Birkhäuser

Trends in the History of Science

Trends in the History of Science is a series devoted to the publication of volumes arising from workshops and conferences in all areas of current research in the history of science, primarily with a focus on the history of mathematics, physics, and their applications. Its aim is to make current developments available to the community as rapidly as possible without compromising quality, and to archive those developments for reference purposes. Proposals for volumes can be submitted using the online book project submission form at our website www.birkhauser-science.com.

More information about this series at http://www.springer.com/series/11668

David E. Rowe · Wann-Sheng Horng
Editors

A Delicate Balance: Global Perspectives on Innovation and Tradition in the History of Mathematics

A Festschrift in Honor of Joseph W. Dauben

Birkhäuser

Editors
David E. Rowe
Institut für Mathematik
Johannes Gutenberg Universität
Mainz
Germany

Wann-Sheng Horng
Department of Mathematics
National Taiwan Normal University
Taipei
Taiwan

ISSN 2297-2951 ISSN 2297-296X (electronic)
Trends in the History of Science
ISBN 978-3-319-36594-7 ISBN 978-3-319-12030-0 (eBook)
DOI 10.1007/978-3-319-12030-0

Springer Cham Heidelberg New York Dordrecht London

Softcover reprint of the hardcover 1st edition 2015

Printed on acid-free paper

Springer International Publishing AG Switzerland is part of Springer Science+Business Media
(www.birkhauser-science.com)

Contents

Preface

The essays in this volume were written to celebrate the seventieth birthday of Professor Joseph W. Dauben, who for four decades has played a pivotal role for historians of mathematics around the world. Many other historians would surely have been pleased to contribute to this Festschrift as well, but the response we received from those initially invited was so unanimously enthusiastic that practical reasons, including a tight time schedule, precluded expanding our original list of invitees. The seventeen papers that appear here, which cover a wide variety of themes relating to the history of mathematics in both western and eastern cultures, reflect many sides of contemporary research that Joseph Dauben helped to pioneer. It would take another whole book to describe all of his many scholarly contributions and organizational activities in the course of a singularly rich career. While we cannot begin to do justice to all that here, we have appended a selected bibliography of his major publications, about which we will restrict ourselves to some brief remarks in the course of discussing his many achievements.

Before commenting further on Professor Dauben's scholarly and professional work, however, a few words must be said about how this Festschrift project first began. Its moving spirit was our dearly missed colleague, Xu Yibao. In July 2013 we met with him in Manchester at the 24th International Conference of the History of Science, Technology, and Medicine. Shortly afterward, the three of us agreed that a Festschrift would be an ideal way to celebrate Joe's forthcoming birthday. Yibao's idea was to present Joe with this volume as a surprise at the 6th International Symposium on Ancient Chinese Books and Records of Science and Technology, which he and Joe were organizing together. This would take place at Yibao's home institution, the Borough of Manhattan Community College, CUNY, from 2–8 October 2014.

So in August 2013, the three of us sent out a letter of invitation to potential contributors, historians of mathematics on the three continents represented in this book. That letter began:

> There is a long tradition in Germany that calls upon former students of a distinguished professor to pay tribute to an esteemed mentor in the form of a Festschrift. This practice is viewed with sympathy in China, where Confucianism teaches people to honor their teachers as sincerely as they honor heaven, earth, the state, and their parents. As former students of Joseph W. Dauben, the three of us would like to follow these traditions by showing our appreciation to Joe through a Festschrift in his honor. . . . We believe a Festschrift would be a meaningful token of our appreciation.

The response we received was overwhelmingly positive, despite the time constraints, so that we were soon able to submit a tentative table of contents to Birkhäuser in Basel. They agreed to publish the book, and so by October Yibao's plan was already off to a smooth start. Then, in early November, we received the tragic news that he had suddenly died after suffering a massive cerebral hemorrhage. This came as a terrible shock, a tragedy especially for his wonderful family, but also a deep loss for all of us who knew him well. He will be memorialized in a special session in his honor at the symposium he helped organize, but was unable to attend.

Yibao had been Joe Dauben's closest collaborator and a dear friend, so with his death we felt both obligated and determined to carry on with the project he initiated. We are especially pleased that this volume contains an English translation of an earlier paper by Xu Yibao that was originally written and published in Chinese. Fittingly, it deals with the interests of an American historian of mathematics, David Eugene Smith, in a collaborative project with Li Yan, an expert on Chinese mathematics. Although he did not live to see the final outcome of his initiative, we feel certain that Yibao would have been very pleased by the response it produced. Now a few words about the man the three of us have known, first as a mentor, but always as a friend for many years.

Joseph Dauben was born in Santa Monica, California on 29 December, 1944. He studied mathematics and English literature at Claremont McKenna College in southern California, graduating in 1966 with a scholarship to study the history of science at Harvard University. He was joined there by another graduate student, Wilbur Knorr, who went on to become a leading authority on the history of Greek mathematics. Both benefited from tutorials with John Murdoch and Judith Grabiner, who prepared them for their oral examinations prior to beginning work for a Ph.D. Joe's dissertation topic, "The Early Development of Cantorian Set Theory," was written under the direction of Erwin Hiebert and Dirk Struik. At Harvard, he also had the opportunity to work with the distinguished algebraist, Richard Brauer, who had been forced to leave Nazi Germany and who took a personal interest in Joe's thesis research.

Thanks to grants from Harvard and the NSF, in 1970-71 Joseph Dauben had the opportunity to spend a year in Europe doing archival research in Germany, Sweden, and Italy. A good deal of that year he spent in the divided city of Berlin, where memories of President Kennedy's visit shortly before his assassination were still vivid. Berlin's mayor then was Willy Brandt, subsequently Chancellor of West Germany, and the architect of the country's new *Ostpolitik*. Joe Dauben's stay there thus came at a perfect time. He could travel freely back and forth, and was soon befriended by the East German historian of mathematics, Kurt R. Biermann, director of the Alexander von Humboldt Forschungsstelle at the Deutsche Akademie der Wissenschaften. In West Berlin, he gained similar support and sympathy from Herbert Meschkowski at the Freie Universität and Christoph Scriba at the Technische Universität. Joe's friendship with Scriba would in the years ahead become especially important for linking the research communities in Europe and North America.

Within the West German community of historians of mathematics, a fairly tight-knit group of researchers gathered for annual meetings at the Mathematisches Forschungsinstitut at Oberwolfach in the Black Forest. Its leading figure was

Joseph Ehrenfried Hofmann, a well–known expert on Leibniz, who established a tradition of seminars focused on what was known as Problemgeschichte der Mathematik. Following Hofmann's death in 1977, Christoph Scriba took over direction of these sessions, which over the years evolved into truly international conferences, though no longer with a core group that received invitations automatically.

Joe Dauben was the first American to attend these meetings on a regular basis in the 1970s and 80s. Through Oberwolfach meetings he soon got to know nearly all the leading Western German historians: Menso Folkerts, Ivo Schneider, Eberhard Knobloch, Herbert Mehrtens, among many others. He and Christoph Scriba also conceived of a project that took advantage of this setting in order to produce a book that aimed to provide an anchor for the modern field of historical studies in the mathematical sciences by presenting a larger picture of the discipline from an historical perspective. Their efforts led to the publication of *Writing the History of History of Mathematics: an Historiography Project of the International Commission on History of Mathematics*, (Basel: Birkhäuser, 2002), edited by Dauben and Scriba. This large-scale undertaking, led by a core group that also included Hans Wussing, Jeanne Peiffer, and Ivor Grattan-Guinness, ultimately involved a large number of other scholars from literally all parts of the world. This novel historiographic venture posed numerous challenges, yet succeeded in gathering together a wealth of documentation relevant for understanding how the history of mathematics emerged as a discipline in various countries at different times, and often influenced by complex cultural and political factors.

Returning to his days as a graduate student, Joe felt lucky back then that his interests in Cantor's life and work came just at the time when Ivor Grattan-Guinness had broken new ground in these areas. The advice and encouragement he received from him initiated a lifelong friendship and collaboration. Needless to say, Ivor's name figures prominently, along with many others, in the acknowledgments at the front of Joe Dauben's groundbreaking study, *Georg Cantor: His Mathematics and Philosophy of the Infinite* (Cambridge, Mass.: Harvard University Press, 1979). From the wider vantage point of historiography, this book marks only the beginning of an ongoing effort to explore Cantor's ideas and their impact in shaping modern mathematics. Joe Dauben's contributions in that regard are amply documented in the selected bibliography appended to this Festschrift. In particular, he took a strong interest in understanding how the "Cantorian revolution" could best be understood in light of Thomas Kuhn's ideas regarding scientific revolutions. Two essays dealing with these themes were written for a volume edited by Donald Gillies, entitled *Revolutions in Mathematics* (Oxford: Clarendon Press, 1992): "Conceptual Revolutions and the History of Mathematics: Two Studies in the Growth of Knowledge" and "Revolutions Revisited." These represent only one aspect within Joe Dauben's larger interest in historiographical issues and problems specific to the history of mathematics.

After taking his doctorate in history of science at Harvard in 1972, Joseph Dauben began his long association with Herbert Lehman College, part of the City University of New York system. He has now taught there for over four decades, first as assistant professor of history, then joining the faculty at the CUNY Graduate Center as associate professor in 1977, and from 2002 till now as a CUNY distinguished professor. Throughout those years, however, he had many visiting appointments elsewhere, far too many to recount here. Already during the academic year 1973-74, he spent a year in Italy as a fellow of the National Endowment for the Humanities. His research project at that time dealt with "Science and Art in the Renaissance, with Special Attention to Mathematics and Perspective." In many subsequent years, he returned to Italy during the summer months, maintaining an affiliation with the American Academy in Rome.

From the beginning and throughout his early career, Lehman College generously supported his research, including the demands of editorial work. More than anyone, Joseph Dauben was responsible for nurturing and sustaining the fledgling journal founded in 1974 by Kenneth O. May, *Historia Mathematica.* Joe entered that venture as managing editor in 1976, and then after May's sudden death the following year, took over as editor. It was under his editorship that *Historia Mathematica* began to be published by Academic Press, from which time it could establish itself as a mature scholarly journal. This year now marks the fortieth anniversary of its founding, and having now established its place in the world of international scholarship, historians of mathematicians can only be grateful to Ken May and Joe Dauben for their foresight and unstinting commitment to this vision. After stepping down as editor, Joe afterward served as Chair of the ICHM from 1986 to 1994.

From his biography of Cantor, Joe turned to the career of Abraham Robinson and the history of nonstandard analysis, or as one might say, he left the mathematics of the infinitely large to pursue the newest chapter in the history of the infinitely small, which Cantor, ironically enough, had completely rejected. In fact, Joe had the ideal background to enter into these waters, which had never been seriously explored by an historian. For he had written his senior thesis as an undergraduate mathematics major on nonstandard analysis, so the topic was a familiar one in which he had a longstanding personal interest. Since many who knew Robinson were still alive, including his widow, he made ample use of oral interviews for this project. Harvard's Garrett Birkhoff proved very helpful, not only because he had known Robinson, but also due to his own interests in the historical implications of nonstandard analysis. Joe also spent time talking with Wim Luxumburg at Caltech and George Seligman at Yale, both former colleagues and good friends of Robinson. In the case of Georg Cantor, the historian faces the problem of a dearth of primary source material to work with, which helps to explain why his life remains a topic of debate. With the life of Robinson, a more contemporary figure, Joe faced just the opposite problem, namely too much information. His definitive biography of 550 pages, *Abraham Robinson. The Creation of Nonstandard Analysis, A Personal and Mathematical Odyssey,* (Princeton: Princeton University Press, 1995), actually involved making deep cuts in the original manuscript. This was not always a labor of love, but the end result was greatly appreciated, above all by the American mathematical community.

Joe Dauben has long played a major part in promoting the history of mathematics within both the American Mathematical Society and the Mathematical Association of America. He has regularly attended the annual joint meetings, and on numerous occasions served as co-organizer of special sessions on history of mathematics. In 2012 Professor Dauben was awarded the American Mathematical Society's Albert Leon Whiteman Memorial Prize for "his contributions to the history of Western and Chinese mathematics and for deepening and broadening the international mathematical community's awareness and understanding of its history and culture." His impact on the American mathematical community has, indeed, been both deep and broad, as was duly noted in the citation:

> In the United States, the history of mathematics, once an excellent but eclectic collection of teaching tools and post-retirement projects, has in the last decades become an integral component of the mathematical community, with well-attended sessions at major meetings. Joe Dauben has spurred this professionalism by his scholarly example and through his service to the profession, which includes organizing international workshops and symposia and editing *Historia Mathematica* for a decade.

In 1988 Joe spent six months in China as part of a program jointly sponsored by the Chinese Academy of Sciences and the U.S. National Academy of Sciences. In Beijing he joined the seminar on history of Chinese mathematics taught by Du Shiran at the Institute for History of Natural Sciences. It was there that he also met Guo Shuchun and Liu Dun, who helped sustain his growing interest in the history of Chinese mathematics. At this time, he had begun an intensive study of Chinese and was working on the *Ten Classics of Ancient Chinese Mathematics.* Thus began a new phase in Joe Dauben's remarkably productive career.

One of the striking qualities of his scholarship that has been deeply appreciated, both in the PRC and Taiwan, is his sensitivity to the distinctive aspects of Chinese mathematics. As a Westerner, someone who thus has the "advantage" of being able to compare very easily Liu Hui's work with Euclid's, he still manages to avoid the standard tendency of portraying Chinese mathematics in a Western light, and thereby losing touch with the intrinsic qualities of the Chinese tradition. This aspect of his research can be seen in his critical investigations of mathematical texts like the *Suanshu shu* and the *Jiuzhang suanshu*, as well as his work on secondary literature dealing with texts from the Qin-Han period, in particular those written by historians of mathematics. These studies are quite unique, and have enabled Joe Dauben to attain a special reputation all his own within the community of historians of Chinese mathematics.

The twenty-first century has made global understanding a major part of (and need within) today's world, and this obviously entails an interest in and respect for the cultural histories of its peoples. Western historians of mathematics have taken new initiatives in this direction, a striking example being the essays in the sourcebook recently edited by Victor Katz, *The Mathematics of Egypt, Mesopotamia, China, India, and Islam* (Princeton: Princeton University Press,

2007). Each was written by an expert on the mathematical cultures represented therein: Annette Imhausen (Egypt), Eleanor Robson (Mesopotamia), Kim Plofker (India), J. Lennart Berggren (Islam), and J. W. Dauben, whose contribution on "Chinese mathematics" provides Western readers with 200 pages of well-chosen source material that reflects the themes and methods of that tradition.

His most recent contribution to Western understanding of Chinese mathematics also represents his single most important achievement in this connection. This is the monumental new edition of the *Jiuzhang suanshu*, undertaken in collaboration with Xu Yibao and Guo Shuchun: *Nine Chapters on the Art of Mathematics. A Critical Edition and English Translation based upon a New Collation of the Ancient Text and Modern Chinese Translation,* Library of Chinese Classics, 3 vols., (Shenyang: Liaoning Education Press, 2013). Thanks to this work, English readers now have ready access to this great classic, and with it they can now truly comprehend how Chinese mathematics, especially as exemplified by Liu Hui, is clearly distinguishable from that of the ancient Greeks or other Western civilizations.

Joe Dauben began his career with special training as an historian of science whose interests from the very beginning went far beyond the realm of modern mathematics. These broader interests are, of course, reflected in his many varied publications, lecture topics, and also the numerous projects he has been pursuing with other scholars. Looking back, his career mirrors some of the major shifts in world affairs over the last four decades. Back in the 1980s, before the fall of the Berlin Wall, he established close friendships with leading historians of mathematics in both East and West Germany. Joe never let politics get in the way, and the same holds for his many friendships with Chinese historians, both in the People's Republic as well as in Taiwan. The breadth of his scholarship has been most impressive, though for us who have enjoyed the rewards of studying under him, his many publications only reflect a small part of his truly astounding range of knowledge and interests. Yet just as impressive, and no less important for the history of mathematics, Joe Dauben has, more than anyone else, helped to establish a professional world-wide community of historians of mathematics through his work as an editor, conference organizer, and tireless spokesman for the discipline. He has been a true global player, who has cultivated a lifetime of friendships around the world like no one else before him. As one of our contributors, the Spanish historian of mathematics Elena Ausejo, so aptly put it, he has been a "bridge-builder of intercultural understanding and friendship." He has created something like a multiplier effect for the history of mathematics, a fragile and rather esoteric field that could never have done so well without him.

The articles this volume contains have been loosely arranged in thematic sections, in part to reflect Joe Dauben's own intellectual journey, starting with his work on Georg Cantor, which broke new ground for the history of modern mathematics. Without any input from us as editors, our authors managed to touch on many of the topics and themes that we have all learned more about by reading Joe Dauben's books and articles. We thank all of them warmly for their steadfast support and cooperation throughout. Our thanks also go to Dr. Anna Mätzener at Birkhäuser in Basel for her help in producing this volume under unusual circumstances. Further thanks go to Philip Wang, for translating Xu Yibao's paper from the Chinese, and to Cheng-Hung CHEN and Jyun-Wei HUANG in Taiwan, for offering technical support in helping prepare the papers by Asian scholars. In Mainz, we were most ably assisted at various stages in the production process by Renate Emerenziani and Matthias Kapffer, who deserve special thanks for their efforts in helping to realize this project.

David E. Rowe
Wann-Sheng Horng

Joseph W. Dauben, June 2014

Set Theory and Foundations of Mathematics

On Cantor's Continuum Problem and Well Ordering: What really happened at the 1904 International Congress of Mathematicians in Heidelberg

Walter Purkert

Cantor on the Continuum Problem and Well Ordering

As a young Privatdozent in Halle, Georg Cantor began to investigate problems involving the uniqueness of trigonometric series espansions, following the lead of his senior colleague, Eduard Heine.[1] In 1870 he was able to prove the following result, known today as Cantor's Uniqueness Theorem: if the series

$$\frac{a_0}{2} + \sum_{n=1}^{\infty}(a_n \cos nx + b_n \sin nx)$$

converges to zero uniformly in $[-\pi, \pi]$, then $a_0 = 0$, and $a_n = b_n = 0$ for all n ([Cantor 1870]). Cantor's researches on trigonometric series took a decisive turn when he began to inquire about the existence of what today are called sets of uniqueness. These are subsets $P \subset [-\pi, \pi]$ which satisfy the condition that if

$$\frac{a_0}{2} + \sum_{n=1}^{\infty}(a_n \cos nx + b_n \sin nx) = 0 \tag{1}$$

for all $x \in [-\pi, \pi] \setminus P$, then all the coefficients must vanish. One can thus regard a set of uniqueness as an allowable set of exceptional points that does not affect the validity of Cantor's Theorem.[2] For such a set of points, the series (1.1) need not converge at all, or it may converge, but not to zero; still, the coefficients in the series will all vanish. After showing in 1871 that every finite subset $P \subset [-\pi, \pi]$ is a uniqueness set, Cantor found one year later that such sets can also be infinite ([Cantor 1872]). In order to characterize these, he introduced the concept of the derivative of a point set: the derivative P' of a point set P is the set of all its limit points. This derivative operation can obviously be iterated finitely many times by successively applying $P^{(n+1)} = (P^{(n)})'$. This process leads to Cantor's theorem: if there exists a natural number n for which $P^{(n)} = \emptyset$, then P is a uniqueness set.

There are several indications that Cantor already by 1870 had the idea of extending the derivation process to transfinite iterations, thereby implicitly introducing transfinite ordinal numbers, though still in the concrete form of orders of

[1] For an analysis of this early work, see [Dauben 1979], 30-46.

[2] The study of such sets has remained an interesting field of active research; see A. S. Kechris and A. Louveau, "Descriptive Set Theory and the Structure of Sets of Uniqueness," *London Mathematical Society Lecture Notes Series*, 122, Cambridge: Cambridge University Press, 1987.

D.E. Rowe and W.-S. Horng (eds.), *A Delicate Balance: Global Perspectives on Innovation and Tradition in the History of Mathematics*,
Trends in the History of Science, DOI 10.1007/978-3-319-12030-0_1

derivations related to point sets.[3] If so, then the actual birth of set theory took place long before 1874, the year of his first publication in this field ([Cantor 1874]), which has often been taken as marking the beginning of set theory. In any case, after 1872 Cantor no longer studied trigonometric series but rather turned instead to work on problems exclusively connected with transfinite sets.

In his aforementioned paper from 1874 Cantor took up an idea from Dedekind to prove that the set of all real algebraic numbers is countable. He showed further (though not yet by using the idea that later came to be called Cantor's diagonal argument) that the set $(0,1)$ is uncountable. This informed the mathematical public for the first time that in the transfinite realm one had to take different quantitative levels into account. Cantor was himself surprised by a result he published in 1878, namely that $[0,1]$ and $[0,1]^n$ have the same cardinality ([Cantor 1878]). It was in this paper that he formulated his continuum hypothesis for the first time:

> According to this, all linear manifolds [sets of real numbers] would consist of two classes, of which the first includes all manifolds that can be placed in the form: functio ips. ν (where ν runs through the natural numbers); whereas the second class consists of those that can be reduced to the form: functio ips. x (where x can take on all real numbers ≥ 0 and ≤ 1). Corresponding to these two classes there would therefore only be two cardinalities for infinite linear manifolds; we postpone for a later opportunity the precise investigation of this question.[4]

In other words: Cantor's continuum hypothesis asserts that every infinite set of real numbers is either countable or has the cardinality of the continuum.

Cantor drew up the outlines for his general theory of transfinite numbers in Part V in his six-part series of articles entitled *Über unendliche lineare Punctmannichfaltigkeiten* ([Cantor 1883]). Part V was also published in 1883 by Teubner as a separate broschure with the title *Grundlagen einer allgemeinen Mannichfaltigkeitslehre – Ein mathematisch-philosophischer Versuch in der Lehre des Unendlichen.* An essential issue for Cantor in the *Grundlagen* concerned the introduction of transfinite ordinal numbers independent of the concept of the order for the derivative of a point set. This required a new concept, the notion of a well ordering, which soon became fundamental for all of set theory. Cantor had arrived at this concept by no later than 1882, as can be seen from a letter he wrote to G. Mittag-Leffler (1846–1927) on 17 December of that year ([Cantor 1991], 100). His original definition in the *Grundlagen* placed transfinite succession in the foreground rather than the property that every non-empty subset of the given linear ordered set has a smallest element. Cantor calls a set M well-ordered if:

> $[\cdots]$ its elements are connected to one another by a definite given succession, according to which there is a *first* element of the set and not

[3] [Purkert/Ilgauds 1987], 39; [Purkert 2000], 193.

[4] [Cantor 1878], 132–133. Here and afterward we cite all of Cantor's works from his Gesammelten Abhandlungen [Cantor 1932].

> only does each individual element have a definite successor (unless it is the last element in the succession) but also for every arbitrary finite or infinite set of elements there belongs a definite element, which is the element that *immediately follows* all of the others (unless there exists no element that succeeds them all)([Cantor 1883], 168).

The possibility of well ordering any given set was for Cantor a matter beyond all doubt, although he never succeeded in proving this. He writes in the *Grundlagen einer allgemeinen Mannigfaltigkeitslehre*:

> The concept of a *well-ordered set* proves to be fundamental for the entire theory of manifolds. In a later publication I will return to a law of thought that is remarkable because of its general validity, namely that it is always possible to exhibit every *well-defined* set in the *form* of a *well-ordered* set.[5]

Cantor introduced the transfinite ordinal numbers in his *Grundlagen* by means of two generating principles: the first insures the transition from α to $\alpha+1$ for every ordinal number α, whereas the second passes from a given transfinite well-ordered set without a last element to its associated limit ordinal.

His so-called *Hemmungsprinzip* leads to certain natural cuts, and thus a segmentation within the ordinal numbers forming different number classes. Cantor designated the set of natural numbers as the first number class with its corresponding cardinality, that is having the power of the first class (later written as $\aleph_0$). He then employed the *Hemmungsprinzip* to generate the next segment by means of the following condition: an ordinal number α, generated by one of the two principles above, will belong to the second number class precisely when the set of ordinals that precedes α has the power of the first number class. Cantor then proved that the power of the second number class (later written as $\aleph_1$) is the next higher cardinality after $\aleph_0$. This was one of the few propositions that he proved in the *Grundlagen*, though it is one that has a fundamental significance for his theory. The third number class, corresponding to the cardinal $\aleph_2$, would follow from the *Hemmungsprinzip* as the set of all α for which $\{\beta \,|\, \beta < \alpha\}$ has the power $\aleph_1$. As Cantor put this: "In the same way the third number class results from the definition of the third power or the power of the third class." ([Cantor 1883], 167). His definition of the series of alephs did not, in fact, go beyond this vague "etc.".

In terms of alephs, Cantor's continuum hypothesis can be formulated simply as $\aleph = \aleph_1$, where $\aleph$ (without index) stands for the cardinality of the continuum. In his *Grundlagen* Cantor referred to this conjecture in connection with the cardinality of the interval $(0,1)$: "[$\cdots$] I hope to be able to answer this [the question of the power of $(0,1)$] soon and by means of a rigorous proof that its power is none

[5] [Cantor 1883], S. 169. Cantor's opinion that the well ordering theorem has the status of a generally valid law of thought was closely connected with his philosophical views, about which see the remarks of U. Felgner in [Hausdorff 2002], 602.

other than that of the *second number class*" ([Cantor 1883], 192). If this conjecture were indeed to be correct, then it would follow immediately that every subset of the continuum is either countable or has a cardinality identical with that of the continuum itself.

Cantor added an interesting note here, in which he even took a first step in the direction of formulating the generalized continuum hypothesis ([Cantor 1883], 207). He asserted, namely, that the set of all real-valued functions has the power of the third number class, or in modern notation: $2^{\aleph} = \aleph_2$. Taking this together with his continuum hypothesis $\aleph = \aleph_1$ leads to the assertion that $2^{\aleph_1} = \aleph_2$. The general equation $2^{\aleph_\alpha} = \aleph_{\alpha+1}$ later became known as the generalized continuum hypothesis, which was formulated for the first time by Hausdorff ([Hausdorff 1914], 182). He, however, characterized the equation $2^{\aleph_\alpha} = \aleph_{\alpha+1}$ as altogether problematic since its very first case ($\alpha = 0$) corresponds to Cantor's continuum hypothesis.

Cantor's hopes for solving the continuum hypothesis rested on progress he had made in connection with point sets in n-dimensional Euclidean space.[6] The central result has come to be known as the Cantor-Bendixson Theorem, which states that every *closed* uncountable point set P can be written in the form

$$P = R + S, \tag{2}$$

where R is at most countable and S is perfect. Cantor could furthermore prove that every perfect set has the power of the continuum, which meant that his continuum hypothesis was indeed correct for *closed* sets: every closed set is either countable or has the power of the continuum.

Starting from the Cantor-Bendixson Theorem, Cantor developed two different strategies for attacking the continuum problem. The first involved constructing a closed point set P with cardinality $\aleph_1$. Since $\aleph_1 > \aleph_0$ the set P would be uncountable and thus, in view of the above theorem, it would have the power of the continuum. On the other hand, by construction it would have the cardinality $\aleph_1$, so that $\aleph = \aleph_1$ would follow immediately. Cantor mentioned this line of argument in a letter to G. Mittag-Leffler from 26 August 1884 where he wrote: "You thus see that everything now reduces to defining a *closed* set of the second power. Once I have brought all that in order I will write you with more details" ([Meschkowski 1967], 243). His second strategy sought to generalize the Cantor-Bendixson Theorem to arbitrary uncountable sets by showing that any such set contained a perfect subset, which would thus lead to a decomposition of type (1.2). His pursuit of this strategy was plainly stated at the end of [Cantor 1884] where he announced:

> We thus have the following theorem:
>
> *An infinite closed point set has either the first power or it has the power of the continuum,* $[\cdots]$ In later paragraphs it will be proved that this remarkable theorem remains valid for *non-closed* linear point sets as well as for all n-dimensional point sets.

[6]The most important of these results were published in Part VI in his series of articles *Über unendliche lineare Punctmannichfaltigkeiten* ([Cantor 1884]).

> This will be shown with the help of the theorems *proved* in Nr. 5 §13 [which show that $\aleph_1$ of the second number class is the power that immediately succeeds $\aleph_0$] that the *linear continuum has the power of the second number class (II)* ([Cantor 1884], 244).

This second strategy thus involved decomposing an arbitrary infinite set into subsets so that the isolated elements form a countable set, whereas the remaining core was dense in itself ([Cantor 1885]). Cantor was unable, however, to draw any conclusions as to the cardinality of the latter. Today it is clear why this approach had no more chance of succeeding than did the first strategy (see 4. below). Nevertheless, Cantor's efforts to extend the Cantor-Bendixson Theorem to a more general class of sets contained the germ for a many-faceted field of present–day research, descriptive set theory. In 1903 W. H. Young (1863–1942) showed that, in analogy with (1.2), every uncountable G_δ-set in the real line contains a perfect subset ([Young 1903]). Hausdorff was able to extend this result in his *Grundzügen der Mengenlehre* ([Hausdorff 1914], 465–466) to $G_{\delta\sigma\delta}$-sets in complete separable metric spaces. Finally, in 1916 Hausdorff and P. Alexandroff (1896–1982) succeeded in proving, independent of one another, that every uncountable Borel set in a complete separable metric space has the power of the continuum ([Hausdorff 1916], [Alexandroff 1916)].

In the years 1895 and 1897 Cantor published a two-part work under the title *Beiträge zur Begründung der transfiniten Mengenlehre* ([Cantor 1895/1897]). In this new presentation of his set theory he avoided all mention of philosophical and historical issues. E. Zermelo, the editor of Cantor's *Gesammelten Abhandlungen*, offered the following characterization of this work:

> ... Cantor's last publication on set theory which form the actual culmination of his life's work. Here, after gradually developing over the course of decades, the fundamental ideas and concepts attain their final form and several of the principal theorems of general set theory now are placed on their classical foundation for the first time ([Cantor 1932], 351).

It would seem even more remarkable that Cantor did not even mention the continuum hypothesis in this work, despite the fact that he had been driven for years by various attempts to prove it. In fact, he could have easily formulated a purely set-theoretic version of the continuum problem in his *Beiträgen* by drawing directly on his exposition of cardinal arithmetic, thereby bypassing any need for a theory of the real numbers. For there he showed that the power of the continuum equals $2^{\aleph_0}$, i.e., the cardinality of the power set of a countable infinite set. This meant that the continuum problem amounted to asking which of the alephs equals the cardinal number $2^{\aleph_0}$. Cantor's continuum hypothesis asserts that $2^{\aleph_0} = \aleph_1$.

The concept of well ordering is defined in the *Beiträgen*, as before, by means of finite and transfinite succession. Cantor then shows that an ordered set is well-ordered precisely when each non-empty subset has a smallest element. He used this property as the definition of a well-ordered set in a letter to Dedekind from 3 August 1899 ([Cantor 1991], 408). Ever since the publication of G. Hessenberg's (1874–1925) *Grundbegriffe der Mengenlehre* this definition ([Hessenberg 1906], S. 538) has been taken as the usual one.

Cantor remained convinced that for every set there exists a well-ordered set having the same cardinality, or in other words, that every power is equivalent to an aleph (the well ordering theorem), as can be seen from passages on pages 295–296 of his *Beiträgen*. We also know from a letter he wrote to Hilbert on 26 September 1897[7] that he planned to add a third installment to the *Beiträgen* in which he hoped to publish a proof of the well ordering theorem. This proof was to have been based on the assertion that one could not conceive of the collection of all alephs as forming a set without falling into a contradiction.[8] Cantor sketched his idea for this proof in a letter to Dedekind from 28 July 1899. The fact that he never published such a proof suggests that he probably felt his argument was not sufficiently convincing.[9]

Following an initial period of skepticism regarding Cantor's general theory of sets, gradual recognition of its importance began to take hold by the end of the century.[10] This led to increasing awareness of and interest in the two fundmental open problems in Cantor's set theory: the continuum hypothesis and the validity of the well ordering theorem. Both of these problems gained prominent attention in the wake of Hilbert's famous lecture at the International Congress of Mathematicians held in Paris in 1900.[11] On that occasion Hilbert placed these two foundational issues (the continuum problem and the question as to whether it was possible to find an explicit well ordering of the real number continuum) at the top of his list of 23 problems, those he recommended to his fellow mathematicians as major challenges for the new century.

Felix Hausdorff's Turn Toward Set Theory

A number of documents from Hausdorff's days as a student (summer semester of 1887 to summer semester 1891) have been preserved, including certificates for courses he attended as well as a fairly complete set of notes he made for the lecture courses.[12] Thanks to these we can form a fairly clear picture of his mathematical background and education at that time. These documents also tell us that it is most unlikely Hausdorff had any exposure to set theory during his years as a student,

[7] Published in [Purkert/Ilgauds 1987], 224–226.

[8] Cantor was well aware of the pitfalls that later became known as the antinomies of set theory; indeed, he was aware of such problems long before they were made known by Burali-Forti and Russell; see [Purkert 1986].

[9] For critique of the argument, see Zermelo's remarks in [Cantor 1932], 451.

[10] On the reception of Cantor's theory see [Dauben 1979], 240-270, and [Hausdorff 2002], 18–27.

[11] [Hilbert 1900]. See also *Die Hilbertschen Probleme.* Ed. H. Wussing and H. Bernhardt. Ostwalds Klassiker, Bd. 252, Geest & Portig Akademische Verlagsgesellschaft, Leipzig 1976, 1979^2.

[12] These documents can be found in the Universitäts- und Landesbibliothek Bonn, Nachlass Hausdorff.

especially since he never took courses in function theory or Fourier series, the two subjects in which set-theoretic ideas might well have been discussed. During his final semester in Leipzig he was working closely with Heinrich Bruns (1848–1919), professor of applied mathematics and astronomy, under whom he wrote his dissertation in 1891 ([Hausdorff 1891]). It dealt with the refraction of light in the atmosphere, a topic he pursued in two further studies ([Hausdorff 1893], in two parts) as well as in his habilitation thesis ([Hausdorff 1895]). Before habilitating Hausdorff was employed from February 1893 to February 1895 as a calculator at the Leipzig astronomical observatory. His other published work up until 1897 dealt with optics ([Hausdorff 1896]) and insurance mathematics ([Hausdorff 1897a]), topics without the slightest connection to set theory.

That a mathematician with this type of applied background should turn to set theory is certainly most surprising. In Hausdorff's case, the explanation has to be sought in connection with his unusually broad intellectual interests and talents. These extended far beyond the realm of mathematics, enabling him to move freely across boundaries that normally separated different cultural spheres. After completing his habilitation, Hausdorff turned to philosophical and literary studies, which he published under the pseudonym Paul Mongré. This activity began with a volume of aphorisms (1897), then a book on epistemology (1898), and a collection of poems (1900). This was followed by a work for the theater (1904) that was performed with great success in numerous cities across Germany. Finally, over a more extended period (1898–1912), he published a series of essays in the leading literary journals of the day.[13] Thus around the turn of the century Hausdorff was better known as a writer and philosopher than as a mathematician.

We will see that Hausdorff's interest in Cantor's set theory was above all awakened by questions of a philosophical nature. In the above mentioned volume of aphorisms, *Sant' Ilario. Gedanken aus der Landschaft Zarathustras* ([Hausdorff 1897b]) one finds a chapter that addresses epistemological issues, "Zur Kritik des Erkennens." In aphorism Nr. 406 (pp. 349 ff.) Hausdorff deals directly with Nietzsche's problematic argument for eternal recurrence (*ewige Wiederkunft*), a central motif in his philosophy, but which Nietzsche also elaborated in notes found among his posthumous papers. Nietzsche had argued therein that time is infinite, but that concretely experienced time (what Kant called the content of time, *Zeitinhalt*) consists in a vast, but nevertheless finite number of atomistic combinations, from which he concluded that an eternal recurrence of the same experience in time must occur. In his aphorism Hausdorff acknowledged the correctness of Nietzsche's pigeon hole principle but not his assumption that the content of time is finite. Indeed, he believed the latter was not only infinite but that it consisted of a much higher order of infinity, an infinity of a *higher dimension*. Thus, he wrote:

> No one should dare to approach epistemological things lacking the mathematical distinction between manifolds of different dimension and infinities of different rank and size; one cannot comprehend the infinite

[13] For a complete list of his writings see [Hausdorff 2002], XI–XVI.

> merely by taking it in opposition to the finite and without noting that within the infinite one can distinguish stages and degrees. ([Hausdorff 1897b], 350).

Around this time, Hausdorff was intensively engaged with problems in the theory of knowledge, and in 1898 he published an epistemological essay entitled *Das Chaos in kosmischer Auslese* ([Hausdorff 1898]). The quotation cited above might suggest that he already had a burning interest in Cantor's set theory, which opened the way to quantifying the various degrees of infinity by means of transfinite numbers. Yet this aphorism nevertheless reveals that at this time Hausdorff was still unfamiliar with Cantor's theory. For here he still views the various degrees of infinity in terms of the dimension of point sets: thus a line contains ∞^1 points, whereas a surface has ∞^2, and a solid figure ∞^3.

Hausdorff completed his *Sant' Ilario* during the winter semester of 1896/97 when he was recovering from an illness on the coast of Liguria. Thus it seems that in early 1897 he was not yet familiar with Cantor's theory. Yet one year later, in *Chaos in kosmischer Auslese*, we find him now fully oriented with regard to the same problems and their connections with set theory. There he wrote the following about Nietzsche's proof:

> In any case, all attempts to decide for or against eternal recurrence on a purely *speculative* basis will run up against the problem that tends to make any apriori deduction dangerous. Nietzsche's own attempt to show the necessity of eternal recurrence must be counted as a similar case. The argument published in his Nachlass is based on the completely groundless assertion that the set of all worldly states (Weltzustände) is finite, whereas the points of time are infinite, so that the finitely many world states (Weltinhalt) must repeat themselves infinitely often over the infinite course of time in order to fill it completely. Time is linear, whereas the collective of all worldly states, already in the simplest purely materialistic conception, can be represented as an infinite dimensional point continuum, and cannot be embedded in time in a *continuous* way. Were one to drop the continuity condition for the correspondence, then one would have to take a further step by showing that the set of all worldly states has a higher *power* than that of the linear continuum in the sense of Cantor's theory of manifolds. Here, incidentally, it suffices to imagine a portion of the world as a fluid body that breaks up in a random fashion however it moves. ([Hausdorff 1898], 193–194).

One finds numerous other places in *Chaos in kosmischer Auslese* where connections are made between philosophical arguments and set-theoretic concepts or propositions (e.g., pp. 81–82, p. 122, pp. 145–146, pp. 157–158); this makes evident that Hausdorff was well versed in Cantor's theory. All indications suggest that by this time Hausdorff had begun to take a truly deep interest in set theory, an interest that was above all motivated by related philosophical problems, in particular epistemological concerns. It was thanks to these extra-mathematical interests that Hausdorff took up set-theoretical studies in the first place, a curious situation, but one that enabled him to do original research in this field to which he would later make fundamental contributions.

This began in 1901, at which time he taught a course on set theory in Leipzig.[14] It was during this time that Hausdorff made his first set-theoretic discovery. He found that the type class $T(\aleph_0)$ of all countable ordinal types had the power $\aleph$ of the continuum. His proof made essential use of the equivalence theorem of Dedekind-Bernstein by showing:

> a) a subset of countable types is equivalent to the continuum.
> b) a subset of the continuum is equivalent to the class of countable types.[15]

Hausdorff penciled in red in the upper margin of page 37:

> a) according to an oral communication from Cantor,
> b) from me myself. Presented on 27 June 1901.[16]

In his lecture course, Hausdorff carefully separated results that were provable and had already received exact proofs from those propositions that Cantor had conjectured, but which he had not been able to prove. Hausdorff called the latter postulates, which he formulated as follows:

> Postulate II: The power of the continuum equals that of the second number class.[17]
> Postulate III: Every well-defined set can be brought into the form of a well-ordered set (*Ibid.*, 440-441).

Postulate I asserted that all cardinal numbers are comparable, which is correct if Postulate III holds. Postulate II is Cantor's continuum hypothesis; this can also be formulated as the continuum problem, which asks for the place of $\aleph$ in the sequence of alephs $\aleph_\alpha$. It was precisely this problem that motivated Hausdorff to undertake a thorough study of ordered sets. Indeed, he conceived a new strategy for attacking the continuum problem by using this theorem concerning the power of $T(\aleph_0)$. For if a countable set has $\aleph$ orderings, but $\aleph_1$ well orderings, then perhaps one could gain a deeper insight into the relationship between $\aleph$ and $\aleph_1$ by investigating new types of ordering structures that were more general than

[14]Nachlass Hausdorff: Kapsel 03: Fasz. 12. Published with commentary by U. Felgner in [Hausdorff 2013], 409-466. Only three auditors attended (Archiv der Universität Leipzig, PA 547, Bll. 10–12). This was one of the first courses on set theory taught anywhere; the only earlier ones were those taught by Schoenflies and by Zermelo in Göttingen (SS 1898 and WS 1900/1901); see [Peckhaus 1991], 82 and [Hausdorff 2013], 452. Cantor himself never taught a course on set theory, though he did occasionally hold seminars that dealt with it.

[15]Nachlass Hausdorff: Kapsel 03: Fasz. 12, Bl. 37. [Hausdorff 2013], 435.

[16]Felix Bernstein (1878–1956) had also proven b) in his dissertation. Hausdorff would later refer to this as the Cantor–Bernstein Theorem.

[17]Nachlass Hausdorff: Kapsel 03: Fasz. 12, Bl. 46. [Hausdorff 2013], 440.

well orderings but not as general as arbitrary orderings. This strategy is already apparent in his first publication in the field of set theory ([Hausdorff 1901]) where he wrote:

> In the field of study of order types opened by G. Cantor it is only in the special area of ordinal numbers that we know anything substantial; with regard to general order types that do not arise from well-ordered sets very little is known. Nevertheless, a precise knowledge of these types and their classification belongs to the central problems of set theory, not least because this could possibly bring us closer to solving the longstanding question of the power of the continuum. For, on the one hand, the Cantor-Bernstein Theorem asserts that the power of the second *type* class (the class of countable types) equals the power $\aleph$ of the continuum, whereas $\aleph_1$ is the power of the second *number* class. Thus, in order to compare $\aleph$ and $\aleph_1$ it is important to consider possible intermediary levels, i. e. to determine a smaller subset of the set of all countable sets for which the countable well-ordered sets form a proper subset. This approach by means of intermediate levels promises to simplify the proof of Cantor's conjecture that $\aleph = \aleph_1$. On the other hand, if $\aleph > \aleph_1$, then this road is likely to turn into a detour ([Hausdorff 1901], 460. [Hausdorff 2013], 5).

In this paper Hausdorff investigated certain types of ordered sets that he called "graded sets" ("gestufte Mengen"). In such graded sets two different initial segments are never similar to one another. This concept is more comprehensive than that of well-ordered sets. Furthermore, the set of all countable graded sets has the power of the continuum, so that Hausdorff could take this as the starting point for the research program described above. Today we know, of course, that the continuum problem cannot be solved using this approach any more than with the strategies that Cantor had pursued.

Set Theory at the 1904 Heidelberg Congress

In 1904 the Third International Congress of Mathematicians took place in Heidelberg from August 8–13. During that week Cantor's set theory, and in particular the continuum problem and the problem of well ordering the continuum, played a major role in the discussions, including some that took place immediately afterward. On 10 August Julius König gave a lecture on the continuum problem in the special section of the congress on arithmetic and algebra. There he proved that the sum $\mathbf{a}$ of an increasing sequence of cardinal numbers satisfies the inequality $\mathbf{a}^{\aleph_0} > \mathbf{a}$. Since $\aleph_{\beta+\omega} = \aleph_\beta + \aleph_{\beta+1} + \cdots$ it follows that:

$$\aleph_{\beta+\omega}^{\aleph_0} > \aleph_{\beta+\omega}. \tag{3}$$

König now made use of a relation in the cardinal arithmetic of alephs that Felix Bernstein had introduced in his dissertation, namely

$$\aleph_\mu^{\aleph_\alpha} = \aleph_\mu \cdot 2^{\aleph_\alpha} \tag{4}$$

Assuming now that $\aleph = 2^{\aleph_0}$ equals an element in the sequence of alephs, say $2^{\aleph_0} = \aleph_\beta$, then setting $\aleph_\mu = \aleph_{\beta+\omega}$ and placing this in the formula (3.2) one obtains

$$\aleph_{\beta+\omega}^{\aleph_0} = \aleph_{\beta+\omega} \cdot 2^{\aleph_0} = \aleph_{\beta+\omega}\aleph_\beta = \aleph_{\beta+\omega} \tag{5}$$

which contradicts (3.1). With that König concluded that the power of the continuum cannot appear in the sequence of alephs, which meant that the continuum cannot be well-ordered. Naturally this would also entail that Cantor's conjecture that $\aleph = \aleph_1$ had to be wrong. The protocol shows that three speakers took part in the discussion afterward: G. Cantor, D. Hilbert, and Arthur Schoenflies (1853–1928).

König's lecture created a great sensation at the time, and little wonder that this was so. For were his argument correct, then this would have demolished one of Cantor's most fundamental beliefs, namely, that every set can be well-ordered. This would also have meant that both parts of the problem Hilbert had formulated four years earlier in Paris had now been answered, though not as he had hoped but rather in the negative and with a simple three-line argument. In his autobiography, *Bestand und Wandel*, G. Kowalewski gave this vivid picture of the events that immediately followed:

> König's lecture thus brought about the refutation of two of Cantor's fundamental assumptions [that every set can be well ordered and that $\aleph = \aleph_1$]. Cantor spoke afterward, deeply moved. He even thanked God in passing for having granted him the chance to witness the refutation of his errors. The newspapers brought out reports on this important lecture by König. The Grand Duke of Baden was informed about the nature of this sensation by Felix Klein.
>
> Thankfully, the next day it turned out that König's proof was unsound. He had made use of a theorem of Felix Bernstein, a result which on closer inspection was shown to be false. Zermelo, an extremely sharp and quick thinker, made this important discovery. In fact, he even found during these days a proof for the possibility of well ordering an arbitrary set, a result that he later confirmed with a second proof that used entirely different considerations ([Kowalewski 1950], 202).

This passage is written in such a striking manner that the reader easily gets the impression these were events Kowalewski witnessed first hand, and yet it is nearly certain that this was not at all the case. He was merely recalling what he had heard about them from others, and furthermore his report was written fifty years after all this happened. In any case, the fact is that the name Kowalewski does not appear in the official list of participants who attended the congress ([Krazer 1905], S. 11–12).

What Kowalewski wrote about Cantor's reaction was surely invented; indeed, his account is flatly contradicted by what Scheonflies reported (see below). Nor could Felix Klein have spoken with the Grand Duke of Baden on this occasion since the latter spent the month of August at a spa in Switzerland. Moreover, it is unlikely that Klein would have been present at König's lecture since he was deeply involved with the section on applied mathematics. Of course he might have spoken with the Grand Duke's son, who held a reception for the participants on the evening of 10 August, but these are just speculations hardly worth mentioning.

The passage that follows this one ("Thankfully ...) in Kowalewski's portrayal is also full of errors. It deserves, however, a more careful refutation because it has often been cited in the literature on the history of set theory where it has for decades colored the picture of what happened after the Heidelberg congress.[18]

In reality it was Felix Hausdorff who first pointed out that König's argument was flawed because he had failed to notice that Bernstein's formula (3.2) was not generally valid. It applies only to those ordinals μ that happen not to be limit numbers. In all likelihood the lively discussions at the congress and afterward in Wengen, a holiday resort in the heart of the Swiss Alps (see below), prompted him to delve into the difficulties surrounding exponentiation with alephs. For only a few weeks later he published his famous recursion formula:[19]

$$\aleph_\mu^{\aleph_\alpha} = \aleph_\mu \aleph_{\mu-1}^{\aleph_\alpha}.$$

This formula loses its meaning if μ is a limit ordinal; when that is not the case, however, then one can derive the Bernstein recursion formula from it. At the end of this paper, Hausdorff made an explicit reference to König's lecture:

> The formula that F. Bernstein obtained by unrestricted recursion (Untersuchungen aus der Mengenlehre, Diss. Halle 1901, S. 50)
>
> $$\aleph_\mu^{\aleph_\alpha} = \aleph_\mu \cdot 2^{\aleph_\alpha}$$
>
> is therefore to be regarded as still unproved. Its correctness appears all the more problematic because from it, as J. König has shown, would follow the paradoxical result *that the power of the continuum is not an aleph and that cardinal numbers exist that exceed every aleph.* ([Hausdorff 1904], 571. [Hausdorff 2013], 33.)

If Zermelo had already pointed out in Heidelberg that Bernstein's formula was unproven for limit ordinals μ and that it was likely to be false in that case, then

[18]It can be found, for example, in [Meschkowski 1967], 165–166; [Dauben 1979], 247-250; [Moore 1982], 87; [Purkert/Ilgauds 1987], 161; [Peckhaus 1991], 83; [Oberschelp 1994], 11; and somewhat relativized in [Ferreirós 1999], 312: " it may well be that Kowalewski misremembered."

[19][Hausdorff 1904]. Republished with commentary by U. Felgner in [Hausdorff 2013], 31-37.

Hausdorff would certainly never have written this passage, or at the least would have mentioned Zermelo in this connection.

Hausdorff's role in resolving the mistake in König's argument also comes plainly into view in A. Schoenflies's recollections of the events in Heidelberg ([Schoenflies 1922]). Schoenflies witnessed the discussions surrounding König's lecture first hand, and so was a far more reliable witness than Kowalewski. For this reason we give his report in its entirety:

> The interest of the mathematical world in set-theoretic problems reached a high point in 1904 at the Heidelberg Congress. The mathematicians were moved by the question as to the power of the continuum as well as the well ordering theorem, in short, the assertion that every set is capable of being well-ordered. For Cantor it was a kind of dogma, central for his set-theoretic knowledge and faith, to require that all sets can be well-ordered and, in particular, that the continuum had the second power. König's lecture, which ended with the climactic result that the continuum could not be an aleph (i.e., that it was incapable of being well-ordered) therefore had an astonishing effect, all the more so because it was based on an extraordinarily thorough and precise argument. Since, however, part of it was based on a relation for alephs found by F. Bernstein that turned out not to be generally valid, König's result had to be restricted to alephs with a transfinite index,[20] and this robbed his argument of any force. Yet characteristically, from the very beginning Cantor thought that König's result could not be correct, despite the exactitude of his proof. It will enhance the picture of Cantor's personality if I mention here a personal recollection. After the Heidelberg meeting a kind of post-congress took place in Wengen. Hilbert, Hensel, Hausdorff, and I myself met there quite by accident. Originally, Cantor was lodging nearby, but he soon joined us because he felt a constant need to be heard. At the heart of these discussions stood once again the theorem of König. It was an altogether dramatic moment when early one morning Cantor came to the hotel where Hilbert and I were staying. He was waiting for us a good long time in the breakfast hall, impatiently ready to pounce on us or anyone else with his new refutation of König's theorem. To Hausdorff, as is known, we owe thanks for undertaking a careful examination of the range of validity of Bernstein's aleph relation.[21] While still in Wengen, he obtained this result, which led to the above mentioned restriction of the theorem, thereby showing that one could draw no consequences from it for the continuum.[22]

Beyond this clear report by an eye witness, it should be mentioned that Schoenflies had already pointed out in 1905 that Hausdorff had one year earlier

[20] Here Schoenflies's assertion goes too far; see below under 4.

[21] Here in a footnote Schoenflies referred to [Hausdorff 1904].

[22] [Schoenflies 1922], 100–101.

shown that Bernstein's aleph relation could be proved if one restricted the index to non-limit ordinals ([Schoenflies 1905], 186).

Should any doubts remain, however, as to which of these two reports, Schoenflies' or Kowaleski's, is more accurate, these can be cleared away by an additional piece of evidence, a letter Hausdorff wrote to Hilbert on 29 September 1904 in which he describes his struggles with this problem:

> After being tormented by the continuum problem nearly like a monomania in Wengen, my first glance here turned naturally to Bernstein's dissertation. The problem lies precisely there where I expected, on page 50: [...] Bernstein's consideration leads to a recursion from $\aleph_{\mu+1}$ to $\aleph_\mu$, but this fails for those $\aleph_\mu$ that have no predecessor, that is precisely for the alephs, the case which Mr. J. König necessarily requires. During my travels, I had already written to Mr. König in the same sense without being able to peruse the work of Bernstein, but without receiving an answer. Now I am more inclined than ever to hold König's proof for false and the theorem for not even remotely probable. On the other hand, you too will certainly have gained the impression that Mr. Cantor was unable in the last weeks to find what he has sought in vain for 30 years, and so it appears that after the Heidelberg congress your problem no. 1 remains exactly where it was when you left it at the congress in Paris.
>
> But perhaps, just as I write this, one of the parties in this dispute is now in possession of the truth. I am very curious about the published proceedings from the congress.[23]

In the written report on his lecture for the congress, König expressly noted that he had withdrawn the final portion in which he had originally tried to make use of Bernstein's aleph relation. He did not mention Hausdorff in this connection ([Krazer 1905], S. 147).

Kowalewski's remark regarding Zermelo (" In fact, he even found during these days a proof for the possibility of well ordering an arbitrary set" ...) suggests that Zermelo found his proof in direct connection with events at the congress. There seems to be no evidence that would support this claim, but what we do know is the following. On 24 September 1904 Zermelo wrote a letter to Hilbert in which he offered a proof for the well ordering theorem. The proof was based explicitly on the axiom of choice, which required the existence of coverings for any infinite collection of sets, so that "there will always exist such correspondences for which each set corresponds to one of its elements" [Zermelo 1904], 516). This letter to Hilbert was soon thereafter published in 1904 in *Mathematischen Annalen* and it is certainly possible that it was stimulated by the discussions at the Heidelberg congress, which Zermelo attended. If so, he made no mention of this and wrote instead that: "The

[23]Niedersächsische Staats- und Universitätsbibliothek zu Göttingen, Handschriftenabteilung, Nachlass Hilbert, Nr. 136. Quoted in full in [Hausdorff 2012], S. 330. Hausdorff appears in the list of participants at the congress.

proof for this [the well ordering theorem] resulted from conversations that I had during the last week with Mr. Erhard Schmidt" ([Zermelo 1904], 514).

Not long ago, Heinz-Dieter Ebbinghaus discovered a postcard that Zermelo sent to Max Dehn on 27 October 1904, in which he mainly described reactions to his proof of the well ordering theorem. In this postcard, however, Zermelo also revealed that he, too, independent of Hausdorff, had found the mistake in Bernstein's dissertation that undermined König's proof. The postcard indicates nothing, however, about when he had noticed this; in all likelihood he had some idea already in Heidelberg as to where the mistake might lie. According to the available sources, Zermelo's discovery of the mistake came later and played no role in the discussions that took place in Heidelberg or afterward in Wengen. Whether or not König himself noticed his mistake, as Zermelo here claimed, must remain open. He could just as easily have been led to this conclusion by Hausdorff's letter, mentioned above in the letter Hausdorff wrote to Hilbert. Zermelo's postcard to Dehn reads as follows:[24]

> Dear D!
>
> Many thanks for [your] friendly agreement. I also received the same the day before yesterday from the "younger Berlin school," i.e. from Landau and Schur in fact. But if you only knew how I am bombarded in the meantime by letters from the experts in set theory, from Bernstein, Schoenflies, and König! Some fine polemic will still unfold in the Annalen; but I am not worried about that. So you still do not know the fate of König's talk? Innocent soul! Nothing else but that was discussed during the entire vacation. K[önig] solemmly informed both Hilbert and me that he revoked the Heidelberg proof. B[ernstein] did the same with his theorem on powers. K[önig] can be happy that the library in Heidelberg had closed so early; otherwise he might have made a fool of himself. Thus I had to wait for checking until my return to G[öttingen]. Then it was instantly obvious. In the meantime, however, König had found that out for himself while writing up his proof. So please try to apply dissertation page 50(?) to the case $\mathbf{s} = \mathbf{m}_1 + \mathbf{m}_2 + \cdots$, the only one treated by K[önig]. You will see the mistake at once. Blumenthal calls it a real "Bernstein ruin," but K[önig] wants to save whatever can be saved. He fantasizes about the set W *of all* order types, which *could* be contained in many sets, which would show that my $L_\gamma = W$ and so my conclusion $L_\gamma = M$ would then be wrong. Strangely enough, he is backed up in this matter by Bernstein and Schoenflies, though the latter only in a quite confused and misunderstood way. The W-believers already deny the definition of subset and even perhaps the principle of the excluded middle!
>
> With heartfelt greetings, Your E. Zermelo

[24]Cited from [Ebbinghaus 2007a], 431; see also [Ebbinghaus 2007b], 50-53.

As this postcard already hints, Zermelo's proof would unleash a long and heated discussion over foundational issues in mathematics, in particular with regard to the axiom of choice, a debate that has received careful analysis in the standard monograph [Moore 1982].

Some Subsequent Developments

These dramatic events from 1904 mark a major turning point in the early history of set theory. Cantor's continuum hypothesis would remain for the next several decades a major open problem that Hilbert and others tried to tackle, but without success. Here I will only make a few remarks that bear directly on the story told here. Reading Zermelo's postcard to Dehn, one might easily be left with the impression that Julius König emerged as the big loser after Heidelberg, but this would be a far too narrow perspective, as I will now indicate.

König had deduced the inequality given above, that is $\mathbf{a}^{\aleph_0} > \mathbf{a}$, when $\mathbf{a}$ equals the sum of an increasing sequence of cardinal numbers, by making use of the following theorem: if $1 < \mathbf{m}_1 < \mathbf{m}_2 < \cdots$, then

$$\sum_{i=1}^{\infty} \mathbf{m}_i < \prod_{i=1}^{\infty} \mathbf{m}_i.$$

This theorem would soon be recognized as one of the fundamental results in set theory. It was generalized in 1907 by the Russian expert in set theory and logic, Iwan Iwanowitsch Shegalkin (1869–1947), and today this result appears in standard textbooks as the Theorem of König ([Shegalkin 1907], 332 ff.). This states that if A is an arbitrary index set and if $\{\mathbf{a}_\alpha\}_{\alpha \in A}$, $\{\mathbf{b}_\alpha\}_{\alpha \in A}$ are sets of cardinal numbers with $\mathbf{a}_\alpha < \mathbf{b}_\alpha$ for all α, then

$$\sum_{\alpha \in A} \mathbf{a}_\alpha < \prod_{\alpha \in A} \mathbf{b}_\alpha.$$

This generalization was also found by Zermelo, independent of Shegalkin, in [Zermelo 1908].

The Theorem of König leads to a partial result for the continuum problem by showing that certain alephs cannot equal the power of the continuum. Consider $\aleph_1 < \aleph_2 < \cdots$ and $\aleph_\omega = \aleph_1 + \aleph_2 + \cdots$. It follows from König's Theorem that

$$\aleph_\omega^{\aleph_0} > \aleph_\omega.$$

For $\aleph$, however:

$$\aleph^{\aleph_0} = \left(2^{\aleph_0}\right)^{\aleph_0} = 2^{\aleph_0 \cdot \aleph_0} = 2^{\aleph_0} = \aleph;$$

from which it follows that $\aleph$ cannot equal $\aleph_\omega$. Schoenflies went too far, on the other hand, when he claimed in 1908 that $\aleph$ could never equal $\aleph_\beta$ when β is a limit ordinal ([Schoenflies 1908], 25). By means of Hausdorff's fundamental concept of cofinality, Lindenbaum and Tarski were able to prove in 1926 that $\aleph \neq \aleph_\beta$ for every limit ordinal β that is cofinal with ω ([Lindenbaum/Tarski 1926], 310).

Cantor's continuum hypothesis led to intensive investigations, especially in the circle of mathematicians who published in the journal *Fundamenta Mathematicae*. Several of these experts sought other equivalent formulations of the continuum hypothesis, or they drew consequences from it for descriptive set theory, point set topology, the theory of real functions, measure theory, as well as functional analysis. As an example of an assertion equaivalent to the continuum hypothesis we mention the following theorem: The statement $2^{\aleph_0} = \aleph_1$ holds if and only if (1) every set $P \subset (-\infty, \infty)$ with power $< \aleph$ has measure 0, and (2) there exists a set $Q \subset (-\infty, \infty)$ with card$(Q) = \aleph$, and for which every set of measure 0 has at most countably many points in Q.

A high point with regard to these efforts came with Wacław Sierpiński's (1882–1969) monograph, *Hypothèse du Continu* ([Sierpiński 1934]), which presented a provisional synthesis of this research. Hausdorff continued to take a deep interest in these developments,[25] but judging at least from the extant material in his posthumous papers, he did not take up the related developments in logic and metamathematics. And it was precisely from this direction that the resolution of the continuum problem was eventually achieved.

Taking (**ZF**) to stand for the Zermelo-Fraenkel system of axioms for set theory and (**CH**) for the assertion that $\aleph = \aleph_1$, Kurt Gödel (1906–1978) proved in [Gödel 1938] that:

(1) If (**ZF**) is consistent, then so is (**ZF**) + (**CH**).

Almost three decades later, Paul J. Cohen (1934 –) showed in two fundamental papers ([Cohen 1963], [Cohen 1964]) that:

(2) If (**ZF**) is consistent, then so is (**ZF**) + non(**CH**). This result and, above all, Cohen's technique of forcing led to an explosion of new metamathematical investigations in this direction that have lasted till the present day.

Finally, we should ask: to what extent can one say that the results of Gödel and Cohen constitute the solution of the continuum problem? From the point of view of research on the foundations of mathematics, the problem is certainly now solved, since it has been proved that the assertion $\aleph = \aleph_1$ is undecidable in (**ZF**). One can thus do mathematics on the basis of (**ZF**) + (**CH**), or one may choose one of the (infinitely many) systems of mathematics in which (**ZF**) + non(**CH**) holds (for example, ((**ZF**) + ($\aleph = \aleph_2$) or (**ZF**) + ($\aleph = \aleph_3$) or $\cdots$). Nevertheless, Gödel and especially Cohen, already took into consideration that research into the further implications of (**ZF**) + (**CH**) on the one hand, and results obtained in systems for which (**ZF**) + non(**CH**) holds, on the other, could lead to a decision after all, though not on purely epistemological grounds. Foundational research, Cohen believed, might lead mathematicians (or at least a majority within the research

[25]His Nachlass contains a 78-page text with excerpts from Sierpiński's book, which Hausdorff reworked in his own notation and terminology, ocassionally simplifying some of the proofs, while adding critical comments; (Nachlass Hausdorff: Kapsel 42: Fasz. 729, dated August, September 1938).

community) to take the standpoint that in order to do reasonable mathematics it would be beneficial to add (**CH**), or an equivalent alternative, to the standard axiom system that mathematicians are accustomed to working with. That would mean reaching a definite decision, but in a very different sense from the one Cantor dreamed of with his conception of a universe of sets as an eternal truth of the divine intellect.[26] Yet this modern point of view, as seen from the philosophical standpoint expressed by Hausdorff in his book *Das Chaos in kosmischer Auslese*, would, in fact, harmonize quite wonderfully.[27]

Acknowledgement

Translated from the German by David Rowe. The author wishes to express special thanks to the North Rhine-Westphalian Academy of Sciences, Humanities, and the Arts for generously supporting the Hausdorff Edition and the many research projects related to it.

Bibliography

[Alexandroff 1916] Alexandroff, Paul, Sur la puissance des ensembles mesurables B., *Comptes Rendus de l'Académie des Sciences Paris*, 162: 323–325.

[Cantor 1870] Cantor, Georg, Beweis, daß eine für jeden reellen Wert von x durch eine trigonometrische Reihe gegebene Funktion f(x) sich nur auf eine einzige Weise in dieser Form darstellen läßt. *Journal für die reine und angewandte Mathematik* 72 (1870), 139–142. Wiederabdruck: [Cantor 1932], 80–83.

[Cantor 1872] Cantor, Georg, Über die Ausdehnung eines Satzes aus der Theorie der trigonometrischen Reihen. *Mathematische Annalen* 5 (1872), 123–132. Wiederabdruck: [Cantor 1932], 92–101.

[Cantor 1874] Cantor, Georg, Über eine Eigenschaft des Inbegriffs aller reellen algebraischen Zahlen. *Journal für die reine und angewandte Mathematik* 77 (1874), 258–262. Wiederabdruck: [Cantor 1932], 115–118.

[Cantor 1878] Cantor, Georg, Ein Beitrag zur Mannigfaltigkeitslehre. *Journal für die reine und angewandte Mathematik* 84 (1878), 242–258. Wiederabdruck: [Cantor 1932], 119–133.

[Cantor 1883] Cantor, Georg, Über unendliche lineare Punctmannichfaltigkeiten. Nr. 5. *Mathematische Annalen* 21 (1883), 545–586. Wiederabdruck: [Cantor 1932], 165–209.

[Cantor 1884] Cantor, Georg, Über unendliche lineare Punctmannichfaltigkeiten. Nr. 6. *Mathematische Annalen* 23 (1984), 453–488. Wiederabdruck: [Cantor 1932], 210–246.

---

[26] See Cantor's letter to Hermite from 30 November 1895, in [Purkert/Ilgauds 1987], 105; on Cantor's theological interests, see [Dauben 1979], 288-291.

[27] See the remarks in [Koepke 1996].

[Cantor 1885] Cantor, Georg, Über verschiedene Theoreme aus der Theorie der Punktmengen in einem n-fach ausgedehnten stetigen Raume G_n. Zweite Mitteilung. *Acta Mathematica* 7 (1885), 105–124. Wiederabdruck: [Cantor 1932], 261–277.

[Cantor 1895/1897] Cantor, Georg, Beiträge zur Begründung der transfiniten Mengenlehre. *Mathematische Annalen* 46 (1895), 481–512; 49 (1897), 207–246. Wiederabdruck: [Cantor 1932], 282–356.

[Cantor 1932] Cantor, Georg, *Gesammelte Abhandlungen mathematischen und philosophischen Inhalts.* Hrsg., Ernst Zermelo, Springer-Verlag, Berlin 1932.

[Cantor 1991] *Cantor, Georg, Briefe.* Hrsg. Herbert Meschkowski und W. Nilson. Springer-Verlag, Berlin 1991.

[Cohen 1963] Cohen, Paul J., The Independence of the Continuum Hypothesis. I. *Proceedings of the National Academy of Sciences* (U.S.A.) 50 (1963), 1143–1148.

[Cohen 1964] Cohen, Paul J., The Independence of the Continuum Hypothesis. II. *Proceedings of the National Academy of Sciences* (U.S.A.) 51 (1964), 105–110.

[Dauben 1979] Dauben, Joseph W., *Georg Cantor. His Mathematics and Philosophy of the Infinite.* Princeton University Press. 1979.

[Ebbinghaus 2007a] Ebbinghaus, Heinz- Dieter, Zermelo and the Heidelberg Congress 1904. *Historia Mathematica* 34 (2007), 428-432.

[Ebbinghaus 2007b] Ebbinghaus, Heinz–Dieter, *Ernst Zermelo. An Approach to His Life and Work.* Springer-Verlag, Heidelberg. 2007.

[Ferreirós 1999] Ferreirós, José, *Labyrinths of Thought. A History of Set Theory and its Role in Modern Mathematics.* Birkhäuser, Basel–Boston–Berlin 1999.

[Gödel 1938] Gödel, Kurt, The Consistency of the Axiom of Choice and of the Generalized Continuum-Hypothesis. *Proceedings of the National Academy of Sciences* (U.S.A.) 24 (1938), 556–557.

[Hausdorff 1891] Hausdorff, Felix, Zur Theorie der astronomischen Strahlenbrechung. I. *Berichte über die Verhandlungen der Königlichen Sächsischen Gesellschft der Wissenschaften zu Leipzig, Math.-phys. Classe* 43 (1891), 481–566. Wiederabdruck: [Hausdorff 2006], 5-90.

[Hausdorff 1893] Hausdorff, Felix, Zur Theorie der astronomischen Strahlenbrechung. II, III. *Berichte über die Verhandlungen der Königlichen Sächsischen Gesellschft der Wissenschaften zu Leipzig, Math.-phys. Classe* 45 (1893), 120–162, 758–804. Wiederabdruck: [Hausdorff 2006], 93-182

[Hausdorff 1895] Hausdorff, Felix, Über die Absorption des Lichtes in der Atmosphäre. *Berichte über die Verhandlungen der Königlichen Sächsischen*

Gesellschft der Wissenschaften zu Leipzig, Math.-phys. Classe 47 (1895), 401–482. Wiederabdruck: [Hausdorff 2006], 215-296.

[Hausdorff 1896] Hausdorff, Felix, Infinitesimale Abbildungen der Optik. *Berichte über die Verhandlungen der Königlichen Sächsischen Gesellschft der Wissenschaften zu Leipzig, Math.-phys. Classe* 48 (1896), 79–130. Wiederabdruck: [Hausdorff 2006], 315-366.

[Hausdorff 1897a] Hausdorff, Felix, Das Risico bei Zufallsspielen. *Berichte über die Verhandlungen der Königlichen Sächsischen Gesellschft der Wissenschaften zu Leipzig, Math.-phys. Classe* 49 (1897), 497–548. Wiederabdruck: [Hausdorff 2006], 445-496.

[Hausdorff 1897b] Hausdorff, Felix (Paul Mongré), *Sant' Ilario. Gedanken aus der Landschaft Zarathustras.* Verlag C. G. Naumann, Leipzig 1897. Wiederabdruck: [Hausdorff 2004], 87-473.

[Hausdorff 1898] Hausdorff, Felix (Paul Mongré), Das Chaos in kosmischer Auslese – Ein erkenntniskritischer Versuch. Verlag C. G. Naumann, Leipzig 1898. Wiederabdruck: [Hausdorff 2004], 589-807.

[Hausdorff 1901] Hausdorff, Felix, Über eine gewisse Art geordneter Mengen. *Ber. über die Verhandlungen der Königl. Sächs. Ges. der Wiss. zu Leipzig, Math.-phys. Classe* 53 (1901), 460–475. Wiederabdruck: [Hausdorff 2013], 5-20.

[Hausdorff 1904] Hausdorff, Felix, Der Potenzbegriff in der Mengenlehre. *Jahresbericht der DMV* 13 (1904), 569–571. Wiederabdruck: [Hausdorff 2013], 31-33.

[Hausdorff 1914] Hausdorff, Felix, *Grundzüge der Mengenlehre.* Veit & Co., Leipzig 1914. Wiederabdruck: [Hausdorff 2002], 93-576.

[Hausdorff 1916] Hausdorff, Felix, Die Mächtigkeit der Borelschen Mengen. *Mathematische Annalen* 77 (1916), 430–437. Wiederabdruck: [Hausdorff 2008], 431-438.

[Hausdorff 2001] Hausdorff, Felix, *Gesammelte Werke.* Band IV: *Analysis, Algebra und Zahlentheorie.* Springer-Verlag, Heidelberg. 2001.

[Hausdorff 2002] Hausdorff, Felix, *Gesammelte Werke.* Band II: *Grundzüge der Mengenlehre.* Springer-Verlag, Heidelberg. 2002.

[Hausdorff 2004] Hausdorff, Felix, *Gesammelte Werke.* Band VII: *Philosophisches Werk.* Springer-Verlag, Heidelberg. 2004.

[Hausdorff 2006] Hausdorff, Felix, *Gesammelte Werke.* Band V: *Astronomie, Optik und Wahrscheinlichkeitstheorie.* Springer-Verlag, Heidelberg. 2006.

[Hausdorff 2008] Hausdorff, Felix, *Gesammelte Werke.* Band III: *Mengenlehre (1927, 1935). Deskriptive Mengenlehre und Topologie.* Springer-Verlag, Heidelberg. 2008.

[Hausdorff 2012] Hausdorff, Felix, *Gesammelte Werke.* Band IX: *Korrespondenz.* Springer-Verlag, Heidelberg. 2012.

[Hausdorff 2013] Hausdorff, Felix, *Gesammelte Werke.* Band IA: *Allgemeine Mengenlehre.* Springer-Verlag, Heidelberg. 2013.

[Heine 1870] Heine, Eduard, Über trigonometrische Reihen. *Journal für die reine und angewandte* Mathematik 71 (1870), 353–365.

[Hessenberg 1906] Hessenberg, Gerhard, Grundbegriffe der Mengenlehre. *Abhandlungen der Friesschen Schule.* Bd. I, Heft 4 (1906), 478–706. Auch selbständig gebunden erschienen bei Vandenhoek & Ruprecht, Göttingen 1906.

[Hilbert 1900] Hilbert, David, Mathematische Probleme. Vortrag, gehalten auf dem Internationalen Mathematikerkongreß zu Paris, 1900. *Nachrichten der Königl. Ges. der Wiss. zu Göttingen. Math.- phys. Klasse* 1900, 253–297. Wiederabdruck: David Hilbert, *Gesammelte Abhandlungen*, Band III, Springer-Verlag, Berlin 1935, 339–364.

[Koepke 1996] Koepke, Peter, Metamathematische Aspekte der Hausdorffschen Mengenlehre. In: Brieskorn, Egbert. (Ed.): *Felix Hausdorff zum Gedächtnis – Aspekte seines Werkes.* Vieweg, Braunschweig–Wiesbaden 1996, 71–106.

[Kowalewski 1950] Kowalewski, Gerhard, *Bestand und Wandel.* Oldenbourg, München 1950.

[Krazer 1905] Krazer, Adolf, (Hrsg.), *Verhandlungen des dritten Internationalen Mathematiker-Kongresses in Heidelberg vom 8. – 13. August 1904.* Teubner, Leipzig 1905.

[Lindenbaum/Tarski 1926] Lindenbaum, Adolf; Tarski, Alfred, Communication sur les recherches de la théorie des ensembles. *Comptes Rendus des Séances de la Société des Sciences et des Lettres de Varsovie,* Classe III, 19 (1926), 299–330.

[Meschkowski 1967] Meschkowski, Herbert, *Probleme des Unendlichen. Werk und Leben Georg Cantors.* Vieweg, Braunschweig 1967, 1983^2.

[Moore 1982] Moore, Gregory H., *Zermelo's Axiom of Choice. Its Origins, Development and Influence.* Springer-Verlag, New York–Heidelberg–Berlin 1982.

[Oberschelp 1994] Oberschelp, Arnold, *Allgemeine Mengenlehre.* BI, Mannheim–Leipzig–Wien 1994.

[Peckhaus 1991] Peckhaus, Volker, *Hilbertprogramm und kritische Philosophie.* Vandenhoek & Ruprecht, Göttingen 1991.

[Purkert 1986] Purkert, Walter, Georg Cantor und die Antinomien der Mengenlehre. *Bulletin de la Société Mathématique de Belgique,* , Ser. A, 38 (1986), 313–327.

[Purkert/Ilgauds 1987] Purkert, Walter; Ilgauds, Hans Joachim, *Georg Cantor 1845–1918.* Vita Mathematica, Band I. Birkhäuser, Basel 1987.

[Purkert 2000] Purkert, Walter, An der Wiege grundlegender Begriffe—Episoden aus der Geschichte der Theorie der trigonometrischen Reihen. In: *Mathematikdidaktik aus Begeisterung für die Mathematik.* Festschrift für Harald Scheidt. Hrsg. von J. Blankenagel und W. Spiegel. Klett- Verlag, Stuttgart 2000, S. 162–199.

[Schoenflies 1905] Schoenflies, Arthur, Über wohlgeordnete Mengen. *Mathematische Annalen* 60 (1905), 181–186.

[Schoenflies 1908] Schoenflies, Arthur, Die Entwickelung der Lehre von den Punktmannigfaltigkeiten. Teil II. *Jahresbericht der Deutschen Mathematiker-Vereinigung,* 2. Ergänzungsband, Teubner, Leipzig 1908.

[Schoenflies 1922] Schoenflies, Arthur, Zur Erinnerung an Georg Cantor. *Jahresbericht der Deutschen Mathematiker-Vereinigung* 31 (1922), 97–106.

[Shegalkin 1907] Shegalkin, Ivan Ivanovich, *Transfinitnyje Tschisla.* (Transfinite Zahlen). Universitätsdruckerei, Moskau 1907.

[Sierpínski 1934] Sierpínski, Waclaw, *Hypothèse du Continu.* Monografie Matematyczne, vol. IV. Garasínski, Warszawa 1934.

[Young 1903] Young, William H., Zur Lehre der nicht abgeschlossenen Punktmengen. *Berichte über die Verhandlungen der Königlichen Sächsischen Gesellschft der Wissenschaften zu Leipzig, Math.-phys. Classe* 55 (1903), 287–293.

[Zermelo 1904] Zermelo, Ernst, Beweis, daß jede Menge wohlgeordnet werden kann. Aus einem an Herrn Hilbert gerichteten Briefe. *Mathematische Annalen* 59 (1904), 514–516.

[Zermelo 1908] Zermelo, Ernst, Untersuchungen über die Grundlagen der Mengenlehre. I. *Mathematische Annalen* 65 (1908), 261–281.

Nonstandard Analysis, Infinitesimals, and the History of Calculus

Craig Fraser

Introduction

Carl Boyer's *The Concepts of the Calculus; a Critical and Historical Discussion of the Derivative and the Integral* was published in 1939 and reprinted in several later editions. Its title harkened back to Ernst Mach's *Science of Mechanics* of 1883 and the conviction that a fuller understanding of a scientific subject was revealed through an examination of its history. Boyer's presentation of the history of calculus encapsulated the prevailing view, one that is still fairly common today. According to this, the first 130 years or so was a "period of indecision," in which infinitesimals were commonly used, series were deployed without attention to questions of convergence and divergence, and new results were prized above all else. In the nineteenth century, this new calculus was then put on a sound basis with the emergence of real analysis, the leading contributors to this development being such figures as Augustin Cauchy, Bernhard Bolzano, P. G. L. Dirichlet, Richard Dedekind, and Karl Weierstrass. The later development embodied an ethos of logical rigor that was absent in the discovery-oriented early years of the subject.

Although infinitesimals and divergent series were apparently banished from mathematics in the post-Cauchy era, they never quite disappeared. A whole new branch of analysis, the theory of summability, emerged at the end of the nineteenth century from the ashes of the older work on divergent series. Attempts to develop a logically rigorous theory of infinitesimals were also pursued by such mathematicians as Giuseppe Veronese, Paul du Bois-Reymond and Otto Stolz. (For a study of these researches see [Fisher 1981] and [Ehrlich 2006].) While the work of the infinitesimalists was noteworthy, they were not as successful as were the inventors of summability theory, and no viable and widely accepted non-Archimedean theory of analysis emerged during the nineteenth century.

About a quarter of a century after the appearance of Boyer's book, new approaches to analysis were advanced that employed infinitesimals. The best known and most influential of these theories was published between 1961 and 1966 by Abraham Robinson, a researcher in model theory in mathematical logic. Robinson's nonstandard analysis was based on a result known as the compactness theorem in model theory. Consider a formalized language that is rich enough to formulate all sentences that are true in the real numbers **R**. The main idea in a somewhat simplified form goes as follows. (For details, see [Enderton 1972], 164-173.) If every finite subset of a given set K of sentences has a model, then by the compactness theorem K itself has a model. For each real number consider the sentence that asserts the existence of a number v such that $c < v$. Let K be the set consisting of all such sentences. It is evident that any finite subset of K will have

D.E. Rowe and W.-S. Horng (eds.), *A Delicate Balance: Global Perspectives on Innovation and Tradition in the History of Mathematics*,
Trends in the History of Science, DOI 10.1007/978-3-319-12030-0_2

R as a model, by selecting v greater than the maximum number in the subset. We conclude by the compactness theorem that K itself has a model $\mathbf{R}^*$, which will be a nonstandard extension of **R**. The element v will be infinite, and an infinitesimal is obtained by taking its reciprocal. $\mathbf{R}^*$ contains infinitesimal numbers, numbers that are smaller than any positive real number but are non-zero.

Robinson's nonstandard analysis embodied a formulation of calculus that used infinitesimals in accord with modern standards of rigor. Authors such as Jerome Keisler [Keisler 1976] wrote calculus textbooks based on non-Archimedean fields that dispensed with the framework of mathematical logic but nonetheless presented infinitesimal procedures in a rigorous way. Since the 1950s researchers have developed other approaches to non-Archimedean analysis, including Curt Schmieden and Detlef Laugwitz's Ω-calculus [Schmieden and Laugwitz 1958], John Conway's [Conway 1976] surreal numbers, and John Bell's smooth infinitesimal analysis [Bell 2008].

In the concluding chapter of his 1966 book Robinson reflected on the history of calculus and the relevance of nonstandard analysis for understanding past mathematics. This line of inquiry was also the subject of his essay "Metaphysics of the calculus," published in the next year. Robinson believed that it was necessary "to redraw" the older history of calculus in light of the rehabilitation of infinitesimals entailed by nonstandard analysis.

Robinson's views were embraced by philosopher Imre Lakatos in an essay written in 1966 and published posthumously in 1978, who used such terms as "revolutionary" and "epoch-making" to describe the significance of nonstandard analysis for understanding the history of calculus. Lakatos was passionately engaged in history and like Robinson believed that one's view of the history of calculus was being distorted through the lens of the modern Weierstrass foundation. Both Robinson and Lakatos focused on Cauchy, who was seen as a more traditional figure than historians such as Boyer had supposed and who continued to deploy infinitesimal methods in his work. According to the conventional historical view Cauchy initiated the modern rigorous formulation of the calculus, a project that was continued and elaborated in more critical detail by Weierstrass. Robinson challenged this view, calling attention to the fact that Cauchy continued to use the language of infinitesimals and employed arguments that were not rigorous by later standards. He maintained that the modern formulation of the calculus that became entrenched by the end of the nineteenth century should be credited to Weierstrass and his contemporaries and not to Cauchy, the latter having been part of the tradition that extended back to Leibniz. Although Lakatos was more qualified than Robinson in developing this argument, he accepted the broad outlines of the thesis. The view that the original formulation by the Leibniz school of calculus in terms of infinitesimals was somehow an unsatisfactory first step on route to the nineteenth-century foundation was of course also rejected.

Other mathematicians were motivated to revisit Cauchy's work on analysis in light of modern non-Archimedean theories. Included in this group were Gordon Fisher [Fisher 1978], and Laugwitz [Laugwitz 1987]. A substantial literature has developed that examines Cauchy's treatment of various results in analysis. The most active worker here has been Laugwitz, who some two decades following the publication by Schmieden and him of the Ω-calculus commenced to publish a series of articles arguing that their non-Archimedean formulation of analysis is well suited to interpret Cauchy's results on series and integrals.[1]

Since the 1960s there has been a new wave of writing about the history of eighteenth-century mathematics. Authors such as Henk Bos, Steven Engelsman, Niels Jahnke, Giovanni Ferraro, Craig Fraser and Marco Panza have charted the development of calculus without interpreting this development as a first stage in the inevitable evolution of an arithmetic foundation. These scholars also have not experienced any particular need to investigate the historical materials in light of recent modern developments in analysis. In their work the dynamics of the growing subject – the "metaphysics of the calculus" – is analyzed on its own terms in its original historical setting. The extent and detailed character of this literature highlight the fairly limited knowledge of the eighteenth-century subject that Robinson and Lakatos brought to their ruminations on the history of analysis.

The new history has focused on uncovering the guiding principles and original outlook of calculus in the period from 1680 to 1820. Of course, classical analysis developed out of the older subject and it remains a primary point of reference for understanding the eighteenth-century theories. By contrast, nonstandard analysis and other non-Archimedean versions of calculus emerged only fairly recently in somewhat abstruse mathematical settings that bear little connection to the historical developments one and a half, two or three centuries earlier. Hence there is a natural tendency either to reject or to overlook nonstandard analysis in arriving at an appraisal of the early calculus. This point has been emphasized by Bos ([Bos 1974], 81-82):

> I do not think that the appraisal of a mathematical theory, such as Leibniz's calculus, should be influenced by the fact that two and three quarter centuries later the theory is "vindicated" in the sense that it is shown that the theory can be incorporated in a theory which is acceptable by present-day mathematical standards.
>
> If the Leibnizian calculus needs a rehabilitation because of too severe treatment by historians in the past half century, as Robinson suggests ([Robinson 1966], 260), I feel that the legitimate grounds for such a rehabilitation are to be found in the Leibnizian theory itself. I believe that, in order to prove its value as a mathematical theory, Leibniz's calculus does not need an adjustment to twentieth century requirements.

---

[1] A balanced discussion of Cauchy historiography is provided by Umberto Bottazzini [Bottazzini 1990] in an introduction to an edited reprint of Cauchy's *Cours d'Analyse*. Concerning Lakatos's claim that the discovery of nonstandard analysis will be "epoch making" for the history of mathematics, Bottazzini writes ([Bottazzini 1990] p. lxxxvii), "Apparently this prophecy has not been fulfilled." Teun Koetsier ([Koestier 1991], Chapters 3 and 4) also gives an illuminating account of the views of Laugwitz and Spalt in his monograph on Lakatos; the penultimate chapter on the equality of mixed partial derivatives is also very worthwhile.

In 1988 Joseph Dauben published "Abraham Robinson and nonstandard analysis: history, philosophy, and foundations of mathematics." Here Dauben accepts the historians' view that nonstandard analysis has limited relevance for an understanding of the early calculus, and he criticizes some of Robinson's claims about Cauchy. He observes that Robinson's larger purpose was to show that model theory and logic were not peripheral parts of mathematics but were linked to subjects of central concern to mathematicians. According to Dauben, Robinson's claim that the history of calculus needed to be rewritten in light of the discovery of nonstandard analysis was simply "propaganda" in his efforts to establish this thesis. (In his biography of Robinson, Dauben ([Dauben 1995], 349-266), provides further discussion of Robinson's views on the history of mathematics.)

In 2013 a group of researchers working on non-Archimedean analysis advocated views similar to those of Robinson, Lakatos and Laugwitz. In their article "Is mathematical history written by the victors," ([Bair et al. 2013], abstract) assert that modern theories of infinitesimals "show the works of Fermat, Leibniz, Euler, Cauchy and other giants of infinitesimal mathematics in a new light," that several of the procedures of the older calculus "were only clarified and formalized with the advent of modern infinitesimals." Although analysis went through several distinct historical stages and involved many different strands of research, the authors believe that modern theories of infinitesimals provide a valuable touchstone to help us understand and interpret this history.

In considering the historiography of analysis we encounter two distinct strands of writings, one that is motivated by present-day mathematical perceptions to look at history, and the other focused primarily on understanding the development of ideas and methods in the context of their time. One must ask whether a given method originally developed in a particular way and with a particular purpose is in need of re-examination in light of modern theories. In the following study we focus primarily on two distinct but related topics. The first concerns the way in which infinitesimals were understood in the early years of the calculus, and the status of the logical critique of the calculus presented by contemporary observers such as Bishop Berkeley. We maintain that an understanding of the early developments requires attention to the dual geometric-algebraic character of the basic processes of the calculus. The question of the logical status of infinitesimals is of secondary interest, from either the perspective of a researcher in the early eighteenth century or an observer today. The second topic concerns the decisive shift to algebraic analysis that occurred in the writings of such figures as Euler and Lagrange in the second half of the century. The transition from algebraic analysis to the modern paradigm of Cauchy and Weierstrass constituted a fundamental change in outlook. Our central claim here is that the emergence of modern analysis in the nineteenth century should be viewed not in terms of the rejection of infinitesimals and divergent series but rather in terms of this broader change in perspective.

Leibniz, l'Hôpital and the early calculus

Infinitesimals in late seventeenth-century mathematical science have been the subject of a large body of recent historical writings, as is evident in the 2008 collection of essays "Infinitesimal Differences Controversies between Leibniz and his Contemporaries" [Goldenbaum and Jesseph 2008] and the references contained therein. We will not attempt to survey the various issues explored by historians, but concentrate on the basic question of how infinitesimals functioned in the logical workings of the new subject. In particular, we examine the first book devoted to the new calculus, the Marquis de l'Hôpital's *Analyse des infiniments petits pour l'intelligence des lignes courbes* of 1696.

Bernard le Bovier de Fontenelle ([Fontenelle 1790-1792] 6:43), the permanent secretary of the Paris Academy of Sciences, had in 1727 included l'Hôpital among the "great" mathematicians who laid the foundations for an "epoch of almost total revolution in science." Two decades earlier in an éloge to l'Hôpital Fontenelle ([Fontenelle 1790-1792], 6:31) observed that the popularity of the *Analyse* showed that "the revolution will become still greater." He ([Fontenelle 1790-1792], 7:67) wrote again in 1720 in connection with this book that "there was in the mathematical world a well-marked revolution [une révolution bien marquée]." This rhetoric of revolution, less common and more significant in l'Hôpital 's time than it is today, indicates the degree of excitement aroused by the new mathematical discoveries. (English translation by Cohen ([Cohen 1985], 213-214). Cohen regards Fontenelle's remarks of note because they are an early explicit statement by a contemporary observer of a revolution taking place in science.)

As innovative as l'Hôpital may have been, the theory he detailed in his book owed a great deal to Johann Bernoulli, who had explained the new mathematical methods to the Frenchman during his stay in Paris in the early 1690s and in letters to him. The contents of the *Analyse* were very indicative of the technical concerns shared by researchers of the period, which remained strongly anchored in the study of curves and their properties. One area of calculus that is particularly revealing in highlighting fundamental conceptions is the investigation of maxima and minima, a subject that illustrates well how new ideas played out in a clearly defined mathematical problem context. This subject was taken up by l'Hôpital in section three of his book. We shall illustrate the theory with the example $y = x(a-x)$, where a is a constant, ([l'Hôpital 1696] §51, Example 4 of the *Analyse*, with $n = m = 1$), and use this example to evaluate some of the logical issues associated with the use of infinitesimals. These issues would later arise in polemical form in Bishop George Berkeley's *Analyst* of 1734.

If $y = x(a-x)$ then we have $dy = (x+dx)(a-x-dx) - x(a-x) = (a-2x)dx - dx^2$. l'Hôpital reasoned that because dx^2 is infinitely small with respect to dx it follows that $dy = (a-2x)dx$. Let us now consider the problem of finding the maximum of the curve given by $y = x(a-x)$. The situation was illustrated in several figures, of which Figure 1 applies in this case. (In the figure MR is dx and Rm is dy.) l'Hôpital reasoned in general that around a value x_0

for which y is a maximum, dy decreases as x approaches x_0, changes sign at x_0, and then increases in absolute value as x increases. At $x = x_0$ the corresponding differential dy must be 0 and at this point y is a maximum. In the example at hand we set $dy = (a - 2x)dx = 0$ and so $2x = a$ and there is a maximum at $x = a/2$.

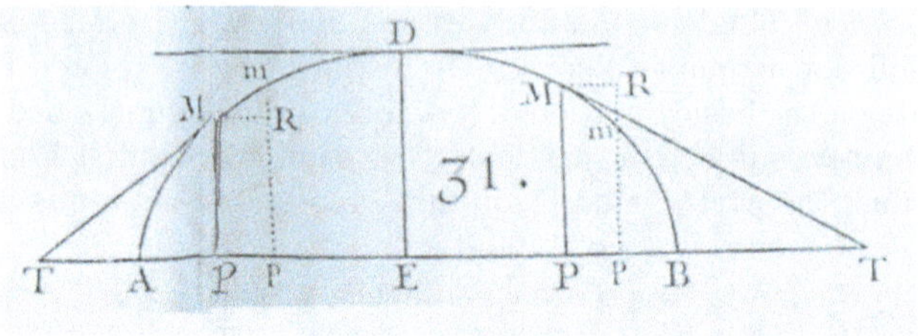

Figure 1: MR is dx and Rm is dy. Not only does Rm tend to zero as M approaches D, Rm also becomes arbitrarily small with respect to MR as M approaches D.

Authors such as Bishop Berkeley [Berkeley 1734] believed that there was a logical defect in the application of the differential calculus to find tangents to curves. The criticism in question would apply to the present example in the following way. Going back to first principles we have $dy = (a - 2x)dx - dx^2 = (a - 2x - dx)dx$. If $dy = 0$ then $(a - 2x - dx)dx = 0$. If $dx \neq 0$ then it follows that $a - 2x - dx = 0$. Setting $dx = 0$ we have $a - 2x = 0$ or $x = a/2$. This derivation seems to require us to suppose that both $dx \neq 0$ and $dx = 0$, an obvious paralogism.

To early practitioners of the calculus this sort of criticism would not have carried much force. In the differential calculus we are positing a partition of the x-axis into intervals of constant length Δx which extrapolates to an infinite partition of intervals of constant infinitesimal length dx. For a series of constant increments Δx the value of the corresponding increments Δy will vary, depending on the value of x. As one approaches a maximum it is evident that the increment $|\Delta y|$ becomes arbitrarily small with respect to Δx . That is, $\Delta y = (a - 2x - \Delta x)\Delta x$ (in absolute value) becomes arbitrarily small with respect to Δx . Suppose $x = c$ is not equal to $a/2$. Then $a - 2c \neq 0$ and $(a - 2x - \Delta x)\Delta x = (a - 2c - \Delta x)\Delta x = (a - 2c)\Delta x - \Delta x^2$ and it is evident that this expression cannot be made arbitrarily small with respect to Δx. On the other hand if $x = a/2$ then $\Delta y = -\Delta x^2$ and $|\Delta y|$ can be made arbitrarily small with respect to Δx by making Δx sufficiently small. Hence at a maximum we must have $a - 2x = 0$ or $x = a/2$.

Moving from increments to differentials, we conclude that at a value x that is a maximum $|dy|$ is less than dx/n for all positive integers n. If $x = c \neq a/2$ then $dy = (a - 2c)dx - dx^2$, and the condition $|dy| < dx/n$ for all n does not hold. On the other hand, if $x = a/2$ then $dy = -dx^2$. Since $dx < 1/n$ for all n it follows that this condition does hold.

In the general case where y is any expression involving x, we have $dy = pdx + (\textit{second order terms})$. At a minimum or maximum we must have $p = 0$. For at a minimum or maximum $|dy|$ becomes arbitrarily small with respect to dx. If $p \neq 0$ then$|dy|$ will not be arbitrarily small with respect to dx. Hence we must have $p = 0$ at a minimum or maximum.

The reasoning underlying the use of infinitesimals was robust, as Jean d'Alembert ([d'Alembert 1754], [d'Alembert 1765]) in his articles in the *Encylopédie* tried to explain in some detail by appealing to the notion of limit. d'Alembert possessed the acutest critical sense of any of the commentators who addressed concerns about the foundations of the calculus. Berkeley's belief that what was actually going on involved a compensation of errors was an idea that lacked any real explanatory power. (For a lucid presentation of Berkeley's idea see [Youschkevitch 1971], 152-153).) The calculus worked not because there was a compensation of errors, but because the difference between $y'(x)dx$ and $y(x + dx) - y(x)$ was infinitesimally small with respect to dx.

On the other hand Berkeley and critics both before and after him did have a point, since the actual infinitesimals dx and dy clearly had a different status from the increments Δx and Δy. A quantity dx that is less than $1/n$ for all n but is not zero clearly is a magnitude very different from the traditional Archimedean magnitudes of classical mathematics.

Leibniz himself was very concerned about the status of infinitesimal entities, a concern that was expressed in greatest detail in his correspondence and manuscript writings. One commentator ([Reyes 2004], 171) has observed that "Leibniz's rhetorical engagement with the issues raised by infinitesimals reaches nearly unmanageable proportions." The large space Leibniz devoted to discussing infinitesimals indicates that he believed there were real foundational issues at stake, although he never developed a coherent theoretical strategy to deal with them. In this respect the modern development of a rigorous theory of actual infinitesimals may be of some relevance to an historical appraisal of the early calculus. If nothing else it provides some pyschological support for the conviction that this calculus should not be viewed as a naïve and logically unsatisfactory form of what later would become real analysis.

Differentials in the eighteenth-century calculus

The eighteenth-century conception of differentials and the rules that governed their use in calculus have been investigated by historians. ([Bos 1974] is the most detailed study of this sort.) In the Leibnizian calculus, the operation of differentiation possessed a dual character: algebraic/algorithmic on the one hand, and geometric on the other. The algebra comprised a set of rules that governed the use of the symbol d and was based on two postulates: $d(x + y) = dx + dy$ and $d(xy) = ydx + xdy$. Accompanying these rules there was also an order principle, according to which higher-order differentials in a given equation were to be neglected with respect to differentials of a lower order.

The differentials that appeared in a given problem could also be understood in another way: as the differences of values of a variable quantity at successive points in the geometric configuration. The differential dx was set equal to the difference of the value of x at two consecutive points infinitely close together; higher-order differentials were set equal to the difference of successive lower-order differentials. Euclidean geometry was used to analyze the properties of the curve in terms of these differentials.

An example of the geometric treatment of infinitesimals is given by l'Hopital's calculution in section 5 of the *Analyse* of the radius of curvature to a curve at a point Let M be any point on the curve AMD (Figure 2). Let m be a point on the curve infinitely close to M. The normals to the curve at M and m intersect at the center of curvature C. The distance MC is the radius of curvature. Suppose $AP = x$ and $PM = y$ are the abscissa and ordinate of M. The lines MR and Rm parallel to AP and PM are the infinitesimal increments dx and dy of x and y. Using the differential algorithm l'Hôpital was able to obtain a formula for the radius of curvature involving the differentials dx, dy and the second order differential ddy. He also derived the same formula in a purely geometric way. Let n be a point on the curve infinitely close to m (Figure 3). L'Hôpital conceived of the portion Mmn as composed of the polygonal segments Mm and mn. The second differential of y, ddy, is given as $ddy = nS - mR = nS - HS = -Hn$. By means of similar triangles l'Hôpital arrived at the same expression for the radius of curvature that he had obtained earlier analytically.

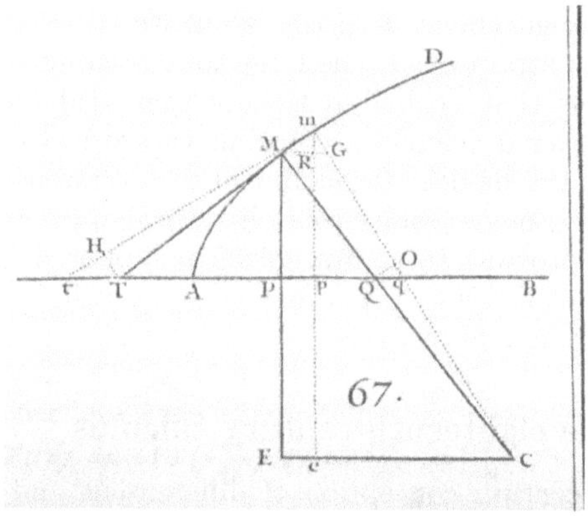

Figure 2: MC is the radius of curvature at the point M of the curve AMD.

In his mid-century treatises Euler, as part of his program of separating analysis from geometry (discussed in more detail below) made the algebraic conception of differentiation fundamental. In so doing he made the concept of the algorithm primary in his understanding of the foundations of the calculus. Some of the issues that arise in this shift in viewpoint are illustrated by his theory of differential expressions set forth in Chapters 8 and 9 of the first part of his 1755 *Institutiones Calculi.* Consider any formula containing $dx, ddx, dy, ddy, \ldots$. Because these

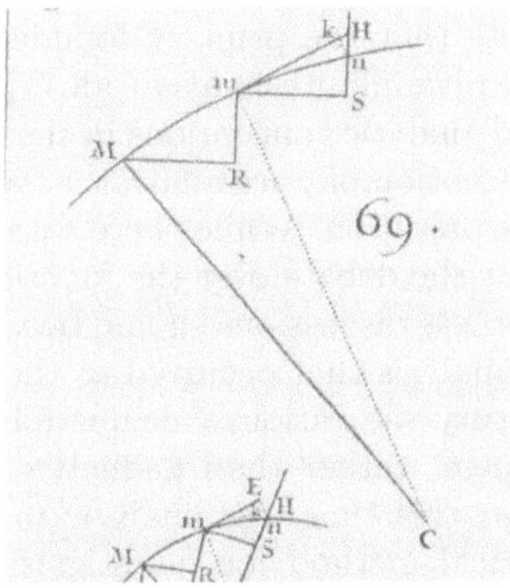

Figure 3: ddy is equal to $-Hn$.

quantities are no longer interpreted geometrically the meaning of the formula is unclear; its value will depend on whether dx or dy is held constant, an assumption that is not evident in the algebra. For example, the quantity ddy/dx^2 is zero if dy is constant; if dx is constant its value will vary according to the functional relation between x and y. Conversely, certain expressions, such as $(dyddx - dxddy)/dx^3$, may be shown to be invariant regardless of which variable is taken to be independent.

Euler's solution to the problem of indeterminacy in differential expressions was to introduce notation that made clear the relations of dependency among the variables. He did so by eliminating higher-order differentials as such, replacing them instead with differential coefficients. Rather than write ddy/dx^2 (dx constant) we define the differential coefficients p and q by the relations $dy = pdx$ and $dp = qdx$; ddy/dx^2 then becomes simply q. Euler provided rules and examples that showed how more complicated expressions can be reduced to ones containing only variables and differential coefficients. In addition to bringing order to the calculus, this emphasis on the differential coefficient was conceptually important in identifying the derivative as an independent object of mathematical study.

Shift to pure analysis

The fundamental notion of the early calculus was the operation of differentiation, understood as an algorithm or a process. As the subject developed, the notion of infinitesimal receded in importance, and the algorithm came to occupy a more prominent place. Johann Bernoulli redefined the integral as the operational inverse of differentiation, a definition that became standard in the eighteenth century. The original Leibnizian concept of an integral as a sum of infinitesimal elements was replaced by an object obtained by the inverse operation of differentiation applied to an analytic expression.

It is important to note that the primary foundational issue in eighteenth-century calculus did not involve infinitesimals at all. This issue concerned the relative place of geometric and analytic conceptions in the foundations of the subject, and the possibility of a conception of pure analysis as a subject autonomous from any geometric basis or interpretation. While there were other more specific foundational discussions, such as the debate over the function concept (which also did not involve infinitesimals), these discussions should be viewed from the perspective of the larger shift to pure analysis that occurred as the century progressed.

In the eighteenth century the various calculus-related parts of mathematics were heavily problem-oriented, rather than deductive or theorem-oriented. The emphasis of mathematics at this time was analytic in the sense that analysis is the appropriate mathematical method for investigating problems. However, by the middle of the century the term analysis had largely lost its original Pappusian meaning of "solution backwards." (For a discussion of analysis and synthesis around 1700 see [Guicciardini 2009].) As is well known, during the early modern period analysis came to denote algebra and the use more generally of symbolic methods in the solution of problems. When one introduces a variable and derives an equation, one is assuming logically at the outset that the thing that is sought is at hand, even if one does know it value. Hence methods in which the existence of the thing sought is first assumed as an unknown variable and its value is derived by means of some mathematical process are analytic. Analysis came to denote algebraic symbolic methods and was contrasted with geometric modes of solution.

In the writings of Euler and Lagrange, analysis and synthesis no longer referred to the kind of mathematical method employed – which could be either geometric or algebraic – in a given mathematical investigation. Analysis was not primarily a method or an art, but rather an autonomous branch of mathematics, a subject area in its own right employing processes that were linguistic or symbolic in character. Above all, analysis avoided geometric modes of representation. Synthesis denoted a geometric conception of the mathematical object in which this object as a whole was taken as given and in which its properties were used in the course of the investigation. (For a more detailed exposition of the Eulerian-Lagrangian conception of analysis see [Fraser 2003].)

In the problems and examples of the early calculus geometric conceptions were ubiquitous. l'Hôpital's book contained 156 figures. In the solutions of Newton, Huygens, and the Bernoullis detailed geometric knowledge of the particular problem situation at hand was often required for its solution. By contrast, analysis – understood as the application of a symbolic process given in terms of variables, operations and coordinate systems – entailed a more general and uniform approach. This fact is noted by Euler in 1736 at the beginning of his book on analytic mechanics:

> But what obtains for all the works composed without analysis holds most of all for mechanics: even if the reader be convinced of the truth of the things set forth, nevertheless he cannot attain a sufficiently clear and distinct knowledge of them; so that, if the same questions be the slightest

> bit changed, he may hardly be able to resolve them on his own, unless he himself looks to analysis and evolves the same propositions by the analytic method. Thus, I always have the same trouble, when I might chance to glance through Newton's Principia or Hermann's Phoronomiam, that comes about in using these, that whenever the solutions of problems seem to be sufficiently well understood by me, that yet by making only a small change, I might not be able to solve the new problem using this method. Thus I have endeavoured for a long time now, to get at the analysis behind those synthetic methods in order to draw out the same propositions that are more readily handled by my own analytic method, and so by working with this latter method I have gained a perceptible increase in my understanding. ([Euler 1736], Preface; translation from [Mahoney 1985], 198) and (Bruce, Euler Archive))

Euler's analytic philosophy of mathematics was embraced and developed to a new level of sophistication and purity by Lagrange. Although Lagrange's analytic tendencies were apparent from the very beginning of his career, his distinctive mathematical style really only became consolidated in the period 1770 to 1776, when he was in his late thirties and comfortably settled at the Berlin Academy. In these years the value of analysis becomes an explicit theme in his writings for the Academy on a range of subjects in pure and applied mathematics. The following is a selection from these writings, and is by no means complete. (Detailed citations for the references which follow are given in ([Taton 1974], 10-20).) In a paper of 1775 on the attraction of a spheroid Lagrange attempted to show that the method of "algebraic analysis" provided a more direct and general solution than the "synthetic" or geometric approach followed by Colin Maclaurin. In his study of the rotation of a solid he advanced an alternative to the mechanical treatment of d'Alembert and Euler, one that was "purely analytic", whose merit consisted "solely in the analysis" that it employed, and which contained "different rather remarkable artifices of calculation." In a memoir on triangular pyramids he noted that his "solutions are purely analytic and can even be understood without figures"; he observed that independent of their actual utility they "show with how much facility and success the algebraic method can be employed in questions that would seem to lie deepest within the province of Geometry properly considered, and to be the least susceptible to treatment by calculation."

The theme of analysis recurred in Lagrange's writings of the later 1770s and 1780s. In a memoir submitted to the Paris Academy in 1778 on the subject of planetary perturbations he offered a method for transforming the equations of motion that would "take the place of the synthetic methods proposed until now for simplifying the calculation of perturbations in regions beyond the orbit" and that "has at the same time the advantage of conserving uniformity in the workings of the calculus". In 1780 he published a memoir on a theorem of Lambert in particle dynamics. The result in question had been demonstrated synthetically, and Lagrange expressed concern that it might be regarded as one of "the small number [of theorems] in which geometric analysis seems to be superior to algebraic analysis". His purpose was to present a simple and direct analytic proof. In the preface to his famous *Traité de la méchanique analitique*, completed around 1783, he announced that in it "no figures would be found", that all would be "reduced to the uniform and general progress of analysis." (See [Fraser 1983] for a discussion of this book.) Directness, uniformity and generality were qualities that Lagrange associated with analysis; he sometimes also mentioned simplicity. Analysis was noteworthy not simply for the results to which it leads, but also for the methods that it offered. In the writings cited above he was affirming the value of analysis in situations where an alternative geometric or mechanical treatment existed; it was the possibility of this alternative that led him to assert his own methodological preferences. One should also note the sheer preponderance of pure analysis in his work of the 1770s and 1780s in such topics as the theory of equations, Diophantine arithmetic, number theory, probability and the calculus, subjects in which explicit questions of approach or methodology did not arise.

In the 1780s a growing interest in the foundations of analysis was reflected in the decisions of the academies of Berlin and Saint Petersburg to devote prize competitions to the metaphysics of the calculus and the nature of the infinite. Lagrange's views on the subject were presented in their most developed form in his *Théorie des fonctions analytiques* of 1797. The subtitle signalled his intentions in this work: "Containing the principles of the differential calculus, freed from all consideration of infinitesimals, evanescent quantities, limits or fluxions and reduced to the algebraic analysis of finite quantities." (For studies of Lagrange's book see [Grabiner 1981] and [Fraser 2005].) Even before the arithmetic revolution of the nineteenth century, Lagrange's analytic approach led him to reject infinitesimals and kinematic conceptions as founding notions of the subject.[2]

Lagrange used the term "algebraic analysis" to designate the part of mathematics that resulted when algebra was enlarged to include calculus-related methods and functions. The central object here was the concept of an analytic function. Such a function $y = f(x)$ was given by a single analytic expression, constructed from variables and constants using the operations of analysis. The relation between y and x was indicated by the series of operations schematized in $f(x)$. The latter possessed a well-defined, unchanging algebraic form that distinguished it from other functions and determined its properties.

Lagrange's contemporary Simon Laplace was also an adherent of algebraic analysis, writing ([Laplace 1835], 465) that "by abandoning oneself to the oper-

---

[2] In his later didactic works Lagrange displayed a certain degree of self-consciousness about the history of the mathematics that he himself had played such a prominent role in shaping. The early geometric calculus had been characterized by a range of detailed procedures and methods that were tailored to the particular problems at hand. In some remarks on the development of the calculus of variations in the preceding century [Lagrange 1806] observed: "But in the current state of analysis we may regard these discussions as pointless, for they concern forgotten methods, which have given way to others that are simpler and more general. However, such discussions may yet retain some interest for those who like to follow step by step the progress of analysis, and to see how simple and general methods arise from particular questions and complicated and indirect procedures."

ations of Analysis . . . one is led, by the generality of this method and by the inestimable advantage of transforming the reasoning into mechanical procedures, to results often inaccessible to synthesis. Such is the fruitfulness of Analysis that it suffices to translate particular truths into this language in order to see emerge from their very expression a multitude of new and unexpected truths." (English translation in ([Hawkins 1977], 121).) [3] The intellectual attitudes of Lagrange and Laplace were historically specific, rooted in modes of reasoning that were common in the second half of the eighteenth century but which gradually lost currency as the nineteenth century progressed.

Algebraic analysis as an approach to mathematics was valued by other prominent researchers of the period as a powerful tool of investigation throughout science. Nicolas de Condorcet ([Condorcet 1847], 467) wrote in 1786 of Euler that he "sensed that algebraic analysis was the most comprehensive and certain instrument one can employ in all sciences, and he sought to render its usage general." (Translation from ([Rider 1990], 115).)[4] The French philosopher Auguste Comte, although a somewhat later figure, revered Lagrange and believed that he had brought mathematics to an almost completed state. In volume one of his *Cours de Philosophie Positive* Comte [Comte 1830] wrote, "Today, in fact, mathematical science is less important even for the very real and very valued knowledge that directly makes it up, than for constituting the most powerful instrument that the human mind can employ in the investigation of the laws of natural phenomena" ([Comte 1830], 112). Comte divided mathematics into an abstract part, which consisted of algebra and all of analysis, and a concrete part and asserted that "only the abstract part is purely instrumental" ([Comte 1830]. 113). According to Comte, "the sense [*l'esprit*] of mathematical analysis consists in considering magnitudes only from the point of their relations, and independently of any idea of determinative value" ([Comte 1830], 215). (For a discussion of Comte's philosophy of mathematics see [Fraser 1990].) In this conception questions about the status of infinitesimals or the meaning of imaginary numbers were very much secondary, and the primary emphasis was on operations, functions, relations and the active process of working to solutions.

[3]Thomas Hawkins ([Hawkins 1977], 122) has referred to the kind of reasoning employed by Lagrange and Laplace as "generic," and contrasted their conception of generality with the more critical approach of Cauchy and Weierstrass. Hawkins does so in a historical study of the work of the four men on systems of linear differential equations with constant coefficients, but the observation would apply to their work taken as a whole.

[4]Historian of physics John Heilbron [Heilbron 1990], looking at the later eighteenth century from the perspective of physical science, has called attention to a distinct instrumentalist orientation to the application of mathematics. Heilbron ([Heilbron 1990], 3) observes that "in the second half of the century, analysis and algebra, which, in contrast to geometry, had an instrumentalist bias, became the exemplar of the mathematical method."

Divergent series and changing philosophical perspectives

For some insight into the historiographical issues raised by modern theories of infinitesimals it is worthwhile to look at other case studies involving the evaluation of older theories in light of later mathematical developments. A good example is Euler's work on divergent infinite series in the 1750s and 1760s and the emergence of summability theory at the end of the nineteenth century. In some respects this is a more promising case study than nonstandard analysis and the early calculus, because the chronological and conceptual separation between the early and later developments is smaller. (The historical thread is well known; for an overview see [Kline 1972], Chapter 47.)

In the eighteenth-century infinite series were conceived of as things that were given as part of objective reality; they were not defined entities. A completely general infinite series $\Sigma_{i=1}^{\infty} = a_i$ where the a_i are arbitrary real numbers was never considered by Euler and his contemporaries. Infinite series were not created through definition but always come from somewhere, generated by an analytic process of some sort. As Euler [Euler 1760] put it, "infinite series find a place in analysis inasmuch as they arise from some closed expression." For example, the infinite series $1 - x + x^2 - x^3 + ...$ is generated from the expression $1/(1+x)$, since $(1+x)(1-x+x^2-x^3+...) = 1$. Setting $x = 1$ Euler concluded that the divergent series $1 - 1 + 1 - 1 + ...$ has the sum $1/(1+1)$ or $1/2$.

The logical difference in outlook between Euler and modern researchers has been recognized by such authorities as G. H. Hardy, who writes in his book *Divergent Series* ([Hardy 1949], 5-6):

> ...it does not occur to a modern mathematician that a collection of mathematical symbols should have a 'meaning' until one has been assigned to it by definition. It was not a triviality even to the greatest mathematicians of the eighteenth century. They had not the habit of definition: it was not natural to them to say, in so many words, 'by X we mean Y'. There are reservations to be made ... but it is broadly true to say that mathematicians before Cauchy asked not 'How shall we define $1 - 1 + 1 - ...$?' but 'What is $1 - 1 + 1 - ...$?',

A similar observation is made by Konrad Knopp ([Knopp 1928], 457), who writes:

> In our exposition, *the symbol for infinite sequences* was created and then worked with; it was not so originally, these sequences *were there*, and the question was, what could be done with them. [italics in the original]

Historian of mathematics Giovanni Ferraro ([Ferraro 2000], 95) has elaborated on this point:

> Today we say that the series $1-1+1-1+...$ is not summable according to the usual definition but is summable by Cesaro's method. This means that the "real" sum of the series does not exist: the sum depends on the chosen definition; one can imagine many different definitions and, therefore, many different sums of the same series. In a sense, modern definitions generate the object of study. Eighteenth century mathematicians viewed the mat-

ter differently. Definitions were clarifications of the unique and necessary "truth" that already existed in Nature. A mathematical theory was not a matter of the logical deduction of axioms and definitions, subject only to the principle of contradiction; rather, it was an idealisation derived from physical reality.

The perspective of eighteenth-century researchers on divergent series reflected the broader conception of the nature of mathematics that was prevalent during this period. Sergei Demidov ([Demidov 1982], 37), writing of the failure of Euler and d'Alembert to understand each other's point of view in the discussion of the wave equation, observes:

> A cause no less important of this incomprehension rests, in our opinion, on the understanding of the notion of a solution of a mathematical problem. For d'Alembert as for Euler the notion of such a solution does not depend on the way in which it is defined ... rather the solution represents a certain reality endowed with properties that are independent of the method of defining the solution. To reveal these properties diverse methods are acceptable, including the physical reasonings employed by d'Alembert and Euler.

A biographer of d'Alembert ([Grimsley 1963], 248) has noted his insistence on "the elementary truth that the scientist must always accept the essential 'giveness' of the situation in which he finds himself." The sense of logical freedom that developed in later mathematics – expressed, for example, in Richard Dedekind's famous statement of 1888 that numbers are free creations of the human mind – reflects aspects of the modern subject that were absent in the eighteenth century.

The situation with respect to the calculus of infinitesimals parallels the history of divergent series. With the establishment of an arithmetic basis for calculus in the nineteenth century, differentials understood as free-standing infinitesimal entities were banished from rigorous analysis. There is a conceptual gulf between Euler's proof of the theorem on the equality of mixed partial differentials and the modern Weierstrassian theorem on the equality of mixed partial derivatives ([Fraser 1989], 319-321). The emergence of nonstandard analysis in the second half of the twentieth century brought with it the resurrection of infinitesimals, and seemed to confer new respectability on the work of the calculus pioneers. Of course, the logical setting of nonstandard analysis was alien to the outlook of the eighteenth century. In the earlier period the calculus of differentials was understood as something that was given to the researcher from without, a domain of objects that could be investigated geometrically or analytically. By contrast, in nonstandard analysis the infinitesimal arises within a system of definitions and imposed axioms. Two of the most prominent features of the early calculus – the tension between analytic and geometric modes of representation and the central place occupied by the algorithm – are not reproduced at all in nonstandard analysis as defining characteristics of the subject.

Functions of a Complex Variable

In the algebraic analysis of Euler and Lagrange no distinction was made between the subjects of real analysis and complex analysis. In his *Introductio in Analysin Infinitorum* Euler ([Euler 1748], §2) wrote, "A variable quantity includes all numbers, positive and negative, whole and fractional, rational, irrational and transcendental. Even zero and imaginary numbers are not excluded in the meaning of a variable quantity." The algebraic understanding of the calculus reinforced the implicit assumption that the extension of the calculus to the complex domain raised no new issues. While there was a substantial amount of work done in the eighteenth century on problems that one today would classify as part of complex analysis, the various results never coalesced into a distinct branch of mathematics. One could by the rules of algebra manipulate expressions containing imaginaries, and it was clear in specific problems that the formalism possessed a geometric interpretation. However, the idea that one must distinguish two theories, for functions of a real and a complex variable, never arose in the 18th century. In reference to algebraic analysis in the eighteenth century, Jahnke ([Jahnke 2003], 107) writes, "...when a formula demanded an interpretation of a variable as being complex this was tacitly assumed. For Euler and Lagrange, a clear distinction between real and complex analysis as developed in the 19th century was unthinkable." (The point in question is also discussed by [Fraser 1989], 327–328).)

Imaginary or impossible numbers, and more generally analytic expressions containing what were sometimes referred to as "imaginary constants," entered calculus at an early stage in its development. A pioneer in this field was Johann Bernoulli, some of whose results were transmitted by Leibniz in an article of 1703 in the *Acta Eruditorum*. In this piece Bernoulli ([Bernoulli 1703], 30) considered an expression of the form $dt/\sqrt{-1}t$, which he referred to as the "differential of an imaginary logarithm." Bernoulli called the integrals of such expressions "imaginary logarithms" or "quadratures of imaginary hyperbolas."

In an article published in 1712 in the *Acta Eruditorum* Johann Bernoulli considered imaginary hyperbolas of the form $y = 1/(x - \sqrt{-1})$ and integrated such expressions to obtain $z = log(x - \sqrt{-1})$. As we noted in §6, Johann redefined the notion of the integral, from a quantity that was a sum of area elements, to something more like what one would call a primitive or antiderivative. It is worth noting that examples such as the one just presented would have provided motivation for such a shift. It is difficult to conceive of the area under the imaginary hyperbola $y = 1/(x-\sqrt{-1})$ or to conceive of the integral of this expression in terms of a summation process. On the other hand from the differential calculus it was known that the differential of $logx$ was dx/x. Hence $z = log(x - \sqrt{-1})$ was an antiderivative of $1/(x - \sqrt{-1})$ and could be taken as its integral, even though $y = 1/(x - \sqrt{-1})$ had no direct geometric meaning.

Cauchy was the single most important figure in the nineteenth century in establishing complex analysis as an independent branch of mathematics. His contributions to the field were developed over thirty years in a series of memoirs devoted to integration in the complex domain. Complex analysis emerged in his work as a subject with its own theorems, problems and applications. Argand and Gauss had shown that the field of complex numbers could be interpreted as points in a two-dimensional continuum. A function of a complex variable became a relation among variables connecting points in two such continua. The framework that Cauchy had developed for real analysis, involving neighborhoods, limits and pointwise definition of the derivative, was immediately generalizable to the complex domain.

With the transition from the algebraic calculus of the eighteenth century to the new analysis of the nineteenth century it became possible for a theory of functions of a complex variable to emerge as a coherent branch of mathematics. The construction of such a theory marked a major event in the history of mathematics. It is worth noting the very minor role infinitesimals and matters of ontology played in this historic shift.[5]

Interlude: Infinitesimals and mathematical science

It is of course important to remember that even today one employs infinitesimals in engineering and physics, where they play a valuable role in the derivation of physical laws and formulae that is free of any concerns in the foundations of mathematics. Even within mathematics proper researchers following Weierstrass continued to use infinitesimals, and one has to be cautious about attributing very much foundational import to this fact. An illustrative example is a paper of Ernst Schröder's from 1870, where he derived the first result in what later became known as the subject of complex dynamics. We follow Daniel Alexander's [Alexander 1994], 6-8) presentation of Schröder's theorem. Let $\phi(z)$ be a complex function which is analytic on a neighborhood of a point x which satisfies $\phi(x) = x$ with $|\phi'(x)| < 1$. Let $\phi^n(z)$ denote the n(th) iterate of $\phi(z)$, that is the *n-fold* composition of $\phi(z)$ with itself. Then for all z in some neighborhood D of x

$$\lim_{n\to\infty} \phi^n(z) = x. \tag{1}$$

In terminology that became standard later, x is an attracting fixed point for the function $\phi(z)$. To prove this result we begin by taking the Taylor expansion for $\phi(z)$ about x:

$$\phi(z) = x + \phi'(x)(z - x) + ..., \tag{2}$$

where $0 < |\phi'(x)| < 1$. If z is sufficiently close to x then the term $\phi'(x)(z - x)$ dominates in this expansion. Hence for infinitely small ϵ

$$\phi(z) = x + \phi'(x)\epsilon. \tag{3}$$

[5]For an indication of Euler's work involving functions and complex variables see [Euler 1983]. The nineteenth-century history of complex analysis is described by Bottazzini [Bottazzini 2003] and Bottazzini and Gray [Bottazzini and Gray 2013]. Smithies [Smithies 1997] is very good on Cauchy's contributions.

Taking ϕ of each side of this equation we have

$$\phi^2(z) = \phi(x + \phi'(x)\epsilon) = \phi(x) + \phi'(x)\phi'(x)\epsilon = x + \phi'^2\epsilon, \tag{4}$$

or

$$\phi^2(z) = x + \phi'^2\epsilon. \tag{5}$$

Continuing this process we obtain

$$\phi^n(z) = x + \phi'^n\epsilon. \tag{6}$$

Because $|\phi'(x)| < 1$ it follows that

$$\lim_{n\to\infty} \phi^n(z) = x. \tag{7}$$

Alexander ([Alexander 1994], 7) suggests that Schröder's use of infinitesimal arguments "is perhaps symptomatic of his isolation from the German mathematics mainstream, since it suggests that Schröder was unaware of Karl Weierstrass' (1815-1897) rigorous delta-epsilon approach to mathematics." While this may be true, it would not be difficult to recast the derivation in a form that satisfies the Weierstrassian criteria for rigor. The appearance of infinitesimals in the derivation was very different from their use by such contemporary mathematicians as Paul du Bois-Reymond, who were consciously attempting to create a viable mathematical theory of infinitesimals. In devising a satisfactory proof that captured the idea of the result, Schröder displayed the pragmatic sense of the working mathematician for whom fine points of mathematical rigor were not of primary concern.

Conclusion

The transition from algebraic analysis of the eighteenth century to Cauchy-Weierstrass analysis of the nineteenth century marked a much greater discontinuity than did the emergence of nonstandard analysis out of classical analysis in the second half of the twentieth century. Nonstandard analysis is an offshoot of modern analysis and sits solidly on the modern side of the conceptual gulf opened up by the Cauchy-Weierstrass foundation. In this respect Robinson and to a lesser degree Lakatos were mistaken in their assessment of Cauchy. Rather than try to understand Cauchy as someone who developed within a given intellectual and historical milieu, they approach the history from an essentially artificial point of view. The claim that Cauchy should be seen as a conservative figure whose approach was aligned with the Leibnizian infinitesimalist tradition constituted a misguided exercise in historical revisionism. In fact, Cauchy was a revolutionary figure in the foundations of calculus, not simply because of his contributions to rigor, but also for developing a more conceptual approach that made the numerical continuum the fundamental object of reference.

The work of Euler, Lagrange and their contemporaries represented a new stage in the history of calculus, one that was very different from both the geometric approach of the pioneers as well as the modern approach initiated by Cauchy. In the writings of Euler and Lagrange analysis was a subject separate from geometry, although it was also a very effective tool for the investigation of curves and surfaces. This separation of the formalism of the calculus from geometry is not part of the logical landscape of the modern subject. A theorem about a function defined on some interval of real numbers under specified conditions of differentiability has a geometric interpretation implicit in its very formulation.

Lagrange's algebraic analysis was made up of functional relations, algorithms and operations on expressions composed of variables and constants. The values that these variables received, their arithmetic or geometric interpretation, were of secondary concern. In real analysis, by contrast, the basic object of study is the numerical continuum. The formalism of the modern subject is in a fundamental sense interpreted – given meaning – as a theory of functions defined on intervals of real numbers. In this respect classical real analysis resembles the calculus of the early eighteenth century, when the calculus was a kind of fine geometry, a way of representing and investigating the curve. The early calculus was interpreted in the geometry of curves in the same way that the modern calculus is interpreted in real analysis.

The geometric continuum of the early calculus possessed a particulate structure, different from that of the numerical continuum of modern analysis, and a calculus of infinitesimals was consistent with such a conception. The various accounts of infinitesimals from the eighteenth century (described for example in [Boyer 1939], Chapter 6) may be viewed as reasonable attempts to formulate a coherent system of magnitudes. The calculus united an algebra of differential expressions and a geometry of discrete infinitesimals, and the resulting conception formed the basis of a powerful and successful mathematical theory.

The relevance of modern non-Archimedean theories to an historical appreciation of the early calculus is a moot point. It is doubtful if it is possible or advisable to reconstruct a past mathematical subject in a way that conforms to modern theories and is also consistent with how a practitioner of the period would have worked. Such a project runs the obvious risk of imposing one's own interests and conceptions on the past subject. It is possible that such an endeavor will end up with something of intellectual interest and mathematical value, but it is unlikely that it will constitute a significant contribution to history. Ivor Grattan-Guinness in his essay "The mathematics of the past: distinguishing its history from our heritage" (2004) has drawn a distinction between history and heritage, between an historical account of the development of past mathematics, and a cultural celebration of past mathematical achievements for today's researchers and students.[6] The early calculus and modern nonstandard analysis both belong to

[6]Referring to the way mathematicians look at the past, Grattan-Guinness observes ([Grattan-Guinness 2004],abstract), "Old results are
modernized in order to show their current place; but the historical context is ignored and thereby often distorted. By contrast, the historian is concerned with what happened in the past, whatever be the modern situation."

the heritage of mathematics, but attempts to connect them as parts of history have been less successful.

Acknowledgment

I am very honored to participate in this tribute to the long and rich career of Joseph Dauben in the history of mathematics, and grateful to David Rowe for the invitation to contribute to the present volume. Thanks go to Noah Stemeroff for his assistance with formatting it for LaTeX.

The research presented in this article has been supported by grants from the Social Sciences and Humanities Research Council of Canada and the University of Toronto.

Bibliography

[Alexander 1994] Alexander, Daniel S. *A History of Complex Dynamics From Schröder to Fatou and Julia.* Braunschweig/Wiesbaden: Vieweg & Sohn, 1994.

[Bair et al. 2013] Bair, Jacques, et al., "Is Mathematical History Written by the Victors?", *Notices of the American Mathematical Society* 60 (2013), 886-904.

[Barbeau and Leah 1976] Barbeau, Edward J. and P. J. Leah, "Euler's 1760 Paper on Divergent Series," *Historia Mathematica* 3 (1976), 141-160.

[Bell 2008] Bell, John L. *A Primer of Infinitesimal Analysis*, 2nd Edition. Cambridge, UK: Cambridge University Press, 2008.

[Berkeley 1734] Berkeley, George. *The Analyst, or, a discourse addressed to an infidel mathematician: wherein it is examined whether the object, principles, and inferences of the modern analysis are more distinctly conceived, or more evidently deduced, than religious mysteries and points of faith*, London, 1734.

[Bernoulli 1703] Bernoulli, Johann. "Problema exhibitum", *Acta Eruditorum* (1703), 26-31.

[Bernoulli 1712] Bernoulli, Johann. "Angulorum arcuumque sectio indefinita per formulam universalem expressa sine serierum auxilio...," *Acta Eruditorum* (1712), 274-277.

[Bos 1974] Bos, Henk J. "Differentials, Higher-Order Differentials and the Derivative in the Leibnizian Calculus," *Archive for History of Exact Sciences* (1974), 1-90.

[Bottazzini 1990] Bottazzini, Umberto. "'Geometric Rigour and Modern Analysis'. An Introduction to Cauchy's Cours d'Analyse," in *Augustin-Louis Cauchy's Cours d'Analyse de l'École Royale Polytechnique Première Partie Analyse Algébrique,* Umberto Bottazzini, ed., Bologna: Cooperativa Libraria Universitaria Editrice, 1990, pp. xi-cxxxix. Volume 6 of the series

Instrumenta Rationis Sources for the History of Logic in the Modern Age, Dino Buzzetti et al., eds.

[Bottazzini 2003] Bottazzini, Umberto. "Complex Function Theory," in *A History of Analysis*, H. N. Jahnke, ed., American and London Mathematical Societies, 2003, pp. 213-260.

[Bottazzini and Gray 2013] *Hidden Harmony—Geometric Fantasies The Rise of Complex Function Theory*, Sources and Studies in the History of Mathematics and Physical Sciences, New York: Springer, 2013.

[Boyer 1939] Boyer, Carl. *The Concepts of the Calculus; a Critical and Historical Discussion of the Derivative and the Integral*, New York: Columbia University Press, 1939. Republished as *The History of the Calculus and its Conceptual Development*, New York: Dover, 1959.

[Cohen 1985] Cohen, I. Bernard. *Revolution in Science*, Cambridge, MA: Belknap Press of Harvard University Press, 1985.

[Comte 1830] Comte, Auguste. *Cours de Philosophie Positive Tome Premier Contenant les Préliminaires Genéraux et la Philosophie Mathématiques*, Paris: Bachlier, 1830.

[Condorcet 1847] Condorcet, Nicolas de. "Discours sur les sciences mathématiques," in Condorcet, *Oeuvres*, vol. 1, A. Condorcet O'Connor and M. F. Arago, eds., Paris, 1847, pp. 453 - 81.

[Conway 1976] Conway, John H. *On Numbers and Games*, London, New York: Academic Press, 1976.

[d'Alembert 1754] d'Alembert, Jean. "Différentiel," in *Encyclopédie, ou Dictionnaire Raisonné des Sciences, des Arts et des Métiers*, vol. 4, Paris, 1754, pp. 985-989.

[d'Alembert 1765] d'Alembert, Jean. "Limite," in *Encyclopédie, ou Dictionnaire Raisonné des Sciences, des Arts et des Métiers*, vol. 9, Paris, 1765, p. 542.

[Dauben 1988] Dauben, Joseph W. "Abraham Robinson and Nonstandard Analysis: History, Philosophy and Foundations of Mathematics," in *History and Philosophy of Modern Mathematics*, William Aspray and Philip Kitcher, eds., Minnesota Studies in the Philosophy of Science, vol. 11, Minneapolis: University of Minnesota Press, 1988, pp. 177-200.

[Dauben 1995] Dauben, Joseph W. *Abraham Robinson The Creation of Nonstandard Analysis A Personal and Mathematical Odyssey*, Princeton: Princeton University Press, 1995.

[Demidov 1982] Demidov, Sergei. "Création et développement de la théorie des équations différentielles aux dérivées partielles dans les travaux de J. d'Alembert," *Revue d'Histoire des Sciences* 35/1 (1982), 3-42.

[Ehrlich 2006] Ehrlich, Philip. "The Rise of non-Archimedean Mathematics and the Roots of a Misconception I: the Emergence of non-Archimedean Systems of Magnitudes," *Archive for History of Exact Sciences* 60 (2006), 1-121.

[Enderton 1972] Enderton, Herbert B. *A Mathematical Introduction to Logic*, New York and London: Academic Press, 1972.

[Engelsman 1984] Engelsman, Steven B. *Families of Curves and the Origins of Partial Differentiation*, Amsterdam: North-Holland, 1984.

[Euler 1736] Euler, Leonhard. *Mechanica sive motus scientia analytice exposita*, St. Petersburg: St. Petersburg Academy of Sciences, 1736. English translation by Ian Bruce, Euler Archive (http://eulerarchive.maa.org/).

[Euler 1748] Euler, Leonhard. *Introductio in Analysin Infinitorum*, Tomus Primus, Lausanne: M.-M. Bouquet & Soc., 1748; reprinted in *Leonhard Euler Opera Omnia*, Series 1, vol. 8, Leipzig and Berlin: B. G. Teubner, 1922.

[Euler 1755] Euler, Leonhard. *Institutiones calculi differentialis cum eius in analysi finitorum ac doctrina serierum*, Impensis Acad. imp. sci. petropolitanæ, ex off. Michælis, Berlin, 1755; reprinted in *Leonhard Euler Opera Omnia*, Series 1, vol. 10, Leipzig and Berlin: B. G. Teubner, 1913.

[Euler 1760] Euler, Leonhard. "De seriebus divergentibus," *Novi Commentarii academiae scientiarum Petropolitanæ* 5 (1760), 205-237; reprinted in *Leonhard Euler Opera Omnia*, Series 1, vol. 14, Leipzig and Berlin: B. G. Teubner, 1925, pp. 585 - 617; partial English translation and commentary in [Barbeau and Leah, 1976].

[Euler 1983] Euler, Leonhard. "Zur Theorie komplexer Funktionen," *Arbeiten von Leonhard Euler,* A. P. Juschkewitsch, ed., German trans. W. Purkert, et al., Darmstadt: Wissenschaftliche Buchgesellschaft, 1983.

[Ferraro 2000] Ferraro, Giovanni. "The Value of an Infinite Sum Some Observations on the Eulerian Theory of Series," *Sciences et Techniques en Perspective* 4 (2000), 73-113.

[Ferraro 2007] Ferraro, Giovanni. *The Rise and Development of the Theory of Series up to the Early 1820s*, Berlin and New York: Springer, 2007.

[Fisher 1978] Fisher, Gordon. "Cauchy and the Infinitely Small," *Historia Mathematica* 5 (1978): 313-331.

[Fisher 1981] Fisher, Gordon. "The Infinite and Infinitesimal Quantities of du Bois-Reymond and their Reception," *Archive for History of Exact Sciences* 24 (1981), 101-163.

[Fontenelle 1790-1792] Fontenelle, Bernard Le Bovier de. *Oeuvres*, vols. 6 and 7, new edition, Paris: Jean-François Baslieu et Jean Servière, 1790-1792.

[Fraser 1983] Fraser, Craig. "J. L. Lagrange's Early Contributions to the Principles and Methods of Mechanics," *Archive for History of Exact Sciences* 28, (1983), 197-241.

[Fraser 1989] Fraser, Craig."The Calculus as Algebraic Analysis: Some Observations on Mathematical Analysis in the 18th Century," *Archive for History of Exact Sciences* 39 (1989), 317-335.

[Fraser 1990] Fraser, Craig."Lagrange's Analytic Mathematics, Its Cartesian Origins and Reception in Comte's Positive Philosophy," *Studies in the History and Philosophy of Science* 21 (1990): 243-256.

[Fraser 2003] Fraser, Craig. "History of Mathematics in the Eighteenth Century", in *The Cambridge History of Science: Volume 4, Eighteenth-Century Science*, Roy Porter, ed., Cambridge University Press, 2003, pp. 305-327.

[Fraser 2005] Fraser, Craig. "Joseph Louis Lagrange, Théorie des fonctions analytiques," in *Landmark Writings in Western Mathematics 1640-1940*, Ivor Grattan-Guinness, ed., Amsterdam: Elsevier, 2005, pp. 208-224.

[Grabiner 1981] Grabiner, Judith V. *The Origins of Cauchy's Rigorous Calculus*, Cambridge, MA: MIT Press, 1981.

[Goldenbaum and Jesseph 2008] Goldenbaum, Ursula and Douglas Jesseph, eds., *Infinitesimal Differences Controversies Between Leibniz and his Contemporaries.* Berlin and New York: Walter de Gruyter, 2008.

[Grattan-Guinness 2004] Grattan-Guinness. "The Mathematics of the Past: Distinguishing its History from our Heritage," *Historia Mathematica* 31 (2004), 163-185.

[Grimsley 1963] Grimsley, Ronald G. *Jean d'Alembert (1717-1783)*, Oxford: Clarendon Press, 1963.

[Guicciardini 2009] Guicciardini, Niccolò. *Isaac Newton on Mathematical Certainty and Method*, Cambridge, MA and London: MIT Press, 2009.

[Hardy 1949] Hardy, Godfrey H. *Divergent Series*, Oxford: Oxford Clarenden Press, 1949.

[Hawkins 1977] Hawkins, Thomas. "Weierstrass and the Theory of Matrices," *Archive for History of Exact Sciences* 17 (1977), 119-163.

[Heilbron 1990] Heilbron, John L. "Introductory Essay," in *The Quantifying Spirit in the Eighteenth Century*, Tore Frangsmyr, J. L. Heilbron, and Robin E. Rider, eds., Berkeley: University of California Press, 1990, pp. 1-23.

[Jahnke 2003] Jahnke, Hans Niels. "Algebraic Analysis in the Eighteenth Century," in *A History of Analysis*, H. N. Jahnke, ed., American and London Mathematical Societies, 2003, pp. 105-136.

[Keisler 1976] Keisler, H. Jerome. *Elementary Calculus*, Boston, MA: Prindel, Weber & Schmidt, 1976.

[Keisler 1976] Keisler, H. Jerome. *Foundations of Infinitesimal Calculus*, Boston, MA: Prindel, Weber & Schmidt, 1976.

[Kline 1972] Kline, Morris. *Mathematical Thought from Ancient to Modern Times*, New York: Oxford University Press, 1972.

[Knopp 1928] Knopp, Konrad. *Theory and Application of Infinite Series*, trans. from 2nd German edition by R. C. Young, London: Blackie, 1928.

[Koestier 1991] Koetsier, Teun. *Lakatos' Philosophy of Mathematics, A Historical Approach*, Amsterdam: North-Holland, 1991.

[Lagrange 1797] Lagrange, Joseph Louis. *Théorie des fonctions analytiques, contenant les principes du calcul différentiel, dégagé de toute consideration d'infiniment petits ou d'évanouissans, de limites ou de fluxions, et réduits à l'analyse algébrique des quantités finies*, Paris: L'Imprimerie de la République. Prairial an V, 1797.

[Lagrange 1806] Lagrange, Joseph Louis. *Leçons sur le calcul des fonctions*, Paris: Courcier, 1806.

[Lakatos 1978] Lakatos, Imre. "Cauchy and the Continuum: The Significance of Non-standard Analysis for the History and Philosophy of Mathematics," in *Imre Lakatos, Mathematics, Science and Epistemology Philosophical Papers*, vol. 2, John Worral and Gregory Currie, eds., Cambridge: Cambridge University Press, 1978, pp. 43-60.

[Laplace 1835] Laplace, Pierre Simon. *Exposition du système du monde*, 6th ed., Paris, 1835; reprinted in *Oeuvres complètes de Laplace*, Tome 6, Paris, 1884.

[Laugwitz 1987] Laugwitz, Detlef. "Infinitely Small Quantities in Cauchy's Textbooks," *Historia Mathematica* 14 (1987): 258 - 274.

[Laugwitz 1989] Laugwitz, Detlef. "Definite Values of Infinite Sums: Aspects of the Foundations of Infinitesimal Analysis around 1820," *Archive for History of Exact Sciences* 39 (1989), 195-245.

[Leibniz 1703] Leibniz, Gottfried Wilhelm,."Continuatio analyseos quadraturarum rationalium," *Acta Eruditorum* (1703), 19-26.

[l'Hôpital 1696] l'Hôpital, Marquis de. *Analyse des infiniments petits pour l'intelligence des lignes courbes*, Paris, 1696.

[Mach 1893] Mach, Ernst. *The Science of Mechanics: A Critical and Historical Account of Its Development*, Chicago: Open Court, 1893. English trans. of 2nd German ed. by Thomas J. McCormack.

[Mahoney 1985] Mahoney, Michael S. "Diagrams and Dynamics: Mathematical Perspectives on Edgerton's Thesis," in *Science and Arts in the Renaissance*, John W. Shirley and F. David Hoeniger, eds., London: Associated University Presses, 1985, pp. 198-220.

[Panza 1996] Panza, Marco. "Concept of Function, between Quantity and Form, in the 18th Century," in *History of Mathematics and Education: Ideas and Experiences*, H. Niels Jahnke et al., eds., Göttingen: Vandenhoeck & Ruprecht. 1996.

[Reyes 2004] Reyes, Mitchell G. "The Rhetoric in Mathematics Newton, Leibniz, the Calculus, and the Rhetorical Force of the Infinitesimal," Quarterly Journal of Speech 90 (2004): 163-188.

[Rider 1990] Rider, Robin E. "Measure of Ideas, Rule of Language: Mathematics and Language in the 18th Century," in *The Quantifying Spirit in the Eighteenth Century*, Tore Frangsmyr, J. L. Heilbron, and Robin E. Rider, eds., Berkeley: University of California Press, 1990, pp. 113-140.

[Robinson 1961] Robinson, Abraham. "Non-Standard Analysis," *Proceedings of the Koninklijke Nederlandse Akademie van Wetenschappen*, series A 64 (1961), 432-440.

[Robinson 1966] Robinson, Abraham. *Non-Standard Analysis, Studies in the Logic and Foundations of Mathematics*, Amsterdam: North Holland, 1966.

[Robinson 1967] Robinson, Abraham. "The Metaphysics of the Calculus," in *Problems in the Philosophy of Mathematics*, Imre Lakatos, ed., Amsterdam: North-Holland, 1967, pp. 28-46.

[Schmieden and Laugwitz 1958] Schmieden, Curt and Detlef Laugwitz. "Eine Erweiterung der Infinitesimalrechnung," *Mathematische Zeitschrift* 69 (1958), 1- 39.

[Schneider 1968] Schneider, Ivo. "Der Mathematiker Abraham de Moivre (1887-1754)," *Archive for History of Exact Sciences* 5 (1968), 177-317.

[Smithies 1997] Smithies, Frank. *Cauchy and the Creation of Complex Function Theory*, Cambridge, UK: Cambridge University Press, 1997.

[Taton 1974] Taton, René. "Inventaire chronologique de l'oeuvre de Lagrange," *Revue d'Histoire des Sciences* 26 (1974), 3-36.

[Youschkevitch 1971] Youschkevitch, Adolf P. "Lazare Carnot and the Competition of the Berlin Academy in 1786 on the Mathematical Theory of the Infinite," in *Lazare Carnot Savant*, Charles C. Gillispie, ed., Princeton, N.J.: Princeton University Press, 1971, pp. 149-168.

Despite Kempe, the Modest Place of Multisets and Multiclasses in Foundational Theories

Ivor Grattan-Guinness

To Joe Dauben, for four decades of multiple relationships

> When mathematicians, logicians, and other exact thinkers think and reason, they think and reason about something. What is that something?
> Opening of A.B. Kempe [Kempe 1890b]

> The notion of a set takes no account of multiple occurrence of any one of its members, and yet it is just this kind of information, which is frequently of importance.
> Opening of Richard Rado [Rado 1975a]

Theories of Collections

Mankind has always dealt with collections of things, and in connection especially with logics, mathematics and philosophy the theories of collections have required explicit attention. Two theories have prevailed. From antiquity part-whole theories emerged in which, say, the collection of Kenyan women is a part (maybe empty, maybe the whole) of the collections of Kenyans and of women. I reserve "classes" to name this type of collection, to which the name "mereology" is often attached. From the late 19th century onwards it was challenged by Georg Cantor's set theory, in which things belonged to sets and their membership was distinguished from inclusion in the form of subsets. A major difference was that in set theory an object is distinguished from its unit set, whereas in part-whole theory this distinction does not hold. When I wish to mention both theories together, I shall speak of "the duo."

Our concern is with a severe limitation of such theories. For the properties of being a part of a class, and being a member of a set, *can obtain only once.* This severely limits the applicability of the duo, for it is very easy to find contexts in which multiple partship or membership is required and no order is imposed upon the repeated occurrences: for example, repeated zeroes of polynomial functions and of latent roots of matrices, or repeated appearance of prime factors for integers such as 16 and 28. We cannot express such properties in terms of the duo; for example, the factors of 16 are $2, 2, 2$ and 2, which forms neither a class nor a set.

Ordinary life also provides many similar cases. Take tallying, for example; you examine the coins in your pocket and find that there are three 50-cent coins, four 1-cents and one 10-cent. Set theory interprets the coins as a set of eight distinct objects; but if you wish to see how much money is there, then you have to treat the collection comprising three 50s, four 1s, and one 10, together totalling 164 cents.

D.E. Rowe and W.-S. Horng (eds.), *A Delicate Balance: Global Perspectives on Innovation and Tradition in the History of Mathematics*,
Trends in the History of Science, DOI 10.1007/978-3-319-12030-0_3

The restriction to single partship or membership undermines any claim that either theory of collections can be a basis of a wide-ranging foundation of mathematics (or of a logic, which we shall also include). The obvious ubiquity of multiple partship or membership makes this limitation rather obvious; however, both properties have been very little recognised. Various names have been proposed for these types of collections; I shall refer to "multiclasses," in imitation of the current preferred name "multisets," which came from the Dutch mathematician and logician N. G. de Bruijn, in correspondence with Donald Knuth [Knuth 1981, 37]. In contrast to the situation in mathematics, the place of multisets are well recognised in computer science, and Knuth has been one of their earliest advocates, even including some historical explorations [Knuth 1973], [Knuth 1981]. A rather scrappy introduction to some of these contexts is given in [Wikipedia. "Set (abstract data type)"].

Most presentations of the duo make no mention of multiclasses or multisets. Nevertheless, some good introductions have been written; see, in particular, [Rado 1975a], [Blizard 1989], [Brink 1987], [Syropoulos 2001] and [Wikipedia "multisets"]. The histories of the duo have been studied quite extensively, but usually with no attention paid to multiclasses or multisets. Nevertheless their histories have been quite well examined, in some of the sources just noted and especially in [Blizard 1991] (building upon his doctoral thesis [Blizard 1986] on multisets). None of these texts is easily accessible; it is high time for someone to write or edit a textbook.

In this article I consider some principal features of the theories and note their (lack of) roles in foundational theories of mathematics, especially the duo, and also in logics and model theory. The underlying theme is bafflement that foundationalists are so unfamiliar with theories of multiclasses or multisets, despite their being more general than the duo. In the final two sections I discuss the work of Alfred Bray Kempe, who advocated multiclasses and recognised their generality from the 1880s onwards far more articulately than anyone else. He did so, however, in the context of quantics and their graphical representation, which was not a foundational topic, so his contributions have hardly ever been noticed, including by later multiple collectionists.

It is especially important to distinguish the histories of multiclasses and multisets from their heritages. There must be numerous cases where the hindsight afforded by heritage allows us to recast theorems, relationships, and proofs in terms of multiclasses or multisets that the original authors did not use. Doubtless they handled the required multiclasses or multisets intuitively without bothering to articulate the straightforward procedures required, for instance, for tallying the variously coloured balls in problems in combinatorial probability. Given the current recognition of multisets in computer science, an interesting example is David Hilbert's metamathematics of axiomatised theories. A well-formed formula in such a "formalism" could be read as a multiset of its signs with their multiplicities tallied; but I see no evidence that Hilbert and his colleagues consciously treated their results in any multisets-like form. The same judgement applies to their immediate successors, such as [Kleene 1952].

Basic Properties of Multisets

Here I relate multisets to set theory; analogous properties hold for multiclasses. A multiset M is an unordered set of ordered pairs $a.Ma$, where a is any member of M and Ma is the positive finite cardinal that indicates the "multiplicity" of its membership. Thus, for example, the multiset $< a, c, a, c, c, b, a, b, d, c >$ can be abbreviated $\{d.1, b.2, a.3, c.4\}$, which displays its "multiplicity profile" (my term). I use corner brackets to mark that M is a multiset. Note the self-reference here; the abbreviated version is itself a tally.

A theory of multisets may be adapted to fit in with an axiomatic version of set theory; [Blizard 1986] and [Rado 1975a] rely on ZF. Paradoxes are available, variant upon those for sets; for example, the set of all multisets with a given multiplicity profile does and does not belong to itself. Some authors have pursued a theory of "hybrid sets," where negative membership of a multiset is permitted; contexts include the formalisation of borrowing money from a bank [Blizard 1990].

Among the properties in which multiset theory differs from set theory, two are worth special attention. Firstly, the essential inattention to the order of the multiple members reduces substantially the status of ordering; in particular, the well-ordering principle is false for multisets. There is a companion theory, called "families of sets," in which the repeated members are interpreted as a vector. However, imposing order on the elements makes the theory different from multisets; [Rado 1975a], [Rado 1975b] deal with both theories, though the latter is not our concern here. Secondly, the laws of complementation of multisets differ from those of sets because of the possible removal of some of the repeated elements. As a result, the algebraic structure underlying multiset theory is that of a lattice rather than a Boolean algebra [Singh 2011 and others].

Multisets and Set Theory

From the mid 1890s onwards set theory became widely recognized as wide-ranging, especially for those who claimed it to be a (the) basis for mathematics. No attention appears to have been paid to the restriction to single membership. This may be due to the strength of the influence of Georg Cantor. Conceiving of a general set as "any collection into a whole M of definite and separate objects m of our intuition or our thought. These objects are called the elements' of M" [Cantor 1895, art. 1], he took for granted that abstraction would collapse any instance of multiple membership to the single case. His friend Richard Dedekind, who helped him to formulate set theory in the first place, presumably took the same view. Yet he ended his booklet [Dedekind 1888] on numbers with a construction that drew upon the notion, very useful in many cases, of systems in which every element is endowed with a certain frequency-number which indicates how often it is to be reckoned as any element of a system, but then added that it did not lie in the line of his memoir to go further into this discussion. Unfortunately he kept his word; the passage remained unaltered in later editions, and he did not devote a manuscript to the topic.

Most later researchers and textbook writers in set theory seem to have ignored multiple membership completely; if two sets overlap, then their union and intersection were specified in the usual ways as sets with the double membership collapsed.[1] A few authors noted it, but discarded it. For example, Erich Kamke explained that "the same element may not be multiply presented. The number complex $1, 2, 1, 2, 3$ will therefore [become] a set only after the omission of the multiply specified elements" [Kamke 1947, 6]. Similarly, for Abraham Fraenkel " a certain object may be contained in a given set, or not, but there is no possibility of its repeated appearance" [Fraenkel 1953, 17]. In a later edition of his book " an object may belong, or not belong, to a set but cannot ' more than belong', for instance belong repeatedly" [Fraenkel 1976, 10]. Doubtless some other authors, who have so far avoided historical detection, considered multiple membership, but I suspect this was a small number; I undertook a moderately substantial trawl of books and papers, but found only the above cases.

Multiclasses in Algebraic Logic

From the 1840s onwards George Boole developed an algebra of logic in which he presumed some universe 1 of discourse and studied the relationships between subclasses $x, y, \ldots$ of 1, including their complements $1 - x$, $1 - y, \ldots$. (In a whiff of model theory unusual at that time he also allowed $x, y, \ldots$ to represent the mental processes of selecting these classes from the universe, but I shall not use it here.) The underlying theory of collections was part-whole theory, and the question of multiple partship arose most prominently in connection with " the laws of the logical process of addition which is symbolised by the sign + [...] the expression $x+y$ seems indeed uninterpretable and that it be assumed that the things represented by x and the things written about y are entirely separate; that they embrace no individuals in common" [Boole 1854, 66; note also pp. 55-56]. He never elaborated reasons for this position, but presumably avoidance of multiple partship was one of them.

The first serious study of Boole's logic was made by the remarkable polymath Stanley Jevons, who wrote a short book outlining an alternative approach to it in his *Pure Logic* [Jevons 1864]. He corresponded with Boole from 1863 until Boole's unexpected death at the end of 1864 [Grattan-Guinness 1991]. Boole did not engage strongly in the exchange because, as he explained to Jevons, he was then preoccupied with revising his textbook of differential equations; but Jevons highlighted the issue concerning the correct way of understanding logical addition. He took the example of the class $x + x$, which for him gave x. By contrast, for Boole $x+x$ was ill-formed, although the equation $x+x = 0$ was legitimate, and his methods of handling the equation gave $x = 0$ as the solution. Even here multiple

[1]This judgement applies also to the exciting new philosophy of mathematics and logics known as "homotopy type theory," in which notions taken from topology form the basis of mathematical logic and axiomatic set theory with applications to theoretical computer science and automated proof checking [Homotopy 2013].

partship did not explicitly arise, because Jevons viewed logic as concerned not with classes or mental processes required to produce them, but with the ' meaning' of terms, where multiple reference did not arise (both "donkey and donkey" and "donkey or donkey" were the same as "donkey").

The next significant British logician was John Venn, who produced a textbook on "symbolic logic" based upon his teaching at Cambridge. He tried to follow Boole's principles whenever possible; for general purposes especially including mathematics he required classes to be disjoint when defining their "aggregation;" however, he allowed that for some " simple forms of language indicative of class aggregation [...] many cases might be conceived in which they do thus doubly reckon the common members" [Venn 1881, 52, 41]. In the second edition he retained this last remark; but he dropped disjointness, " partly, I must admit, because the voting has gone this way, and in a matter of procedure there are reasons for not standing out against such a verdict" [Venn 1894, 46, 43]. Between Boole and Venn comes a major figure, and the first from the USA. From the mid 1860s Charles Sanders Peirce began publishing in algebraic logic using a Boolean algebra. Unaware of Jevons's work at that time, he also adopted the same reading of addition, but put forward a perverse interpretation of Boole's position. Let classes " $a+b$ denote all the individuals containing a and b together, that is, the class of those things which are a not$-b$, b not$-a$ or both a and b". Then the operation here performed will differ from [Boole's] arithmetical addition because " what is common to a and b is not taken into account twice over, as it would be in arithmetic" [Peirce 1867, 3-4]. It is a mystery that Peirce could have thought that Boole had advocated double partship.

Peirce's version of algebraic logic was based upon his reading of the union of multiclasses; it drew upon hair-raising analogies with the algebras of the binomial theorem, the differential calculus and Taylor's series [Walsh 2012, chs. 3-4]. He focussed upon the logic of relations, where he drew to some extent on Augustus de Morgan's contributions; he highlighted "relatives," the classes of "individuals" that satisfied a relation such as "square of." (The word "quantifier" is his, from 1883.) Occasionally he considered multiple partship [Peirce 1870, 64-65], and he also studied certain aspects of Cantor's theory, although outside the framework of his logics, and multiple membership was not one of his concerns [Dauben 1977]. But he was aware of it, enough to appreciate its next advocate, who came from an unlikely source.

Kempe's Initiative

The Englishman Alfred Bray Kempe, by profession a lawyer, was a mathematician of considerable ability who contributed to several topics, most famously with an attempt to prove the four-colour conjecture about the colouring of maps. The paper of concern here has never received the attention that it deserves: an elaborate theory [Kempe 1886] on "mathematical form," published with the Royal Society of London, of which he had been made a Fellow in 1881.

Kempe's motivation came from the graphical representation of quantics,[2] but his conception was quite general. He took any thing as composed of "separate entities or units," which " come under consideration in a variety of garbs—as material objects, intervals or periods of time, processes of thought, points, lines, statements, relationships, arrangements, algebraical expressions, operators, operations, & c., & c., occupy various positions, and are otherwise variously circumstanced." Units occurred not only singly but also in "collections" of "pairs, triads, ... n-ads".... Every collection of units has a definite *form* due (1) to the number of its component units, and (2) to the way in which the distinguished and undistinguished units, pairs, triads, etc., are distributed through the collection. For example, the graph ⅄ contained a trio of undistinguished end units and one distinguished central unit. He defined properties corresponding in class theory to inclusion ("component"), disjointness ("detached") and intersection ("connection") [Kempe 1886, 3-5].

Kempe presented a remarkably wide-ranging ensemble of forms, each one based upon its distribution of distinguished and undistinguished units; he was much concerned with their "graphical representation." His stress upon undistinguished units not only makes him a multiclassist but also gave his philosophy a structuralist character; however, his forms were drawn more from geometry and graph theory than from abstract algebra(s).[3] However, he treated substitution groups by tabulating all the possible cases up to 12 units, calculating the number of forms in each case [Kempe 1886, 35-47]. His Royal Society referee Arthur Cayley was stimulated to rework these tables in [Cayley 1889] to express the results in terms of his own previous investigations of this kind.

In 1890 Kempe published an essay in *Nature* on "the subject-matter of exact thought," distilling the philosophical content of his forms paper " in as simple and non-technical manner as possible." He wondered if the focus of exact thought lay in the " infinite variety" of its products, or in " a harmonious whole, admitting of a definition and treatment which, though perfectly general, we had preserve the essential characteristics of its component parts?" He preferred the second option, since it highlighted "the essential or necessary matter of exact thought" rather than "merely the dress in which the necessary matter is clothed" [Kempe 1890a, 156]. He most detailed claim was that two very different collections took the same form. The first was a "geometrical theory of points," which was based upon collinear and non-collinear triads of points, where c is (or is not) on the line ab joining a and b. He drew upon the construction of the quadrilateral in projective geometry to lay out laws for their distribution "through the whole system of points." He laid down the laws that " distinguished and undistinguished triads" satisfied, such as " if the two triads of [points] a, b, c and b, c, d are collinear triads, so also are the

[2]An algebraic formula E in variables and constants is a quantic if, when its variables are linearly transformed, an expression associated with E is left basically unchanged. A simple case is the discriminant of a quadratic form.

[3]On the philosophical aspects of Kempe's work see [Vercelloni 1989, preface]. The history is given in more detail in [Grattan-Guinness 2002]. Various letters in Kempe's *Nachlass* may be consulted at the West Sussex Record Office in Chichester.

triads a, c, d and a, b, d''."

The other collection was "the logical theory of classe." Kempe considered "statements" to be subjects of the exact thought of the logician, "[that] compose a system of entities which are undistinguished from each other," so that "these mutual relations" between statements are " the object of the logician to investigate and define". Usually statements " are conceived of [...] as either true or false," but he also admitted special kinds, " *truisms* and *falsisms*," [that] " must always be regarded as always true, and others as always false."[4] Then he expressed the implication "if a then b" between statements as a *three*-statement relationship that manifests the same form as that for points; for example, if a was assumed to be a truism and c true, then b was also true. He explained other cases, and laid down laws for statements to satisfy [Kempe 1890a, 159-160]. But this construal of implication is limited and the analogy with points rather forced. In addition he did not engage with the predicate calculus; the closest approximation to it was a Boolean algebra for "classes," which he did *not* explicitly define but presumably allowed for multiple membership [Kempe 1886, 63-64]. In a much more elaborate account of the two collections he devoted only a few pages to logic, and added only a clearer account of the logical connectives between statements [Kempe 1890b, 169-174].

Later Kempe published a few papers on his theory. In particular, in his Presidential Address to the London Mathematical Society he claimed that mathematics be defined as *"q the science by which we investigate those characteristics of any subject-matter of thought which are due to the conception that it consists of a number of differing and non-differing individuals and pluralities"* [Kempe 1894, 15]. But reaction was always modest.

Some Responses to Kempe

On publication Kempe's original paper [Kempe 1886] significantly influenced only Peirce. He corresponded at once with Kempe, who sent to the Royal Society a revised version of two of its sections [Kempe 1887]. Early in 1889 Peirce seized upon Kempe's use of graphs to develop the first version of what he later called "existential graphs," a topological representation of the syntax of propositions. He published improved versions from [Peirce 1897] in the philosophical journal *The Monist*, where he praised Kempe's theory; but he understood it as contributing to his own logic of relatives, whereas in reply [Kempe 1897] rejected the implied subsidiary status of his theory of forms. These graphs became a major concern of Peirce, and later an important source of his later influence; multiclasses do not feature.

[4]It is striking that Kempe gave primacy in logic to the statement rather than to the term; he also anticipated Ludwig Wittgenstein's concern in the *Tractatus* [Wittgenstein 1922] with tautologies and contradictions. The Cambridge logician W.E. Johnson was to speak of truisms and falsisms [Johnson 1921, 223-224]; but he did not cite Kempe, and may not have known of his work.

In connection with his Lowell Lectures at Boston in 1903, Peirce alerted the Harvard philosopher Josiah Royce to this work, and Royce later gave a lengthy account of Kempe's theory in the context of the foundations of geometry [Royce 1905], preferring it to Russell's use of relations.[5] He became interested in mathematical logic, and in his final years he supervised several students who became prominent logicians or philosophers: Henry Sheffer, Morris Cohen, C. I. Lewis and C. J. Ducasse (in the 1930s the chief founder of the Association of Symbolic Logic).[6] For various reasons Peirce was not impressed with Royce.[7]

But apart from this and a few other ripples, Kempe's work (and also Royce's) was overlooked or ignored by the principal set theorists, logicians and foundationalists, apart from Peirce, and so the undoubted significance of "undistinguishableness" was missed.[8] Why?

Presumably if a class or set theorist looked at Kempe's theory at all, he might have taken it to be a treatment of graphs and quantics, thereby overlooking the significance of undistinguished units. Further, multiple partship was not always prominent when others made use of the theory; for example, Peirce's existential graphs, where the topology was the prime feature. (One modification for this interest was its possible use in solving the map-colouring problem, an obsession for Peirce to which Kempe had made an important contribution.) While Kempe called a collection a "discrete heap" when each unit occurred only once (corresponding to a Cantorian set), he did not invoke set theory itself; indeed, he was probably unaware of it, like most British mathematicians before the 1890s. (His own definition of "set" was quite different from Cantor's, concerning the availability in such a collection of n-ads [Kempe 1886, 23-24].) Further, his examples were always of finite collections, whereas the infinite(s) was a major concern of set theory. Finally, the treatment of logic, while ingenious, might have seemed limited and artificial with its forced analogy with points.

So Kempe's theory would not have seemed relevant to foundations; but then everyone missed the important lesson about the place of multiple partship and membership in mathematics and theories of collections, whether or not they supported his theory of forms. Among apparent non-readers, Whitehead is particularly

[5]Many pertinent manuscripts are held in Royce's *Nachlass* at the Pusey Library in Harvard University.

[6]Another student would have been Norbert Wiener, who wrote his doctoral thesis in 1913 on a comparison of mathematical logic in Russell and algebraic logic in Ernst Schröder [Grattan-Guinness 1975]; but by then Royce was too ill to act, dying in 1916.

[7]On Peirce at this time see his letters to Royce, manuscripts on graphs and comments on Royce, and the letters (one gigantic) to Royce's colleague professor William James [Peirce 1976, 956-972, 405-413, 815-876]. In 1900 he also wrote to Cantor [Peirce 1976, 767-779], who seems not to have answered but in 1901 enquired about him to Philip Jourdain [Grattan-Guinness 1971, 113-115]. Peirce himself corresponded with Jourdain in 1908 [Peirce 1976, 879-888]. Although the specification of sets was considered in these sources, multisets did not feature.

[8]For more details on these later stages see [Vercelloni 1989] and [Grattan-Guinness 2002]. Kempe's work seems to have been ignored also by invariance theorists and their historians; for example, only an earlier paper on quantics is mentioned, in a footnote, in the extensive history [Meyer 1892, 196].

noteworthy. Kempe's main paper [Kempe 1886] appeared when Whitehead had recently graduated from the Mathematics Tripos; so it is not surprising that he did not seize on it, although as an algebraist he would have appreciated its motivation from quantics. When Kempe read his presidential address [Kempe 1894], Whitehead was deeply immersed in Grassmannian algebras and so was not much concerned with kinds of collection. But when he and his former student Bertrand Russell developed their logicist thesis that mathematical logic (including set theory and relations) sufficed as their basis for mathematics during the 1900s, they never considered multiple membership; I have not seen a manuscript or published reference to it by either man.

In 1906 Kempe did address Russell's mathematical logic when he refereed a paper that Russell had submitted to the London Mathematical Society. As usual Russell subsumed sets, but instead of drawing upon propositional functions and relations he went for propositions and their truth-values in a "substation" theory. Kempe doubted the merit of subsumption and thought that the mind was capable of forming "aggregates" (then a frequently used name for sets); but he did not raise the possibility of multisets, and when Russell withdrew his paper soon afterwards it was because of his own quite different doubts about this theory.[9]

Another opportunity arose in 1909, when they submitted the manuscript of their *Principia Mathematica* to the Cambridge University Press and sought a grant from the Royal Society to help cover the considerable printing expenses [Grattan-Guinness 2000, 581-584]. Kempe had been the Treasurer of the Royal Society since 1898, and so presumably gained a good idea of the content of their project, especially the central role of set theory; but he does not seem to have invited them to consider his alternative option of multi-parts. Their method of constructing mathematics out of mathematical logic including set theory, can be imitated using multisets [Grattan-Guinness 2011].

Kempe's theory seems also not to have influenced those who have reinvented multiset theory in modern times. Rado told me that he had never heard of Kempe. The American logician Ted Hailperin reworked Boole's algebra as a ring structure on signed multisets [Hailperin 1976]; unaware of the name "multisets" he used "heaps" at first, but when I asked him if he had taken it from Kempe, he responded "Who?"[10] Without having to adopt Kempe's theory of mathematical forms, will others acquire his level of recognition of the ubiquity of multiclasses and of multisets? Could such awareness even become commonplace?

Bibliography

[Blizard 1986] Blizard, W. D. 1986. Generalizations of the Concept of Set: a Formal Theory of Multisets, Doctoral thesis, University of Oxford.

[9]For Kempe's report and further discussion by G.H. Moore, see [Russell 2014, 236-239].

[10]Further, Peirce's pioneering manuscripts on existential graphs were omitted from volume 6 (2000) of the new edition of his writings! They are transcribed in [Grattan-Guinness 2002].

[Blizard 1989] Blizard, W. D. 1989. Multiset theory, *Notre Dame Journal of Formal Logic*, 30, 36-66.

[Blizard 1990] Blizard, W. D. 1990. Negative Membership, *Notre Dame Journal of Formal Logic*, 31, 346-368.

[Blizard 1991] Blizard, W. D. 1991. The Development of Multiset Theory, *Modern Logic*, 1, 319-352.

[Boole 1854] Boole, G. 1854. *An Investigation of the Laws of Thought*, Cambridge: MacMillan; London: Walton and Maberly. [Several reprs.]

[Brink 1987] Brink, C. 1987. Some Background on Multisets, TR - ARP-2/87, *Research School for Social Science*, Australian National University, 1-11.

[Cantor 1895] Cantor, G. F. L. P. 1895. Beiträge zur Begründung der transfiniten Mengenlehre, pt. 1, *Mathematische Annalen*, 46,481-512. [Also in *Gesammelte Abhandlungen mathematischen und philosophischen Inhalts*, Berlin: Springer, 1932, 282-311.]

[Cayley 1889] Cayley, A. 1889. On the Theory of Groups, *American Journal of Mathematics*, 11, 139-157. [Also in *The Collected Mathematical Papers*, vol. 12, 639-656.]

[Dauben 1977] Dauben, J, W. 1977. C. S. Peirce's Philosophy of Infinite Sets, *Mathematical Magazine*, 50, 123-135.

[Dedekind 1888] Dedekind, J. W. R. 1888. *Was sind und was sollen die Zahlen?*, Braunschweig: Vieweg. [Various reprs. and eds.]

[Fraenkel 1953] Fraenkel, A. A. H. 1953. *Abstract Set Theory*, 1st ed., Amsterdam: North-Holland.

[Fraenkel 1976] Fraenkel, A. A. H 1976. *Abstract Set Theory*, 4th rev. ed. (rev. A. Levy), Amsterdam: North Holland.

[Grattan-Guinness 1971] Grattan-Guinness, I. 1971. The Correspondence between Georg Cantor and Phillip Jourdain, *Jahresbericht der Deutschen Mathematiker-Vereinigung*, 73 (1971), part 1, 111-130.

[Grattan-Guinness 1975] Grattan-Guinness, I. 1975. Wiener on the Logics of Russell and Schröder. An Account of his Doctoral Thesis, and of his Subsequent Discussion of it with Russell, *Annals of Science*, 32, 103-132.

[Grattan-Guinness 1991] Grattan-Guinness, I. 1991. The Correspondence between George Boole and Stanley Jevons, 1863-1864, *History and Philosophy of Logic*, 12, 15-35.

[Grattan-Guinness 2000] Grattan-Guinness, I. 2000. *The Search for Mathematical Roots, 1870-1940. Logics, Set Theories and the Foundations of Mathematics from Cantor through Russell to Gödel*, Princeton: Princeton University Press.

[Grattan-Guinness 2002] Grattan-Guinness, I. 2002. Re-interpreting ”⅄”: Kempe on Multisets and Peirce on Graphs, 1886-1905, *Transactions of the C. S. Peirce Society*, 38, 327-350.

[Grattan-Guinness 2011] Grattan-Guinness, I. 2011. Numbers as Moments of Multisets. A new-old Formulation of Arithmetic, *The Mathematical intelligencer*, 33, no. 1, 19-29.

[Hailperin 1976] Hailperin, T. 1976. *Boole's Logic and Probability*, 1st ed., Amsterdam: North-Holland. [2nd ed. 1986.]

[Homotopy 2013] Homotopy 2013. *Homotopy Type Theory: Univalent Foundations of Mathematics*, Princeton: The Univalent Foundations Program, Institute for Advanced Study. [Multiple authorship; no named editor.]

[Jevons 1864] Jevons, W. S. 1864. *Pure Logic*, London: Stanford. [Also in Pure Logic and other Minor Works, London: MacMillan, 1890 (repr. Bristol: Thoemmes, 1991), 1-77.]

[Johnson 1921] Johnson, W. E. 1921. *Logic*, pt. 1, Cambridge: Cambridge University Press.

[Kamke 1947] Kamke, E. 1947. *Mengenlehre*, 2nd ed., Berlin: de Gruyter [English trans. *The Theory of Sets*, New York: Dover, 1950.]

[Kempe 1886] Kempe, A. B. 1886. A Memoir on the Theory of Mathematical Form, *Philosophical Transactions of the Royal Society of London*, 177, 1-70.

[Kempe 1887] Kempe, A. B. 1887. Note to a Memoir on the Theory of Mathematical Form, *Proceedings of the Royal Society of London*, 42, 193-196.

[Kempe 1890a] Kempe, A. B. 1890a. The Subject Matter of Exact Thought, *Nature*, 43, 156-162.

[Kempe 1890b] Kempe, A. B. 1890b. On the Relation between the Logical Theory of Classes and the Geometrical Theory of Points, *Proceedings of the London Mathematical Society*, (1)21, 147-182.

[Kempe 1894] Kempe, A. B. 1894. Mathematics, *Proceedings of the London Mathematical Society*, (1)26, 5-15.

[Kempe 1897] Kempe, A. B. 1897. The Theory of Mathematical Form. A Correction and Explanation, *The Monist*, 7, 453-458.

[Kleene 1952] Kleene, S. C. 1952. *Introduction to Metamathematics*, Amsterdam: North Holland; Groningen: Noordhoff. 2nd ed., Reading, Mass.: Addison-Wesley.

[Knuth 1973] Knuth, D. E. 1973. *The Art of Computer Programming*, vol. 3, *Sorting and Searching*, Reading, Mass.: Addison-Wesley.

[Knuth 1981] Knuth, D. E. 1981. *The Art of Computer Programming*, vol. 2, *Seminumerical Algorithms*, 2nd ed., Reading, Mass.: Addison-Wesley.

[Meyer 1892] Meyer, F. 1892. Bericht über den gegenwärtigen Stand der Invariantentheorie, *Jahresbericht der Deutschen Mathematiker-Vereinigung*, 1, 79-292.

[Peirce] Peirce, C. S. CP1, CP3. *Collected Papers*, vol. 1, vol. 3, Cambridge, Mass.: Harvard University Press, 1931, 1933.

[Peirce 1867] Peirce, C. S. 1867. On an Improvement in Boole's Calculus of Logic, *Proceedings of the American Academy of Arts and Science*, 7, 250-261. [Also in CP1, 3-15.]

[Peirce 1870] Peirce, C. S. 1870. Description of a Notation for the Logic of Relatives, *Memoirs of the American Association for the Advancement of Science*, 9 (1873), 317-378. [Offprints printed 1870. Also in CP1, 27-98.]

[Peirce 1897] Peirce, C. S. 1897. The Logic of Relatives, *The Monist*, 7, 161-217. [Also in CP3, 288-345.]

[Peirce 1976] Peirce, C. S. 1976. *The New Elements of Mathematics*, vol. 3 (ed. C. Eisele), The Hague: Mouton; and Atlantic Highlands, New Jersey: Humanities Press.

[Rado 1975a] Rado, R. 1975a. The Cardinal Module and some Theorems on Families of Sets, *Annali di Matematica pura ed applicata*, (4)102, 135-154.

[Rado 1975b] Rado, R. 1975b. Families of Sets, in *Proceedings of the International Congress of Mathematicians*, Vancouver 1974 (ed. R. D. James), Vancouver: Canadian Mathematical Congress, vol. 2, 491-496.

[Royce 1905] Royce, J. 1905. The Relations of the Principles of Logic to the Foundations of Geometry, *Transactions of the American Mathematical Society*, 6, 353-415. [Also in his Logical Essays, Dubuque: Brown, 1951, 379-441.]

[Russell 2014] Russell, B. A. W. 2014. *Collected Papers*, vol. 5 (ed. G. Moore), London (Routledge).

[Singh 2011] Singh, D. and others 2011. Complementation in Multiset Theory, *International Mathematical Forum*, 6, no. 38, 1877-1884.

[Syropoulos 2001] Syropoulos, A. 2001. Mathematics of Multisets, in C. S. Calude and others (eds.), Multiset Processing: *Mathematical, Computer Science, and Molecular Computing Points of View*, Berlin: Springer-Verlag (LNCS 2235), 347-358.

[Venn 1881] Venn, J. 1881. *Symbolic Logic*, 1st ed., London: Macmillan.

[Venn 1894] Venn, J. 1894. *Symbolic Logic*, 2nd ed., London: Macmillan. [Repr. New York: Chelsea, 1970.]

[Vercelloni 1989] Vercelloni, L. 1989. *Filosofia delle Strutture*, Florence: Nuovo Italia.

[Walsh 2012] Walsh, A. 2012. *Relations between Logic and Mathematics in the Work of Benjamin and Charles S. Peirce*, Boston: Docent Press.

[Wittgenstein 1922] Wittgenstein, L. 1922. *Tractatus Logico-Philosophicus*, London: Kegan, Paul. [Various reprs. and eds.]

On the Boundaries of Mathematics and Physics

Proofs and Contexts: the Debate between Bernoulli and Newton on the Mathematics of Central Force Motion

Niccolò Guicciardini

Interpreting Mathematical Debates

In this essay I analyze the solutions given by Isaac Newton and Johann Bernoulli to a well-posed mathematical problem known in the eighteenth century as the 'inverse problem of central forces'. This work prompted polemical exchanges between two rival groups: the first, whose leader was Newton, was based in Oxford, Cambridge, and London; the second, whose leading representatives were Leibniz and Bernoulli, was scattered througout Europe, though centered in Basel and Paris. The problem they were concerned with remains of fundamental importance for Newtonian mechanics, so it comes as no surprise that its solution is still taught, albeit in modernized form, at our universities today. It may be stated, in brief, as follows.

The inverse problem of central forces:

- A body moving in space without any external resistance is acted upon by a central force.
- If its initial position and velocity are given, determine its trajectory.[1]

'Determining the trajectory' can mean either finding the plane curve the body will traverse, or locating not only this curve but *also* the body's position as a function of time.[2]

This said, the task of the historian would seem to be quite straightforward. By reading published and perhaps unpublished texts, one should search for evidence that would clarify when, where, and by whom this problem was first stated and then properly solved. Naturally, one would expect that certain difficulties might complicate the details of this story. For example, this particular problem might have been stated in different, but essentially equivalent ways. Or perhaps the first attempted solutions failed or, on closer inspection, some lacuna was found, or maybe an undetected lapse in the reasoning. Still, one might take for granted that after several aborted proposals this central problem was at last stated and solved in a form that we recognize as valid today, that is, a solution fairly close to the one presently found in modern textbooks on mechanics.

[1] Today this problem is usually called the 'direct' problem of central forces: to determine the motion of a particle under the influence of a single central force.

[2] As we shall later see, Hermann and Bernoulli set themselves the task of determining the curve alone under the hypothesis of an inverse-square force.

D.E. Rowe and W.-S. Horng (eds.), *A Delicate Balance: Global Perspectives on Innovation and Tradition in the History of Mathematics*,
Trends in the History of Science, DOI 10.1007/978-3-319-12030-0_4

Nevertheless, as this essay will demonstrate, things were and are not nearly so simple. Indeed, the complications and nervous debates surrounding the evaluation of evidence over who first solved this mathematical problem, or when and how it got solved, suggest that the historian's task is far more difficult than mathematicians might assume. Indeed, this case study might be taken as a reminder of the kinds of problems historians of mathematics often have to face. Mathematics is rightly regarded as the discipline that best epitomizes objectivity; its truth claims are open to evaluation through highly rigorous, intersubjective standards that aim to miminize human error in mathematical proofs. This image of mathematics should not, however, be thought of as somehow guiding the interpretative work of historians who read texts about past mathematics. Historical studies should strive to judge older work in the light of what we know about its contemporary context. This necessarily involves gauging the state of knowledge and understanding of the mathematicians who wrote and read these works.[3]

Before delving into the details of the eighteenth century debates, I would like to mention a related debate from a few decades ago. This modern clash was sparked by a paper published in 1982 by Robert Weinstock, an American physicist based at Oberlin College. His paper bore the incendiary title 'Dismantling a Centuries-Old Myth: Newton's *Principia* and Inverse-Square Orbits' (1982).[4] Soon thereafter a firestorm broke out in the pages of usually rather algid journals, such as the *American Journal of Physics* or the *Archive for History of Exact Sciences*, apparently rekindling some gunpowder left behind by the older combatants, namely, Newton, Bernoulli, and their respective followers and supporters. I believe it will be instructive to embark on this short preliminary detour, which gives me a chance to expand on the above remarks in regard to the complexities of historical interpretation of past events and arguments.

Weinstock reconsidered Newton's solution of the inverse problem for the special case of greatest significance in the *Principia*, namely when the central force varies as the inverse square of the distance from the center. Newton had famously argued that a body acted on by such a force will – under arbitrary initial conditions, so long as the velocity vector does not lie on the line joining the mass to the force centre – trace out a conic section. His solution can be found in Corollary 1 to Propositions 11–13 and in Proposition 17 of the first Book of his masterpiece. This, however, says Weinstock, is just a 'myth' that simply evaporates if one reads carefully; so he denies that Newton produced a valid proof that the orbits for an inverse-square force field must be conic sections. Indeed, he claims that Newton tacitly assumed what he wanted to prove, namely that the orbit is a conic sec-

[3]An historian of medicine will certainly find it easier to offer contextualized readings of ancient texts than an historian of mathematics simply because of the great conceptual sea-changes that have occurred in Hippocrates' discipline as compared with Euclid's. On the notion of revolution in mathematics, see Joseph Dauben, 'Are there revolutions in mathematics?' (1992a) and 'Conceptual revolutions and the history of mathematics' (1992b).

[4]See also, Robert Weinstock, 'Long-Buried Dismantling of a Centuries-Old Myth: Newton's *Principia* and Inverse-Square Orbits' (1989) and 'Inverse-Square Orbits in Newton's *Principia* and Twentieth-Century Commentary Thereon' (2000).

tion, thereby committing a patent *petitio principii.*[5] Weinstock averred that the correct approach required integrating the pertinent differential equation to obtain the general solution, in which case the integral curve will vary with the force law and need not be a conic. Furthermore, the integral calculus provides a proof that a particular solution exists and is unique whenever the initial conditions are specified. So if Weinstock's arguments are correct, then he has discovered a huge lacuna in Newton's mathematical treatment of gravitation. Yet, as we shall see, criticisms of Newton's mathematical treatment of central force motion were advanced three centuries ago by Johann Bernoulli. Nevertheless, Bernoulli's criticisms were somewhat different in nature from Weinstock's, and it would be interesting to put these differences into relief (as I intend to do in a subsequent work).

Here, again, one might imagine that it should be a straightforward matter to gauge the validity of Weinstock's thesis. If so, we would expect that competent readers of Newton's *magnum opus* should have been able to reach a quick consensus with regard to his pointed claims. Yet nothing of the kind took place; on the contrary, some were convinced that Weinstock was right, while others disagreed with him. Most notable amongst his critics were Bruce Brackenridge, Herman Erlichson, Bruce Pourciau, and Michael Nauenberg.[6] They claimed that Weinstock had failed to understand the geometrical, and therefore constructive nature of Newton's demonstration. Newton's aim – according to this viewpoint – was not the integration of a differential equation (as Weinstock required), but rather to geometrically construct the trajectory as a function of the initial conditions. Since, for any given initial condition, a conic could be constructed that satisfied the equation of motion, and since only one trajectory could be given, Newton's method, as expounded in the *Principia*, provides precisely what the geometrician required. The eminent mathematician Vladimir Arnold, backed by the Nobel prize winner Subrahmanyan Chandrasekhar, went even further, arguing that Weinstock had failed to appreciate another central result in the *Principia.* In the subsequent Lemma 28 of Book 1, Newton proved geometrically that oval curves cannot be squared algebraically, thereby anticipating (according to Arnold) a fundamental result in the theory of Abelian integrals.[7]

Yet, notwithstanding such formidible and authoritative defense of Newton's method of solution, there are not a few experts who claim that Johann Bernoulli should be credited as the first to give a complete proof of the inverse problem for inverse square forces. Two competent historians who have expressed views

---

[5]In §3, we will examine Newton's proof in Corollary 1 to Propositions 11–13. For reasons of space, I will only make brief remarks about his Proposition 17. For more information, see Niccolò Guicciardini, *Reading the Principia* (1989), pp. 56–8.

[6]See, Herman Erlichson, 'Comment on "Long-buried dismantling of a centuries-old myth: Newton's *Principia* and inverse-square orbits" by Robert Weinstock' (1990); Bruce Pourciau, 'On Newton's Proof that Inverse-Square Orbits Must Be Conics' (1991); Bruce J. Brackenridge, *The Key to Newton's Dynamics* (1995); Michael Nauenberg, 'Newton's *Principia* and Inverse Square Orbits' (1994).

[7]Vladimir I. Arnold, *Huygens and Barrow, Newton and Hooke* (1990); Subrahamyan Chandrasekhar, *Newton's Principia for the Common Reader* (1995).

very much in line with Weinstock's are François De Gandt and, especially, Pierre Costabel. Referring to Corollary 1, Proposition 11, and Proposition 17, Book 1 of the *Principia*, De Gandt writes: 'All of this reasoning is restricted to the case of an inverse-square force, and presupposes that the solutions are conics. The only question is that of determining precisely the conic agreeing with initial conditions. This is, therefore, an altogether partial and imperfect response to the general inverse problem.'[8] Costabel went even further, referring to Corollary 1 (as emended in the second edition by Newton) he presents an argument that accords perfectly with Weinstock's views.[9]

So how should we view this modern debate? Should a historian consider taking sides, perhaps hoping to adjudicate the conflict in favor of one of the contending parties? Or would it be better to dismiss these differences as merely part of a sterile dispute caused by anachronistic readings of past texts? I personally do not favor either of these two options. The first would amount to casting aside the principles of 'impartiality' and 'symmetry' that the historian should insist upon in a case like this. One ought to make every effort to read the texts of Newton and Bernoulli from a neutral standpoint, which means, in particular, avoiding any preconceived notion of what a solution of the inverse problem of central forces should look like.[10] Rather, I suggest that both the modern and the old debates should be taken as a chance to reconsider how these technicalities, which were so passionately discussed by the actors involved, should be seen in relation to the broader context of the respective research agendas endorsed by the principal parties. I will elaborate this viewpoint in section §4, in which I will claim that it would be reductive to consider the disagreement between Newton and Bernoulli as merely a matter of certain technicalities in their demonstrations. Newton and Bernoulli viewed their mathematical arguments differently because of factors that concern what might loosely be called the 'intellectual context' in which they set forth their results.[11]

Something quite similar might be said about the case of the more recent debate. Indeed, the mathematicians who participated in it interpreted Newton's and Bernoulli's texts from the standpoint of different (perhaps only tacitly assumed) research programs for the mathematization of rational mechanics. Most notably, Arnold's campaign in favor of a qualitative approach to the theory of dynamical systems quite explicitly structured the way he understood the development

[8]François De Gandt, *Force and Geometry in Newton's Principia* (1995), p. 246.

[9]Costabel (1989), pp. 23–4. Further, we can cite Speiser's opinion: 'Newton solved what was called afterwards for a short time "the direct Kepler problem" ("le problème direct"): given a curve (e.g. an ellipse) and the center of attraction (e.g. the focus), what is the law of this attraction if Kepler's second law holds? The "problème inverse" (today: the "problème direct") was attacked systematically only later, first by Jacob Hermann, then solved completely by Johann Bernoulli in 1710 and following Bernoulli by Pierre Varignon'. David Speiser, 'The Kepler Probem from Newton to Bernoulli' (1996), p. 103. We shall review Bernoulli's solution in §2

[10]The principles of impartiality and symmetry were first introduced in David Bloor, *Knowledge and Social Imagery* (1976), p. 7. I find it quite reasonable to adopt these principles in this case.

[11]I will have more to say on the notion of context in section §4.

of rational mechanics from Newton to Poincaré. His admiration for the geometrical style in both the *Principia* and in Huygens's *Horologium* clearly reflected the research approach he endorsed as a practicing mathematician. It would be interesting to compare Arnold's opinions on the development of geometry and dynamics with Truesdell's authoritative efforts aimed at 'rediscovering the rational mechanics of the age of reason'. As in Arnold's case, Truesdell's historical studies were influenced in large part by his own research in rational thermodynamics, the work for which he was best known as a scientist.[12] For Truesdell, Newton's *Principia* belonged to a preliminary stage in the mathematical science of motion. This stage served as the background that eventually led to the advent of rational mechanics as epitomized in the work of Leonhard Euler. In the case at hand, Truesdell considered Bernoulli's calculus approach to central force motion to be a major breakthrough rather than as simply a translation into symbols of Newton's geometrical reasoning, as Arnold would claim.

Mathematicians, not suprisingly, look at the past through the lens of their own research interests, which reflect a diversity both in their aims and methods. As a historian, I am interested in how they go about reconstructing the past, and for this reason I cannot take the purist position, the second option indicated above, which would dismiss the debate surrounding Weinstock's paper as an instance of amateurish anachronism. The views of mathematicians turned historians are often prejudiced by their research interests, but I do not share the skepticism, sometimes even derision, so commonly felt by professional historians when they accuse mathematicians who turn to history of producing work that is hopelessly naïve with respect to hermeneutic methodology. This rather comfortable stance is essentially self-defeating because the mathematicians' point of view has its own value, and this for two reasons.

First, anachronism is a way – to be sure, not the only way – to make past texts come alive for us today.[13] In the hands of scientists of the caliber of Chandrasekhar and Arnold, Newton and Bernoulli become our virtual contemporaries. We enter into a conversation with their texts; so much so that we can ask whether their demonstrations meet our standards of rigor, or even our esthetic standards. Was Newton's demonstration cogent? Was it more elegant than Bernoulli's? Questions like these can lead to fruitful discourse in the modern classroom, particularly if the appropriated texts in question have been carefully analyzed by a competent contemporary reader. This didactical purpose has the beneficial effect of keeping older texts in active use, making them both deployable and even enjoyable.

Second, we can better appreciate the mathematicians' approach to past texts by thinking about another genre of writing that might be called works by connois-

[12] Clifford Truesdell, 'A Program toward Rediscovering the Rational Mechanics of the Age of Reason' (1960).

[13] For two defenses of anachronism, see Nicholas Jardine, 'Uses and Abuses of Anachronism in the History of the Sciences' (2000) and, from a different viewpoint, Dominick LaCapra, *Rethinking Intellectual History* (1983).

seurs.[14] One of the aims of this genre is to re-examine classical or now forgotten scientific texts in order to assess and to explain what is valuable in them and why. I would invite the reader to consider – before raising the charge of Whiggism against this genre of writing – that few readers would dismiss the value of connoisseurs' assessments in scholarly works on the history of music or the other arts. Should we not listen with interest and respect to what a renowned pianist has to say about the use of chromatic harmony in Chopin's Ballades? Would we not do the same, if a famous painter spoke about Titian's techniques for employing combinations of colors? Surely the opinions and interpretations of connoisseurs contribute greatly to our appreciation of past works, making them seem alive and meaningful for us. Are we, as historians of mathematics, really so different from historians of music or art that we can simply treat the views of practitioners as irrelevant for our own work?

It is not as a connoisseur that I write, but rather as a member of a more humble profession. As a historian, I will attempt in what follows to convey a clear reading of the intellectual contexts that framed the mathematical arguments set forth by the British and Continental camps. My aim will be to show that the confrontation that arose between Newton and Bernoulli was guided by the differing agendas these two mathematicians endorsed. By 'agenda' I refer to a component within what Marc Bevir has called a 'web of beliefs'.[15] This refers to the complex, sometimes even contradictory ways in which a mathematician conceives his profession and role, including convictions held about the significance of past developments and hopes for the future evolution of the discipline. Mathematicians look to the past for authoritative masters; very often these are authors who enjoyed exemplary success that their successors hope to imitate or build upon. Such hopes and convictions are often inscribed in their reconstructions of the past, which imbue the discipline with a sense of collective identity. These incentives, in turn, motivate the production of participants' histories: the writing and rewriting of different mathematical genealogies or mythologies. Newton pointed to Pappus, Apollonius, and Huygens as models to his acolytes; Bernoulli taught and defended Leibniz's mathematical heritage. And — in a similar vein — Arnold praised Newton, Hooke, Barrow, and Huygens; Truesdell's hero was Euler. Such genealogies are by no means extraneous to the historian's mind-set either. Surely we, as historians of mathematics writing in the twenty-first century, have to take account of all these factors in our contextualized interpretations of past texts. And these will, of course, reflect our own webs of beliefs, our aims and agendas, coloring and influencing our readings of the works of past actors. In some ways, our views now are as contingent as theirs were then.

---

[14] On the notion of connoisseurship see Hasok Chang, 'The Historian of Science: Painter, Guide, or Connoisseur' (2008). Chang's paper is a comment on Clifford Truesdell, 'The Scholar's Workshop and Tools' (1973).

[15] Marc Bevir, *The Logic of the History of Ideas* (1999), p. 5ff.

Johann Bernoulli's letter of 1710

The inverse problem of central forces

On October 7, 1710, Johann Bernoulli sent a letter to the Académie des Sciences that was to mark a turning point in the development of the applications of mathematics to the science of motion. This was written in reply to a letter from Jacob Hermann, then Professor of Mathematics in Padua, and is a pleasure to read, being a characteristic example of Bernoulli's crystal-clear style. In this 'Réponse à M. Herman', he dealt with the 'inverse problem of central forces' by obtaining a general solution and then applying it to an inverse-square force.[16]

Hermann's approach

Hermann's short letter began with the area law, which since the publication of Newton's *Principia* was known to hold for any central force. Following a few algebraic manipulations, he obtained a second-order differential equation, which he could integrate under the hypothesis of an inverse-square force. This led to the equation for a conic section in Cartesian coordinates.[17]

Hermann's reasoning, as Bernoulli was quick to spot, was incomplete since a constant of integration is missing. But he also leveled other criticisms:

1. Hermann's solution is 'faite à dessein': that is, according to Bernoulli, Hermann was able to integrate the differential equation only because he knew in advance what the solution should be.
2. Furthermore, Hermann did not know how to apply his procedure to a different force law.

In short, according to Bernoulli, Hermann was not in possession of a general method for solving the inverse problem of central forces. As we shall see, these criticisms amounted to a rehearsal for the long polemic that Bernoulli launched against Newton's methods.

Bernoulli's approach

Bernoulli approached the problem from a viewpoint quite different from Hermann's. Drawing inspiration from the Propositions 39–41 in Book 1 of Newton's *Principia*, he introduced polar coordinates and employed two conservation laws,

---

[16] Johann Bernoulli, 'Extrait de la Réponse de M. Bernoulli à M. Herman, datée de Basle le 7 Octobre 1710' (1710) and Jacob Hermann, 'Extrait d'une lettre de M. Herman à M. Bernoulli' (1710).

[17] On Bernoulli's solution, see Michael Nauenberg, 'The Early Application of the Calculus to the Inverse Square Force Problem' (2010). In this masterful paper, Nauenberg also provides an in-depth analysis of the approaches of Leibniz, Hermann, and Newton, while showing that the algebraic manipulations performed by Hermann and Bernoulli were inspired by Newton's geometrical constructions in the *Principia*.

rather than just one: the conservation of areal velocity (equivalent to what nowadays is understood as 'conservation of angular momentum') and a formula for the speed equivalent to what nowadays is understood as 'conservation of mechanical energy'. For a body of unit mass, Bernoulli could now rewrite what Newton had stated in Propositions 39-40 as:

$$v = \sqrt{ab - \int \phi \, dx}, \tag{1}$$

where a and b are constants, x denotes the body's distance from the force center, v its speed, and ϕ the strength of the force acting on the body (a factor of $1/2$ is omitted, as Bernoulli noted, 'pour éviter les fractions').[18] It should further be noted that a denotes the initial distance of the body from the force's center, whereas ab corresponds to what in modern terms would be called twice the total energy. As noted below, the constants in these equations (such as (1)) have been taken so that geometrical homogeneity between the left and right terms is guaranteed.[19]

Bernoulli expressed these two conservation laws (for angular momentum and mechanical energy) in algebraic terms through an easy translation of Newton's geometrical language in Propositions 39-41. This enabled him to obtain the first-order differential equation (2), thereby reducing the inverse problem of central forces (for any central force ϕ) to an integration problem. Bernoulli's differential equation for this general case is:

$$dz = \frac{a^2 c dx}{\sqrt{abx^4 - x^4 \int \phi \, dx - a^2 c^2 x^2}}, \tag{2}$$

where the constant ac denotes twice the areal velocity (i.e., angular momentum per unit mass).

Today we are accustomed to writing polar coordinates with the symbols r and θ, whereas Bernoulli's coordinates are $x = r$ and $z/a = \theta$.[20] If we introduce this more familiar notation, taking l to be the angular momentum and assuming the body has unit mass,[21] then we can rewrite Bernoulli's equation (2) as follows:[22]

[18]We would write $E = mv^2/2 + V(r)$.

[19]Note that in (1) and (4) the dimension of the speed and the force's strength is 1. The necessity of preserving geometrical homogeneity leads Bernoulli to introduce what, from a modern point of view, is a redundant constant. In what follows a modern mathematician, freed from the worries of geometrical representability that still burden Bernoulli's mathematics, could set, for instance, $a = 1$.

[20]It was customary to consider an angle as a dimensionless ratio between two lines, i.e., a ratio between an arc of a circle and its radius, therefore Bernoulli denotes the polar angle as z/a.

[21]We are considering a one-body problem, therefore the mass should not concern us.

[22]A modern reader might encounter some difficulty in making sense of Bernoulli's constants. In modern notation we set: $x = r$, $z/a = \theta$, $\phi = -2F$, $ab = 2E$, $ac = l$, where F is the force's strength, E is the total energy, and l is the angular momentum.

$$d\theta = \frac{l dr}{\sqrt{2Er^4 + 2r^4 \int_r^\infty F dr - l^2r^2}}. \tag{3}$$

For an inverse square force

$$\phi = a^2 g/x^2, \tag{4}$$

(where g is a constant[23]) Bernoulli obtained the following differential equation:

$$dz = \frac{a^2 c dx}{x\sqrt{abx^2 + a^2gx - a^2c^2}}, \tag{5}$$

which he could integrate by just substituting variables. As I remarked earlier, this is essentially the method still used in textbooks on Newtonian mechanics (a somewhat ironic name for this discipline from the standpoint of the present story). It was by this means that Bernoulli was able to show, of course, that a body acted upon by an inverse-square force necessarily moves along a conic section. Robert Weinstock would surely have nodded in approval, so let us consider this argument more closely.

Bernoulli began by introducing a change of variables via:

$$x = \frac{a^2}{y} \tag{6}$$

followed by the transformation:

$$y = a^2g/2c^2 - t. \tag{7}$$

Then, 'pour abréger', he introduced a new constant h such that:

$$c^2h^2 = a^3b + a^4g^2/4c^2. \tag{8}$$

which leads to the differential equation:

$$\frac{dz}{a} = \frac{dt}{\sqrt{h^2 - t^2}}, \tag{9}$$

which he could easily integrate.[24]

One can then verify that this leads to the equation for a conic section written in the form:

$$x = \frac{c_1}{1 - c_2 \sin((z - z_0)/a)}, \tag{10}$$

[23]In contemporary notation, $a^2g = 2k$ for a force $F = -k/r^2$.

[24]Bernoulli writes :'qui [dz/a] est une différentielle d'arc de cercle (dont le rayon est $= h$, & le sinus $= t$) divisé [sic] par son rayon'. We translate as $z/a = \arcsin(t/h) + z_0/a$. Therefore, $t = h\sin((z - z_0)/a)$. Therefore, because of (7): $a^2g/2c^2 - y = h\sin((z - z_0)/a)$. And, because of (6): $x = (2c^2/g)(1 - (2c^2/a^2g)h\sin((z - z_0)/a))^{-1}$. Because of (8), $h = \sqrt{a^4g^2 + 4a^3bc^2}/(2c^2)$. Elimination of h leads to equation (10).

c_1 and c_2 constants. We note that the constants c_1 and c_2 above depend on the initial conditions as follows: $c_1 = 2c^2/g$ and $c_2 = \sqrt{1 + 4bc^2/ag^2}$ (recall that Bernoulli used a redundant constant for the sake of preserving the geometrical homogeneity of the equations). We should further note that Bernoulli did not make use of a notation for trigonometric functions but rather expressed (10) in geometric terms.[25] Finally, Bernoulli changed back to the Cartesian coordinates x and y in order to write the final result as the general equation for conic sections in this familiar form.

In sum, we should see Bernoulli's methods as part of the larger project of Continental work with the new calculus that had taken place over the last two decades. Indeed, his result was the outcome of a protracted and successful research program in the application of Leibniz's calculus and, most notably, the related theory of differential equations to problems in natural philosophy. This program had been initiated by Leibniz himself, but was soon pursued by several others, most notably Pierre Varignon, Hermann, and Johann Bernoulli. This takes us into the larger methodological conflicts that separated these practioners from their counterparts in the circle around Newton.

Bernoulli's first assault

When the long-simmering debate between Newtonians and Leibnizians reached a boiling point in 1710 , Johann Bernoulli tried to come to Leibniz's aid. He did so by claiming that the absence of calculus in the *Principia* was proof positive that in 1687, three years after Leibniz's initial publication of the differential calculus, Newton was still ignorant of these new methods. This, wrote Bernoulli, was evident from Newton's treatment of the inverse problem, a criticism he took to be devastating.[26] Even more, according to Bernoulli, Newton had not proven that a body accelerated by an inverse-square force must necessarily have a conic section as its orbit. Thus, we read in his letter of 1710:

> In order to determine that this curve *ABC* under this hypothesis [namely, the force is inverse-square] is always a conic section, which Newton supposed on page 55 Corollary 1 without proving it, takes much more skill.[27]

[25]In modern notation ($ab = 2E$, $ac = l$, $ga^2 = 2k$): $c_1 = l^2/k$ and the eccentricity is $c_2 = \sqrt{1 + 2El^2/k^2}$, where E is the energy, l the angular momentum, and $F = -k/r^2$. Thus, taking into consideration that $\sin(z/a) = \sin(\theta) = -\cos(\theta + \pi/2)$, the integral (10) is:

$$1/r = (k/l^2)(1 + \sqrt{1 + 2El^2/k^2}\cos(\theta + \pi/2 - \theta_0)),$$

where $\theta_0 = z_0/a$. See Herbert Goldstein, *Classical Mechanics*, eq. (3–46).

[26]Another incendiary spark was set off by John Keill in another paper on central force motion presented to the Royal Society in 1708, where one could read the offending lines: 'All these things follow from the nowadays highly celebrated arithmetic of fluxions, which Mr Newton beyond any shadow of doubt first discovered, ..., and yet the same arithmetic was afterwards published by Mr Leibniz in the *Acta Eruditorum* having changed the name and the symbolism'. John Keill, 'Epistola ad Clarissimum Virum Edmundum Halleium' (1710). English translation in Hall, *Philosophers at War* (1980), p. 145.

[27]'Pour voir presentement que cette Courbe *ABC* dans cette hypothese est toûjours une Section

This first brief public thrust against the *Principia* marked, as we shall see, the beginning of a prolonged battle between John Keill (acting as Newton's watchdog) and Bernoulli, a war that produced an extensive trail of papers, letters, and polemical pamphlets in its wake.

Newton on the inverse problem of central forces

Corollary 1 to Propositions 11–13, Book 1

Let us now turn to Newton's *Principia*, beginning with the matters discussed on page 55 as referred to by Bernoulli, in order to see how Newton dealt with the inverse problem of central forces. Indeed, what we find there is difficult to comprehend. When one opens the *Principia* in search of a solution of the inverse problem, what one first encounters in Book 1 are the three Propositions 11–13, followed by a Corollary 1, referred to by Bernoulli in the letter cited above.

In those preceding propositions, Newton proves that if the orbit is a conic section and the area law is valid for a focus S, then the force acting on the body is inversely proportional to the square of the distance from S and it is directed toward or away from S. Then follows Corollary 1 which reads:

> From the last three propositions [i.e., Propositions 11–13] it follows that if any body P departs from the place P along any straight line PR with any velocity whatever and is at the same time acted upon by a centripetal force that is inversely proportional to the square of the distance of places from the center, this body will move in some one of the conics having a focus in the center of forces; and conversely.[28]

Quite understandably, this terse statement was subjected to criticism; certainly it appears that Newton here conflated the main conclusion from the preceding propositions with its converse. Newton himself, while preparing the second edition of the *Principia*, emended Corollary 1 by adding a sketch of what he considered to be a valid proof. In a letter, dated October 11, 1709, he instructed Roger Cotes to complete Corollary 1 by adding the following lines:

> For if the focus and the point of contact and the position of the tangent are given, a conic can be described that will have a given curvature at that point. But the curvature is given from the given centripetal force [and velocity of

Conique, ainsi que M. Newton l'a supposé pag. 55 Corol. 1 sans le démonstrer; il y faut bien plus d'adresse'. Johann Bernoulli, 'Extrait de la Réponse de M. Bernoulli à M. Herman, datée de Basle le 7 Octobre 1710' (1710), p. 527.

[28]Isaac Newton, *The Principia: Mathematical Principles of Natural Philosophy* (1999), p. 467. 'Ex tribus novissimis Proportionibus consequens est, quod si corpus quodvis P, secundum lineam quamvis rectam PR, quacunque cum velocitate exeat de loco P, & vi centripeta quæ sit reciproce proportionalis quadrato distantiæ a centro, simul agitetur; movebitur hoc corpus in aliqua sectionum Conicarum umbilicum habente in centro virium; & contra'. Newton, *Philosophiae Naturalis Principia Mathematica* (1972), p. 125.

the body]; and two different orbits touching each other cannot be described with the same centripetal force [and the same velocity].[29]

What Newton cryptically states here can perhaps be elaborated in modernized language as follows:[30]

1. For a body of mass m moving with known velocity $\vec{v}$ at the initial point P in a central force field of strength F and with center S (where it is assumed that $\vec{v}$ is not to be directed in line with the force center), Newton proposed to exhibit a geometrical technique for the construction of the body's orbit.
2. The initial position and velocity thus determine not only the point P on the trajectory of the body but also the tangent to the trajectory at P.
3. Since the strength and direction of the force at P are given, the curvature of the trajectory at P is also uniquely determined from the initial conditions ($F_N = mv^2/\rho$, where F_N is the normal component of force at P and ρ the radius of curvature at P).
4. In Proposition 17, Book 1, Newton showed how to construct a unique conic from the known tangent and curvature at P (determined in steps 2 and 3) and with one focus located at the force center S.[31]
5. These initial conditions also determine the areal velocity, which, since the force is central, must be a constant of motion (Propositions 1 and 2, Book 1).[32]
6. Because of Propositions 11–13, Book 1, the Keplerian orbit determined from step 5 satisfies the equations of motions.
7. Since for any initial conditions—thus given both position and velocity—a Keplerian orbit can be constructed that satisfies the equations of motions, one concludes (assuming that only one such orbit is possible) that Keplerian orbits are necessary in a central inverse-square force field.

Newton expressed the assumption of uniqueness (implicit in point 7) in the final lines of the revised corollary ('[T]wo different orbits touching each other

[29] Isaac Newton, *The Principia: Mathematical Principles of Natural Philosophy* (1999), p. 467. The text in square brackets was added in the third (1726) edition. 'Nam datis umbilico, & puncto contactus, & positione tangentis, describi potest sectio conica, quae curvaturam datam ad punctum illud habebit. Datur autem curvatura ex data vi centripeta, [& velocitate corporis]: & orbes duo se mutuo tangentes eadem vi centripeta [eademque velocitate] describi non possunt'. See Isaac Newton, *Correspondence*, 5, p. 5.

[30] For further details, see Bruce Pourciau, 'On Newton's Proof That Inverse-Square Orbits Must Be Conics' (1991) and 'Newton's Solution of the One-Body Problem' (1992).

[31] Proposition 17 would deserve a separate treatment, see Niccolò Guicciardini, *Reading the Principia: the Debate on Newton's Mathematical Methods for Natural Philosophy from 1687 to 1736*, Cambridge, 1999, pp. 56–8.

[32] In Section 6, Book 1, Newton showed how a motion satisfying Kepler's area law along a given conic can be approximated.

cannot be described with the same centripetal force and the same velocity'). Thus, it seems Newton thought that, for any given initial conditions, the uniqueness of the orbit was intuitively obvious. Still, he might have deduced this property from Proposition 41, Book 1, where the inverse problem is reduced to quadratures.[33]

Criticisms leveled at Corollary 1

Corollary 1 never ceased to be an object of criticism and debate, and as noted in the introduction, it has even recently found its critics and defenders. It is not my intention to enter into these matters further, but rather I wish to consider instead what Bernoulli had to say against Corollary 1, and how Newton and his acolytes responded to that challenge. Indeed, in the course of their passionate confrontation, the combatants on these two sides expressed quite different views about the role of mathematics in natural philosophy, and this is what especially interests me here.

At this juncture we might summarize this long, protracted quarrel by saying that:

- Newton and his acolyte Keill defended the cogency of the 1713 version of Corollary 1. Most interestingly, they claimed that Corollary 1 was both more elegant and more concise than Bernoulli's solution in terms of differential equations.[34]
- Bernoulli, like all eighteenth-century mathematicians up until Laplace, also accepted the cogency of Corollary 1 in the emended form of 1713. This, in fact, is a major difference between this older dispute and the contemporary debate that we reviewed in §1. However, Bernoulli also claimed that only by means of differential equations could one obtain a solution that was truly general and 'a priori'. This, as we saw already in his letter to Jacob Hermann, was simply because the procedures of the integral calculus are universal and can thus be applied in a systematic way to different particular cases, without knowing the solution in advance.[35]

[33]For a recent discussion of the uniqueness assumption in Corollary 1, see Bruce Pourciau, 'Proposition II (Book I) of Newton's *Principia*' (2009), p. 164–166. On page 166, Pourciau expresses the uniqueness assumption as follows: '[T]wo different motions, urged by the same centripetal force throughout, cannot pass through the same given point P at the same time and with the same speed and direction'.

[34]For instance, Keill writes to Newton on November 9, 1713: 'Since I left London I have considered Mr Bernoulli's solution of the Inverse Problem of Central Forces, and I am amused at his impudence [...] In his application of it to the particular case [of an inverse square force] he has with a great deal of labour showed that the curve described must be a Conick-Section when the thing may be demonstrated in a few lines'. Isaac Newton, *Correspendence*, 6, p. 37.

[35]For instance, Bernoulli writes to Newton on December 10, 1719: 'I only said something against the form of that assertion [namely, Corollary 1], and wished that someone would give an analysis that led a priori to the truth of the inverse [proposition] and without supposing the direct [proposition] to be already known. This, indeed, which I would not have said to your displeasure, I think was first put forward by me, at least so far as I know at present'. Isaac Newton, *Correspendence*, 7, p. 78, translation from Latin by A. Rupert Hall and Laura Tilling.

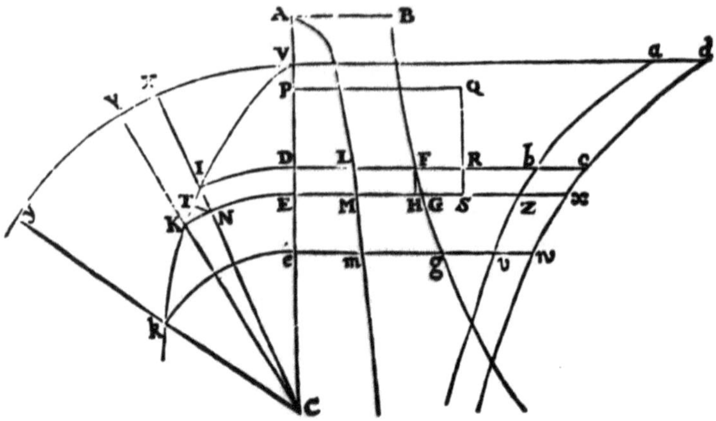

Figure 1: Diagram for Proposition 41, Book 1, of the *Principia* (1687). Source: Newton, *Philosophiae Naturalis Principia Mathematica* (1972), p. 219.

Proposition 41, Book 1

Before discussing the debate between Bernoulli and Newton any further, we need first to turn again to the *Principia.* In Sections 7 and 8 of Book 1, we discover that Newton had something more to offer that was highly significant for the inverse problem. There, in Proposition 41, we find a more general treatment than that indicated in Corollary 1. This reads:

> Supposing a centripetal force of any kind and granting the quadratures of curvilinear figures, it is required to find the trajectories in which bodies will move and also the times of their motions in the trajectories so found.[36]

This proposition is clearly based on the assumption that we are in possession of a method for finding the 'quadrature of curvilinear figures' (i.e., an integration method). In his youth, Newton had developed several quadrature techniques as part of the 'inverse method of fluxions'. However, in the *Principia* Newton chose not to make his mathematical discoveries in this field entirely explicit.

In the present context, Newton began with two conservation laws, which can be identified somewhat anachronistically with conservation of angular momentum and mechanical energy. These two laws, as we noted earlier, were the starting point for Bernoulli's 1710 solution. Newton next considers (see figure 1) a 'body' that has a given initial speed at point V and moves in a direction perpendicular to CV

[36]Isaac Newton, *The Principia: Mathematical Principles of Natural Philosophy* (1999), p. 529. 'Posita cujuscunque generis vi centripeta & concessis figurarum curvilinearum quadraturis, requiruntur tum Trajectoriæ in quibus corpora movebuntur, tum tempora motuum in Trajectoriis inventis.' Newton, *Philosophiae Naturalis Principia Mathematica* (1972), p. 218.

while acted upon by a centripetal force directed towards the center of force C. He writes VIk for the curve sought and lets IK be an infinitesimal arc traversed by the body in a tiny interval of time.[37] Further, KN is drawn perpendicular to CI. Since Kepler's area law holds, the infinitesimal interval of time is proportional to the area of the sector CIK, that is to $(CI \times KN)/2$. Newton denotes the constant areal velocity by $Q/2$; so that the arc IK is traversed in the infinitesimal time interval measured by the ratio $(CI \times KN)/Q$. Since the speed v along IK can be assumed constant, this will be measured by the ratio between the arc IK and the measure of the time:

$$v = \frac{IK \times Q}{CI \times KN}. \tag{11}$$

Note that on the right side of figure 1 Newton has drawn the curves that must be squared in order to solve the inverse problem (he corrected the drawing in the second edition). Thus the curve BFG gives a geometrical representation of the strength of the force as a function of the radial displacement (DF is thus proportional to the force's strength at I).

This relates to Propositions 39 and 40 of Book 1, where Newton proves that the square of the speed at I is proportional to the area of the surface $ABFD$ under the curve, which represents force as a function of radial displacement CI. This result, which he formulates in prose, can be rendered algebraically as

$$v \propto \sqrt{ABFD}.[38] \tag{12}$$

Newton could thus write (we employ the symbol '$\propto$' where Newton would use 'is proportional to')

$$\frac{IK \times Q}{CI \times KN} \propto \sqrt{ABFD}. \tag{13}$$

By setting $CI = A$ and $Q/A = Z$, he then states:

$$\frac{IK}{KN} \propto \frac{\sqrt{ABFD}}{Z}.[39] \tag{14}$$

The proportion (14) is the basic result that allows Newton to tackle the inverse problem of central forces in Proposition 41.[40]

---

[37] 'Let the points I and K be very close indeed to each other'. Isaac Newton, *The Principia: Mathematical Principles of Natural Philosophy* (1999), p. 530. 'Sint autem puncta I & K sibi invicem vicinissima'. Isaac Newton, *Philosophiae Naturalis Principia Mathematica* (1972), p. 219.

[38] Propositions 39 and 40 can be read in modern terms as the expression of the conservation of mechanical energy.

[39] We are here using a somewhat more symbolical formulation. What Newton actually writes is: '$\sqrt{ABFD}$ will be to Z as IK to KN'. Isaac Newton, *The Principia: Mathematical Principles of Natural Philosophy* (1999), p. 530. '& erit semper $\sqrt{ABFD}$ ad Z ut IK ad KN'.

[40] In order to facilitate understanding of the geometrical proportionality (14), we can translate this into our more familiar Leibnizian symbolic language. Substituting $Q = l$, $IK = \mathrm{d}s$, $IN = \mathrm{d}r$, $KN = r\mathrm{d}\theta$, $CI = r$, $CV = r_0$, and furthermore, following Propositions 39 and 40, we put

Throughout the *Principia*, Newton consistently represented all of the concepts and propositions in geometrical form, and here he introduces an auxiliary circle $VXYR$. From (14), after some simple manipulations with ratios, he deduces that

$$\text{area } CVI \left[= \frac{Q}{2} t \right] = \text{area under curve } abz = VabD, \tag{15}$$

where the ordinate of the curve abz is

$$Db = \frac{Q}{2\sqrt{(ABFD - Q^2/A^2)}}. \tag{16}$$

Furthermore

$$\text{area } CVX \left[= \frac{CX^2}{2}\theta \right] = \text{area under curve } dcx = VdcD, \tag{17}$$

where the ordinate of the curve dcx is

$$Dc = \frac{Q \times CX^2}{2A^2\sqrt{(ABFD - Q^2/A^2)}}.^{41} \tag{18}$$

If one 'squares' abz, this establishes a functional dependence between time $[t]$ and distance $[r]$. Similarly, by 'squaring' acx, one obtains a relation between the polar angle $[\theta]$ and the distance $[r]$. Note that instead of working directly with polar coordinates, as in the Leibnizian translation that we proposed in footnote 40 and inserted in the square brackets above, Newton relates the distance CI to geometrical quantities (areas CVI and CVX) proportional to time t and to the polar angle θ respectively, which can be visualized in figure 1.[42]

$ABFD = v_0^2 + 2\int_r^{r_0} F\mathrm{d}r = v^2$, where v is the speed at I, v_0 the initial speed at V. Assuming the mass $m = 1$, we derive from (14)

$$\frac{\mathrm{d}s^2 (= \mathrm{d}r^2 + r^2\mathrm{d}\theta^2)}{r^2\mathrm{d}\theta^2} = \frac{r^2}{l^2}\left(v_0^2 + 2\int_r^{r_0} F\mathrm{d}r\right).$$

By means of simple algebraic manipulations, we now obtain the differential equations for the orbit in polar coordinates:

$$\mathrm{d}t = \frac{\mathrm{d}r}{\sqrt{(v_0^2 + 2\int_r^{r_0} F\mathrm{d}r - l^2/r^2)}}.$$

and

$$\mathrm{d}\theta = \frac{l\mathrm{d}r}{r^2\sqrt{(v_0^2 + 2\int_r^{r_0} F\mathrm{d}r - l^2/r^2)}},$$

It should be stressed that this translation, which is very close to what Johann Bernoulli did in 1710 and while helpful for the modern reader, does not convey accurately what one finds in the *Principia*. This is hardly surprising when we keep in mind that Newton felt a strong need to retranslate algebraic reasoning into geometry.

[41] In order to facilitate the understanding I have added in square brackets symbolic expressions in terms of polar coordinate θ and time t that do not appear in the *Principia*.

[42] The role of visualization has been underlined by Herman Erlichson in 'The Visualization of Quadratures in the Mystery of Corollary 3 to Proposition 41 of Newton's *Principia*' (1994).

Translating Proposition 41 into calculus

It is noteworthy that in his 1710 letter Bernoulli explicitly cited Propositions 39, 40, and 41, 'translating' these results into algebra before deriving his differential equation (2). Indeed, the first part of his letter merely renders Newton's geometrical proportionalities into the language of calculus, a translation that is quite straightforward.

So the question naturally arises, was Newton able to perform a comparable translation? To this, the answer I would give is yes and no. Newton, of course, did not use, as Bernoulli did, Leibniz's version of the calculus. He had long before developed his own methods for handling infinite series and fluxions, methods that were mathematically equivalent to and yet different from Leibniz's algorithms. On the other hand, as we shall now see, Newton was able to translate Proposition 41 into the terminology of his fluxional method.

Inverse-cube forces

In Corollary 3 to Proposition 41, Newton applies his general result to an inverse-cube central force rather than the familiar case of an inverse-square. The question he posed was thus: which orbits are described by a body which is acted on by an inverse-cube central force? In Corollary 3 Newton gives the solution only in the form of a geometrical construction, to which he then adds this remark:

> All this follows from the foregoing proposition [41], by means of the quadrature of a certain curve, the finding of which, as being easy enough, I omit for the sake of brevity.[43]

This is one of the most famous instances of those gaps in the demonstrative structure of the *Principia* that aroused both interest and frustration among its readers.[44]

Newton's reply to Gregory

One such curious reader was David Gregory, who paid a visit to Newton in May 1694 during which he asked about the mysterious quadrature method applied in the construction of Corollary 3. The esteemed Lucasian Professor answered by translating the basic result of Proposition 41 into two 'fluxional' ('differential')

[43]Isaac Newton, *The Principia: Mathematical Principles of Natural Philosophy* (1999), p. 532. 'Consequuntur hæc omnia ex Propositione præcedente, per Curvæ cujusdam quadraturam, cujus inventionem ut satis facilem brevitatis gratia missam facio'. *Philosophiae Naturalis Principia Mathematica* (1972), p. 223.

[44]On Corollary 3, Proposition 41, Book 1, see Derek T. Whiteside, 'The Mathematical Principles Underlying Newton's *Principia Mathematica*' (1970); Herman Erlichson, 'The Visualization of Quadratures in the Mystery of Corollary 3 to Proposition 41 of Newton's *Principia*' (1994); Bruce Brackenridge, 'Newton's Easy Quadratures "Omitted for the Sake of Brevity"' (2003); and Niccolò Guicciardini, *Isaac Newton on Mathematical Certainty and Method* (2009), pp. 271–8.

equations.[45] He then applied these equations to the case of an inverse-cube force, thereby obtaining the construction given in Corollary 3 of the *Principia*. As Gregory remarked in a memorandum recounting this visit:

> on these [quadratures] depend certain more abstruse parts in his philosophy as hitherto published, such as Corollary 3, Proposition 41.[46]

Newton thus used a mixture of geometric and algorithmic methods that was less algebraic than the approach of Bernoulli. Nonetheless, there can be no doubt that Newton relied to a certain extent on quadrature techniques that he expressed in algebraic language in order to solve the inverse problem of central forces in the case of an inverse-cube central force. Indeed, the fluxional solution he revealed to Gregory in 1694 was, in certain key respects, the same as the one Bernoulli gave in 1710.

In his letter to Gregory, Newton denotes by x the distance between the body and the center of force, which he takes to be

$$F = a^4/x^3, \tag{19}$$

where a is a constant.[47] In Propositions 39 and 40, Newton proves that the square of the speed at distance x is proportional to the area $ABFD$ under the curve BFG, whose ordinate DF represents the strength of the central force (see figure 1 and formula 12). For an inverse-cube force satisfying (19) he then deduces:[48]

$$ABFD = 2a^4/x^2 - 2a^4/c^2. \tag{20}$$

After some algebraic calculation, Newton shows that Corollary 3 can be written in terms of two differential equations, so the solution now reduces to squaring the curves whose ordinates Db and Dc are given by:

$$Db = \frac{Q}{2\sqrt{(2a^4 - Q^2)/x^2 - 2a^4/c^2}} = \frac{1}{2}\frac{Qx}{\sqrt{2a^4 - Q^2 - 2a^4x^2/c^2}}, \tag{21}$$

and

$$Dc = \frac{Q \times CX^2}{2x^2\sqrt{(2a^4 - Q^2)/x^2 - 2a^4/c^2}} = \frac{Q \times CX^2}{2x\sqrt{2a^4 - Q^2 - 2a^4x^2/c^2}}. \tag{22}$$

[45] Royal Society Library. Gregory M.S.: Folio 163.

[46] Memorandum, July 1694 (Edinburgh University Library. M.S. Gregory C 42). Translation from Latin in *Correspondence*, 3, p. 386. For a discussion of D. Gregory's comments to the *Principia*, see Guicciardini, *Reading the Principia* (1999), pp. 179–84.

[47] Like Bernoulli, constants are employed so that geometrical homogeneity is preserved. Here, in the numerator the constant a is raised to the fourth power, so that the intensity of the force is one dimensional.

[48] We note that this result has the incorrect factor 2 in the numerator. In modern notation, note that c is chosen so that $-2a^4/c^2 = v_0^2 - 2a^4/x_0^2$, where x_0 and v_0 are the initial position and velocity (mind again the erroneous factor of 2).

The quadrature of the curve given by (21) is elementary and results in a function that describes the distance x in relation to time. In order to square the second curve associated with (22), Newton had recourse to rather advanced quadrature techniques. Essentially, he proceeded by making a substitution of variables similar to (6), as employed by Bernoulli in his letter to Hermann from 1710.

Warfare between Newton and Bernoulli

A larger context: the calculus controversy

Now that we have reviewed the methods used by Newton and Bernoulli in their treatment of motion under central forces, we can turn to reconsider the debate that ensued between the two of them on this issue. Newton did not enter directly into this debate, but rather remained in the background, issuing his replies to Bernoulli either indirectly or through intermediaries. We now know that Newton was the author of the anonymously published review from 1715 of the *Commercium Epistolicum* of 1713, the Royal Society's official report that upheld his priority as the inventor of the differential and integral calculus, while effectively dismissing Leibniz's claims as bogus and smacking of palgiarism. Otherwise, he left the battle field to his acolytes, in particular John Keill, Brook Taylor, and Abraham De Moivre.[49] Bernoulli's position also circulated via anonymous pamphlets and correspondence, but his views received additional backing from certain members of his family, such as Nicolaus I, who served as proxies.[50]

Thus, the debate over the proper method for solving the inverse problem was deeply embedded in the more general dispute between Newton and Leibniz that produced their infamous priority controversy. This, of course, was a decisive issue within the larger context of a conflict that led to a deep rift between British and Continental mathematical methods over the course of the eighteenth century. Johann Bernoulli was quick to recognize that the old-fashioned geometrical style of the *Principia* might be used as proof of Newton's lack of ability to make use of calculus in 1687. This alone must have seemed like telling forensic evidence that could be deployed against Newton's claim of priority. After all, if Newton already had the calculus well before Leibniz published his first paper in 1684, why had he not used it in the *Principia*? But probably even more problematic for Newton's case, in Bernoulli's mind, were the issues raised above in connection with the inverse problem of central forces, since these seemed to reveal a number of telling

[49] John Keill embarked on a campaign against Bernoulli in the pages of the *Journal litéraire*. In his papers he repeatedly crossed swords with Bernoulli. The French in which his papers were written was recognized by Bernoulli as typical of the Huguenot refugees, thereby exposing De Moivre's participation.

[50] In September 1712 Nicolaus I Bernoulli, Johann's nephew, visited London. He met Newton and informed him that his uncle had detected a mistake in Proposition 10, Book 2, of the *Principia*. Johann and Nicolaus cooperated by publishing a series of papers in the *Acta Eruditorum* on motion in resistant media; these were directed against Newton, and then against Keill and Taylor, who attempted a defense.

mistakes and lacunae in the *Principia* that would not have occurred had Newton been a mathematician versed in Leibnizian methods.

The larger context framed by the calculus controversy also had many other ramifications. These lead into the tangled issues concerning philosophical and even political matters that heightened the conflict between Newton and Leibniz. They disagreed over physical questions, such as the nature of gravity, time and space, atoms and the void, but also the fundamental problem of deciding which quantites are properly conserved in accordance with the laws of nature. In the realm of metaphysics, they held opposing views with regard to issues such as free will and God's providence, the religious significance of miracles, and the relation between body and soul. In short, they were not just mathematical opponents: they were philosophical enemies as well. Even in the political arena, in which both took on leading roles, their activities only served to fuel their intellectual disputes. Little wonder, then, that their purely mathematical differences could ignite so much fury.

Newton had been elected to the Convention Parliament (1689). He was very close to the entourage of Charles Montagu (the 1st Earl of Halifax), to the Whigs who had promoted the Glorious Revolution. In 1696 he left Cambridge for London, where he became rich and powerful as Warden of the Mint. In the meantime, Leibniz had served as a counsellor to the Emperor and to the Czar, and he had long been in the service of the Dukes of Hanover. It was also thanks to Leibniz's diplomatic efforts that Duke Ernst August gained the status of Elector. When his son ascended to the throne of Great Britain and Ireland in 1714 as George I, Leibniz hoped to cross the Channel as well. The prospect of having Leibniz — an experienced diplomat and towering metaphysician who actively pursued an ecumenical policy of reconciliation between the Christian Churches — as Royal Historian in London must have been a daunting one for Newton's party, which rather favoured anti-Catholicism and a Protestant interpretation of Anglicanism. Leibniz had been on close terms with the king's mother Sophia, but George I had little sympathy with his polymath courtier.[51] So the king refused his request to join the English court, leaving Leibniz stranded in Hanover, where he died in bitter isolation in 1716. Newton thus triumphed during his lifetime, and yet after his death in 1727 Leibnizian mathematicians on the Continent proved to be far more productive than their Newtonian counterparts.

In reflecting on the larger dimensions of this conflict, we might ask whether we should take this larger context into account when examining the history of the inverse problem for central forces? This might seem to be a most promising enterprise that accords well with the desiderata of other historians, most notably those influenced by the strong program of the Edinburgh school or by the Sussex

[51]After Sophia's death in June 1714, Leibniz had the Princess of Wales initially on his side in London. He hoped to be called to the court in London as "Historiographer of Great Britain", but the Newtonian entourage (most notably, Samuel Clarke) succeeded in diverting Caroline's sympathies. The Leibniz-Clarke correspondence has its origin here, and — I would also claim — the calculus controversy became intertwined with these maneuvers at the royal court.

school in intellectual history.[52] In fact, there are a number of intellectual historians, such as Stefan Collini, who argue forcefully in favor of synchronic over diachronic approaches. As Collini so aptly expresses this in describing the methodology of the Sussex school:

> Instead of works which cut a 'vertical' (and often teleological) slice through the past [...] the tendency of recent works has been towards excavating a more 'horizontal' site, exploring the idioms and preoccupations of a past period as they manifest themselves in thought and discussion about various issues that cannot readily be assigned to current academic pigeon-holes.[53]

I find Collini's ideas attractive, and the Sussex school's strong predilection for local, trans-disciplinary studies in intellectual history has produced magisterial work. Collini, in particular, represents an extreme, though wonderfully learned example of what Albert William Levi once called the 'molder'. Levi observed that historians sometimes approach their subjects more like sculptors, who work in clay or plaster, rather than carvers, whose art proceeds by cutting away at a block of wood that must be 'pared down to size'. These historians practice a technique of 'maximum incorporation' that attempts to integrate all relevant data into the picture, including much that might appear superfluous, In the present case, this would include the none-too-well-shaped material coming not just from science but also from its applications, or even from extra-scientific traditions like literature or religion.[54] Historians of science who approach their topics not as 'molders' but rather as 'carvers', who strip away 'superfluous' details in order to put their diachronic narratives into relief, risk removing the very human agents whose ideas and motivations lie at the heart of the story.

Still, I have reservations about following Collini's lead too enthusiastically when it comes to this case. I fear that an excessive broadening of the context, interesting as this might be, would render the interpretation of the texts I am considering far too opaque. Or, what would be even worse: the exploration of the entanglement between text and context might tempt me to indulge in formulating generalizations thar are far too facile (like those we sometimes find in the worst examples of Marxist or SSK studies, for example). Placing technical mathematics into a larger intellectual context is a matter of combining the *esprit de géométrie* with the *esprit de finesse*: we have to open our perspective without losing our primary focus on the problems, formulas, and techniques that we are trying to interpret as historians of mathematics. There are no simple rules for us to follow

[52] One can also draw here on prior studies of the controversy between Newton and Leibniz that have paid close attention to the philosophical and political contexts, above all the classic study by A. Rupert Hall, *Philosophers at War* (1980). The Edinburgh school methodology is expounded in Steven Shapin, 'Of Gods and Kings' (1981). A more convincing study along these lines is Domenico Bertoloni Meli, 'Caroline, Leibniz, and Clarke' (1999).

[53] Stefan Collini, 'General Introduction' (2000), p. 3.

[54] Albert W. Levi, *Philosophy as Social Expression* (1974), p. 3.

here, of course, so historians who wish to face the challenge of interpreting mathematical texts will always have to struggle with the problem of striking some kind of balance between a conceptual reading aimed at carving diachronic developments of concepts and theories, and a contextual description aimed at synchronic molding. The only way to face this dilemma – which presents itself in different forms in relation to different historical case studies – is by learning from the masters of our discipline (including the one celebrated in this volume). History is a craft learnt in the workshop, it needs laws and rules about as much as a multinational corporation needs an expert in Aramaic. So, with this caveat duly noted, let us embark on the present challenge of reading Newton's and Bernoulli's proofs in context. Let us humbly hope that we cast our hermeneutic net with elegance and precision!

Newton's methodological views

In his confrontation with Bernoulli, Newton adopted a viewpoint that deserves our close attention. When Bernoulli began the debate on the treatment of central force motion in the *Principia*, Newton's response gave pride of place to the (emended) Corollary 1, which he saw as the correct solution to the inverse problem. Those in his circle, for example John Keill, were well aware that Proposition 41 enabled Newton to reduce the inverse problem of central forces to quadratures in a general way. Keill also underscored its equivalence with the first part of Bernoulli's 1710 letter, and even claimed that Newton's argument was superior in its elegance.[55] Yet Keill's intervention notwithstanding, the predominant Newtonian response to Bernoulli's criticism was to point to the geometrical construction alluded to in Corollary 1. To be sure, Continental mathematicians were still in the dark when it came to Newton's beautiful solution (1694) of Corollary 3 to Proposition 41, which remained in private circulation and was given no publicity. Still, we must try to answer two related, but central questions: first, why did Newton prefer to meet Bernoulli's critique with the construction argument in Corollary 1 rather than the far more powerful Proposition 41? Second, why did he choose to avoid using the method of fluxions in his *Principia*?

---

[55]For instance we read in a paper published in 1716: 'Since the solution of M.Bernoully differs from that of M. Newton only in the characters or symbols'. = 'La Solution de M. Bernoully ne differant de celle de M. Newton, que dans les Caractéres [sic] ou Symboles'. John Keill, 'Défense du Chevalier Newton' (1716), p. 418. Keill had already stressed the equivalence between Bernoulli's differential equation and Newton's Proposition 41 in 1714: 'I see that Bernoulli's formula completely coincides with the Newtonian; it differs only in the notation of quantities. In fact if one puts $ab - \int F\dot{x}$ for $ABGE$ [$\approx ABFD$], ac for Q, & x for A, a for CX, & $\dot{x}$ for IN [...] it will result that that formula does not differ from the Newtonian, more than the same words when written in Latin and in Greek'. = 'vidi Bernoullianam formulam omnino cum Newtoniana coincidere; nec nisi in notatione quantitatum ab ea differre. Nam si pro $ab - \int F\dot{x}$ ponatur $ABGE$, pro ac ponatur Q, & x pro A, a pro CX, & $\dot{x}$ pro IN [...] unde constat formulam illam non magis a Newtoniana discrepare, quam verba Latinis literis expressa differunt ab iisdem verbis scriptis in Graecis characteribus'. John Keill, 'Observationes ...de Inverso Problemate Virium Centripetarum' (1714), p. 94.

There can be no simple answer to these questions. It seems to me fruitful, however, to view the different strategies followed by the British and the Continentals in their debate over the inverse problem of central forces as reflecting different attitudes about how mathematics should be employed by natural philosophers. One might express this by saying that these two camps had different research agendas that reflected fundamental differences in their respective webs of belief, perhaps even two different philosophies of mathematics. This is the (partial) explanation I shall offer here, being conscious that history is a complex discipline (certainly not a nomological discipline) so that no simple causal law (of the kind that would explain why A happened because of B) can be discerned. Indeed, we can in this case note at the outset that Newton's preference for geometry in natural philosophy was influenced by at least two other very important factors. First, he was certainly aware of the mathematical competence of his most interested readers, many of whom had a thorough background in geometry but who had virtually no idea of the calculus in 1687; this surely had an impact on the mathematical style he chose in writing the first edition of the *Principia*. Second, Newton was interested in harnessing the power of geometry in order to provide a visualization of real accelerations in the physical context fundamental to his new world system based on absolute and true space and time. For reasons of space, let us leave these two themes aside, notwithstanding their interest, in order to devote our attention to Newton's views on mathematics and its role in natural philosophy.[56]

Newton on the role of mathematics in natural philosophy

In order to answer the first question raised in the previous section, let me begin by considering a pronouncement Newton made when he was still very young. In his 1670 Lucasian *Lectiones Opticae*, he explained why he valued the use of mathematics in natural philosophy so highly in these words:

> Thus although colors may belong to physics, the science of them must nevertheless be considered mathematical, insofar as they are treated by mathematical reasoning. Indeed, since an exact science of them seems to be one of the most difficult that philosophy is in need of, I hope to show—as it were, by example—how valuable is mathematics in natural philosophy. I therefore urge geometers to investigate nature more rigorously, and those devoted to natural science to learn geometry first. Hence the former shall not entirely spend their time in speculations of no value to human life, nor shall the latter, while working assiduously with an absurd method, perpetually fail to reach their goal. But truly with the help of philosophical geometers and geometrical philosophers, instead of the conjectures and probabilities that are being blazoned about everywhere, we shall finally achieve a science of nature supported by the highest evidence.[57]

[56] I deal with these themes in Niccolò Guicciardini, *Reading the Principia* (1999) (e.g. on pp. 95–8) and *Isaac Newton on Mathematical Certainty and Method* (2009) (e.g. on pp. 252–7).

[57] Isaac Newton, *The Optical Papers* (1984), pp. 87, 89 and 437, 439. Translation by A. Shapiro.

Here Newton was positioning himself against both Cartesian hypothetico-deductive modeling as well as the Baconian empiricism 'blazoned about' at the Royal Society by virtuosi like Boyle and Hooke. Most notably, the mitigated skepticism endorsed by eminent fellows of the Royal Society tended to exalt the value of fallible and/or probable outcomes in natural philosophy (as can be verified by even a cursory reading of the Preface of Hooke's *Micrographia* (1665)). Such skepticism was quite in vogue in Restoration England for both cultural and political reasons that have been studied in great detail by Barbara Shapiro and Richard Popkin. Indeed, the probabilism and skepticism that were so widespread in Newton's England were part of a much larger movement toward tolerance endorsed by the Fellows of the Royal Society.[58] Newton, however, would have no truck with mere probabilites: he wanted certainty. Thus he sought to validate his natural philosophy mathematically, while at the same time outstripping methodologies based on probabilism, a trend quite rampant at this time and about which he sometimes complained.

So the program Newton delineated in his *Lectiones Opticae* from 1670 would remain a cornerstone for his agenda as a natural philosopher. Indeed, he famously reformulated it in the 'Praefatio ad Lectorem' that opens the first edition (1687) of his *Principia*, subscribing to it again in the preface to the second edition (1713) that was signed by Roger Cotes, and he elaborated on it in the 'Scholium Generale' that appeared in that edition. There are a number of other relevant manuscripts and letters that are worthy of attention in this connection. It would be fascinating to follow the development of Newton's ideas on the role of mathematics in natural philosophy from his early thoughts through to the more mature formulations, many of which bear on the context of reception of his *Principia* on the Continent.[59] More than anyone else, Alan Shapiro has taught us to appreciate the pervasive role of the quest for certainty in Newton's thought. He convincingly demonstrates that Newton came to recognize the limits of knowledge quite early in his intellectual career. Thus, Newton conceded that natural philosophy, even when approached mathematically, could never attain the absolute certainty of geometrical truth. Nevertheless, he was convinced that geometers who practiced natural philosophy could go far beyond the modest aims (conjectures and probabilities) of the experimental naturalists.[60]

[58]Barbara J. Shapiro, *Probability and Certainty in Seventeenth-Century England* (1983) and Richard H. Popkin, *The History of Scepticism: From Savonarola to Bayle* (2003), pp. 112ff.

[59]See Niccolò Guicciardini, *Newton on Mathematical Certainty and Method* (2009).

[60]We also learn from Alan Shapiro that in the last years of his life Newton changed his mind concerning this matter, especially with respect to the possibility that philosophical geometers and geometrical philosophers can actually approach absolute certainty. As Shapiro remarks: 'Although he [Newton] soon modified his excessive claims and more carefully distinguished mathematical demonstrations from less than certain experimental conclusions, he continued to prize mathematical theories and seek 'truth' and 'certainty', words that appear frequently in his writings. Only in the last decades of his life did he accept the probabilism of his contemporaries'. Alan Shapiro, *Fits, Passions, and Paroxysms* (1993), p. 14. See also Alan Shapiro, 'Newton's "Experimental Philosophy"' (2004).

A tension between method and practice

Such a methodological stance generated a tension between Newton's views on mathematicized natural philosophy and his own practices as a mathematician. For if mathematics is to inject certainty into natural philosophy, it must be practiced according to criteria that guarantee the certainty of its methods. Yet the new algebraic methods that Newton mastered so well, methods often taken to be the salient characteristic that distinguished seventeenth-century mathematics from earlier mathematical achievements, also appeared to many of his contemporaries as being far from rigorous. Indeed Newton himself, a master in algebra and calculus, came to believe that the use of such symbolism constituted part of the heuristic stage of mathematical research. He thus regarded algebra and calculus as non-demonstrative in nature, despite their utility as tools of discovery.

Newton's reflections on the relationship between geometry and algebra can be seen already from the annotations he made in the 1670s on Kinckhuysen's textbook on algebra. Indeed, the decade of the 1670s proved to be crucial in shaping Newton's mature views as a mathematician. It was then that he began reading Pappus's *Collectio*, an experience that caused him to turn enthusiastically toward an attempt to recover the lost ancient method of discovery and proof described by Pappus as analysis and synthesis. After his appointment to the Lucasian Chair in 1669, Newton was required to submit the manuscripts from his lectures, and this compelled a very different approach to mathematics than the unsystematic forays into new territory that he earlier could afford to undertake as a young *protégé* of Isaac Barrow. Then came the encounter with Huygens's *Horologium Oscillatorium* in 1673, which was quite probably another decisive factor that influenced Newton's views regarding the efficacy of geometry, particularly when it came to dealing with difficult physical problems. As a result, he gradually came to accept the epistemic superiority of classical geometrical methods over mechanical calculations that used Cartesian symbolic algebra. Huygens' work showed him how cutting-edge natural philosophy could be pursued in terms consonant with the standards of mathematical certainty and elegance that Newton had come to associate with the methods of ancient Greek geometers. In this decade we thus begin to witness the conflict that Newton faced when he compared the mathematical practices of the moderns, in which he so clearly excelled, with the role he attributed to geometry in his philosophical agenda. Furthermore, Newton began elaborating on his deep distaste for all things Cartesian, and he did so for a whole series of reasons that went far beyond purely mathematical enquiry. It was during this period that he began to develop that veneration for the ancients that pervades all his later alchemical and theological work. It seems that almost everything in Newton's mind was pointing him against Cartesianism.

After composing the *Principia*, he began a deep reconsideration of Greek geometry in a series of geometrical works that would remain unpublished until the appearance of volume 7 in Whiteside's edition of the *Mathematical Papers*. In these pages, laden with philosophical reflections, Newton expressed his conviction that the ancients possessed a method of discovery that was, compared with the Cartesian, not only more elegant but also more powerful. He also believed this ancient geometrical analysis could be recovered by undertaking a reconstruction of the lost books of Euclid's *Porisms.* By this time, his pronouncements on algebra were laced with trenchant criticisms like this: '[algebraic calculus is] so tedious and entangled as to provoke nausea'.[61]

Newton also encouraged his acolytes to pursue researches in ancient analysis and synthesis, and he never missed an opportunity to praise those who continued to write their works in geometrical language. These included writers from the great Huygens to the obscure and long forgotten Antonio Hugo de Omerique, both of whom had resisted the prevailing taste for mathematics couched in the symbolism of the moderns, those 'bunglers in mathematics', as he famously labeled them.[62] Little wonder that when the polemic with Leibniz later exploded, Newton took up these same anti-Cartesian views and turned them against his German rival. Thus, seen within this larger intellectual context, one can easily understand the dilemma Newton faced. For his mature mathematical interests and publications, including a number of anonymously written polemical pieces, were driven by a philosophical agenda that was difficult to reconcile with his own mathematical practice. Indeed, his earlier mathematical works, many also unpublished, were usually based on formal manipulations of algebraic equations, infinite series, or infinitesimal magnitudes, that is, on techniques that belonged to what Michael Mahoney identified as the 'analytical school' in seventeenth-century mathematics.[63]

Newton on algebra

Newton's geometrical writings from the 1690s also provide a rich trove of information for discerning his attitude toward algebra. One of his typical pronouncements was reported by Gregory in a *memorandum* from 1694:

> Our specious algebra is fit enough to find out, but entirely unfit to consign to writing and commit to posterity.[64]

Here we can begin to discern the consequences such a view surely had for Newton's own published work. What to us seems so important, the application of calculus to dynamics, has been downgraded to the level of a mere heuristic technique, a method of reasoning not fit 'to consign to writing and commit to posterity'. Thus, according to Newton, algebra and its extension to infinitesimal magnitudes in the

[61]From a manuscript written in the late 1670s in Isaac Newton, *Mathematical Papers*, 4, p. 276.

[62]Walter G. Hiscock, *David Gregory, Isaac Newton and Their Circle* (1937), p. 42.

[63]Michael Mahoney, *The Mathematical Career of Pierre de Fermat (1601-1665)* (1973), pp. 1–14.

[64]'Algebram nostram speciosam esse ad inveniendum aptam satis at literis posterisque consignandum prorsus ineptam'. University Library Edinburgh MS Gregory C42, translation by D. T. Whiteside in Isaac Newton, *Mathematical Papers*, 7, p. 196. See also *Correspondence*, 3, p. 385.

calculus should not appear in a published proof (the 'composition' or 'synthesis') since these methods lack the certainty of geometrical demonstration. Symbolism can be used in the process of discovery (called 'resolution' or 'analysis' in Newton's time), and it can be communicated informally to interested parties, but such arguments should not appear in print, especially when one is constructing a mathematicized natural philosophy. As Newton pointed out in the 1690s:

> if a question be answered by the construction of some equation, that question is resolved by the discovery of the equation and composed by its construction, but it is not solved before the construction's enunciation and its complete demonstration is, [a]from beginning to end excluding analysis perfectly[a] with the equation now neglected, composed.[65]

Here Newton alludes to Descartes' work on the 'construction of equations'. Descartes, as Henk Bos has shown in his important study, developed a technique aimed at constructing the roots of algebraic equations via intersections of algebraic curves.[66] Combining such geometrical constructions with algebraic methods proved to be one of the most important innovations in seventeenth-century mathematics. Descartes thus used algebra in order to resolve a geometrical problem, for example the trisection of an angle, which enabled him to reduce it to the problem of solving the corresponding algebraic equation (for the trisection of an angle a certain cubic rquation). But it did not suffice to find the roots of the equation algebraically (for example by invoking Cardano's formula for solving the cubic equation): a solution of a geometric problem could only be achieved by means of a geometrical construction, hence the need to devise techniques for constructing segments whose lengths correspond to the roots of a given equation.

There is an interesting connection here with Newton's approach to central force motion. As we have seen, unlike Bernoulli, Newton interpreted the inverse problem of central forces as requiring a construction of the curve of motion of the body. Thus, a calculus solution via the integration of a differential equation clearly failed to provide the kind of geometrical result he demanded. For Newton, a true solution had to be constructive in nature; only then could it meet his desiderata as expressed in the passage quoted above. Such statements insisting on the primacy of constructive geometry, in fact, abound in Newton's writings.[67] The invective one finds in his tirades against the use of algebraic symbols clearly informed much of his opposition to the Leibnizian calculus.

These issues point then to the larger project Newton had in mind, namely to restore the lost art of geometrical analysis as practiced by the ancient Greeks,

---

[65] [a a] cancelled. Isaac Newton, *Mathematical Papers*, 7, p. 307. 'si quaestioni per constructionem aequationis alicujus respondeatur, quaestio illa resolvitur per inventionem aequationis, componitur per constructionem ejusdem, sed non prius solvitur quam constructionis enunciatio ac demonstratio tota [a]a principio ad finem exclusa omni analysi perfecte[a] componitur, aequatione neglecta'. *Mathematical Papers*, 7, p. 306.

[66] Henk Bos, *Redefining Geometrical Exactness* (2001).

[67] For more details, see Niccolò Guicciardini, *Isaac Newton on Mathematical Certainty and Method* (2009).

a project that led him to develop important new ideas in projective geometry. I believe that Newton pursued that program with a twofold conviction and purpose: he was at once a follower of the ancients and an innovator who conceived of his work just as any creative mathematician might have. He was following the 'practice of the ancients' in order to obtain new results in the spirit of that venerable tradition, thereby inscribing himself into a mathematical genealogy that reached back to Apollonius and Archimedes. Thus seen, we should not view Newton's defense of geometry as motivated simply by his conservatism, much less a merely antiquarian interest in the tradition of the Greeks. Identifying the trend toward algebraization in seventeenth-century mathematics as the engine of progress, so to speak, fails to do justice to the achievements of Huygens, Barrow, and Newton, whose work reflected the fertility of geometrical approaches. Moreover, such a one-sided, or asymmetrical, view fails to capture the essential tension that informed the confrontation between these traditionalists and their counterparts in the other camp – Descartes, Wallis, Leibniz, and others – who placed their faith in the power and efficacy of algebraic techniques.

Newton's agenda

For Newton, mathematics played a central role in his broad-ranging program as a seventeenth-century natural philosopher. We have to appreciate, however, that his agenda for a mathematicized natural philosophy – which he spelled out in many pronouncements (as, for instance, in the Lucasian *Lectiones Opticae*) – was at odds with deeply held values endorsed by eminent fellows of the Royal Society. Newton's statement that he could exploit the certainty of mathematics by importing it into natural philosophy was surely perceived by many as an arrogant claim. Such being the case, the Lucasian Professor had to cushion himself against potential criticisms that might be leveled at his mathematical methods. One way to do that was by portraying himself as an heir to the venerable ancient tradition of the Greeks, while consciously downplaying his own accomplishments as a virtuoso in the new symbolical methods associated with the differential and integral calculus. This, I believe, helps to explain what I described above as the seemingly irrational views adopted by Newton and his acolytes in their debate with Bernoulli over the proper method for handling the inverse problem of central forces. It might also explain Newton's reluctance to print the method of fluxions and his preference for communicating it via correspondence and by circulating manuscripts.

Bernoulli's agenda

Just as in Newton's case, so with Bernoulli: to understand his position in this methodological debate one has to take into account his broader views on the role and nature of mathematics in natural philosophy. In order to compare Bernoulli's agenda with Newton's, one would do well to start with his anonymously authored 'Epistola pro Eminente Mathematico' from 1716, in which he with due immodesty identified himself merely as an 'eminent mathematician'. Such self-praise must have seemed to him befitting, since in the very first lines of the text he stakes his claim to the invention of the integral calculus. (He could not have known of those quadratures that Newton had long kept as a 'secret fountain' in his drawer and reluctantly revealed only to a handful of acolytes who approached him either through letters or who visited him in his rooms.)[68] The 'eminent mathematician' had this to say about respective priority rights to the calculus:

> Whoever told you that the invention of the integral calculus was obtained by Johann Bernoulli by his own strengths does not pronounce anything that is alien to the truth, especially if we wish to distinguish this calculus from the differential calculus, which also according to Bernoulli himself is without any doubt entirely due to Leibniz.[69]

Clearly, Johann Bernoulli was proud of his integration technique based on substitution of variables, which he used with such success in solving the inverse problem for central force motions. Thus, he stressed in the polemic with Keill (as well as his letter to Hermann) that he was in possession of a method that allowed him to formulate that problem in the form of a first-order differential equation. Thanks to the integral calculus, of which Bernoulli considered himself the master, he was able to find a general solution that could then be used to determine the orbit of a body for some specific force law. The power of this method lay in its generality, which was precisely what Bernoulli aimed to achieve. He sought a method for handling a whole class of problems at once; from the general solution he could then write down a specific solution in which the integration constants were expressed as functions of the given initial conditions.[70] His approach thus stood in sharp contrast with that of Newton, who emphasized intelligibility rather than generality.

In fact, for Benoulli the inverse problem of central forces was not a fundamental one for natural philosophy; his interest in it, at any rate, was almost entirely mathematical.[71] We can be sure that he had no wish to subscribe to a Newtonian natural philosophy with its mysterious and seemingly occult notion of a gravitational force that acts at great distances across empty space. For him, the *Principia* was essentially just a repository of interesting mathematical problems; what Newton defended there was, from a physical standpoint, sheer nonsense. Moreover, Bernoulli had a perception of his role in the arena of European mathematics that differed markedly from Newton's. The latter thought of himself as a 'geometrical

---

[68] As Newton wrote to John Collins on November 8, 1676. Isaac Newton, *Correspondence*, 2, p. 180.

[69] [Johann Bernoulli]. 'Epistola pro Eminente Mathematico', p. 296. The translation from Latin is mine.

[70] It was only later that mathematicians realized that exact quadratures in terms of elementary functions are possible only for a handful of very special force laws.

[71] In short: Bernoulli thought that the inverse-square action at a distance that Newton had 'deduced from the phenomena' was a manifestation of some deeper contact action mechanism that had to be discovered.

philosopher'; consequently, his mathematical research was embedded in the broad ranging scope of an early-modern natural philosopher. Bernoulli, by contrast, saw himself as embarking on a new professional role as a mathematician; his prestige and income as a professor of mathematics and frequent winner of academic prizes were fundamentally due to his competence and authority as an expert in the new calculus. This does not mean that natural philosophical interests were extraneous to Bernoulli's intellectual biography. Quite the contrary, his interests spanned from medicine to cosmology, including, of course, fields like mechanics, hydrodynamics, ship maneuvering, barometric light, and optics.[72] Nevertheless, Bernoulli was inclined to bracket these interests out when he practiced mathematics, and by so doing he was, in effect, defining a new professional role for the mathematician. So what I am suggesting here is that Bernoulli's dissociation of mathematics from natural philosophy – as exemplified in writings such as the 'Epistola pro Eminente Mathematico' – marks him as a leading representative of a new generation of Continental mathematicians who would soon come to dominate the scene in the Continental academies; by the middle of the eighteenth century these analysts would attain a level of prestige that was simply unthinkable in Britain.

The prominent social status of these Continental mathematicians was consonant with the notion that mathematics could be conceived of and practiced as an autonomous discipline. This generation of mathematicians was eager to receive and accept Leibniz's doctrine according to which mathematics could be understood as a *cogitatio caeca*, a kind of blind reasoning that proceeds by manipulating symbols, which are themselves free from metaphysical commitments, but open to various specific interpretations. Leibniz — notwithstanding his life-long interest in the metaphysics of the calculus — left his disciples the choice of maintaining, *philosophice loquendo*, different foundational positions, for instance, regarding ontological questions with respect to the existence of infinitesimals or the roots of negative numbers. What he wished to defend was the utility of symbols for mathematical calculation. Time and again Leibniz encouraged his mathematical disciples to ignore interpretive metaphysical questions when practicing mathematics. While it would be highly misleading to think of all Continental mathematicians as devout followers of Leibniz's philosophy, they did tend to share a pragmatic Leibnizian faith in algorithmic manipulations. This standpoint gained additional adherents through the influence of Etienne B. Condillac, and it was put into practice by Euler, Lagrange, and Laplace.

This new orientation toward pure analysis in mathematical research was further reinforced by growing skepticism with regard to traditional metaphysics. Thus, as Mary Terrall has shown, 'one feature of the [French] academy's growing commitment to analytic methods in physics over the course of the eighteenth century was the erasure of teleological metaphysics from rational mechanics'.[73] According to Terrall, the polemics that led to the marginalization of Pierre-Louis

[72]For instance, Bernoulli wrote essays on fermentation, muscular motion, and nutrition.

[73]Mary Terrall, 'Metaphysics, Mathematics, and the Gendering of Science in Eighteenth-Century France' (1999), p. 247.

Moreau de Maupertuis, a powerful advocate of a broad teleological approach in connection with his controversial law of least action, represents just one step in a more general development that eventually eliminated metaphysical explanations from science in general and mathematics in particular. Indeed, the language of algebraic analysis – as elaborated by leading figures such as D'Alembert, Lagrange, and Laplace – served this larger strategy of purification in that enlightened mathematics could act as an antidote to imprecise metaphysical thinking. Thus, Continental academicians saw the pursuit of mathematics as an enterprise whose success could be entirely defined in its own terms, without reference to theology, metaphysics, or natural philosophy.

This does not imply that they considered foundational questions irrelevant. Nevertheless, such questions did not tend to interfere with their work; they could thus be bracketed when practicing mathematics. Indeed, such bracketing held sway not only when dealing with foundational issues but also when it came to applications of mathematics. In fact, when mathematicians like Bernoulli, Varignon, or Euler worked on mechanical problems, they did so primarily for the sake of their mathematical interest, even when the original motivation came from an applied context. It has often been said—and with good reason—that analytical mechanics provided the main stimulus for eighteenth-century mathematical research. Still, the famous problems that aroused such interest on the Continent – whether the brachistochrone and vibrating string problems in analytical mechanics, or orthogonal trajectories in geometry – had only faint connections with applications; they were pursued because of their intrinsic mathematical interest. As Terrall notes:

> The most rigorous analytical mathematics that filled the pages of the [French] academy's journal was not necessarily useful in terms of practical applications or public displays. The equations of celestial mechanics or hydrodynamics did not translate easily into engineering design or industrial production.[74]

When considering eighteenth-century analytic mechanics we tend to forget how remote this sort of pure mathematics was from significant physical applications. As Thomas Kuhn long ago remarked, before the advent of the mathematical theories of heat, magnetism, light, etc., the domain of advanced mathematical physics had little relevance except for the astronomer.[75]

Concluding remarks

Apart from the technical, philosophical, and methodological differences that separated Newton and Bernoulli in their debate over the inverse problem in mechanics, it is important to recognize that these protagonists also represented different eras in the history of science. With only slight exaggeration, one might say

[74]Mary Terrall, 'Metaphysics, Mathematics, and the Gendering of Science in Eighteenth-Century France' (1999), pp. 256–7.

[75]Disciplines that fell within the province of mathematical astronomy were also cartography and navigation.

that in this debate Newton, acting as the last great seventeenth-century natural philosopher, crossed swords with Johann Bernoulli, the first mathematician of the Continental Enlightenment. In their confrontation they were unable to reach an agreement not only because of the passions and rivalries that dominated their cantankerous minds, but also because they were committed to fighting different kinds of battles. I take this to be a case in which the historiographic importance of controversies is particularly manifest. Indeed, the most interesting controversies in the history of science are neither longstanding disputes – such as that between science and religion, or debates over whether the fundamental constituents of matter are particles or waves – nor broad conceptual sea-changes or scientific revolutions, like those epitomized by the overturn of Aristotelianism in the seventeenth century or the demise of Newtonian mechanics in the early twentieth century. I find that the most interesting debates, as the works of Joe Dauben on the history of set theory and nonstandard analysis have taught us, are those controversies that unfold over a limited span of time and concern a well-localized group of actors. Being bounded in space and time means they can be approached from a meaningful historiographic point of view. Only in such cases can the actors' specific linguistic conventions be detected with some precision, and with that the possibility of gaining a closer understanding of the motivations behind their work. Only then can the broad agendas behind their disagreements emerge with sufficient clarity.[76]

Acknowledgment

I presented an earlier version of this paper at the symposium *'Unreasonable effectiveness'? Historical origins and philosophical problems for applied mathematics* organized by Benjamin Wardhaugh at All Souls (Oxford) on 16–17 December 2008. I thank the participants for their comments and, in particular, Michael Nauenberg for sharing his ideas with me in an ensuing extensive correspondence. I also thank Massimo Mori for allowing me to participate at his doctoral seminars on the 'development and problems in the history of ideas' held at the University of Turin (2013), and David Rowe for his most enjoyable and enlightening editorial guidance.

I gratefully acknowledge financial aid from the Gerda Henkel Stiftung.

Bibliography

[Arnold 1990] Arnold, Vladimir I., *Huygens and Barrow, Newton and Hooke: Pioneers in Mathematical Analysis and Catastrophe Theory from Evolvents to Quasicrystals.* Translated from the Russian by Eric J. F. Primrose. Basel, Boston, Berlin: Birkhäuser, 1990.

[76] I draw from Trevor Pinch and Christine Leuenberger, 'Studying Scientific Controversy from the STS Perspective', Paper presented at the EASTS Conference Science Controversy and Democracy http://sts.nthu.edu.tw/easts/conference.htm (accessed on June 7, 2012).

[Bernoulli 1710] Bernoulli, Johann, "Extrait de la Réponse de M. Bernoulli à M. Herman, Datée de Basle le 7 Octobre 1710". *Mémoires de l'Académie des Sciences* (1710): 521–33.

[Bernoulli 1716] Bernoulli, Johann, "Epistola pro Eminente Mathematico, Dn. Johanne Bernoullio, Contra Quendam ex Anglia Antogonistam [sic] Scripta". *Acta Eruditorum* (1716): 296–315.

[Bertoloni Meli 1999] Bertoloni Meli, Domenico, "Caroline, Leibniz, and Clarke". *Journal of the History of Ideas* 60 (1999): 469–86.

[Bevir 1999] Bevir, Marc, *The Logic of the History of Ideas*. Cambridge: Cambridge University Press, 1999.

[Bloor 1991] Bloor, David, *Knowledge and Social Imagery*. Chicago: The University of Chicago Press, 1991 (first ed. 1976).

[Bos 2001] Bos, Henk J. M., *Redefining Geometrical Exactness: Descartes' Transformation of the Early Modern Concept of Construction*. New York: Springer, 2001.

[Brackenridge 1995] Brackenridge, Bruce, *The Key to Newton's Dynamics*. Berkeley, Los Angeles, London: California University Press, 1995.

[Brackenridge 2003] Brackenridge, Bruce, "Newton's Easy Quadratures: 'Omitted for the Sake of Brevity"'. *Archive for History of Exact Sciences* 57 (2003): 313–36.

[Chandrasekhar 1995] Chandrasekhar, Subrahmanyan, *Newton's Principia for the Common Reader*. Oxford: Clarendon Press, 1995.

[Chang 2008] Chang, Hasok, "The Historian of Science: Painter, Guide, or Connoisseur". *Centaurus* 50 (2008): 37–42.

[Collini 2000] Collini, Stefan, "General Introduction". In *Economy, Polity, and Society: British Intellectual History 1750–1950*. Edited by Stefan Collini, Richard Whatmore, and Brian Young. Cambridge: Cambridge University Press, 2000.

[Costabel 1989] Costabel, Pierre, "Courbure et Dynamique: Jean I Bernoulli Correcteur de Huygens et de Newton". Symposium Leibniz et Bernoulli (Basel, June 15-17, 1987). *Studia Leibnitiana* Sonderheft 17 (1989): 12–24.

[De Gandt 1995] De Gandt, François, *Force and Geometry in Newton's Principia*. Princeton: Princeton University Press, 1995.

[Dauben 1992a] Dauben, Joseph, "Are There Revolutions in Mathematics?" In Echeverria, J., Ibarra, A., and Mormann, T., editors. *The Space of Mathematics, Philosophical, Epistemological, and Historical Explorations*, pp. 205–229. Berlin: Walter de Gruyter, 1992a.

[Dauben 1992b] Dauben, Joseph, "Conceptual Revolutions and the History of Mathematics: Two Studies in the Growth of Knowledge". In Gillies, D., editor.*Revolutions in Mathematics*, pp. 49–71. Oxford: Clarendon Press 1992b.

[Erlichson 1990] Erlichson, Herman, "Comment on 'Long-buried Dismantling of a Centuries-old myth: Newton's *Principia* and Inverse-square Orbits' by Robert Weinstock". *American Journal of Physics* 58 (1990), 882–4.

[Erlichson 1994] Erlichson, Herman, "The Visualization of Quadratures in the Mystery of Corollary 3 to Proposition 41 of Newton's *Principia*". *Historia Mathematica* 21 (1994): 148–61.

[Guicciardini 1995] Guicciardini, Niccolò, "Johann Bernoulli, John Keill and the Inverse Problem of Central Forces". *Annals of Science*, 52 (1995): 537–575.

[Guicciardini 19999] Guicciardini, Niccolò, *Reading the Principia: The Debate on Newton's Mathematical Methods for Natural Philosophy from 1687 to 1736.* Cambridge: Cambridge University Press, 1999.

[Guicciardini 2009] Guicciardini, Niccolò, *Isaac Newton on Mathematical Certainty and Method.* Cambridge, Mass.: MIT Press, 2009.

[Goldstein 2001] Goldstein, Herbert, *Classical Mechanics*, 3d edition. Addison Wesley, 2001.

[Hall 1980] Hall, Alfred R, *Philosophers at War: The Quarrel between Newton and Leibniz.* Cambridge: Cambridge University Press, 1980.

[Hermann 1710] Hermann, Jacob, "Extrait d'une Lettre de M. Herman à M. Bernoulli, datée de Padoüe le 12. Juillet 1710". *Mémoires de l'Académie des Sciences* (1710): 519–521.

[Hiscock 1937] Hiscock, Walter George, *David Gregory, Isaac Newton and Their Circle: Extracts from David Gregory's Memoranda 1677–1708.* Oxford: for the Editor, 1937.

[Jardine 2000] Jardine, Nicholas, "Uses and Abuses of Anachronism in the History of the Sciences". *History of Science* 38 (2000): 251–270.

[Keill 1708] Keill, John, "Epistola ad Clarissimum Virum Edmundum Halleium Geometriae Professorem Savilianum, de Legibus Virium Centripetarum". *Philosophical Transactions* 26 (1708): 174–8.

[Keill 1714] Keill, John, "Observationes ... de Inverso Problemate Virium Centripetarum". *Philosophical Transactions* 29 (1714): 91–111.

[Keill 1716] Keill, John, "Défense du Chevalier Newton". *Journal Litéraire* 8 (1716): 418–33.

[LaCapra 1983] LaCapra, Dominick, *Rethinking Intellectual History: Texts, Contexts, Language.* Ithaca, NY: Cornell University Press, 1983.

[Levi 1974] Levi, Albert W., *Philosophy as Social Expression.* Chicago: The University of Chicago Press, 1974.

[Mahoney 1973] Mahoney, Michael S., *The Mathematical Career of Pierre de Fermat (1601–1665).* Princeton: Princeton University Press, 1973. (2d rev. ed., 1994)

[Nauenberg 1994] Nauenberg, Michael, "Newton's *Principia* and Inverse Square Orbits". *The College Mathematics Journal* 25 (1994): 211–222.

[Nauenberg 2010] Nauenberg, Michael, "The Early Application of the Calculus to the Inverse Square Force Problem". *Archive for History of Exact Sciences* 64 (2010): 269–300.

[Newton 1715] Newton, Isaac, "An Account of the Book Entituled Commercium Epistolicum". *Philosophical Transactions* 29 (1715): 173–224. [Facsimile in Hall, *Philosophers at War* (1980), pp. 263–314].

[Newton 1959-1977] Newton, Isaac, *The Correspondence of Isaac Newton.* Edited by Herbert W. Turnbull, John F. Scott, A. Rupert Hall, and Laura Tilling. 7 vols. Cambridge: Cambridge University Press, 1959–1977. (cited as *Correspondence*)

[Newton 1967-1981] Newton, Isaac, *The Mathematical Papers of Isaac Newton.* Edited by Derek T. Whiteside. 8 vols. Cambridge: Cambridge University Press, 1967–81. (cited as *Mathematical Papers*)

[Newton 1972] Newton, Isaac, *Philosophiae Naturalis Principia Mathematica, The Third Edition (1726) With Variant Readings.* Edited by Alexandre Koyré and I. Bernard Cohen, assisted by Anne Whitman. Cambridge: Cambridge University Press, 1972.

[Newton 1984] Newton, Isaac, *The Optical Papers of Isaac Newton. Vol.1. The Optical Lectures 1670–1672.* Edited by Alan E. Shapiro. Cambridge: Cambridge University Press, 1984.

[Newton 1999] Newton, Isaac, *The Principia: Mathematical Principles of Natural Philosophy ... Preceded by a Guide to Newton's Principia by I. Bernard Cohen.* Translated by I. Bernard Cohen and Anne Whitman, assisted by Julia Budenz. Berkeley: University of California Press, 1999.

[Popkin 2003] Popkin, Richard H., *The History of Scepticism: From Savonarola to Bayle.* Oxford: Oxford University Press, 2003.

[Pourciau 1991] Pourciau, Bruce, "On Newton's Proof that Inverse-Square Orbits Must Be Conics". *Annals of Science* 48 (1991): 159–72.

[Pourciau 1992] Pourciau, Bruce, "Newton's Solution of the One-Body Problem". *Archive for History of Exact Sciences* 44 (1992): 125–46.

[Pourciau 2009] Pourciau, Bruce, "Proposition II (Book I) of Newton's *Principia*". *Archive for History of Exact Sciences* 63 (2009): 129–167.

[Shapin 1981] Shapin, Steven, "Of Gods and Kings: Natural Philosophy and Politics in the Leibniz- Clarke Disputes". *Isis* 72 (1981): 187–215.

[Shapiro A 1993] Shapiro, Alan, *Fits, Passions, and Paroxysms: Physics, Method, and Chemistry and Newton's Theories of Colored Bodies and Fits of Easy Reflection.* Cambridge: Cambridge University Press, 1993.

[Shapiro A 2004] Shapiro, Alan, "Newton's 'Experimental Philosophy"'. *Early Science and Medicine* 9 (2004): 185–217.

[Shapiro B 1983] Shapiro, Barbara J., *Probability and Certainty in Seventeenth-Century England: A Study of the Relationships Between Natural Science, Religion, History, Law and Literature.* Princeton: Princeton University Press, 1983.

[Speiser 1996] Speiser, David, "The Kepler Problem from Newton to Bernoulli". *Archive for History of Exact Sciences* 50 (1996): 103–116.

[Terrall 1999] Terrall, Mary, "Metaphysics, Mathematics, and the Gendering of Science in Eighteenth-Century France". In *The Sciences in Enlightened Europe.* Edited by William Clark, Jan Golinski and Simon Schaffer. Chicago: University of Chicago Press, 1999.

[Truesdell 1960] Truesdell, Clifford, "A Program toward Rediscovering the Rational Mechanics of the Age of Reason". *Archive for History of Exact Sciences* 1 (1960): 3–36.

[Truesdell 1973] Truesdell, Clifford, "The Scholar's Workshop and Tools". *Centaurus* 17 (1973), 1–10.

[Weinstock 1982] Weinstock, Robert, "Dismantling a Centuries-Old Myth: Newton's *Principia* and Inverse-Square Orbits". *American Journal of Physics* 50(7) (1982): 610–617.

[Weinstock 1989] Weinstock, Robert, "Long-Buried Dismantling of a Centuries-Old Myth: Newton's *Principia* and Inverse-Square Orbits". *American Journal of Physics* 57(9) (1989): 846–849.

[Weinstock 2000] Weinstock, Robert, "Inverse-Square Orbits in Newton's *Principia* and Twentieth-Century Commentary Thereon". *Archive for History of Exact Sciences* 55(2) (2000): 137–162.

[Whiteside 1970] Whiteside, Derek T., "The Mathematical Principles Underlying Newton's *Principia Mathematica*". *Journal for the History of Astronomy* 1 (1970): 116–38.

Saturn's Rings from Laplace to Poincaré

Tom Archibald

Introduction

The term "applied mathematics" has always had many senses, like the term "mixed mathematics" commonly used in the early modern period. Now, "mixed" sciences were long understood as those that study more than one of the sciences in the Aristotelian classification of subject-genera, even if Aristotle himself was a bit unclear on the issue of the proper subject of mathematical inquiry. From very early days, this notion of mixing carried with it the idea of taint, just as the word "pure", used for the unmixed kind of mathematics, suggests. These terms carried connotations that continued to be felt well into the early modern era.

By the seventeenth century, however, the power of mathematics for studying the natural world had come decisively to the fore, as witnessed by the works of Galileo, Descartes, Newton, and many others. The prestige of mathematics as an intellectual endeavour doubtless received a major boost from Newton's harnessing of mathematical tools to create the theory of the universe expounded in the *Principia.* This prestige was to transform in the nineteenth century into a prestige for the discipline and those associated with its professional practice. We see from the careers of Lagrange, Laplace, and Gauss that the practice of mathematical astronomy was key to their advancement. Indeed, the extension of the methods associated with this work, notably the theory and solution of partial differential equations, was a key feature of the development of mathematical physics throughout the course of the century, and the body of shared techniques grew into a structured set of methods that were learned and applied in standard ways. These varied from one community to another, not only at the national or linguistic level but also locally. While this did not change the association of applied mathematics with things more base in a permanent way—think of Hardy's *Mathematican's Apology*—it roiled the waters of scientific practice in such a way that older rankings were no longer obvious. As a result we see a multitude of attitudes among practitioners, not always completely clear, as to the role of mathematics in scientific inquiry.

It is the purpose of this paper to take a look at some of these attitudes as they developed between the late eighteenth and early twentieth centuries. To do so, I will focus attention on the special context of issues associated with a set of problems in celestial mechanics which have to do with the rings of Saturn. Those who have puzzled over the use of terms such as mathematical physics, applied mathematics, and theoretical physics will probably not finish reading this essay with a feeling of sudden illumination. However, I would like to draw attention to certain distinctions in the nature of research in the area of applied mathematics that I hope may be helpful at some level in understanding, historically, the relationship between practices and labels. The cases we will look at concern very different writers:

D.E. Rowe and W.-S. Horng (eds.), *A Delicate Balance: Global Perspectives on Innovation and Tradition in the History of Mathematics*,
Trends in the History of Science, DOI 10.1007/978-3-319-12030-0_5

Pierre-Simon de Laplace, James Clerk Maxwell, Sofia Kovalevskaya, and Henri Poincaré, as well as a supporting cast of varying importance and influence. These writers exemplify varying national and period styles, and various ideas of what mathematics is for are embodied in their work, even though their views were, in the main, not explicitly stated.

The issue at hand has to do with figures of equilibrium of rotating fluid masses. This was a problem that presented numerous interesting challenges ever since it was first tackled by Sir Isaac Newton. It was Newton, who by using the mechanics of his own devising had first famously claimed that the earth was flattened at the poles, this due to the equilibrium between its gravitational attraction and the centrifugal forces. Newton's approach required, however, that the eccentricity of the figure be small, whereas Colin Maclaurin later established that more general oblate spheroids would satify such equilibrium conditions. Legendre argued, however, that for small eccentricities the Newtonian spheroid was the only admissible solution. Then, in 1834, Carl-Gustav Jacobi showed that general ellipsoids, those that are not necessarily figures of revolution, could also be equilibrium figures. The subject was subsequently taken up by Liouville, Dirichlet, Dedekind, and Riemann, among others, so that by mid-century it had gained an important mathematical pedigree quite independent of its cosmological interest.

In what follows, we will see that by around 1800 Laplace's work had begun to recast Newtonian mechanics by idealizing its problems in such a way that they could be expressed by partial differential equations. Indeed, the chief aim of this idealization was to make the equations themselves useful, that is not only for describing idealized problems, but also for solving them.[1] In Laplace's idealizations the tenability of the hypotheses and their simplifications, including the question of whether they imply changes to the outcome that are of the same order as the phenomena being described, are not taken up seriously as issues. The mathematics (including a basic mechanical equilibrium argument based on Newtonian ideas) drives the entire enterprise in the sense that the aim of idealization is to make the mathematical method possible. Thus, Laplace's idealizations serve as a reference point against which other writers react in their various ways. In Maxwell's work, carried out between 1856 and 1859, I will argue that his mathematics is approximative in character, a conclusion in part indicated by the fact that Maxwell's model is made to serve a much more restricted physical analysis. By his time the standards and habits of precision measurement had become widespread, and we see their influence in the course of the story. The point of Maxwell's approach is to thoroughly exclude both solid and liquid rings without having to make any assumptions about the cross-sectional form of the ring. This left the field open for an analysis of the possibility that equilibrium resulted from the local structure of the ring (or rings), which could now be conceived as consisting of a very large

[1]Here I follow John Norton [Norton 2012] in distinguishing between models that serve as idealizations or as approximations. For Norton, an approximation is "an inexact description of a target system," whereas an idealization is a distinct description, *some* of whose properties are those of the system being modelled.

number of small satellites, interacting only gravitationally. The ground for further research on such questions changed somewhat in 1882 with the announcement by Thompson and Tait of new families of mathematically possible equilibrium figures. The combination of the appearance of annular shapes in their list, together with the absence of an explanation for their methods, acted as a direct stimulus for Poincaré's research into stability. Meanwhile, Kovalevskaya's approach, which originated from her still unpublished thesis studies from the 1870s, followed Laplace's lead of idealization, while developing the mathematical tools more fully. It is not clear whether she knew initially of Maxwell's work, but she referred to it later when she published her findings in 1885. In the meantime, Poincaré, who was apparently at first unaware of Kovalevskaya's paper, attacked the problem of stability of the rings in the context of his own extensive mathematical work on equilibrium figures. He saw Saturn's rings both as an interesting case not only for studying the equilibria of rotating fluid masses but also for applications of his qualitative theory of differential equations. His specific approach may, in fact, be seen as reflecting classical "mathematical" idealization, mixed with methods associated with the mathematics of approximations.

By this time, theoretical physics had begun to emerge as a distinct discipline, whereas the trend toward pure mathematics would soon lead to the formation of national societies of research mathematicians, few of whom followed recent developments in physics. Far more striking was the new linkage between pure mathematics and developments in philosophy centered on a newly revitalized academic logic. Philosophers continued to teach Aristotelian syllogisms, of course, but the seeds for what would what later become mathematical logic had already been sown. This convergence of interests had deep long-term implications. Soon the notion arose that mathematics could in some sense emerge from logic and philosophy to become the central tool for describing the eternal laws of the universe, a view that gave apt expression to the power that resided in mathematical learning. As far as the study of the physical world was concerned, this was above all a power that arose through the use of the tools of analysis for continuous modelling. The idea that a deterministic model could accurately predict the future of the universe was interesting in a broad cultural sense, though in some quarters, notably in Catholic circles, such a strong determinism had to be reconciled with issues having to do with human free will and the responsibility of the individual for making moral judgements. Yet, even decades before the issues arising from the quantum came to the forefront, the idea that such models possessed the character of a final reality was widely questioned. Indeed, Poincaré's explicitly conventionalist approach to the nature of the relationship between mathematical models and physical reality became very well-known, and this viewpoint may be seen as extending to a characteristic eclecticism in his mathematical physics. Moreover, Poincaré's popular writings from around the turn of the century were read and discussed in circles that extended far beyond France. Indeed, members of the educated public were increasingly drawn to the many fascinating scientific issues that had a direct bearing on the present state and future fate of the universe. Poincaré's discussions of the stability of the solar system, or the possibility of modelling the creation of the planets and the earth-moon system, as well as many other issues of this kind, touched the imaginations of a broad readership. Jules Verne's *De la terre à la lune* (appearing in 1865, when Poincaré was about 11 years old) and the 1902 George Méliès film *Le voyage dans la lune*, provide evidence of a broad general interest in such topics. We might also recall that Schiaparelli's Martian canals were from this very same period.

The rings of Saturn may thus likewise be seen against this background of general fascination with things astronomical that was in full blush during the latter part of the nineteenth century. For mathematicians, these rings posed a striking challenge given their seemingly simple form, for the problem of explaining how they persisted in place clearly resisted any kind of simple theoretical treatment. Little wonder that these difficulties attracted the attention of astronomers and mathematicians alike.[2] So let us now turn to the problem in question, the mathematical description or explanation of the rings of Saturn and their stability. The mathematics in what follows will, to use a phrase of Robert Langlands, be provided only in homeopathic doses. We should take note, however, that this subject is even now of unusually current interest as a new moon of Saturn appears to be in the process of formation as we write, joining its over 200 siblings, as seen by the Cassini orbiter.

Background: the Role of Laplace

From the moment of their first observation, the rings of Saturn fired imaginations by their oddity, their almost monstrous character. It was Christiaan Huygens who first decided that the peculiar and variable appearance of the planet was due to a ring around the planet – something he first recorded in 1656 in an anagram

> a a a a a a a c c c c c d e e e e e g h i i i i i i i l l l l m m n n n n n n n
> n n o o o o p p q r r s t t t t t u u u u u.[Huygens 1659, 47]

As Huygens was to explain later, this means: Annulo cingitur, tenui, plano, nusquam cohaerente, ad eclipticam inclinato. (It is surrounded by a thin flat ring, nowhere touching, and inclined to the ecliptic.)

But even before that, the "handles" of Saturn provoked imaginative explanation, for example by Roberval, who posited exhalations of vapors from the equatorial regions of the planet. Huygens imagined a single solid ring, though G. D. Cassini's 1675 observation of the dark band that we now call the Cassini gap led him to speculate that the ring was not solid at all, but made of a large number of small satellites.

[2]Maxwell's contributions have been discussed in [Brush et al. 1983] and the original version of Maxwell's Adams Prize Essay is contained in Harman's edition of Maxwell's letters and papers ([Maxwell Letters, v. 1, 438-479]). [Cooke 1984] provides a detailed discussion of Kovalevskaya's work.

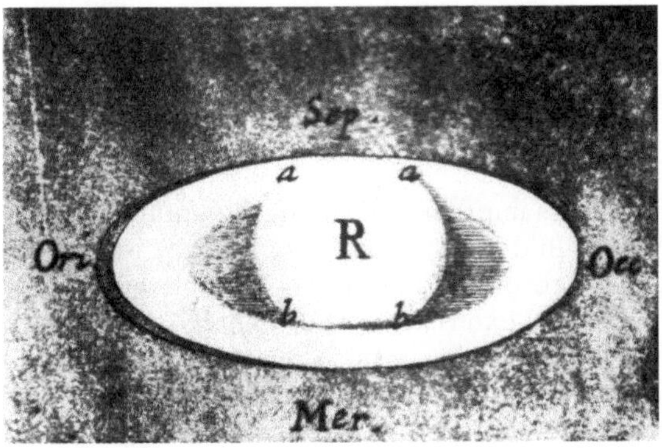

Figure 1: Robert Hooke's image of Saturn, Phil. Trans., 1666

The question of the possibility of describing the ring mathematically in a way that accounts for its properties was obvious after Newton. The first to discuss the matter was Laplace, who ruled out the notion that the ring could be solid on the grounds that the slightest instability would send it crashing to the surface. Laplace then stated an assumption about the actual state of affairs, without arguing why it should be so: he concluded that there were several rings (for which astronomical evidence was becoming clearer at his time) and that they had to be irregular in shape. This appeared to overlook other options, such as that of a large number of independent small satellites, or that of liquid rings, oversights noted by the British Astronomer Royal George Biddle Airy among others. It was James Clerk Maxwell who addressed the problem of proving that the ring was made of a large number of independent satellites, a difficult argument presented in his Adams Prize essay of 1856, and one of his first independent sustained pieces of mathematical work.

Laplace had long been interested in the figures of the earth and planets, and he sought to derive the conditions for an ellipsoidal shape as early as 1773. The argument involves the requirement of equilibrium between the gravitational and centrifugal forces, already used by Newton in the *Principia.* Saturn comes up in several places in Laplace's *Mécanique céleste*; already in the second volume from 1799 a dozen pages are devoted to the figure of the rings and to the question of their stability [Laplace An VII, 155-166]. He announced the recent observations of Short, to the effect that there were several rings, and then proceeded to use a standard equilibrium argument to deduce the figure of each of them. It should be stressed that the telescopes of the day made it difficult to distinguish between differences in illumination and irregularities in the cross-sections of the rings.

The discussion is, to modern eyes, a curious mixture of ad hoc assumptions and firm conclusions, but in all it fits very well into the "idealization" category. Indeed, typical of Laplace's strategy in celestial mechanics, the core of the method consists in making simplifying assumptions that allow the mathematical tools he develops to be deployed successfully.[3] The discussion is quite explicit about this ploy. Laplace posits that the ratio of the diameter of the ring to the diameter of the planet is taken as 2 for the interior ring, "ce qui s'éloigne peu de la vérité". On this basis, he is able to argue that the mean density of the planet would be 1.3 times that of the ring.

The fundamental assumption of the stability argument is that the rings consist of a homogeneous fluid mass. From this, Laplace investigated the idea that each ring is a figure of revolution generated by a closed curve about an axis in the plane of the curve and external to it, ultimately settling on an ellipse as the generating figure. One of his aims was to relate the radius of the planet to the radii of the rings, respectively. Once determined, this would enable him to calculate the ratio of the mean densities of the rings and the planet. Here the accuracy of observation was an obstacle, however, since dispersion made it difficult to get sufficiently good data for the radii.

The basic idea behind Laplace's approach is an equilibrium method. One assumes a ring filled with homogeneous matter, another evident simplification, and likewise posits a shape that permits the calculations to be carried out successfully. This shape consists of the rotation of an ellipse about an axis lying in its plane but not passing through it. On the surface of the solid of rotation we imagine an infinitely thin layer of fluid which is attracted both by the ring and by the attraction of a central body (Saturn) with centre of gravity coinciding with the centre of the ring. The question, then, is to determine whether the angular velocity of rotation and the distances can be specified in such a way that the fluid is in equilibrium.

This is equivalent to the requirement that the following equation obtain between the variables:

$$V + \frac{M}{\sqrt{r^2 + z^2}} + \frac{1}{2}n^2r^2 - C = 0.$$

Here V is the potential due to the ring at a point of the surface (recall that the potential satisfying what is now known as the Laplace equation was a central tool in Laplace's theories), M is the mass of the planet (taken as being concentrated at its centre of gravity), r is the distance from a point on the surface to the axis of rotation, and z is the distance of the point from the equatorial plane. The rotational speed is n, and C is a constant. To solve this equation, Laplace further assumed that the radius of the ring (the distance from its centre to the centre of the ellipse that generates the ring by its rotation about the axis of the planet) was very large compared to the semi-axis of the generating ellipse. This permitted Laplace to ignore the effect of distant particles of the ring on the points of its surface, so that to calculate the potential of the the ring on a point of the surface he treated the ring as an infinite cylinder with elliptical cross-section. This enabled him to get

[3] On Laplace's use of such simplifications early in his career, the reader is referred to [Lysne 2010].

the relationship in which stability will obtain (as it evidently does empirically by virtue of the persistence of the rings) employing an argument based on equilibrium obtaining between centrifugal and gravitational forces at the surface of the ring. On the basis of these numerous simplifying assumptions Laplace concluded that the ring can't be stable if the ratio of the density of the planet to that of the ring exceeds 1.3.

Underlining his own awareness of the assumptions he is making, Laplace goes on to state that the assumption of homogeneity would appear to be invalid:

> These inequalities [in the cross-section of the rings] are necessary to maintain the ring in equilibrium around Saturn, for, if it were perfectly similar in all its parts, its equilibrium would be disturbed by the slightest force, such as the attraction of a comet or a satellite, and the ring would wind up falling onto the surface of Saturn.[4]([Laplace An VII, 164])

To support this claim that a slight perturbation would lead to a fatal separation of the centres of the planet and the ring, he presents another mathematical argument. Laplace calculates that on the assumption of homogeneity the attractive force from the centre of the planet to the centre of the ring will be consistently negative, meaning that the ring must crash into the planet.

This much is based on simplifying assumptions that are, of course, not precisely met, but nevertheless the conclusion Laplace drew from them comes as a surprise. He claimed that his analysis meant that the various rings must be irregular solids of variable breadth at different points of their circumferences, and furthermore, that the centres of gravity of the rings would never coincide with the geometric centre. However, he offers no mathematical arguments to support these conclusions, and his subsequent discussion of the form and stability of the rings passes over these difficulties. The idealization that Laplace works with here, in fact, makes strong assumptions about the shape of the rings and the overall geometry. He invokes these, it seems, in order to employ the Newtonian equilibrium argument together with his own favorite mathematical tool for this context, the potential function V. No perturbing effects external to the planet-rings system are considered, so this method is clearly not intended to be approximative in any serious way. For example, while three decimal places are given in the expression involving the density ratios and the radii, taking the radius of the ring to be equal to twice the radius of the planet is another simplifying assumption. Since his assumptions seem to be pulled from thin air, what the mathematical model illustrates is simply that the Laplacian tools provide a means of addressing the problem, adding to what might be termed the completeness of the *Mécanique céleste*. What we see, and what were to be seen by later writers, as physically implausible assumptions, were to Laplace an essential part of creating his model.

[4]Ces inégalités sont nécessaires pour maintenir l'anneau en équilibre autour de Saturne; car, s'il était parfaitement semblable dans toute ses parties, son équilibre serait troublé par la force la plus légère, telle que l'attraction d'une comète ou d'un satellite, et l'anneau finirait par se précipiter sur la surface de Saturne.

Aftermath of Laplace to Mid-Century

The type of equilibrium argument given by Laplace for the figures of the planets and the rings of Saturn came to be important set pieces for those apprenticing in celestial mechanics. The adaptation of methods of partial differential equations to the task was broadly learned, and Laplace's original techniques were improved until they became standard routine. One example of a writer in this vein was George Biddell Airy, an early product of the reforms in Cambridge mathematics that began with the introduction of continental methods. In 1826, Airy wrote a paper on the figure of the earth that was to become standard reading for later Cambridge Tripos men, via his *Mathematical Tracts* ([Airy 1826], [Airy 1826a]). Airy became Astronomer Royal in 1835, a post he was to occupy until 1881, during which period he gained notoriety for his failure to discover Neptune. Airy's careful observational work was complemented by both physical experimentation and mathematical modelling, and he was not infrequently in the public eye. For example, his pendulum experiments of 1854 to determine more precisely the density of the earth were reported in the popular press ([Illustrated London News 1854]). Airy's earlier efforts in this regard were first undertaken in 1828 by the same method, namely measuring differences in the motion of pendula at the top and bottom of a mine shaft. This initial effort foundered on various grounds, the most important being the impossibility of determining synchronicity at the top and bottom of the shaft, a problem that was overcome in the later experiment through the possibility of telegraphic signals. Airy is thus representative of the shift toward precise measurement and its integration with theory mentioned in the introduction.

Airy's authoritative position makes his views on Laplace's theory of Saturn's rings particularly apposite. In 1859, reviewing Maxwell's Adams Prize paper on the subject, he wrote:

> Probably no competent mathematician has ever read this paper [by Laplace] without remarking on the want of a ground for the last conclusion. Professor Maxwell has well given the following as the conclusion that ought to have been drawn: "If the rings were solid and uniform, their motion would be unstable, and they would be destroyed. But they are not destroyed, and their motion is stable; therefore they are either not uniform or not solid."[Airy 1859, 159]

Indeed, since Saturn continued to attract much interest, a number of observational astronomers and theorists continued to make critical remarks relating in one way or another to Laplace's account. In the late 1840s and early 1850s, for example, observers noticed various bands or gaps in the rings, and in 1852 it was widely remarked that the visible edge of the planet's surface (the "limbs" of Saturn) were visible through the rings, indicating that the rings could hardly be solid. It thus became widely assumed that they were in fact liquid, though the differences between a truly liquid ring and a ring made up of a very large number of small satellites were probably not grasped by some, like William and George Bond around 1850, who proposed this view [Bakich 2000, 257]. The Bonds were working with one of the two largest refractors in the world, a 15-inch telescope at Harvard. Some writers entered the field with evidently speculative proposals, for example that set forth by Harvard's Benjamin Peirce (1809-1880), who thought that the moons of Saturn might be responsible for the stability of its rings. In the course of these discussions, a decisive contribution was made by the young James Clerk Maxwell.

Maxwell, Thomson and Tait

Maxwell

Maxwell was Second Wrangler in 1854 at Cambridge, and became a Fellow at Trinity College in 1855. Soon thereafter, a vacancy at Marischal College in Aberdeen became available, and he applied for and was awarded the position, which he took up in November of 1856. In the meantime, the Adams Prize problem for 1857 had been announced, calling for a study that would account for the motion of Saturn's rings. In December of 1856, Maxwell submitted his essay for the prize; details may be found in [Brush et al. 1983] and in [Maxwell Letters, v. 1]. It turned out to be the only submission, and Maxwell was awarded the prize in May of 1857. He revised his essay substantially for the 1859 monograph published by Macmillan under the title "On the Stability of the Motion of Saturn's Rings" [Brush et al. 1983, 69-158]. For this work he was awarded a prize of £130.

Maxwell sought to account for the shape and stability of the rings—in his view it was established that there were only three—by appealing to mechanical principles. His stated aim was to achieve this or to "admit that, in the Saturnian realms, there can be motion regulated by laws which we are unable to explain"[Maxwell 1859, 73]. Calling the rings the "most remarkable bodies in the heavens, except, perhaps, those still less *useful* bodies—the spiral nebulæ", Maxwell made careful note of the recent observational history and their probable dimensions. His figure of the rings is at the scale of one inch to 100 000 miles, with the diameter of the outer ring being 176,000 miles, while the thickness was given to be not more than 100 miles. His aims, doubtless reflecting the remarkable recent discovery of Neptune, included being able to learn more about these bodies "than the telescope can yet ascertain" [ibid. p. 74]. This, then, is a regime in which phenomena are to be discovered on the basis of models, a situation in which the approximative or simulating role of the model plays the key role.

After a careful examination of Laplace's work, Maxwell accepted his first conclusion that stability would be impossible for a uniform solid ring, but then he drew the appropriate inference from this in the words quoted above, as cited by Airy. Maxwell then proposed to "determine the amount and nature of the irregularity which would be required to make a permanent rotation possible." Here he noted already from the beginning that if the rings were solid, then to be stable they had to be highly non-uniform. In particular, under certain assumptions, a small (roughly satellite-sized) region in the ring would have to contain (almost exactly) 4.5 times the mass of the rest of the ring. As there was nothing in the observational data to suggest that this was the case, Maxwell accordingly abandoned the theory of solid rings.

This negative conclusion was accompanied by a much subtler positive reflection. Since the rings cannot be solid, they must either be formed of a liquid substance or else composed of many small individual satellites. In the latter case, the action of the ring will be small with respect to that of the planet, so there should be no possibility of the small pieces ceasing to be satellites. But there is nothing (without further reflection) to guarantee the stability of the ring as a whole. Maxwell argues, in fact, that there will be a tendency for the pieces to collapse into larger satellites, but that "by the revolution of the ring it is converted into the condition of dynamical stability."[Maxwell 1859, 77].

The methods by which Maxwell came to these conclusions have been described in [Brush et al. 1983, 12-21]. He solved the initial problem of the motion of the ring by calculating the potential of the ring at the centre of the planet, which leads to conditions on the form of the ring due to its uniform rotation. Stability conditions are obtained by an equilibrium argument, and these, in turn, can be used to derive conditions on the shape of the ring to ensure stability. Instead of assuming the elliptical cross-section, as Laplace had done, Maxwell gives a Fourier expansion for the form of the ring and then obtains its coefficients from the constraints—a nice example showing how the data and dynamical theory enter into the model to eliminate ad hoc geometric assumptions. Establishing that rigidity is impossible, Maxwell then looks at the propagation of waves through the rings and finds that there are several different kinds, for which he determines bounds on their periods. Finally, he argues that for a rotating ring the instability with respect to tangential displacements will tend to be damped as a consequence of the motion of the ring and by virtue of the coupling of radial and tangential displacements.

This model leaves much unsaid; it does not explain, for example, why three rings are observed, and it leaves many other questions open. But the picture provided is one in which several rings of small satellites, each rotating with a speed appropriate to its radius, may be stable subject to density and velocity constraints. Using arguments similar to those of Laplace, Maxwell was able to show that semi-rigid or liquid rings could be ruled out as they appeared to be unstable. Rather remarkably, he even produced a demonstration device with a system of 36 satellites in order to illustrate the kinds of wave actions predicted by his model.

The monographic publication of Maxwell's treatise, the density of its mathematics, and the rather arcane nature of the subject may have condemned it to a small readership, despite Maxwell's hopes that it might "redeem the character of mathematicians, and make it intelligible," as he wrote to Thomson in November 1857 ([Maxwell Letters, v. 1, p. 557]). Clearly both Thomson (one of the original posers of the problem, and a correspondent during the revision for the book) and Airy would have read it carefully. Still, its methods, being tightly tied to the particular features of the problem, limited the direct influence of the work. Nevertheless, some have speculated that Maxwell's later work on statistical mechanics may have found part of its genesis in these efforts, and this is certainly plausible. These kinds of stability results surely also had held much interest for Thomson, particularly when he took up his later considerations of equilibrium figures, which we briefly describe below.

Thomson and Tait

The *Treatise on Natural Philosophy* of William Thomson and Peter Guthrie Tait first took form as a project in the early 1860s. It was originally intended as a compendious work that aimed to provide a unified textbook treatment of applications of mathematical analysis to all areas of physical inquiry, but with due attention to empirical support for the theoretical results. The first volume was originally published it 1867, and dealt with the statics and dynamics of particles as well as rigid bodies and fluids. It was to become a standard feature of the education of those training for the Cambridge Tripos examination, and hence had broad influence within the British community. It was also much appreciated by Hermann von Helmholtz, and at his instance was translated into German as *Handbuch der theoretischen Physik*; this appeared during the years 1871-1874, but it would only have a limited influence in Germany. The extension of this project to cover the entire gamut of physical theory was ultimately abandoned, but a revision of the first volume was published in two parts, in 1879 and 1883, respectively. The portion of the second part pertaining to the equilibrium figures of rotating fluids is that which interests us here because of the new findings annouced there. G. H. Darwin worked on this revision, and it may be that portions of the work we mention below bear his imprint, though the sections covered are not among those credited to him. Be that as it may, we can read that:

> During the fifteen years which have passed since the publication of our first edition [in 1867] we have never abandoned the problem of the equilibrium of a finite mass of rotating incompressible fluid. Year after year, questions of the multiplicity of possible figures of equilibrium have been almost incessantly before us, and yet it is only now, under the compulsion of finishing this second edition of the second part of our first volume, with hope for a second volume abandoned, that we have succeeded in finding anything approaching to full light on the subject [Thomson and Tait 1882, v. 1, 332].

Before this, prior investigators were concerned with the question of finding conditions (involving radii, densities, and angular velocity) under which a spheroid would be in equilibrium. The stability of this equilibrium had been investigated via what we would now term energy considerations, stated in terms of simple geometry. Here, Thomson and Tait establish conditions for the rotating ellipsoid that show that for stability the angular velocity ω should be proportional to the square root of the density ρ, and they provide a table relating the eccentricity of the ellipsoid to the ratio $\omega^2/2\pi\rho$.

i.	ii.	iii.	iv. see § 775.	v. see § 776.
eccentricity $e = f/\sqrt{(1+f^2)}$	$f = e/\sqrt{(1-e^2)}$.	$\frac{\omega^2}{2\pi\rho}$	Rotational period, in mean solar seconds, for case of density equal to Earth's mean density.	$\frac{\mu^2}{k} = \frac{(1+f^2)^{\frac{4}{3}}\omega^2}{2\pi\rho}$ where μ is moment of momentum, and k a constant *.
0	0	0	∞	0
0·093	·0934	·0023	86,164	·0023
·1	·1005	·0027	79,966	·0027
·2	·2041	·0107	39,397	·0110
·3	·3145	·0243	26,495	·0258
·4	·4365	·0436	19,780	·0490
·5	·5774	·0690	15,730	·0836
·6	·7502	·1007	13,023	·1356
·7	·9804	·1387	11,096	·2172
·8	1·3333	·1816	9,697	·3588
·8127	1·3946	·1868	9,561	·3838
·9	2·0648	·2208	8,804	·6665
·91	2·1949	·2225	8,759	·7198
·92	2·3474	·2241	8,729	·7813
·93	2·5304	·2247	8,718	·8533
·94	2·7556	·2239	8,732	·9393
·95	3·0423	·2213	8,783	1·045
·96	3·4282	·2160	8,891	1·179
·97	3·9904	·2063	9,098	1·350
·98	4·9261	·1890	9,504	1·627
·99	7·0175	·1551	10,400	2·113
1·00	∞	0·0000	∞	∞

Figure 2: Table of Eccentricities and Angular Velocities, [Thomson and Tait 1882, v. ii, 327]

This research is certainly in the approximative mode, and in that respect it is very similar to what Maxwell had done. For example, the authors draw the conclusion that if the angular velocity is greater than the value calculated from $\omega^2/2\pi\rho = 0.2247$, "equilibrium is impossible in the form of an ellipsoid of revolution" [Thomson and Tait 1882, 327]. Note though the words "in the form of an ellipsoid of revolution." For right after this, Tait and Thomson turn to consider the problem of finding other possible figures of equilibrium for a rotating fluid, a much more difficult problem than demonstrating stability for a given figure. Along with their new results, they also list the known possibilities ([Thomson and Tait 1882, 330-335]); these included the regular Newtonian flattened spheroid and the ellipsoidal solution discovered by Jacobi in the 1830s.

It seems remarkable that they give no demonstrations for any of the cases beyond the ellipsoidal ones, which had already been studied earlier. This, in fact, suggests that rather than a general method they had simply done "experimental" investigations, including in some cases attempts to calculate conditions of stability for various other kinds of figures. The language they employ also implies that their efforts were not entirely conclusive. We quote two examples to give an idea of the nature of this incomplete achievement:

(b) If the condition of being a figure of revolution is imposed [as in the

> Laplace case], without the condition of being an ellipsoid, there is, for large enough moment of momentum, an annular figure of equilibrium which is stable, and an ellipsoidal one which is unstable. It is probable, that for moment of momentum greater than one definite limit and less than another, there is just one *annular* figure of equilibrium, consisting of a *single ring.*
>
> (c) For sufficiently large moment of momentum it is certain that the liquid may be in equilibrium in the shape of two, three, four or more separate tings, with its mass distributed among them in arbitrary portions, all rotating with one angular velocity, like parts of a rigid body. It does not seem probable that the kinetic equilibrium in any such case can be stable. [Thomson and Tait 1882, v. ii, 333].

Not only is this discussion entirely descriptive, it is also devoid of any information about how these results were found. In some cases, parameter values are specified, but these surely involved carrying out lengthy calculations. We may speculate that these were found to be either too long for the *Treatise* or of marginal relevance to the overall project; they may also have been insufficiently conclusive to warrant independent publication at the time. Since this project had remained largely dormant for the preceding fifteen years, the authors may well have felt reluctant to go beyond their preliminary findings, especially in the absence of a clear method for resolving the many remaining technical difficulties. In any event, their remarks stimulated the interest of Poincaré, who in the course of his work on this theme became aware of other relevant recently published work by Kovalevskaya, to which we now turn.

Kovalevskaya

The Russian mathematician Sofia Kovalevskaya began work on a mathematical description of the shape of Saturn's rings back in 1873 when she was studying under Weierstrass in Berlin. She completed her paper on this topic in the spring of 1874, and it was included as part of the work she submitted for her doctorate, which she was awarded *in absentia* from Göttingen. The circumstances involved in this unusual procedure are somewhat cloudy, but it seems likely that Weierstrass was advised by Göttingen's Lazarus Fuchs that she submit this paper along with her other work. Fuchs apparently thought it wise that she provide a physical subject that could be judged positively by the renowned physicist Wilhelm Weber.[5] There appears to be no extant documentation that might shed light on her original reasons for selecting this theme, though her earlier work on the potential of an annulus helped prepare her to take it up. Moreover, physical themes occur frequently in her research, albeit with a strong mathematical bent that makes it not

[5] I am grateful to Eva Kaufholz for this information, which is based on a letter from Weierstraß to Fuchs of July 19, 1874.

at all similar to Maxwell's work. The specific topic of stability is also prefigured in a letter she received from Weierstrass, written August 20, 1873.

Kovalevskaya's paper did not appear in print, however, until 1885, when it was forwarded to the *Astronomische Nachrichten* by the Swedish astronomer Hugo Gyldén. This was no doubt linked to her efforts to solidify her employment at Stockholm's newly founded university, where Gyldén was part of Mittag-Leffler's network. Kovalevskaya also spent at least part of that year in Paris, a point to which we shall soon return. It seems she recognized that her results from a decade earlier were unlikely to interest astronomers because she points out that Maxwell's work had in the meantime shown the implausibility of Laplace's original hyotheses regarding the structure of Saturn's rings. This may well explain why she did not push her own calculations further. We can not be sure whether or not she knew Maxwell's work when she first undertook her own investigations, but it seems more likely that she became aware of it later, for reasons we will return to below.

What we can say with assurance is that Kovalevskaya's research was directly inspired by the Laplacian approach. It thus represents an idealization of the same kind, though she began with weaker hypotheses than those assumed by Laplace. In particular, he had assumed that the opening in the ring (the distance from its centre to the centre of the ellipse that generates the ring by its rotation about the axis of the planet) was very large compared to the semi-axis of the generating ellipse. As noted earlier, this permitted Laplace to ignore the gravitational effect of distant particles of the ring on the points of its surface, thereby enabling him to calculate the potential of the ring on a point of the surface by treating the ring as an infinite cylinder with elliptical cross-section. This particular idealization then made it possible to solve the equation. But this assumption is a very strong one, since, as Kovalevskaya pointed out, one cannot tell without further consideration whether the error due to the assumption is really less than the order of the solution.

She therefore proceeded with a far more general model: instead of assuming an elliptical cross-section, she parametrized this curve with an arbitrary odd function written in a Fourier sine series with coefficients β_i. While a similar strategy was employed in Maxwell's work, this was by now a standard technique in dynamics so there is no particular reason to think she took this approach from Maxwell. Using an energy conservation argument, she then expresses the potential in terms of the angular momentum and the β_i coefficients. This, of course, leads to an infinite system of equations, from which she obtains an expression involving complete elliptic integrals, which she then evaluated for the first few terms. The calculations required here are, however, heroic, so she chose to break off further work on them. Here she merely remarks:

> I have been deterred from a yet more precise determination of the cross-section of the ring not only by the difficulty of the calculation, but also by the fact that, due to Maxwell's research ... it has become doubtful whether Laplace's view of the structure of the rings of Saturn is acceptable.[Cooke 1984, 81](Cooke's translation)

As noted above, it seems at least plausible that she had been unaware of Maxwell's work when she first took up this problem over a decade earlier. One can also easily imagine why she, as one of Weierstrass's most gifted pupils, would have been drawn to this topic initially. Both the difficulty of the problem and the obvious limitations of Laplace's approach, in addition to the general prestige of celestial mechanics, made it ideally suited for her technical abilities. Yet once she learned that Maxwell's work had largely undermined the whole Laplacian scheme, she clearly had good reason not to bury herself in fruitless calculations. On the other hand, she could still draw on this work to strengthen her professional status by publishing her paper in an astronomical journal. The approach through idealization employed by Kovalevskaya was entirely standard for the German mathematics of this period, as was the approach of generalizing an existing idealization by dropping some of the restrictions or otherwise improving the model. [Archibald 2001]. More significant still, her publication attracted the attention of Poincaré, both because of its subject matter and in connection with the difficulties involved in solving infinite systems of equations.[6].

Poincaré's work on Stability and Saturn's Rings

The work of Kovelevskaya and Poincaré represents a different tack from that of the British natural philosophers. Poincaré's interest in stability issues dates from around 1884.[7] We note in passing that the year 1883, which saw the appearance of the second edition of Thomson and Tait's *Treatise*, was also the year of Poincaré's first publication in celestial mechanics, a paper on the three-body problem. In his 1885 papers, described below, he made clear that he had read the relevant portions of the *Treatise* by Thomson and Tait.

Poincaré's first paper on stability appeared as a 10-page note in the *Bulletin astronomique* [Poincaré 1885a]. This was undertaken independently of Kovalevskaya's paper, though he became aware of her work while preparing his own, perhaps through Mittag-Leffler or Hermite. In any event, he added a footnote to the first page of his publication, stating that:

> I just received news of a memoir by Mme de Kowalewsky that I did not know of at the moment when I wrote the present work ... Although the problem treated by the learned *mathématicienne* was not exactly the same as the one I worked on, her analysis is very close to mine and I have added only a little to the results that could be deduced easily from her memoir.[8][Poincaré 1885a, 109]

---

[6]Both the content of this paper and its role in Kovalevskaya's career have been discussed in detail in [Cooke 1984, Chapter 4]

[7]Poincaré's work in this area has been discussed recently and cogently in [Gray 2013], to which the following account is indebted. Gray's scientific biography helps chart the complex trajectory of Poincaré's work on various subjects.

[8]Je viens de recevoir communication d'un Mémoire de Mme de Kowalewsky dont je n'avais pas connaissance au moment où j'ai rédigé le présent travail ... Bien que le problème traité par

The question of the generosity of these remarks may be left to the reader. Before he published the longer version of this paper, Poincaré had the opportunity to discuss these mattesr with Kovelevskaya, since she was in Paris in early 1885. In this paper for *Acta* he notes:

> In this study I met with Mme Kowalevski who had already used the same procedures in a memoir on the rings of Saturn. (Dans cette étude je me suis rencontré avec Mme Kowalevski qui avait déjà employé les mêmes procédures dans un mémoire sur l'anneau de Saturne.) [Poincaré 1885d, 260]

Poincaré on Stability

Poincaré's papers were, however, above all a response to the recent findings by Thomson and Tait on new equilibrium figures. He thus sought, first, in [Poincaré 1885d] to provide a proof for one of their new cases, the annulus (quoted as (b) above). This gave him an opportunity to use what he termed "continuity" methods, which he had already employed in his work on the three-body problem. Such methods are an extremely important feature of Poincaré's work up to at least 1896; thiey involve taking a solution to a differential equation in a similar but "nearby" case, and then extending that solution to a solution of the perturbed equation. This focus on the relation between a differential equation and the functions it defines goes back to Poincaré's thesis, notoriously awarded the "trois boules blanches", the minimum necessary for success. This work attempted to extend ideas of Fuchs and Flocquet on linear differential equations to the case of several variables. In his new work on stability, Poincaré noted that the usual successive approximation methods (which go back to Liouville, but were used in some sense by Schwarz and Neumann and later popularized by Picard) were not adequate to establish the existence of a solution in the cases he was considering. In striking contrast to the decimal places in Maxwell's work, he writes his numerical values in rational units.

To obtain an equation for stability, Poincaré takes W as the potential energy of the system (that is, taking into account the action of one part of the toroid on another) and I as the moment of inertia with respect to the axis of revolution. Setting

$$U = W + \frac{\omega^2 I}{2},$$

where ω is the angular velocity, the condition for equilibrium is then $\delta U = 0$. This expresses the physical property that small perturbations in the system will lead to a nearby stable state. Two types of perturbations in the form of the ring need to be taken into account. The first consists of wobbles, as it were, in the surface

la savante mathématicienne ne soit pas tout à fait le même que celui dont je me suis occupé, son analyse se rapproche beaucoup de la mienne et je n'ai ajouté que peu de chose aux résultats qu'on pourrait facilement déduire de son Mémoire.

of the ring that leave the mean meridional cross section unchanged. This Poincaré handles in the same way as Kovalevskaya did, that is, by expressing the cross-section via a Fourier series with "stability coefficients" β_i. Here Poincaré created new tools to solve the problem rigorously by diagonalizing an infinite quadratic form. The second kind of perturbation involves an inflation or deflation of the entire meridional cross-section of the ring.

As for Saturn's rings, [Poincaré 1885c] gives an argument to rule out the possibility that a fluid ring can rotate as a single piece. This was in part a footnote to the work of his friend Tisserand, who in 1882 had carried the analysis of Laplace still further to show even more convincingly that stability in the case of a solid ring was impossible. Tisserand had also used Laplacian arguments in order to give stability conditions for the maximum thickness of a ring with a given density and mean radius [Poincaré 1885c, 507]. Poincaré now gave a new stability criterion in the form of a lower bound for the density of a fluid ring. His arguments for this came directly from [Poincaré 1885a], where he showed that a requirement for stability was that the angular velocity

$$\omega < \sqrt{2\pi\rho f},$$

where f is the gravitational attraction of two mass points a unit distance apart and ρ is the density of the ring. For the rotational speed, $\omega^2 r = fM/r^2$, where M is the mass of Saturn and r the radius of the ring. Using the above inequality and the expression for the mass of Saturn in terms of its density, he could easily show that

$$\frac{\rho}{\rho'} > \frac{2}{3}\left(\frac{r}{r'}\right)^3,$$

where the right side of this inequality varies between 1/5 and 1/16. (Here the primed variables refer to the density and radius of the planet.) This shows that the density of the outer ring, for example, would have to be at least 1/16 that of the planet. This, however, was clearly impossible in the light of Maxwell's work, for he had argued that this density should be at most 1/300 that of the planet in order to attain stability. Thus, Poincaré's analysis provided an additional argument against a liquid ring and in favour of a large number of small satellites.

This brief description might leave the impression that these are simple arguments; they are anything but that. For although Poincaré presented the overall arguments in only a few pages to the less mathematically inclined readership of the *Bulletin Astronomique*, the full version in *Acta Mathematica* filled 122 pages. Moreover, this long paper is an astonishing display of his mathematical powers and virtuosity. Let us briefly consider some of the main features of this body of research, including its aims, both as mathematical methods and the applications thereof.

To begin with, these papers present a striking combination of fairly standard methods and substantially new techniques. At their core are the tools of the Laplacian tradition in celestial mechanics, namely perturbation methods, combined with those of Lagrangian mechanics, in particular the method of "imposition and removal of constraints." A potential, or force function, depending on n generalized coordinates, is in equilibrium when the first derivatives of this potential are all zero, and this equilibrium state will be stable if this is a relative maximum. Indeed, Poincaré's notation and style of presentation directly recall Laplace and Lagrange, which is merely to say that he adhered to the notation that was usually taught. Newer is his use of diagonalization of a quadratic form for the energy to get its "stability coefficients"— the β_i of Kovalevskaya—and the generalization to the case of infinitely many coefficients. There is no reason to doubt that Poincaré's work on this was independent of Kovalevskaya's, though her research preceded his. He was able to resolve some of the difficulties she had struggled with by placing the force function, whose variation must be shown to vanish, in an infinite-dimensional configuration space. The condition for stability could then be analyzed in terms of properties of the trajectory of this force function, as this depends on a single parameter (such as the angular velocity), so that conditions can be specified as to whether we are in a parameter domain where equilibrium obtains or otherwise. The points of transition between these domains—such as Thomson and Tait's 0.2247— now become, explicitly, bifurcation points. Indeed, the phrase "équilibre de bifurcation" is the heading of the second section of the paper in *Acta*.

So was this idealization or approximation? Poincaré actually addresses this issue rather directly. His main aim was, he claims, a rigorous demonstration of the possibility of a solution for the annular case, as first proposed by Thomson and Tait. Once this is done, "on se sert des méthodes habituelles pour la calculer approximativement" [Poincaré 1885a, 110]. Now, existence results for solutions to a particular model were not new, though they were relatively recent. The issue of the Dirichlet principle, the criticism it had attracted, and the work-arounds of such writers are Carl Neumann and Hermann Amandus Schwarz to guarantee existence of solutions in certain well-behaved cases, all these were common knowledge in 1870s in Germany, Italy and France. Poincaré's creation of a new conceptual apparatus to obtain a workable method thus followed in these footsteps, no matter how innovative the method itself happened to be. But the aim of the method was clearly approximative. One seeks values of the parameters that specify stability, or transition from one regime to another. Nevertheless, without the existence proof, there is no way of knowing whether one is approximating an artifact of the model under study. Poincaré's well-known and much-evidenced conventionalism, the notion that the nature of a model did not matter as long as the results were correct, would seem to put him firmly in the approximation camp. Regarding the truth value of such models he remained quite indifferent.

Concluding Remarks

Problems in celestial mechanics deal with the structure of the universe, and solving such problems in a way that allows an apparently correct (if approximate) prediction of phenomena presented evidence of the ability of the human mind to grasp the workings of that universe in a fruitful way. Despite the increasing prestige of the experimental physical sciences in the late nineteenth century, such questions as the stability of the solar system exercised a strong power over the human imagination. The selection of such problems by leading scientists shows the enduring attraction of this æsthetic over the course of the century, both among scientists as well as the cultivated public. Poincaré taught these methods, and indeed they were widely studied, even if the direction they sought to go turned out not to provide the key to understanding the formation of planetary systems, something we still strive to understand.

Poincaré's power as a mathematician gave credibility to what turned out to be a suggestive, but ultimately unfruitful model for the formation of the planets and planetary systems. As the distinguished astrophysicist, Subrahmanyan Chandrasekhar, was to write,

> Poincaré conjectured that the bifurcation of the pear-shaped body leads onward stably and continuously to a planet attended by a satellite, the bifurcation into the fourth zonal harmonic probably leads unstably to a planet with a satellite on each side ... The grand mental panorama thus created was so intoxicating that those who followed Poincaréé were not to recover from its pursuit. ... Finally in 1924 Cartan established that the Jacobi ellipsoid becomes unstable at its first bifurcation point .... at this point the subject quietly went into a coma.[Chandrasekhar 1969]

This survey was intended to explore differences in using mathematics to understand the rings of Saturn. These various approaches may be characterized as differences that have to do with mathematical "style", without attempting to formulate a precise meaning for that term. Choice of subject, choice of method, the precise mix of methods, the media employed for transmission of new and received knowledge, all form part of this picture. In the present case, we see stylistic features that have some bearing on the issue of shifting national styles, at least on the level of preliminary evidence. The work in the tradition of Laplace, which has often been described as Newtonian, reflected, in fact, a mixed set of tools that were largely absent on the British scene. Its practitioners combined algebraic analysis with a *vis viva*-based mechanics, using potentials as a key tool. The idealization that this approach requires was common to much French and German mathematical work on physical subjects of the nineteenth century, though the German methods, which we have only touched on, generally had a more geometric flavour by virtue of their focus on boundary value problems.

Poincaré's continuity with Laplace lies to a considerable extent in his ability to push these older methods in new directions, both algebraically and analytically. Indeed, while retaining the centrality of celestial mechanics for this enterprise, Poincaré was able to contend with the fertile offspring of a very different style, that exemplified by Maxwell, Thomson and Tait. Their approximative approach, with its use of Lagrangian mechanics and a correspondingly freer relationship between mathematical variables and physical phenomena, blended well with Poincaré's conventionalist stance. This helps to explain how he so easily integrated this British work into the array of methods he used, transformed, and taught. This picture aligns very well with a larger point that has been cogently expressed by Jesper Lützen: it is not (usually) the case that new mathematics is prefabricated as "pure" mathematics and then made to fit onto physics like a ready-to-wear outfit. On the contrary, mathematics is almost always developed along with its applications; it emerges in accordance with the nature of the model used in the application and reflects the features that are highlighted or ignored, idealized or approximated, just as the model can likewise be changed by the choice of mathematical tools—or, particularly, by the success of the researcher in developing them for the task.

The qualitative differences between the Laplacian model and those developed later by Poincaré and Maxwell seem very clear, and these point to the dramatic changes that abstract mathematical science brought to the modelling of physical phenomena over the course of the nineteenth century. Joseph Dauben has frequently drawn attention to Georg Cantor's remark to the effect that the essence of mathematics lies in its freedom. We see this freedom at work in the multiplicity of approaches we have just examined, and it lies in the power of such freedom that mathematics is able to shed light on arcane and difficult matters that arise as mathematicians look at beautiful and perplexing things, while trying to uncover their secrets.

Bibliography

[Airy 1826] Airy, George Biddle. "On the Figure of the Earth", *Philosophical Transactions*, 116, 548-578.

[Airy 1826a] Airy, George Biddle. *Mathematical Tracts on Physical Astronomy, the Figure of the Earth, Precession and Nutation, and the Calculus of Variations.* Cambridge, "Printed by J. Smith, Printer to the University".

[Airy 1859] Airy, George Biddle, "Review of Maxwell's Essay on Saturn's Rings", *Monthly Notices of the Royal Astronomical Society* 19, 297-304. In [Brush et al. 1983, 297-304].

[Archibald 2001] Archibald, T. "Images of Applied Mathematics in the German Mathematical Community: Authors of Physico-Mechanical Papers in Crelle-Borchardt, 1860-1875" in *Changing Images of Mathematics in History*, eds. A. Dahan and U. Bottazzini, p 49-67. London: Gordon and Breech. 2001.

[Brush et al. 1983] *Maxwell on Saturn's Rings*, eds. S. Brush, C. W. F. Everitt, and E. Garber. Cambridge Mass., MIT Press, 1983.

[Bakich 2000] Bakich, Michael, E. *The Cambridge Planetary Handbook*, Cambridge, Cambridge University Press, 2000.

[Chandrasekhar 1969] Chandrasekhar. *Ellipsoidal Figures of Equilibrium*, New York, Dover, 1969.

[Cooke 1984] Cooke, Roger. *The mathematics of Sonya Kovalevskaya.* New York, Springer, 1984.

[Gray 2013] Gray, Jeremy. *Henri Poincaré: a Scientific Biography.* Princeton, Princeton University Press 2013.

[Huygens 1659] Huygens, Christian, *Systema Saturnium, sive De causis mirandorum Saturni Phaenomenon, et Comite eius planeta novo.* The Hague, Vlacq, 1659.

[Illustrated London News 1854] "The South Shields Pendulum Experiments by the Astronomer Royal." *Illustrated London News* [London, England] 4 Nov. 1854: 446+. Illustrated London News. Web. 30 June 2014.

[Kovalevskaya 1885] Kovalevskaya, Sofia, "Zusätze und Bemerkungen zu Laplace's Untersuchung über die Gestalt der Saturnsringe", *Astronomische Nachrichten* 111, 1885, 38-48.

[Laplace 1776] Laplace, Pierre-Simon. "Mémoire sur l'inclinaison moyenne des orbites des comètes, sur la figure de la terre, et sur les fonctions." *Mémoires de l'Acad. Royale des Sci., Savants Étrangers,* [pour l'année 1773], t . VII, 1776, 279-

[Laplace An VII] Laplace, Pierre-Simon. *Traitéde Mécanique céleste,* Tome 2, Imprimerie de Crapelet, Paris, An VII [1799].

[Lysne 2010] Lysne, Menolly, *Scientific patronage and rivalry in Laplace's early career,* Unpublished MSc thesis, Simon Fraser University, Burnaby, available at http://summit.sfu.ca/item/11312.

[Maxwell 1859] Maxwell, James Clerk. *On the Stability of the Motion of Saturn's Rings: an Essay which obtained the Adams Prize for the year 1856, in the University of Cambridge.* Cambridge, Macmillan and Co., 1859.

[Maxwell Letters] *The scientific letters and papers of James Clerk Maxwell,* ed. P. M. Harman. Cambridge, Cambridge University Press, 3 v. in 5, 1990-2002.

[Norton 2012] Norton, John D. "Approximation and Idealization: Why the Difference Matters", *Philosophy of Science* 79 (2):207-232 (2012).

[Poincaré 1885a] Poincaré, Henri. "Sur l'équilibre d'une masse fluide animée d'un mouvement de rotation." *Bulletin astronomique* 2, 1885, 109-118.

[Poincaré 1885b] Poincaré, Henri. "Sur l'équilibre d'une masse fluide animée d'un mouvement de rotation." *Bulletin astronomique* 2, 1885, 405-413.

[Poincaré 1885c] Poincaréè, Henri. "Note sur la stabilité de l'anneau de Saturne." *Bulletin astronomique* 2, 1885, 507-508.

[Poincaré 1885d] Poincaréè, Henri. "Sur l'équilibre d'une masse fluide animée d'un mouvement de rotation." *Acta Mathematica* 7, 1885, 259-380.

[Poincaré 1902] Poincaré, Henri, *Figures d'équilibre d'une masse fluide: Leçons professés la Sorbonne en 1900.* Rédigés par L. Dreyfus. Paris, C. Naud, 1902.

[Thomson and Tait 1882] Thomson, William, and P. G. Tait *Treatise on Natural Philosophy*, New (second) edition, Cambridge, Cambridge University Press, 1879-1883.

Henri Poincaré and Hermann Weyl on the Foundations of Mathematics

Jeremy Gray

I first learned of Joe Dauben through his biography of Cantor; it is a pleasure to bring him an essay on two of Cantor's most fascinating critics.

Introduction

Henri Poincaré and Hermann Weyl are two of the mathematicians in the period from the 1890s to the 1930s who thought most deeply about the philosophy of mathematics and the relationship of mathematics to philosophy. They were also, as is well known, two of the most gifted mathematicians of their time who worked creatively in the fields of both mathematics and mathematical physics. This led them to deep, and conflicting, reflections on several topics that are discussed in this essay. One is the precise relationship of mathematics to truth. Are mathematical statements true – and if so, in what sense? Or are they merely proved, according to some agreed system of logical rules? Is there one mode of mathematics for numbers and simple arithmetic, which both men saw as providing certain knowledge, and another for geometry and our knowledge of space? A second topic is the relation of mathematics to physics. For Poincaré the two disciplines were almost inseparable, mathematics being, as he put it, the only language the physicist speaks, but the only reliable parts of physics were the confirmed observations and the proven theorems – the rest was a matter of conventions and theories that, most likely, time would reveal to be flawed. For Weyl, if mathematics could not be confined to the rather small realm of meaningful statements but was essential to the growth of physics then the hope was that somehow the truth of the whole of physics would confer truth on the mathematics it required.

The deepest question for both men was that of explaining why the human mind was capable of doing mathematics at all. Poincaré was close to being a Kantian about arithmetic and famously advocated conventionalism about geometry and, in a different form, in physics. His fundamental state was pragmatic, even sceptical. Weyl all his life sought certainty, even if he was not to find it. He responded more to the appeal of Leibniz, whose vision was of a divine rational knowledge of the world to which humans could at least aspire. As a result, Poincaré and Weyl saw the life of the creative mathematician in different terms. Poincaré called upon the good mathematician to be productive, to find and exploit analogies, and to be prepared to be less than rigorous at times the better to be understood. Weyl was never less than rigorous, he saw the mathematical ego, which has subjective but absolute knowledge, striving to discover truths through action and vanishing in the act of creating a theory that was both objective and relative. Poincaré saw the frailty of human work in the positive light of discovery;

D.E. Rowe and W.-S. Horng (eds.), *A Delicate Balance: Global Perspectives on Innovation and Tradition in the History of Mathematics*,
Trends in the History of Science, DOI 10.1007/978-3-319-12030-0_6

Weyl in more anguished terms. Both knew whereof they spoke.

Historians as well as philosophers have often discussed the foundational ideas of Poincaré and Weyl, both of whom resisted the trend toward modernism associated with Cantor's theory of sets and its first axiomatization in the hands of Zermelo. Poincaré, who was some three decades older than Weyl, began to address foundational issues in mathematics shortly before the turn of the century. His views were published in widely read journals and books, and what he had to say about these matters remained fairly constant throughout his life, though he reacted to new developments as the controversies over foundations gathered momentum. Weyl, on the other hand, grew up mathematically during the period when Poincaré had already established himself as the leading figure of the age. He, unlike Poincaré, would witness the battles over foundations that ensued in the wake of the Great War; indeed, for a time he was one of main figures involved in the fighting. During the course of the 1920s, however, his philosophical views shifted, and they would continue to undergo various subtle changes afterward. These circumstances explain why his views are far more difficult to pin down than those of Poincaré. And for this reason, we will begin this essay with a reconsideration of Weyl's views as expressed in a work that he published twice, though with a time lag of more than twenty years.

Weyl's classic book, the *Philosophy of Mathematics and Natural Science*, is the English edition (1949) of a German original, the *Philosophie der Mathematik und Naturwissenschaft* (1927).[1] Most of the translating was done by a professional translator, Olaf Helmer, two chapters were translated by Hermann Weyl's son Joachim Weyl, and an appendix consisting of six new chapters was written by Weyl, who also extensively altered Sections 13–15, and made some further unremarked changes.[2] One purpose of this essay is to compare some aspects of these two texts. The earlier one was written when Weyl was negotiating his way to a workable compromise between the ideas of Brouwer, which he had so passionately espoused but come to realise put too much mathematics out of reach, and the symbolic mathematics and metamathematics of Hilbert that seemed to deprive mathematics permanently of meaning. The later text offers the reader an opportunity to see how much of his old ideas Weyl now wished to change, and how much he still adhered to.[3] There is a considerable degree of overlap between the two versions, so it will be necessary to survey them in tandem before we can appreciate their differences. The second purpose is to give a more systematic comparison of the ideas of Poincaré and Weyl on mathematics, physics, and the relations between them than has been given hitherto.[4] It is not the case that Poincaré was a

[1]I shall refer to them as *Philosophie* (1927) and *Philosophy* (1949), and drop the dates when the reference is clear.

[2]The appendices, which will not be discussed here for reasons of space, caused him more trouble than the book had, and he often contemplated throwing the manuscript of the appendices on the fire when it was finished; see (Weyl 1955, 211).

[3]For an account of the last phase of Weyl's intellectual life, see (Scholz 2005).

[4]See, however, (Feist 2002), which concentrates on Weyl's appropriation of some of Poincaré's

major influence on Weyl; indeed, Weyl came only gradually to appreciate some of the points his French predecessor had made. Rather, the strengths and weaknesses of each man's thinking allow us to gain an insight into these difficult matters and into the contexts in which they were put forward.

Hermann Weyl: *Philosophie* (1927) and *Philosophy* (1949)

It is well-known that Weyl's intellectual life went through several phases. Starting in 1918, the year he turned 33 and by which time he was a professor at the ETH in Zürich, he began to express his philosophical ideas about mathematics and physics in language drawn from the philosophy of Husserl, which he knew through his wife, who had studied under Husserl in Göttingen. In that year he published the first edition of his *Raum–Zeit–Materie* and his book *Das Kontinuum*, in which he gave a proposal for a meaningful mathematics, in which its objects are accessible to the mind. As a consequence, he rejected all the higher reaches of what he referred to as Cantor–Dedekind–Zermelo set theory and offered alternative foundations of his own. These, as he admitted, could not reach as far as much of modern analysis, including the customary definition of the real numbers, but he thought that there was no alternative other than to abandon what could not be saved.

This radical phase was transformed by his personal acquaintance with Brouwer and his adoption of Brouwer's even more dramatic ideas about the continuum, as famously espoused in his paper (Weyl 1921), in which he proclaimed "Brouwer – that is the revolution."[5] But his enthusiasm for Brouwer's ideas faded as his involvement in what became his best mathematical work—the representation theory of semi-simple Lie groups—deepened. Eventually, he came to feel that Brouwer's ideas would never provide foundations for mathematical physics, which therefore, by Weyl's lights, made that subject impossible.

The *Philosophie* and the *Philosophy* both present two philosophies of mathematics. In so doing, they reflect the ambiguous feelings Weyl had in 1927 and, it would seem, continued to have two decades later. These philosophies – which are present in both editions – share an appreciation of the logical, formal, or axiomatic side of mathematics, but one retains sympathy for his radical ideas while the other turns to the ideas of Hilbert that Weyl had once opposed. Accordingly, we can follow Weyl's two texts noting that the opening accounts of the role of formalism in mathematics, and the way meaning can enter mathematics, largely agree. Next, we can note that the sections on intuitionism and Hilbert's philosophy of mathematics and metamathematics also agree. The crucial differences between the *Philosophie* and the *Philosophy* come in sections 13 to 15, which deal with space and geometry, and the role played by mathematics in the construction of space-time. Here we can see that although Weyl had noted, in the discussion of meaning in both the *Philosophie* and the *Philosophy*, that mathematical structures

ideas about the natural numbers, impredicative definitions, and infinite sets.

[5]See (Weyl 1921, 56). Weyl did not entirely adopt Brouwer's ideas. For an account of the similarities and differences, see (van Dalen 1965).

are determined only up to isomorphism, his appreciation of the implications of this insight for any mathematical account of space deepened significantly between 1927 and 1947.

Formalism and Mathematics

In the opening section ('Mathematical logic, Axiomatics') Euclidean geometry is presented as the prototypical deductive science. Yet while it is instructive regarding the processes of deduction and inference, it does not draw its conclusions impeccably, and so Weyl turned to a modern model, namely the theory of relations. By means of this formalism, for example, one can make inferences about when one number is less than another, or infer that if one proposition happens to be true , then another will be true, and so on. This was a way of presenting mathematics that Weyl had made use of in his Habilitationsschrift (Weyl 1910) and more substantially in his first major venture into the philosophy of mathematics, *Das Kontinuum* (1918).

There, as is well known,[6] Weyl had claimed that the house of analysis was to a large degree built on sand. So he pleaded that its foundations should not be covered with a fake wooden structure of formalism— by which he meant the set-theoretic approach to the continuum associated with the work of Cantor, Dedekind, and Zermelo— and that what could be saved had to be placed on pillars of enduring strength, while giving up the rest. These pillars rested on a view that there are nested families of types of proposition: there are grammatical propositions, some of which may also be meaningful, and among the meaningful propositions some may also be true. The mathematical statements that Weyl proposed as meaningful were those about the natural numbers; true statements about them can be seen to be so, he suggested, by individual acts of the mind. The pillars themselves were constructed by means of relations, and the algebra of relations was what Weyl relied on to extend mathematics beyond the natural numbers and as far as it would go— but no further. He was able to save a considerable amount of mathematics by the use of sequences indexed by the natural numbers, but he could not produce the real number continuum within his theory (only a countable subset of it).

In the *Philosophie*, and again in the *Philosophy*, Weyl provided a quick tour through the machinery of relations. But he dropped the formalism governing the construction of meaningful sentences, and with that omission he, in effect, dropped its function of guaranteeing that valid mathematical statements have a meaning. What was left amounted only to a way of producing valid statements. Then Weyl discussed what he meant by a "constructive mathematical definition," giving among other examples that of a function. As in (Weyl 1910), he emphasised that mathematicians do not say what their objects *are*, but rather they specify what we can say with them or do with them. "Nobody can explain what a *function* is,

[6]The literature here is large; let me merely cite (Feferman 2000) and (Mancosu 1998).

but this is what really matters in mathematics: A function f is given whenever with every real number a there is associated a number b .... b is then said to be the *value* of the function f for the argument value a."[7]

So far, so formal, but at this stage in the *Philosophie* (and also in the *Philosophy*) Weyl presented his updated versions of meaningful mathematics along with his theory of sets:[8]

> With every property $E(x)$ which is meaningful for the objects x of a given category we correlate a set, namely 'the set of objects x having the property E'. ...To say that we know a set means only that we are given a property characteristic of all its elements.

The phrase "meaningful for the objects ...of a given category" was Weyl's way of keeping the far less restrictive demands of Cantor–Dedekind–Zermelo set theory at bay. It implies that the category of objects is surveyable, and therefore at most countable; the key example Weyl had in mind was, of course, the natural numbers, which he had taken as fundamental in *Das Kontinuum.*

He then discussed logical inference and proofs, introduced truth tables, and mentioned that the truth of formally admissible formulae can be determined mechanically (the completeness of the propositional calculus). Then, he added, this is not the case when existential quantifiers—'There exists' and 'For all'—are admitted. Now step by step demonstration is necessary, and at this point he quoted Fries on the "primordial obscurity of reason," remarking that "We do not possess the truth. It wants to be attained by action."[9] It will be worth returning to this remark later when we discuss his attitude toward creation in mathematics.

Next Weyl returned to the axiomatic method, in which, he explained, one attempts to write down the basic concepts and facts from which all concepts and theorems of a science can be deduced. When this is done, this science, following Husserl, is said to be definite. Euclid's *Elements*, he remarked, is unsuccessful in this respect, but the goal of defining all the concepts of Euclidean geometry and proving all its theorems solely from the definitions was reached completely at the turn of the century by Hilbert in his *Grundlagen der Geometrie.* Other than this, Weyl did not have much to say about axiom systems. He observed that they must be consistent, and that preferably the axioms should be independent. He also noted that the consistency of all but the simplest axiom systems is established by means of models. Euclidean geometry, for example, is consistent because Cartesian coordinate geometry shows it to be so; but this has nothing to do with its truth. It is also interesting to know whether an axiom system is complete, where completeness, with respect to pertinent, general propositions, means that if a proposition a is pertinent and general, then one of a and $\neg a$ ('Not a') can be established. If, on the other hand, the notion of completeness is taken to mean the establishment of procedural rules that would resolve this question for every a, then mathematics

[7]Quotations are from *Philosophie* (1927, 8) and *Philosophy* (1949, 8).
[8]Quotations from *Philosophie* (1927, 11) and *Philosophy* (1949, 12).
[9]Quotations from *Philosophie* (1927, 14) and *Philosophy* (1949, 16).

would be trivialised, to which he quickly added that "such a philosopher's stone has never been discovered and never will be discovered." Analogy, experience, and intuition are required.[10]

Weyl next observed that a domain of scientific investigation is determined only up to isomorphism, a concept "of fundamental importance for epistemology." Two domains of objects are said to be isomorphic with respect to a system of relations if the objects may be put in a one-to-one correspondence in such a way that relations hold for the one system of objects if and only if they hold for the corresponding objects of the other system. "Isomorphic domains may be said to possess the same structure," so that "an axiom system can be conceived of as a 'logical mold' of possible sciences."[11] Only in the *Philosophy* (1949) did he discuss systems in which any two models are isomorphic; such axiom systems are said to be *categorical*. Even a categorical axiom system does not allow one to say that it describes a unique set of objects, although any two collections of objects thus described are equivalent for all logical purposes. Non-categorical systems, on the other hand, may admit very different models. In this way, Weyl built up a modern picture of mathematics: a hypothetico-deductive theory of relations indifferent to any interpretation. As we shall see, it is this system that Weyl will later compare with theories in physics.[12]

Intuitionist Mathematics and Metamathematics

In Section 8, Weyl discusses set theory before turning, in Section 9, to Brouwer's intuitionist mathematics. He comments that while modern set theory produces "an almost unbearable awkwardness," the Brouwerian alternative forces the mathematician to "watch with pain the larger part of his towering edifice which he believed to be built of concrete blocks dissolve into mist before his eyes."[13] This last, much-quoted remark is addressed to the reader, and in particular to the mathematician, who is persuaded of the Cantor–Dedekind–Zermelo orthodoxy, but has now read Weyl's attempt to convince him or her of Brouwer's radical insight. Such persons, it seems, were now confronted with the problem that their remarkable subject ultimately was unable to formulate meaningful propositions that the mind could assess.

However, in writing this, Weyl was himself travelling in the other direction; indeed, he was learning to come to terms with an interpretation of the orthodox view that he had spent almost a decade denouncing. In the wake of the "new foundations crisis" (Weyl 1921a), he spent six months searching through philosophical

[10]Quotations from *Philosophie* (1927, 20) and *Philosophy* (1949, 24). This is not to be read as a foreshadowing of Gödel's later results; it was a commonplace feeling among mathematicians at the time.

[11]Quotations from *Philosophie* (1927, 21) and *Philosophy* (1949, 25).

[12]Chapter Two, 'Number and Continuum, the Infinite' differs little from (Weyl 1910) and *Das Kontinuum*; moreover, *Philosophie* and the *Philosophy* differ little as well, so I shall say nothing about it here.

[13]Quotations from *Philosophie* (1927, 44) and *Philosophy* (1949, 54).

writers from Leibniz to Husserl, a search that was surely conducted in part from a wish to find a way out of a philosophical position that had become intolerable for him. So in Section 10 Weyl turned, one may suppose with some regret, to what he called symbolic mathematics: Hilbert's game of symbols and the metamathematics that governed it. Here he presented Hilbert's philosophy as a game with fixed rules that govern the use of empty symbols, but he now conceded that the best one could hope for would be to show that mathematics is consistent, not that it is true. Hilbert's metamathematics was designed, of course, to investigate the rules of the "mathematical game" and to show that they can never lead to a contradiction.

In Weyl's opinion it is only in the realm of metamathematics that Hilbert required the finite meaningful mode of thought, and at this level intuitive inference via the principle of complete induction is allowed. Yet, he still thinks metamathematics has its limitations. Weyl quoted Brouwer's opinion that the possibility of a defence of the rules—the choice of these rules and not others—is rendered unanswerable. Brouwer believed that the only way answers could be found, on Hilbert's plan, was in philosophy, psychology, or anthropology.

Unlike Brouwer, who took no interest in physics, Weyl was unhappy that so much of mathematical physics might be out of reach, and on this point he quoted Hilbert's *Axiomatisches Denken* (1917): "Every potential subject of scientific thought, as soon as it is ripe for the formation of a theory, is bound to fall under the axiomatic method and, therefore, indirectly to the lot of mathematics," to which he added that Hilbert's remark "reminds us that it is the function of mathematics to be of service to the natural sciences."[14] This led to some further remarks about the relationship between mathematics and physics. He pointed out that, because individual propositions in mathematics are meaningless, it can only be the whole of science that contributes any kind of meaning to the whole of mathematics. Symbolic mathematics of this kind Weyl regarded as a priori, and as superior to any natural language. This claim stands in sharp opposition to Poincaré's views about mathematical physics, as we shall see below.

Weyl's hope at the time he wrote *Das Kontinuum* had been that mathematics could be presented as a meaningful activity and that mathematical cognition could be treated on a par with other kinds of human knowledge. He retained this ambition when he transferred his allegiance to Brouwer's approach, but during the mid-1920s gradually relinquished this faith in the face of the demands it made on advances in mathematics. The Husserlian underpinnings remained, however, and they in turn had traces of Kantian ways of thinking. This is clear in Section 11, where Weyl gave a qualified endorsement of Kant's fundamental viewpoint, namely, that construction is essential in mathematics.[15] In this context, he summarised his position by labelling modern set theory a naive realism, noting that Brouwer responded with an idealist position, and that Hilbert's axiomatic for-

[14]Quotations from *Philosophie* (1927, 48, 49) and *Philosophy* (1949, 60, 61).
[15]Quotations from *Philosophie* (1927, 52) and *Philosophy* (1949, 63).

malism attempts to represent the transcendant by appealing to the pure symbol. Regarding the latter, Weyl asserted that the theoretical desire within us for totality teaches us that we can only be content on one condition, namely "that we are satisfied with the symbol and renounce the mystical error of expecting the transcendent ever to fall within the lighted circle of our intuition."[16]

This dense remark responds to the principal thrust of Kant's *Critique of Pure Reason*. Kant had set up the interaction of intuition and understanding as a way to obtain reliable synthetic knowledge. Above understanding sits reason, which has a sound regulative function and follows the laws of logic. Yet reason has a desire for totality that is responsible for unsound, totalising ideas, and it must therefore be held in check. For Kant it was a form of mysticism—and a trend that the scientific turn of mind was to reject for ever—to imagine that reason works constitutively; this was, for him, exactly what was wrong with metaphysics. Weyl agreed; what is transcendent may not be intuited. Mathematical knowledge is not, directly, knowledge of the world. From this lofty height, there followed some rather unremarkable comments on different kinds of geometries, and then he turned to Chapters 13, 14, and 15, the sections of his text from 1927 that were extensively revised for the *Philosophy of Mathematics and Natural Science*.

Revisions concerning Space and Geometry

In the *Philosophie*, Chapter 13, Weyl took up issues associated with the new space-time physics. He began with the observation that when dealing with its constituent parts, one can only distinguish one point of space or space-time from another by means of a coordinate system. As he had done in *Raum–Zeit–Materie* a decade earlier, he adopted a view of coordinates that allowed for Husserlian ideas about intuition:[17]

> But the objectivisation conferred by the exclusion of the ego and its data derived directly from intuition, is not entirely satisfactory: the coordinate system which can only be specified by an individual act (and then only approximately) remains as the unavoidable residuum of the ego's annihilation [das unvermeidliche Residuum der Ich-Vernichtung].

The relativity problem then appears in two parts: first, establish what must be shown for an arbitrary entity (such as a point P) to be determined by introducing a coordinate system, which provides a conceptual definition for these fundamental entities. Second, establish the lawful connections between the coordinates of the same entity in two such systems. Here Weyl gave as an example the geometry on a line. To impose Euclidean, metrical geometry on the real line, one needs to choose an origin and a scale of measure. One then has a coordinate system for every point

[16]Quotations from *Philosophie* (1927, 53) and *Philosophy* (1949, 66).

[17]See *Philosophie* (1927, 57). Although, as Erhard Scholz has pointed out to me, Husserl's 'Weltvernichtung' by the knowledge acquiring subject has been turned into an 'Ich-Vernichtung'.

on the line, and the coordinate transformations $x \mapsto ax + b$ correspond to changes of the origin and the scale. Weyl noted that these transformations form a group.

In Section 14 he presented "Groups of motions as the fundamental idea of geometry" while briefly surveying rigid body motion in the fashion of Helmholtz and the the Helmholtz-Lie space problem. He concluded by noting the surprising realisation that all the relevant groups were already known. The following Section 15—"Riemann's standpoint, analysis situs"—considered Riemann's purely infinitesimal analysis based on a metric of the form $ds^2 = \sum_{j,k} g_{jk} dx_j dx_k$. This, Weyl observed, had allowed Riemann to show that Euclidean geometry is only a special case of within the theory of geometries as three-dimensional Riemannian manifolds. As almost every writer since Helmholtz had gone on to observe, metric homogeneity leads to constant curvature and free mobility, but, said Weyl, Riemann allowed the metric to vary and to be influenced by matter.

That matter affects the metric of a space seems to have been an idea that Riemann somewhat obscurely expressed in his writings on *Naturphilosophie*.[18] But here, as elsewhere in his work on the general theory of relativity, Weyl elided Riemann and Einstein. He argued instead that Einstein's theory of gravity applied Riemann's ideas to space-time—and went on to remark that the metric field makes itself known through its physical effects (on rigid bodies, paths of light, etc.): "But whatever acts must suffer too; it must itself be something real and cannot be enthroned in unattackable 'geometric' rigidity above the forces of matter."[19] This allowed Weyl to conclude that free mobility of bodies without deformation takes place even in inhomogeneous spaces, because the presence of a body will deform the field. We now turn to consider how he dealt with these matters some two decades later.

In the *Philosophy* Weyl made much greater use of the ideas he developed in earlier chapters. Thus, in Section 13 he explained that the mathematical concepts necessary for an objective account of space are:

- an axiomatic science, such as Euclidean geometry after Hilbert (expressed in terms of relations), but any axiomatised domain would do. Individual elements are called points and any subset a figure.
- the automorphisms of the domain. Figures are called similar if there is an automorphism taking one to the other. E.g. all points of Euclidean space are similar. A relation among figures is said to be *objective* if it is invariant with respect to every automorphism of the domain.

Of course, real space does not present itself this way. Neither the axioms nor the basic relations are given to us; we begin rather with a group of transformations. This establishes the degree of homogeneity of the domain. So Weyl proceeded to give a vigorous presentation of the Kleinian view of geometry as the way to analyse a problem geometrically.

[18]For a careful recent investigation, see (Bottazzini and Tazzioli 1995).

[19]Quotation from *Philosophie* (1927, 62) and *Philosophy* (1949, 87).

As the next step, one must define the domain and distinguish points within it. This forces us to employ a coordinate system (once again, the inevitable residue of ego-extinction), and here Weyl repeated the paragraph about geometry on a line. He then discussed classes of frames and conjugate subgroups, noting that geometrical properties are identical with the group invariant properties: these are the objective and relative properties of space. As an example, he noted that when rigid body motion is taken as establishing the concept of congruence, this leads to the invariant theory of the group $\Delta^+ = SO(3)$ of direct Euclidean isometries.

At this point, but only in the *Philosophy*, Weyl made some interesting remarks about the distinction between left and right, noting that:

> Kant finds the clue to the riddle of left and right in transcendental idealism. The mathematician sees behind it the combinatorial fact of the distinction between even and odd permutations. The clash between the philosopher's and the mathematician's quest for the roots of the phenomena which the world presents to us could hardly be illustrated more strikingly.[20]

In the *Philosophy*, when he came to Section 15—now entitled "Riemann's point of view. Topology"—Weyl began right away with n-dimensional manifolds, while explaining carefully that a *differentiable* transformation of the manifold gives rise to a linear transformation on the tangent space. This shift reflects how the concept of a manifold had taken on much greater sophistication during the intervening years. Earlier, it had been customary to treat the transformations of general theory of relativity as both entirely general (meaning merely continuous) and yet sufficiently differentiable, an inconsistency that was finally disappearing. Weyl noted that "While topology has succeeded fairly well in mastering continuity, we do not yet understand the inner meaning of the restriction to differentiable manifolds. Perhaps one day physics will be able to discard it"—though at present, he said, it seems to be indispensable.

Now, but not in 1927, Weyl drew out the idea that the true lawfulness of nature, according to Leibniz's continuity principle, resides in the infinitesimal: "only in the infinitely small may we expect to encounter the elementary and uniform laws." This was a familiar Weylian theme, "near-geometry" takes precedence over "far-geometry", or theories based on action at a distance. He then repeated a passage from 1927 on the metrical inhomogeneity of space, which by interacting with matter conditions the "metric field," calling this Einstein's adaptation of Riemann's ideas to space-time. Writing in the *Philosophy* in 1949, he identified the automorphisms of space-time as the group Ω of all differentiable transformations of the manifold.[21] Both here, and also in the earlier *Philosophie*, Weyl spoke of the metric of space-time varying from point to point as the infinitesimal rigid

---

[20]Quotation from the *Philosophy* (1949, 84). Weyl added that he regarded the distinction between odd and even as the second most fundamental fact in mathematics, after counting.

[21]'Diffeomorphism" is the modern term, but it was taken up later.

Euclidean cubes soften into fluid Riemannian ones. Weyl's use of a "Flüssigkeit" (fluid) surely reflected his field-theoretic sensibilities.

It had been one of Weyl's successes in (Weyl 1921b, 1923) to show that the transformations in Ω split into a translational part that moves points in space-time determining all possible transformations of the coordinates, and a rotational part, in which Euclidean rotations are anchored at each point of space-time. In 1929, he had also been able to rework, and reconceptualize, his unsuccessful gauge theory of the electro-magnetic field in order to apply it to Dirac's relativistic field theory of the electron. Here, Weyl noted, but did not explain the connection between the so-called group of spinors and the electron wave field.[22] He also raised the question as to why the Euclidean group turns up, but without giving an answer.

From this point on, the two texts begin to agree again. In Section 17 ("Subject and object: the scientific implications of epistemology"), Weyl aimed to show that a series of insights, going back to Locke's distinction between primary and secondary qualities, had now reached the point where one could regards the notions of space and time as secondary qualities within the symbolic construction of space-time. This led Weyl to consider the familiar problem of idealism: How do we construct the world from our subjective impressions?

In the general theory of relativity, he argued, an objective world is constructed from what is given in intuition by a symbolic construction in the fashion of Hilbert. The general theory of relativity has enabled physicists to emancipate themselves entirely from intuitive space and time in the construction of the objective world. One can say that in the objective world, the instantaneous light cone of the moving observer "crawling upward along the life line of their body" determines what is available to their consciousness and brings a section of the world fleetingly to life; the objective world, however, "simply *is*, it does not *happen*."[23] The distinction Weyl now drew between what he called the subjective-absolute and the objective-relative, struck him as particularly important: this was his way of at once separating lived experience from the impersonal accounts of it that form physics while bridging the gap between idealism and realism.[24]

There are rules governing the construction of the objective world. Following views he attributed to Lambert and Helmholtz, Weyl insisted that"a difference in perceptions is always founded in a difference in the real conditions," and so the objective world must be constructed in such a way that differences in it manifest

[22] See (Scholz 2004) for an account of the revised gauge theory and its differences with the earlier attempt.

[23] Quotations from *Philosophie* (1927, 82) and *Philosophy* (1949, 115).

[24] As noted above, the body-ego, or its residue in the form of a coordinate system, is the place that both has the experience of space and time and transforms it into an objective, scientific account. The instantaneous experience of the ego is, for Weyl subjective and absolute: subjective because it is that ego's experience and cannot be the experience of any other ego; absolute because it cannot be denied (once deceptions are excluded). The scientific account is objective and relative because it is presented as the account from one point of view (in the coordinates of the ego's system) that can be transformed into the account (and coordinates) of any other ego.

themselves in differences in perceptions.[25] As a consequence, there should be no natural processes in the objective world which remain forever undetectable. As an example, Weyl discussed how one should not interpret the Schwarzschild metric. One could, he said, try to claim that space in the region around a star is Euclidean, though measuring rods it the star's vicinity behave in a strange way (why not?, he asked, after all, they change with temperature).[26] But to this Weyl countered that such an interpretation introduces a principle that contains "an arbitrary element that has no perceptually performable consequences and must therefore be eliminated."[27] Indeed, this elimination was accomplished by Einstein insistence that physical geometry is determined simply by what the measuring rods indicate (in this case, the Schwarzschild metric describes a space that is not Euclidean). But we also construct space as infants in a much more elementary way, and Weyl next turned to this much-discussed topic.

Our concept of space is constructed from our sensations. In fact, we construct several spaces—by sight, by touch, and by our motions. The two-dimensional, purely visual field, coupled together with eye movement, constitute "the centered three-dimensional space in which the body of the ego finds its position, though still the distinguished position of the centre."[28] Indeed, only now does a body-ego occur. The senses of touch and the motion of the limbs also fit in at this level. Weyl referred approvingly to Husserl's observation that all these things, alien to "true," "objective" space, are essential necessities (*Ideen* (1916), 315). Our conception of homogeneous space arises as a result of our own free motibilty, which allows us to recognise our body as an equal among other spatial objects: "Only this space can be conceived as one and the same for several subjects; it is the presupposition for the construction of the intersubjective world."[29] Here Weyl referred positively to Poincaré's mental compensation theory based on the notion of rigid bodies. As pointed out in (Sieroka 2009), consciousness and mental activity are not passive for Weyl, who agreed with Kant and Fichte that "I am originally not only sentient but also intuiting" and therefore "the penetration of the This (here-now) and the Thus is the general form of consciousness."[30]

It is interesting to see that Weyl believed the general theory of relativity ascribes both *a priori* and *a posteriori* properties to space-time. He felt that Helmholtz had largely succeeded in separating the Kantian idea of space as a pure form of intuition from the conclusion that Euclidean geometry holds *a priori*. Yet, from a modern perspective, an axiomatic account of "mathematical space" distinguishes it from "physical space," where one is concerned with an ordering scheme for the real things integral to the theoretical construction of the world. This led him

[25]Quotations from *Philosophie* (1927, 83) and *Philosophy* (1949, 116).

[26]Surely he had Poincaré's ball model of non-Euclidean geometry in mind here, which was explained in precisely these terms, see (Poincaré 1895).

[27]Quotations from *Philosophie* (1927, 84) and *Philosophy* (1949, 118).

[28]Quotations from *Philosophie* (1927, 92) and *Philosophy* (1949, 128).

[29]Quotations from *Philosophie* (1927, 92) and *Philosophy* (1949, 129).

[30]Quotations from *Philosophie* (1927, 93) and *Philosophy* (1949, 130).

to make some remarks about the principles of theoretical construction, from which we learn that he saw Poincaré's conventionalism as the most extreme standpoint. Weyl eventually concluded that among the *a priori* features of space-time must be counted the absolutely given Euclidean–Pythagorean *nature* of the metric, and its topological connectivity and dimension. Among its *a posteriori* features are the quantitative course of the metrical field, which is dependent on the distribution of matter.[31]

In summary, we can see that for Weyl in 1927, and even still in 1949, there are two main approachs to the philosophy of mathematics: the pure intuitionistic mathematics of Brouwer, or Hilbert's formalist game, played according to metamathematical rules. They are distinguished, however, on philosophical grounds —no longer did he decry formalism as a meaningless game—and in scope. The most significant difference between Weyl's *Philosophie* and his *Philosophy* is that in 1949 he gave a more thorough account of the role played by mathematics in shaping the construction of space and space-time. An ultimately axiomatic mathematics deals with objects known only up to isomorphism, and in the case of space-time the mathematical apparatus of diffeomorphisms of a four-dimensional manifold was found to be sufficient to allow for the construction of a homogeneous objective-relative world. It would also seem, as Scholz and Sieroka have noted, that the influence of Fichte and Leibniz on Weyl increased, whereas that of Husserl declined, as reflected in the role of the ego as a more active agent in Weyl's later work.

A Comparison with Poincaré's Philosophical Views

Writing *Das Kontinuum* seems to have been one occasion when Weyl began to consider some of Poincaré's views. Yet, as he himself said, it was only after finding his own way out of the vicious circle of foundational confusion that he discovered the ideas of Frege and Russell, and "Poincaré's very uncertain remarks about impredicative definitions." At this point, he went on to say, "I became firmly convinced (in agreement with Poincaré, whose philosophical position I share in so few other respects) that *the idea of iteration, i.e., of the sequence of the natural numbers, is an ultimate foundation of mathematical thought*."[32]

In the mid-1920s Weyl, as we have seen, made a serious investigation of different philosophies of mathematics, a study by no means restricted to Brouwer's ideas: he drew on a German tradition that extended, as he saw it, from Leibniz to Husserl. Poincaré, on the other hand, had much less of a French tradition to draw upon, and although he was influenced by some of his brother-in-law Émile Boutroux's ideas, and had some acquaintance with the ideas of Kant, he was a more isolated figure when it came to philosophical matters. Nevertheless, both men

[31] Weyl also speculated about how these *a priori* possibilities could arise.

[32] *Das Kontinuum* pp. 47–49, Weyl's italics. Feist (2002, 287) notes how Weyl gave the distinction between the natural numbers and reasoning by recurrence a Husserlian setting.

took philosophy seriously; they expected it to shape their work and not simply conform to it.

It is easy enough to list some of the points of close similarity between the ideas of the two men. Famously, both were willing to put their trust in the natural numbers as mathematical objects intelligible to the human mind, and neither was willing to accept the new set-theoretic formulation of mathematics. Both placed considerable emphasis on physics and the broader implications this had for mathematics due to its central role in scientific work. As for geometry, the oldest of the natural sciences, both were comfortable in general with the modern notion of implicit definitions, but they also sought more than axiomatics alone could deliver, namely a system of geometry that might account for our experience of space and time.

It is just as easy to state some significant differences, some of which can be traced to the fact that they came from two different generations. Weyl lived through, and helped advance, the great transformations in physics that took place during the early decades of the twentieth century—the general theory of relativity and quantum mechanics—whereas Poincaré sensed these were coming but did not live to see these revolutionary developments. Indeed, Weyl was strongly drawn to general relativity, whereas Poincaré only knew the special theory and had an ambiguous relationship with it. Moreover, Weyl alone lived to see Hilbert's metamathematics and the mainstream responses to the paradoxes of set theory as well as Brouwer's radical alternative; we simply do not know whether or how Poincaré might have adapted his ideas to the mathematical and scientific world of the 1920s.

Among the intriguing points of comparison, however, is the nature of human agency in the practice of mathematics. This issue, often forgotten in discussions of the philosophy of mathematics, was central to both men's views, although in quite different ways. Another, though somewhat related, aspect concerns their attitude toward rigour and its role in the acquisition of knowledge. Let us address this issue first. For Poincaré, rigour was the least interesting part of proof; what mattered to his mind was the role a proof plays in understanding a piece of mathematics. Of course, as he explicitly said, mathematics has to be rigorous. As he put it in his address on the future of mathematics: "Rigour is not everything – but without it there is nothing." But rigour can nonetheless be obstructive. Proofs can easily become unwieldy, and rigorous proofs must be prevented from becoming almost incomprehensibly long. Moreover, calculations should be held to an irreducible minimum, and they should never be blind.[33]

Proofs can also be wrong in kind, as Poincaré felt they were in potential theory. The rigorous arguments for the existence of solutions, some of which he had produced himself, depend on convergence arguments that cannot yield effective

[33]For the quote, see Poincaré 'L'avenir des mathématiques' (1908), in *Science et méthode* (1908, 22). The most accessible text and its English translation differ markedly from Poincaré's text; for a full translation and commentary, see (Gray 2012b). Moreover, there is no standard edition of Poincaré's essays, which makes page references to them inevitably erratic.

numerical procedures. Moreover, they do not mimic the actual processes involved, whereas the arguments that did, the more-or-less intuitive proofs based on an appeal to Dirichlet's principle, had little value for the mathematician, although they could well satisfy physicists because they left the mechanism of the phenomena transparent. More generally, Poincaré felt that rigour alone was not all one should ask of an argument. For, on occasion, there were legitimate demands from philosophy that a purely formal view of mathematics cannot respond to. Poincaré thus pointed out in 1905 that even though Hilbert had uncovered the formal character of reasoning in geometry, this did not mean that formalism could be extended to all mathematical knowledge. Indeed, even if the same could be done for arithmetic and analysis this did not imply that all of mathematics could somehow be reduced to a purely logical system without mutilating it. After all, he insisted, the origin of the axioms would still have to be investigated, however conventional they were taken to be.[34]

This leads to the question of agency and the active part played by the mathematician. For Poincaré, as we noted, mere accuracy and logical correctness were not the crucial elements in a mathematical argument. A lengthy calculation may lead to a striking result, but this was not satisfying, he said, until we understand at least why the characteristic features of the result could have been predicted. And because it is not order per se, but only unexpected order that has a real value, the mechanical pursuit of mathematical knowledge would be an almost pointless pursuit; or, as he put it in a striking phrase, *"A machine can take hold of the bare facts, but the soul of the fact will always escape it".*[35] Isolated facts had no appeal for Poincaré, but a class of facts held together by analogy brings us into the presence of a law, and following Mach he asserted that "The importance of a fact is measured by the return it gives – that is, by the amount of thought it enables us to economise." This led him to opinions that some have regarded as aesthetic. The elegance of a good proof, he said, reflects an underlying harmony that in turn introduces order and unity, and this "enables us to obtain a clear comprehension of the whole as well as its parts. But that is also precisely what causes it to give a large return."[36] So this aesthetic response to mathematics was taken by Poincaré to be a sign of its efficacy. Evidently he saw the latter as the more potent virtue, rather than beauty, although this pair of ideas—elegance and efficiency—clearly went hand in hand.

Weyl did not express his opinions about agency as a reflection on rigour. In his *Das Kontinuum*, and in more Brouwerian terms in (Weyl 1921a), he posed a stark dichotomy between humanly intelligible mathematics and a meaningless and valueless version, a formal and 'fake' mathematics. The latter, he forcefully asserted, had to be abandoned. When he later advocated the Hilbertian system of mathematical games regulated by metamathematics, however, he never discussed

[34]See (Poincaré 1905) reprinted in *Science et méthode*, (1908, 158).
[35]*L'Avenir*, 1908, quoted in *Science et méthode*, p. 27.
[36]*L'Avenir*, 1908, quote in *Science et méthode*, p. 23.

the implications of trying to work literally in that way. That proofs might be too long, wrong in kind, or unproductive, these were matters of no real concern to him; the clear implication is that, for Weyl, a mathematical argument simply had to be rigorous. Yet, in his view of mathematics, the ego is nonetheless actively converting the subjective-absolute into the objective-relative. Mathematics, for him, is not merely a network of definitions, theorems, and proofs, woven together by a correct use of the theory of relations and simple logical inference, however much it might seem that way in the opening pages of the *Philosophie* and, even more so, in the *Philosophy*.[37] The difference in the way he and Poincaré expressed their various concerns reflects, in fact, the very different contexts in which they worked in several ways.

Poincaré wrote not only for mathematicians but also for more general audiences. His essays appeared in magazines with a wide circulation, and there he was concerned with mathematical issues that both illuminated but also justified its pursuit to this readership, some of whom surely thought of mathematics as an arid subject. Poincaré was therefore keen to present it as a creative subject, but also a useful one, a field of inquiry that addressed unexpected, but palpable concerns that non-mathematicians could identify with. Weyl wrote initially for mathematicians, then for a multi-volume German encyclopedia, and only after arriving in the United States in the early 1930s did he address more general, non-specialist audiences.[38] After he gave up trying to get mathematicians to change their ways, he still took it for granted that they would want to be rigorous in their work; his struggle was thus far more a personal one that mainly involved his coming to terms with Hilbert's approach to foundational issues. It is also true that Weyl personally had no affinity for mathematics that was not rigorous. He did not see it as his business to offer approaches that fell short of complete correctness, whereas on several occasions Poincaré did. Indeed, Poincaré was often informal even in his strictly mathematical papers because he wanted his readers to appreciate how he was thinking and to give them the chance to think along with him. He advocated conducting research by pursuing the right analogy, and analogies are inevitably vague, often at the crucial point. This is not to endorse the popular, though incorrect view of Poincaré's work as frequently unrigorous, but only to note that he introduced less than rigorous arguments into his papers for a purpose. This was something Weyl never did.

All philosophies try to keep psychological issues at bay. They claim in various ways not to be interested in actual human minds with all their oddities and failures, preferring to discuss cognition, or any other issue, at a level of depth and generality that would apply to any other intelligence in the universe. Their mathematician, if he or she enters their purview at all, has an arbitrary large memory, an instant

[37]Much that is wrong with philosophy of mathematics today stems from a view of mathematics as an already existing body of knowledge of a perhaps obscure kind, a tradition that derives from the successful work in mathematical logic of Gödel and Tarski.

[38]Somewhat exceptional were his several essays written for *Die Naturwissenschaften*, but in most cases these were still somewhat technical and, hence, not intended for a broader readership.

response time, and infallible logical precision. Poincaré and Weyl largely share in this approach. Curiously, Weyl is more intense about the place of the creative mathematician, but ultimately has less to say, whereas Poincaré has more advice to offer.

It is well-known, and has been noted above, that the opening pages of Weyl's *Raum–Zeit–Materie* present coordinate systems in the general theory of relativity as the residue of ego-annihilation. But in the course of having experiences and creating knowledge, this almost extinct ego acts: it intuits, it suffers. The language is dramatic, as so often with Weyl, but other egos caught up in this drama receive little advice. Poincaré, on the other hand, does not speak to the anguish of the mathematician. If anything, doing mathematics is almost too easy; the task is to do good mathematics. The aesthetic component is not suppressed, but rather rapidly subjected to the requirement that it comes with understanding, which enables a person to do more mathematics. This is partly why Poincaré wrote in his discursive manner: to enable others to acquire an overview of the subject that gives it its meaning.

Weyl, as we saw above, hoped that physics would give meaning to mathematics, just as mathematics would bring organisation to physics. How this interplay was meant to operate was, unfortunately, never precisely articulated. Each individual mathematical statement, being a combination of symbols, was in itself meaningless. Taken together, the logical form they create, the theory, could itself be understood as merely a logical mold in which the objects are determined only up to isomorphism. Among the objects described in *Raum–Zeit–Materie*, for example, are the length and inertia fields, which direct the motion of matter in the universe. These fields are not detectable directly, so we know them only through their effects. How, then, do the truths of physics, or any science, permeate the body of mathematics? Weyl expressed the problem this way in *Raum–Zeit–Materie*: "The problem of space is at the same time a very instructive example of that question of phenomenology that seems to the author to be of greatest consequence, namely, in how far the delimitation of the essentialities perceptible in consciousness expresses the structure peculiar to the realm of presented objects, and in how far mere convention participates in this delimitation."[39]

In his later writings, Weyl moved away from phenomenology and tried to grasp the problem of space from a different angle. Consciousness of space-time has experiences that it also begins to process, through the use of coordinates, until it arrives at numbers or measurements that can be compared with those from other conscious experiences, and the invariants under the appropriate automorphisms are the objective facts. This Kleinian approach makes the place of mathematics clear, but this gives the mathematician no further guidance. The Hilbertian view, according to which every well-grounded branch of science would fall to the axiomatic method, was not only naive, it failed even to account for the development of the general theory of relativity. It would seem, in fact, that Einstein's own

[39] *Raum–Zeit–Materie* , English translation of the 4th edition (1952, 148).

path to that theory, and Weyl's prompt exploration of it, conform rather better to Poincaré's views on the relationship between mathematics and physics. He had argued that mathematics and physics are inseparable, and wrote that "the first reason why the physicist cannot give up mathematics is: it provides him with the only language he can speak."[40] This was not a simple endorsement of a view popularly attributed to Galileo, according to which the book of nature is written in the language of mathematics. Indeed, for Poincaré, science and mathematics were constantly linked in discourse.

Research in physics typically ends, sooner or later, in what is called a law of physics, and Poincaré suggested that this process of discovery generally involved three assumptions. First, the researcher assumes that a physical process can be related to factors that can be traced to the immediate past. Second, the explanation focusses on small, phsically accessible regions of space. Third, these localised analyses usually lead to fairly simple quantitative relationships; typically they are expressed mathematically by a differential equation. The success of mathematical physics can then be explained by the fact that the same well-understood type of mathematics is involved in many different branches of this science.[41]

In this way, he argued, the physicist begins to distil physical laws from experimental work. However, all is not so simple: experiments are particular, but laws are general, and experiments are approximate, whereas laws are precise. To pass from one to the other requires an appeal to analogy, and Poincaré regarded this capacity to formulate such analogies as the real contribution of mathematics to physics. "The aim of mathematical physics," he wrote, was "to reveal the hidden harmony of things ...", by which he meant a harmony of relations between facts (experimental results and mathematical theorems).[42] This harmony he regarded as lying beyond any mind that conceives it; rather "what we call objective reality is, in the last analysis, what is common to many thinking beings; ...this common part, ...can only be the harmony expressed by mathematical laws. It is this harmony then which is the sole objective reality."[43] So for Poincaré the "universal harmony [which] is the source of all beauty" is a condition for the intelligibility of the world.

Poincaré also considered the extent to which these laws might be arbitrary. He argued that there was a creative element in the formation of a scientific law (particularly those laws that have been raised to the status of principles). In his view, as expressed in (Poincaré 1900), the equality of action and reaction, the definition of force, and other fundamental laws have the status of first principles. This means they could no longer be viewed in purely empirical terms: such laws

[40]See (Poincaré 1897, 141).

[41]It is interesting to note that Poincaré believed that the range of topics accessible to this approach was running out, and the student of natural science would be forced to find other modes of generalisation.

[42]See (Poincaré 1897) reprinted in *La valeur de la science* 1905, p. 147, English translation (Poincaré 2001, 276).

[43]See the introduction to *La valeur de la science* 1905 p. 9, English translation (Poincaré 2001, 191).

are not increasingly well confirmed experimental results—for as such they would be open to refutation by new experiments—rather they have been elevated to the status of conventions. That is to say, they are rules that we agree to abide by when conducting work in the relevant branches of physics. Indeed, on those rare occasions when the truth status of a fundamental scientific laws might be brought into question, substantial disagreements[44] will not be settled by appeal to empirical evidence but rather by means of conventions. For Poincaré, principles are neither true nor false, they are conventional and convenient (Poincaré 2001, §4).[45]

Moreover, it was Poincaré, far more than Weyl, who emphasized that theories change and die, and that this phenomenon constituted a serious challenge for any philosophy of science. It is not difficult to talk about how scientists got something right; it is far harder to explain how, although they got it wrong, they can be trusted to do better next time. Poincaré believed that what survived in the transition from one theory to the next were the relations in which the theory was expressed, which is the reason why his name is often invoked in structural realist accounts in the philosophy of science.[46] Yet when this harmony of relations between facts seemed to break down, and one contemplated the melancholy remains of failed theories, he argued that:[47]

> The equations in which the old theories were expressed are still true, the relations they capture preserve their reality. But as for what is related, these are merely names of the images we substituted for the real objects which Nature will hide for ever from our eyes. The true relations between these real objects are the only reality we can attain, ...When theories seem to contradict each other, it is likely that it is the images we have supplied which stand in contradiction.

Here Poincaré gave the examples of billiard ball atoms and the fluids of Coulomb, once old-fashioned "yet here they are re-appearing under the name of electrons."

As a result of this dynamic view of science, Poincaré seldom spoke about truth in science. He preferred to argue that we rely on the testimony of experts and on a shared communication with others, furthermore that we speak a shared family of languages, natural, scientific, mathematical which work because of a shared set of conventions, and finally that we have ideas about what we would do if our statements conflict or communication failed. As he put it: "No discourse —no objectivity."[48] For this reason, already in his day Poincaré was frequently accused of scepticism or idealism (as opposed to realism). The implication is that

[44]For Poincaré on the advent of Minkowski's space-time see his (1912).

[45]Poincaré was by certainly not the first to consider the status of the so-called laws of physics, and specifically of mechanics; Mach was also critical of loose talk of these 'discoveries'.

[46]I have argued for a language game interpretation of Poincaré's ideas in (Gray 2012). Poincaré's conventionalism is surely much more akin to a Wittgensteinian language game than a structural realist position.

[47]See (Poincaré1900), in *La Science et l'hypothèse*, 174.

[48]See (Poincaré 1902b) in *La Valeur de la Science*, 262.

he was a relativist about truth, and would allow faultless disagreement, and indeed he would. His refusal to adopt Minkowski's space-time (see (Poincaré 1912)) at the end of his career and his geometric conventionalism—his explicit argument that the decision between Euclidean and non-Euclidean geometry is a matter of convention—are all evidence of that.

But Poincaré was not a sceptic when it came to mathematics, as his views on the natural numbers attest. In his view, reasoning by recurrence is an activity of the mind that is indubitable, creates knowledge, and therefore allows for the growth of knowledge.[49] On these philosophical grounds, Poincaré believed that we *know* what reasoning by recurrence is in an almost Kantian fashion. If we now recall that, for Poincaré mathematics and physics are inseparable, we confront a tension in his thought that conventionalism in physics tended to reduce by locating an arbitrary element in the production of scientific theories. But, as we have seen, Poincaré's views on proof also had a creative element. Poincaré saw mathematics as a practice, and mathematical understanding as enabling. A good proof rested on general ideas that are capable of wider application (by analogy). In mathematics it explained why some things are the case; in physics it dealt in relations that could survive changing beliefs or practices about objects. Thus, to summarise one of his arguments in "L'Avenir," for Poincaré a good proof is a new and valid use of the terms it involves.

The nearest Weyl came to discussing some of these issues was in his Terry lectures on "The Open World" at Yale University in 1930. There he said that a scientific law holds not between events but between types; it may only be revealed after the perceptual qualities of things have been resolved and 'concealed characters' revealed; and it passes from the given and special to the general and ideal—what Weyl called the constructive generation of the possible. Moreover, a useful law must be mathematically simple enough for us to use it, and it should continue to hold in new circumstances (such as much greater levels of accuracy).

As for whether such laws are found or invented, Weyl claimed that:

> Constructive natural science is confronted with the general problem of assigning to objects such constructive characteristics that their behavior under circumstances described by the same kind of characteristics is entirely determined and predictable by means of the natural laws. The implicit definition of the characteristics is bound to these laws. The fact that we do not find but enforce the general principles of natural knowledge was particularly emphasized by the conventionalism of H. Poincaré. But I believe one may also consider the hastily sketched developments just completed as an interpretation of Kant's doctrine of the categories.[50]

Still, the idea that physics and mathematics were engaged in a perpetual discourse is wholly lacking in Weyl's work. The closest he came to grappling with

[49]See (Poincaré 1902a) in *La Valeur de la Science*, p. 21.
[50]See (Weyl 2009, 58).

the inter-subjective world was, as we saw above, in discussing how we acquire a sense of the homogeneity of space. Otherwise, he seems to have adhered to the view that somehow the best science always surfaces,[51] and that it survives until it is refuted. Even his unsuccessful first introduction of gauge–theoretic ideas in 1918, and their later more successful reworking in 1929, failed to prompt any reflections beyond the usual confrontation of a mathematical interpretation with experiment.

Weyl's remarks about "near-geometry" in the *Philosophy* may seem close to Poincaré's views on mathematical modelling as focussing on localized phenomena, but those remarks were not in themselves very original. The more significant point is that Weyl went back to Leibniz. As has been pointed out elsewhere (see, for example (Scholz 2012)), Leibniz was a major influence on Weyl, and he is cited more than any other author in (Weyl 1949). With the steady publication of his work in the opening decades of the 20th century, Leibniz came to be seen as a thinker who posed a serious alternative to Kant rather than a philosopher whose ideas had been surpassed. Weyl's increasing interest in Leibniz reflects this trend, and in the context of "near-geometry" he claimed that the laws of physics must be about the behaviour of infinitely near points, and not merely on pragmatic grounds. It is Leibnizian, too, even to believe that there are laws of physics. Leibniz believed that what seems contingent and synthetic to us, due to our limited minds, was analytic to God. If we look closely at the lectures on 'Mind and nature' we see that Weyl invoked God as the ultimate reason for the mathematical lawfulness of nature, and devoted two lectures to the idea that "recognition of the world as it comes from God ...can be gained only by symbolical construction."[52] So although Weyl took a constructive view of laws in physics, he was also able to contemplate the idea that there are fixed, eternal laws of physics, some of which we may have been able to obtain, although more accuracy is always desirable. Poincaré, on the other hand, saw only conventions which, ultimately, might have to change.

Weyl, as we noted above, was willing to declare some aspects of the physics of space-time *a priori* and others *a posteriori*. In particular, the geometry of space-time was given locally by a varying Minkowskian metric, the precise nature of which was determined by the distribution of matter. Poincaré, with only two alternatives to consider (Euclidean and non-Euclidean geometry), had famously argued that there was no way a decision could be reached about which was the true geometry of space. Weyl, faced with the infinite flexibility of Riemannian–Einsteinian geometry, had been forced to give a reason why the behaviour of measuring rods unambiguously determines the nature of space-time.

The final comparison I wish to make concerns the use of language. Poincaré wrote elegant, though somewhat impersonal French prose. Even though he was expressing personal opinions, he seldom took a strong partisan stance when arguing against those who held contrary views. He made his points elliptically, so that it is possible to misunderstand him or even to think that he has nothing substantial

[51]As when he quotes Einstein's remark to Planck in 1918 that evolution has shown that there is always one theory far superior to all the others, see "Mind and nature" (Weyl 2009, 115).

[52]See (Weyl 2009, 50).

to say, a quite mistaken view, though Minkowski expressed a similar opinion in a letter to Hilbert (4 May 1908). It is, however, true that Poincaré was not entirely consistent, and he did not always think his arguments through or explore issues in depth.

Weyl was brought up in a more rigorous tradition, and he applied himself with more determination to overtly philosophical problems. What is striking, however, is the number of times he—who did not see himself as a leader of men and yet struggled with the burden of knowing that he was practically predestined to become Hilbert's successor—adopted minority positions which led him to voice strong criticisms of others.[53] His remarks in *Raum–Zeit–Materie*, *Das Kontinuum*, and the (1921a) espousal of Brouwer's foundational ideas are of this kind, but they are so well-known that they need not be repeated here. Rather, we may note the Weylian manner in which the ego, the consciousness, acts, and is treated in return, on its way through life. As quoted above, in Weyl's view (quoting Fries) we are confronted with "the primordial obscurity of reason—we do not possess the truth. It wants to be attained by action." This is not the researcher's habitual lament that doing good original work is difficult. It is offered here as an archetypal and painful truth, and a hope at best.

Even more striking is the remark that occurs only in the *Philosophy* (1949), where Weyl wrote that the life of the mind "is light and darkness, contingency and necessity, bondage and freedom, and it cannot be expected that a symbolic construction of the world in some final form can ever be detached from it." It would take another paper to explore the differences between Poincaré's elegant concerns and Weyl's gravity, but it is clear that both men saw mathematics as a life of action, in which one strives to do new and valuable things. Poincaré's charmed life brought him few defeats, but Weyl had to abandon the hopes that had sustained him in his thirties. Both men offer the prospect of thinking through the creation of mathematics and physics as uncertain, existential acts of commitment that historians and philosophers today, well after they wrote, could profitably embrace.

Acknowledgements

I would like to thank my hosts and the participants in several seminars at Sphere, Université Paris–Diderot, for the opportunity to present talks and join discussions that helped me write this paper. David Rowe's advice has been particularly helpful in shaping the presentation of this paper. The literature on Weyl is large and expert, and several books and articles have had an impact on this paper without being cited explicitly. I would like to note here the helpful writings of Paolo Mancosu, Richard Tieszen, Thomas Ryckman, and Erhard Scholz.

[53]On Weyl's complicated relationship with Hilbert, and how this intertwines with Weyl's philosophy and his physics, see (Rowe 2003).

Bibliography

[Bottazzini and Tazzioli] Bottazzini, Umberto, and Tazzioli, Rossana, 1995. *Naturphilosophie* and its Role in Riemann's Mathematics, *Revue d'Histoire des Mathématiques* 1: 3–38.

[Feferman 2000] Feferman, Solomon, 2000. The Significance of Hermann Weyl's *Das Kontinuum*, in *Proof Theory: History and Philosophical Significance*, V. F. Hendricks, et al., eds., Dordrecht, Kluwer, pp. 179–194.

[Feist 2002] Feist, Richard, 2002. Weyl's Appropriation of Husserl's and Poincaré's Thought, *Synthese* 132: 273–301.

[Gray 2012a] Gray, Jeremy J., 2012a. *Henri Poincaré: a Scientific Biography*, Princeton University Press.

[Gray 2012b] Gray, Jeremy J., 2012b. Poincaré replies to Hilbert: on the Future of Mathematics, ca. 1908, *Mathematical Intelligencer* 34: 15–29.

[Hilbert 1917] Hilbert, David, 1917. Axiomatisches Denken, Vortrag gehalten in der Schweizerischen mathematischen Gesellschaft, am 11. Sept. 1917 in Zürich, *Mathematische Annalen* 78: 405–415.

[Mancosu 1998] Mancosu, Paolo, ed., 1998. *From Brouwer to Hilbert. The Debate on the Foundations of Mathematics in the 1920s*, Oxford University Press.

[Poincaré 1895] Poincaré, Henri, 1895. L'espace et la géométrie, *Revue de métaphysique et de morale* 3: 631–646; modified in *La Science et l'hypothèse*, Paris, Flammarion, 1902.

[Poincaré 1897] Poincaré, Henri, 1897. Sur les rapports de l'analyse pure et de la physique mathématique, *Acta Mathematica* 21, 331–341; in *La Valeur de la Science*, Paris, Flammarion, 1905, 136–155.

[Poincaré 1900] Poincaré, Henri, 1900. Sur les rapports de la physique expérimentale et de la physique mathématique, *Rapports présentés au congrés international de physique*, C.-E. Guillaume and L. Poincaré, eds., vol. 1, Paris, Gauthier-Villars, pp. 1–29; reprinted as 'Les hypothèses en physiques' and 'Les théories de la physique moderne' in *La Science et l'hypothèse*, Paris, Flammarion, 1902, pp. 157–172, 173–190.

[Poincaré 1902a] Poincaré, Henri, 1902a. Du rôle de l'intuition et de la logique en mathématiques, in *Comptes rendus du IIe Congrès international des mathématiciens*, E. Duporcq, ed., Paris, Gauthier-Villars, pp. 115–130; reprinted in *La valeur de la science*, Paris, Flammarion, 1905, pp. 11–34.

[Poincaré 1902b] Poincaré, Henri, 1902b. Sur la valeur objective de la science, *Revue de métaphysique et de morale* 10: 263–293; reprinted with modifications as 'La science est-elle artificielle?' and 'La Science et la Réalité' in *La valeur de la science*, Paris, Flammarion, 1905, pp. 213–247 and 248–276.

[Poincaré 1905] Poincaré, Henri. 1905. Les mathématiques et la logique, *Revue de métaphysique et de morale* 13: 815–835; reprinted with modifications in *Science et méthode*, Paris, Flammarion, 1908, pp. 152–171.

[Poincaré 1908] Poincaré, Henri, 1908. L'avenir des mathématiques, *Revue générale des sciences pures et appliquées* 19: 930–939; partial reprint in *Science et méthode*, Paris, Flammarion, 1908, pp. 19–42, and in English trans., The Future of Mathematics, *Monist*, 20, 1910, 76–92 and in *Science and method.*

[Poincaré 1910] Poincaré, Henri, 1910. Über transfinite Zahlen, in *Sechs Vorträge über ausgewählte Gegenstände aus der reinen Mathematik und mathematischen Physik, Mathematische Vorlesungen an der Universität Göttingen*, Leipzig, Teubner, pp. 33–41.

[Poincaré 1912] Poincaré, Henri, 1912. L'espace et le temps. *Scientia (Rivista di Scienza)* 12, 159–170; reprinted in *Dernières Pensées*, Paris, Flammarion, 1913, pp. 97–109.

[Riemann 1990] Riemann, Bernhard, 1990. Naturphilosophie, in *Gesammelte Mathematische Werke*, R. Narasimhan, ed., Berlin, Springer, pp. 558–570.

[Rowe 2003] Rowe, David E., 2003. Hermann Weyl, the Reluctant Revolutionary, *Mathematical Intelligencer* 25: 61–70

[Scholz 2001] Scholz, Erhard, ed. 2001. *Hermann Weyls* Raum–Zeit–Materie *and a General Introduction to his Scientific Work*, Basel, Birkhäuser.

[Scholz 2004] Scholz, Erhard, 2004. Hermann Weyl's Analysis of the "Problem of Space" and the Origin of Gauge Structures, *Science in Context* 17: 165–197.

[Scholz 2005] Scholz, Erhard, 2005. Practice-related Symbolic Realism in H. Weyl's Mature View of Mathematical Knowledge, in *The Architecture of Modern Mathematics: Essays in History and Philosophy*, J. Ferreirós, J..J Gray, eds., Oxford University Press, pp. 291–309.

[Scholz 2012] Scholz, Erhard, 2012. Leibnizian Traces in H. Weyl's *Philosophy of Mathematics and Natural Science*, in *New Essays on Leibniz Reception*, Ralf Krömer, Y. Chin-Drian, eds., Basel: Birkhäuser, pp. 203–216.

[Sieroka 2009] Sieroka, Norman, 2009. Husserlian and Fichtean Leanings: Weyl on Logicism, Intuitionism, and Formalism, *Philosophia Scientiæ* 13(2): 85–96.

[van Dalen 1995] van Dalen, Dirk, 1995. Hermann Weyl's Intuitionistic Mathematics, *Bulletin of Symbolic Logic* 1: 145–169

[Weyl 1910] Weyl, Hermann, 1910. Über die Definitionen der mathematischen Grundbegriffe, *Mathematisch- naturwissenschaftlcieh Blätter* 7: 93–95, 109–113; reprinted in (Weyl 1968, vol. I, 298–304).

[Weyl 1917] Weyl, Hermann, 1917, *Raum–Zeit–Materie*, Springer.

[Weyl 1918] Weyl, Hermann, 1918. *Das Kontinuum. Kritische Untersuchungen über die Grundlagen der Analysis*, Leipzig, Veit. English translation *The Continuum: A Critical Examination of the Foundation of Analysis*, S. Pollard and T. Bole, eds., New York, Dover, 1994.

[Weyl 1921a] Weyl, Hermann, 1921a. Über die neue Grundlagenkrise der Mathematik, *Mathematische Zeitschrift* 10: 39–79; reprinted in (Weyl 1968, vol. II, 143–180); English translation in (Mancosu 1998, 86–118).

[Weyl 1921b] Weyl, Hermann, 1921b. Das Raumproblem, *Jahresbericht der Deutschen Mathematiker–Vereinigung* 30: 92–108; reprinted in (Weyl 1968, vol. II, 212–228).

[Weyl 1923] Weyl, Hermann, 1923. *Mathematische Analyse des Raumproblems, Vorlesungen gehalten in Barcelona und Madrid*, Berlin, Springer; reprinted Darmstadt, Wissenschaftliche Buchgesellschaft, 1963.

[Weyl 1927] Weyl, Hermann, 1927. *Philosophie der Mathematik und Naturwissenschaft, Handbuch der Philosophie*, Abt. 2A, München, Oldenbourg.

[Weyl 1932] Weyl, Hermann. 1932. *The Open World. Three Lectures on the Metaphysical Implications of Science*, London, H. Milford; reprinted in (Weyl 2009, 34–82).

[Weyl 1933] Weyl, Hermann, 1933. *Mind and Nature*, Philadelphia, University of Pennsylvania Press; London, Oxford University Press; reprinted in (Weyl 2009, 83–161).

[Weyl 1949] Weyl, Hermann, 1949. *Philosophy of Mathematics and Natural Science*, 2nd ed. 1950. Princeton University Press.

[Weyl 1954] Weyl, Hermann, 1954. Erkenntnis und Besinnung (Ein Lebensrückblick), *Studia Philosophica*; reprinted in (Weyl 1968, vol. IV, 631–649); translated as Insight and Reflection, in (Weyl 2009, 204–221).

[Weyl 1968] Weyl, Hermann, 1968. *Gesammelte Abhandlungen*, 4 vols., K. Chandrasekharan, ed., Springer, Berlin.

[Weyl 2009] Weyl, Hermann, 2009. *Mind and Nature: Selected Writings on Ohilosophy, Mathematics, and Physics*, Peter Pesic, ed., Princeton University Press.

European Mathematics in Transition

Yixing (一行) in Paris

Jean-Claude Martzloff

The Pantheon

The Pantheon in Paris is a Catholic temple that was transformed into a secular mausoleum dedicated to the memory, glorification, and commemoration of the most illustrious French citizens. This transformation took place in 1791 by a decree of the revolutionaries under the short-lived First French Republic (1792-1804), which was established after the Revolution of 1789 but then overthrown by Napoleon. Erected on top of the montagne Sainte-Geneviève on the left bank of the river Seine, this neo-classical mausoleum occupies a highly symbolic position, not only because Sainte-Geneviève is the name of a Parisian 'mountain' (in fact a modest hill, something like one hundred meters above sea level). It is most noteworthy, however, because of a shepherdess named Geneviève (ca. 422-512), who was credited with having repelled the attack of Attila's Huns against this original Parisian settlement, which would later become the cultural heart of Paris during the Middle Ages. Beyond this important "battle feat", this heroic maiden was admired for her piety, devotion, works of charity, and her prophecies. Some of these, as the hagiography has it, came to pass, and she thus became the living symbol of the then nascent Catholicism. Even after her death, her relics were believed to be propitious for the realization of further miracles, such as the warding off of epidemics. Therefore, she was sanctified in 512 and an abbey bearing her name was created by Clovis I, the king of the Franks, under the authority of Saint Benoit (born in 480 and the founder of the monastic orders in the Western world).[1] Then, despite many ups and downs (most notably the plundering of the Abbey by the Normans during the IXth and Xth centuries), a school of copyists and a library were founded there in 1148. After this, the cult of Sainte-Geneviève was maintained over many centuries and was still alive more than a thousand years later.

In 1744, the King of France, Louis XV, went on a pilgrimage to the church of this abbey, where St. Geneviève's tomb was still preserved; for he believed that the holy maiden had enabled his recovery from a life-threatening disease.[2] At the same time, the local priests beseeched him to reconstruct a church devoted to her at the expense of the crown; their wish was fulfilled not long thereafter. The construction was thus undertaken in 1757 and had still not come to an end when in 1789 the French Revolution burst forth and brought down the monarchy.[3]

Two years later, the desire of the revolutionaries to eradicate all symbols

[1]See [GDU, tome 8, p. 1146-1149]. This very interesting article has many features not often found in more recent sources, notably references to contemporary sculptures, paintings and the like.

[2][Perouse 1994, p. 384.]

[3]Ibid.

D.E. Rowe and W.-S. Horng (eds.), *A Delicate Balance: Global Perspectives on Innovation and Tradition in the History of Mathematics*,
Trends in the History of Science, DOI 10.1007/978-3-319-12030-0_7

of the defunct system, including in the first place those of the Catholic religion, resulted in the decree that brought this plan to a sudden halt. The Church Sainte-Geneviève was to be assigned another purpose than the worship of a Catholic cult: it would serve as a mausoleum devoted to the great men of France. At the same time, the new institution was intended to extol universal values of freedom, science, humanity, and democracy, in conformance with the ideals of the revolution. Despite their rejection of religion, however, the revolutionaries deified their secular heroes by calling the former church 'Pantheon' from the ancient Greek word *πάνθειον*, meaning literally 'a temple devoted to all gods'. The building itself also had parallels in antiquity, notably with the Pantheon in Rome, which is still visible.

Not long afterward, with the advent of Napoleon, the cult of Sainte-Geneviève was restored. After Napoleon's fall from power, however, the cult of great men was alternatively reestablished or abolished according to the vicissitudes of revolutionary turmoil in France, marked by the political upheavals that took place in 1830, 1848 and 1871. But then in 1885 the Pantheon was reinstituted once and for all. For in that year the great republican novelist and poet Victor Hugo (1802-1885) was 'pantheonized', immediately after his death.

Nowadays, this lofty Parisian building evokes the illustrious memory of two famous philosophers from the French Enlightenment: Voltaire (1694-1778) and Jean-Jacques Rousseau (1712-1778). But it also honors the mathematicians Nicolas Condorcet (1743-1794), Joseph-Louis Lagrange (1736-1813), and Gaspard Monge (1746-1818); the man who abolished slavery, Victor Schoelcher (1804-1893); the influential pacifist socialist leader Jean Jaures (1859-1914), assassinated just before the beginning of the First Word War; the two Nobel laureates, Pierre Curie (1859-1906) and Marie Curie (1867-1934), famed for their research on radioactivity; as well as the mathematician Paul Painlevé (1863-1933), who was active in the development of aviation in France, to mention only a few names.

Beyond this highly symbolic commemorative function, the Pantheon can also be regarded as the geographical center of gravity of a large number of prestigious French institutions of higher learning, all located no more than a few hundred meters away from the 'Place du Panthéon' (the Pantheon Square). Just south of it, we find the École Normale Supérieure and the Institute Henri Poincaré (a famous research center for mathematics). Eastward, we meet the site of the ancient 'École Polytechnique' (an Institute devoted to education in science and technology at the highest level) and the Asian Institutes (devoted mainly to the study of China, Japan, India, and Tibet, with their corresponding specialized libraries). Northward, we stumble upon the Collège de France, where the most prestigious professors offer lectures and seminars in all domains of learning for anyone who feels inclined to attend them, without restrictions or any kind of final examination. Toward the northwest, we find the Faculty of Medicine, and in the immediate vicinity of the Pantheon, the Sainte-Geneviève Library and the Sorbonne University one cannot miss, to mention only some of the most prestigious French institutions devoted to the pursuit of knowledge in this area of Paris.

Hence the well-known appellation 'Quartier latin' (Latin Quarter) has been given metonimically to this area of Paris. For the study of the Latin language has been the backbone of learning in France over a long series of centuries, from the Middle Ages to well after the beginning of the twentieth century. (Until around 1900, French doctoral theses were still often written in Latin and these were by no means equivalent to our present Ph.Ds. They were much more demanding, if only because *two* theses were required before the diploma could be validated).

The Bibliothèque Sainte-Geneviève or the Unknown Pantheon

Whereas the cult of the great men held at the Pantheon was typical of the lofty ideals of the newly born First French Republic, it also happens that another much less well-known Parisian building, but also historically connected with the cult of Sainte-Geneviève, was likewise dedicated to similar ideals. Its range is much wider, however, for it does not merely concern Frenchmen, but also illustrious personalities and covering all historical periods, from the earliest recorded origins to the time it was inaugurated in December 1850. The building in question is the Bibliothèque Sainte-Geneviève (Sainte-Geneviève Library), located along the north side of the Pantheon Square and running parallel to the left wing of the Pantheon (Fig. 1). Historically, it derives from the former library of the Abbey Sainte-Geneviève. During the French Revolution it was temporarily called the 'Bibliothèque du Panthéon' (Library of the Pantheon), following the definitive disappearance of the Abbey Sainte-Geneviève in 1796.[4]

On its front and two adjacent walls no less than eight-hundred and ten easily legible names are carved in vermilion. These cover twenty-seven huge panels, nineteen on the front wall and four on each lateral wall.[5] No matter where they are located, all these names can be marveled at from the street. Remarkably, they are harmoniously placed midway between the base of the building and its top, just below its upper circular windows; the panels extend over a distance of approximately one hundred meters. Given the sizeable dimension of the letters, even the casual passer-by cannot fail to miss these names and to identify them readily without effort. They might even suggest to some kind of books they might find inside, if ever they desired to enter. One easily notices too, that these names belong to historical personalities from all branches of learning, forming a sort of encyclopedia. Beginning with Moses, the Hebrew prophet, and ending with the Swedish chemist Berzelius (1779-1848), the list includes theologians, monks, philosophers, musicians, painters, playwrights, poets, travelers, mathematicians, astronomers, and many other categories of illustrious personalities. Many of them are doubtless familiar to practically everyone: Jesus, Hipparchus, Ptolemy, Alcuin, Roger Bacon, Marco Polo, Copernicus, Tycho Brahe, Kepler, Galileo, Descartes, Gassendi, Pas-

[4]From the official internet site of the Bibliothèque Sainte-Geneviève : http://www.bsg-univ-paris1.fr/; [Perouse 1994, p. 502-505], also provides a brief overview of its history.

[5][Levine 2012, p. 92] notes that this practice follows from similar models attested in the Roman Empire, where many monuments bear numerous legible inscriptions.

cal, Molière , Fermat, Spinosa, Leibniz, Newton, Euler, Diderot, Buffon, Goldoni, Lavoisier, Condorcet, Monge, Laplace, Beethoven, Goethe, Hegel. Others, including numerous monks and theologians from the Middle Ages are certainly unfamiliar, if not completely forgotten. This is certainly also the case, regarding a Chinese-looking name, curiously spellt Y HIANG.

Figure 1: The Pantheon (on the right) and the Bibliothèque Sainte-Geneviève on the left.

This name appears on the immediate left of the main entrance of the library, alongside some forgotten monks from the Middle Ages, and just above what is now a bus stop. But is it really a Chinese name? And if so, to whom does it refer? And how is it that it was integrated into such an impressive, but heteroclite list?

The identification of the name Y HIANG

When I first noticed the name Y HIANG on the front wall of the Bibliothèque Sainte-Geneviève (Fig. 2), I was baffled. For I could absolutely not figure out who this was, not even whether it was a Chinese name. Still worse, it occurred in the same column along with a number of medieval Europeans, that I could not identify at first (Fig. 2). Many Chinese, have names beginning with 'y', but what Chinese syllable could correspond to 'hiang'?

An obvious place where such information could be found sought was, of course, the Great Chinese-Latin-French dictionary, published in 1813 by order of Napoleon, one of the prestigious scholarly achievements of early nineteenth century Chinese studies.[6] But whereas the pronunciation 'y' was associated with various

[6][De Guignes 1813]. However, it turns out that, despite its 14,000 Chinese characters, this dictionary is more prestigious than really useful. It contains long lists of rare characters and rare meanings, but no compounds at all. In fact, it is essentially a slightly improved copy of a much earlier work due to the Franciscan Basilio Bollo da Glemona (1648-1704).

Figure 2: The panel of the Bibliothèque Sainte-Geneviève where the name Y HIANG is carved (upper right column, beginning with 'Eugène, Frédégaire, etc.). Apart from 'Béde' (in fact Bède, the 'é' is erroneous; latin: Beda Venerabilis, Bede the Venerable) which is a still well-known figure, famed for his wide learning, including an history of England and, above all, theological works covering the Old and the New Testament. Most other names from the same column as Yixing refer to generally forgotten monks.

Chinese characters such as 一, 乙 or 佾, this dictionary does not list 'hiang' as a possible Chinese syllable.[7] This resource thus left the name Y HIANG remained as mysterious as ever.

Some years ago, while we were walking together along the Bibliothèque Sainte-Geneviève, Jacques Gernet, former professor of Chinese at the Collège de France, suggested to me that Y HIANG could perhaps be the astronomer Yixing (一行). So it then occurred to me that Antoine Gaubil's pioneering history of Chinese astronomy, published in Paris in 1732, might confirm this hypothesis.[8] But once again, whereas passages of this famous work where Yixing is mentioned were easily recognizable,[9] the famous monk and astronomer was never referred to as Y HIANG, only as 'Y-hang', and so neither of us could reach a definite conclusion.

Later, it occurred to me that Louis-Gabriel Michaud's (1773-1868) outstanding, but by now often overlooked biographical dictionary,[10] seldom used by scholars

[7] Ibid. The table of phonetical transcriptions of the Chinese language listed at the beginning of this dictionary, p. xiii and xiv, does not mention any syllable spelt 'hiang'.

[8] [Souciet 1732]. See also [Dehergne 1944], [Dehergne 1945] and [Dehergne 1973, p. 106-107].

[9] Almost everything found in Gaubil is taken either the Chinese official histories or from Xing Yunlu 邢雲路 (c.1580)'s *Gujin lülikao* 古今律曆考 (*Researches into ancient and modern astronomical canons and musical tubes*). Gaubil mentions Yixing (in the form Y-hang) very often from the former source, notably p. 124, 125, 129, 131, 163, 168, 170, 171 (these page numbers are not too easily retrieved for Gaubil does not follow the chronological order). As for the latter source, see [Chen Meidong 2003, p. 618 sq].

[10] [Michaud]. On Louis-Gabriel Michaud and his biographical dictionary, see the biographical notice in [GDU, tome 11, p. 215].

(except for those involved with the history of nineteenth century France) could perhaps provide another source for identifying the name Y HIANG. Indeed, from the long article [Michaud, vol. 45, p. 259-260] devoted to Yixing 一行 (683-727) in this monumental work it turned out this was the man hidden under this obsolete transcription.[11] Moreover, by way of further reward, the Michaud also provided a direct and easy identification of all the other names on the walls of the Pantheon library, not only those listed together with the famous Tang astronomer but also in the other panels. So here was another unexpected clue to the still unknown source of these illustrious and lesser known names: Michaud's biographical dictionary.

As can easily be checked, the generally excellent articles in this outstanding dictionary were all written by eminent scholars of the first half of the nineteenth century, and their scholarly standards ranked very high. In particular, the content of the entry concerning Yixing is well-informed, given the early period of its composition. In particular, it reports Yixing's long distance expedition from the north to the extreme south of China in order to determine the variations in the meridian length of a gnomon. In addition, other details also reveal a working knowledge of the relevant chapters of the *Xin Tangshu*,[12] most notably Yixing's *Dayan li* 大衍曆. This is correctly regarded as an astronomical system and not as a calendar, a mistake found in many recent articles (*Dayan li* literally means "the Dayan astronomical canon,"[13] *dayan* being the name of a well-known *Yijing* divination technique).

So who was the author of this lengthy and substantial notice, signed 'KL—H', an unusual abbreviated form from the two first letters and the last of the name in the author in question? As indicated at the end of volume 45, this corresponds to 'Klaproth'. Although this name is rare, two famous scholars from the eighteenth and nineteenth century went by it. The first was Martin-Heinrich Klaproth (1743-1817), a German Professor of chemistry at the Berlin University. The second was his son, Julius Klaproth (1783-1835) (Fig. 3).
The latter became a renowned sinologist, with strong links to France where he lived from 1815, so he was doubtlessly the author of the notice.[14] As the catalog of his numerous and wide ranging publications extending far beyond the 'mere' knowledge of the Chinese language clearly shows, he also mastered Manchu and other oriental languages; the list of his publications is truly impressive.[15] Clearly, he was a professional of the first magnitude during this early era of European sinology. Unfortunately, however, his untimely controversy with Champollion over

[11]Numerous recent biographies of Yixing are available, notably [Chen Meidong 1998] and [Hockey 2007] but ancient ones, notably Ang Tian Se's unpublished Ph.D. dissertation is still precious [Ang Tian Se 1979].

[12][XTS, j. 3a-4b, p. 587-694]

[13]It is true that *li* 曆 means 'calendar' in general, but in a technical sense, it also designates collections of mathematical and astronomical procedures intended not only for the calendar but, more generally for the construction of planetary ephemerides, or for the prediction of eclipses, normally never mentioned in calendars.

[14][Walravens 1999].

[15]*Ibid.*

Figure 3: Julius Klaproth (From [Walravens 1999])

the deciphering of the Egyptian hieroglyphs put him on the wrong side of history, which incurred him the disfavor of later scholars.

Now that we have identified Y HIANG, a final piece of the puzzle is still missing, namely the question of the origin of the extraordinary list of personalities that adorn this building. In this respect, a detailed knowledge of certain aspects of the history and purpose of the construction of the Bibliothèque Sainte-Geneviève and of the intellectual French context during that time is in order.

The Bibliothèque Sainte-Geneviève and Auguste Comte's Religion of Humanity

The original project of construction of the Bibliothèque Sainte-Geneviève goes back to 1836 under the July monarchy. However, construction of the library was launched only two years later and it was only completed long afterward in December 1850, after the fall of Louis-Philippe's regime and the proclamation of the second French republic (1848-1852), a consequence of the revolution of 1848.[16]

The library had been specially designed for the benefit of the students of the Latin Quarter and it was kept open until 10 PM. There, despite the state of technology around 1850, they could work by night with the help of gas lighting and a comfortable room temperature of not less than 18 degrees Celsius, maintained in winter by a system of central heating.[17] In other words, the revolution had not only political but also intellectual and social implications, all the more so considering that the Bibliothèque Sainte-Geneviève ranked among the greatest of the time.

The encyclopedic list of names on the walls of the Bibliothèque Sainte-

[16] [Leniaud 2001], [Labrouste 2012].
[17] [Levine 2012]

Geneviève was drawn up by its architect Henri Labrouste (1801-1875)[18] in July 1848, and it was officially approved on August 24th. The engravings were realized between September 30 and November 21 of the same year.[19] Here, we note in passing that Labrouste had established his lengthy list in a strikingly short span of time. Therefore, he must have had recourse to a convenient and handy source rather than relying on a wide variety of documents. Although we do not know exactly what he did in this respect, it seems probable that he merely consulted Michaud's biographical dictionary, for, as mentioned above, he could have found everything he needed there.

As for the intellectual climate, it is quite strikingly, that the ideas of the French founder of positivism, Auguste Comte (1798-1857), were at their zenith at this same time. The renowned philosopher had developed his idea of a systematic cult or of a religion of humanity in a booklet entitled *Culte systématique de l'humanité; l'amour pour principe, l'ordre pour base et le progrès pour but. Calendrier positiviste ou système general de commemoration publique, destiné surtout à la transition finale de la grande république occidentale [···] (A Systematic Cult of Mankind ; love for principle, order for basis and progress for goal. Positivist Calendar or General System of Public Commemoration [···]).*[20]

As Comte explains, this system aimed to commemorate great men of the past, independent of theological considerations and nationality in order to prepare future progress. Also quite significantly, the famous philosopher names his system "Le Panthéon positiviste" ("the positivist *Pantheon*" (my emphasis)).[21] Explaining all the minutiae of Auguste Comte's ideas would lead us too far, but it will suffice here to note that the list of names commemorated in his positivist calendar is quite similar to those in the panels of the Bibliothèque Sainte Geneviève. In both cases, the progress of humanity from the theological age to the philosophical and positivist periods is obvious, and both lists are similarly encyclopedic in character. They are not identical, however, because the number of names included in the 365 days of the calendar year is, of course, much smaller than the eight-hundred and ten names of the Bibliothèque Sainte-Geneviève, but also because not exactly the same names have been retained. For instance, the positivist calendar (see Fig. 4) celebrates Confucius, Mozi, Laozi, and even the mythical Chinese emperor Fuxi 伏羲, but not Yixing. Nevertheless, Confucius, Mozi and Laozi are present in the Sainte-Geneviève list. It remains, however, that Yixing is the unique Chinese scientist. Obviously Labrouste had composed his lengthy list in a hurry, though both lists were certainly inspired by similarly lofty ideals.

---

[18]All the needed biographical information concerning Labrouste is easily found in [Leniaud 2001] and [Labrouste 2012]. He was one of the leading French architect of the XIXth century and, besides the Bibliothèque Sainte-Geneviève, he has also been responsible of the renovation and construction of the ancient buildings of the Bibliothèque nationale, located not far from the Opera in Paris.

[19][Leniaud 2001, p. 92]

[20][Comte 1849], [Leniaud 2001, p. 92-93]

[21][Comte 1849, p. 18]

CALENDRIER POSITIVISTE,

POUR UNE ANNÉE QUELCONQUE.

JANVIER,

PREMIER MOIS, CONSACRÉ A LA THÉOCRATIE INITIALE.

MOÏSE.

1 —	Lundi.	Prométhée.	
2 —	Mardi.	Hercule.	*Thésée.*
3 —	Mercredi.	Orphée.	
4 —	Jeudi.	Ulysse.	
5 —	Vendredi.	Lycurgue.	
6 —	Samedi.	Romulus.	
7 —	DIMANCHE.	NUMA.	
8 —	Lundi.	Bélus.	*Sémiramis.*
9 —	Mardi.	Sésostris.	
10 —	Mercredi.	Menou.	
11 —	Jeudi.	Cyrus.	
12 —	Vendredi.	Zoroastre.	
13 —	Samedi.	Les Druides.	*Ossian.*
14 —	DIMANCHE.	BOUDDHA.	
15 —	Lundi.	Fo-Hi.	
16 —	Mardi.	Lao-Tseu.	
17 —	Mercredi.	Meng-Tseu.	
18 —	Jeudi.	Les théocrates du Thibet.	
19 —	Vendredi.	Les théocrates du Japon.	
20 —	Samedi.	Manco-Capac.	*Tamehamea.*
21 —	DIMANCHE.	CONFUCIUS.	
22 —	Lundi.	Abraham.	
23 —	Mardi.	Joseph.	
24 —	Mercredi.	Samuel.	
25 —	Jeudi.	Salomon.	*David.*
26 —	Vendredi.	Isaïe.	
27 —	Samedi.	Saint-Jean-Baptiste.	
28 —	DIMANCHE.	MAHOMET.	

Figure 4: The month of January of Auguste Comte's perpetual positivist calendar. His Chinese "saints" Fo-Hi (Fuxi), Lao-Tseu (Laozi), Meng-Tseu (Mengzi) and Confucius are celebrated on January 15, 16, 17 and 21. The latter, Confucius, is inscribed in capital letters, like Buddha and Mahomet, and occurs on a Sunday (from [Comte 1849 p. 23])

Bibliography

[Comte 1849] Comte, Auguste. *Calendrier positiviste ou système général de commémoration publique destiné surtout à la transition finale de la grande république occidentale*, Paris, 1849 (quoted here from its reedition, Paris, Fontfroide, Fata Morgana, 1993).

[De Guignes 1813] *Dictionnaire chinois, français et latin, publié d'après l'ordre de sa Majesté L'Empereur et Roi, Napoléon le Grand ; par M. De Guignes, Résident de France à la Chine*, Paris, Imprimerie Impériale, 1813.

[GDU] Larousse, Pierre. *Grand dictionnaire universel français, historique, géographique, biographique, mythologique, bibliographique, littéraire, artistique, scientifique* [⋯], Paris, Administration du Grand dictionnaire universel, 1867-1890.

[Michaud] Michaud, *Biographie universelle, ancienne et moderne.* Second edition. Paris, C. Desplaces; Leipzig, Brockhaus, 1843-1865, Tome 45.

[Souciet 1732] Souciet, Le P. Étienne, S.J. *Observations mathématiques, astronomiques, géographiques, chronologiques et physiques, tirées des anciens livres chinois ou faites nouvellement aux Indes, à la Chine et ailleurs, par les Pères de la Compagnie de Jésus. Rédigées et publiées par le P. Etienne*

Souciet de la même compagnie. Tome III, contenant un traité de l'astronomie chinoise par le P. Gaubil de la même compagnie. Paris, Rollin Père, 1732.

[XTS] *Xin Tangshu* 新唐書 (*New Tang History*) by Ouyang Xiu 歐陽脩 et al., Beijing, Zhonghua shuju. 1975.

[Ang Tian Se 1979] *I-Hsing (683-721 A.D.): His Life and Scientific Work, A Thesis Submitted for the Degree of Doctor of Philosophy to the Faculty of Arts and Social Sciences*, University of Malaya, Kuala Lumpur. 1979.

[Chen Meidong 1998] Chen Meidong 陳美東, Yixing *in* Lu Jiaxi 盧嘉錫 (ed.) *Zhongguo Kexue jishu shi, renwu juan* 中國科學技術史人物卷, Beijing, Kexue chubanshe (1998), pp. 278-294.

[Dehergne 1944] Dehergne, Joseph S.J. Le père Gaubil et ses correspondants (1689-1759), *Bulletin de l'Université Aurore*, 1944, ser. 3, tome 5, n. 2, p. 354-392.

[Dehergne 1945] Dehergne, Joseph S.J. Gaubil historien de l'astronomie chinoise, *Bulletin de l'Université Aurore*, ser. 3, tome 6, n. 1, p. 168-227.

[Dehergne 1973] Dehergne, Joseph S.J. *Répertoire des jésuites de Chine de 1552 à 1800.* Roma, Institutum Historicum S. I. ; Paris, Letouzey et Ané, 1973.

[Fierro 1996] *Dictionnaire historique de Paris.* Paris, Robert Laffont, 1996.

[Hockey 2007] Yixing in Hockey, Thomas (ed. in chief). *Biographical Encyclopedia of Astronomers*, Berlin, Heidelberg, New-York, Springer, 2007, p. 1252-1253.

[Labrouste 2012] Bélier, Corinne ; Bergdoll, Barry and Le Coeur, Marc (dir.). *La structure mise en lumière, Labrouste (1801-1875) architecte.* Nicolas Chaudun, Cité de l'Architecture et du Patrimoine, Paris; Museum of Modern Art (New York) and Bibliothèque Nationale de France, 2012. [This document has been released at the occasion of an exhibition held at the Cité de l'Architecture et du Patrimoine, Paris, 11 october 2012-7 january 2013 and a few months later at New York, Museum of Modern Arts, 10 march-24 june 2013. An English version has also been published under the following title: *Henri Labrouste: Structure Brought to Light*, MoMA Publications, 2013. The present article refers only to the French version].

[Leniaud 2001] Leniaud, Jean-Michel (dir.). *Des palais pour les livres, Labrouste, Sainte Geneviève et les bibliothèques*, Paris, Maisonneuve Larose, Bibliothèque Sainte Geneviève, 2001.

[Levine 1975] Levine, Neil Arthur. *Architectural reasoning in the age of positivism : the neo-grec idea of H. Labrouste's Bibliothèque Sainte-Geneviève.* New Haven, Yale University, 1975, 3 vol.

[Levine 2012] La bibliothèque publique à l'aune de la bibliothécomomie, in [Labrouste 2012, p. 164-179].

[Pérouse 1994] Pérouse de Montclos, Jean-Marie (dir.). *Le guide du patrimoine, Paris.* Paris, Direction du Patrimoine, Hachette, 1994.

[Walravens 1999] Walravens, Hartmut. *Julius Klaproth (1783-1835) Leben und Werk.* Wiesbaden, Harrassowitz Verlag, Orientalistik Bibliographien und Dokumentationen, 1999.

Needham's 19 (j) and Fibonacci's *Liber Abaci*: A New Approach to the Communication and Transmission of Mathematical Knowledge from China to the West

JI Zhigang

Introduction: The Significance of Needham's 19(j)

In his monumental *Science and Civilisation in China*, Joseph Needham focuses on mathematics in subsections (a) to (k) of section 19 in volume III. After evaluating the characteristics and contributions of ancient Chinese mathematics, Needham turns in the tenth subsection: (j) Influences and Transmissions. Here, he writes, "When, however, we ask what mathematical ideas seem to have radiated from China southwards and westwards, we found a quite considerable list." ([Needham 1959], 146) Below we have compiled Needham's "quite considerable list" in the form of a concise table.

Needham then goes on to say: "Thus in spite of all the 'isolation' of China, and all the inhibitory social factors which we shall presently examine, it seems probable that between –250 and +1250 a good deal more came out than went in." ([Needham 1959], 148) This is the general picture drawn by Needham regarding the exchange and transmission of mathematical knowledge between the West and the East. But we should also note that not every historian of mathematics agrees with what Needham wrote about this. For example, Professor Jean-Claude Martzloff has remarked:

> This type of argumentation is controversial because of its chronological and methodological imprecision; there is nothing to prove that concomitant developments could not have taken place. Precisely which algebraic and geometric relationships are referred to in point 9? Does Needham mean those which may be used to deduce algebraic formulae from geometric figures? Which "Pascal triangle" is referred to in point 14? What form did it have? What was it used for? Was it used to solve combinatorial problem, as describe by historians of Indian mathematics? Or was it used to extract roots, as in China? Or, again, was it used, as Pascal used it, to calculate probabilities?([Martzloff 1997], 91)

Martzloff's arguments are worthy of careful consideration. In fact, the exchange and transmission of mathematical knowledge between West and East has now become a hot issue, as can be seen from publications such as *From China to Paris: 2000 Years Transmission of Mathematical Ideas* ([Dold-Samplonius 2002]). In addition, three symposia held at the XXII International Congress of History of Science (ICHS) were especially concerned with the transmission of mathematical knowledge (SC13: *Along the Silk Road*; and SC14: *Multicultural Transmission of*

D.E. Rowe and W.-S. Horng (eds.), *A Delicate Balance: Global Perspectives on Innovation and Tradition in the History of Mathematics*,
Trends in the History of Science, DOI 10.1007/978-3-319-12030-0_8

Table 1: Summary of Needham's List ([Needham 1959], 146-147)

Subjects in China	Related Items in Other Areas
1. A numeral notation: nine figures, blank space, place-values	India
2. The extraction of square roots and roots of higher degree by Chia Hsien[1]	Brahmagupta (630) , al-Kashi (15th c.)
3. Rule of three	India
4. A way to express fractions	India used the same method as in China
5. Negative numbers	In India by the time of Brahmagupta
6. Hsuan Thu[2] method of proving the Pythagorean Theorem	Bhaskara (12th c.)
7. Geometrical surveying	Mahavira (9th c.)
8. The area of a circle	Mahavira
9. The fundamental identity of algebraic and geometrical relations	al-Khwarizmi to Khazaria
10. The rule of false position	Regola elchataym known in Italy from the Islamic world
11. Indeterminate analysis	Ayabhata (5th c.), Brahmagupta (7th c.), Issac Argyros (14th c., a Byzantine monk)
12. Indeterminate equations	Mahavira (9th c.), Bhaskara (12th c.)
13. Cubic numerical equations	Fibonacci (13th c.) was the first to give a solution
14. Binomial coefficients (Pascal's Triangle)	Persia, Europe in late 16th c.

Mathematical Knowledge; S26: *Mathematical Practitioners and the Transformation of Natural Knowledge in Early Modern Europe*) [Book of Abstracts of ICHS 2005: 289-295, 296-301, 114-118]. Just after the close of this Congress, the 1st International Conference on History of Exact Sciences Along the Silk Road was held in Xi'an. All this indicates that Needham's 19(j) still has significance today, as delicately indicated by a metaphor that Dr. Christopher Cullen used in speaking of *Influences and Transmissions of Mathematical Knowledge*, namely that after a long accumulation of massive energies, "a new mountain range" seems to be emerging from "the Tectonic Plates" of Needham's 19(j) ([Cullen 2005], 4-5).

In this paper, we hope to find a route via Fibonacci's *Liber Abaci* to climb this "new mountain range." First, we will give a brief introduction to Fibonacci and his famous *Liber Abaci*. Then, drawing on the Chinese version of *Liber Abaci*, we will show that some of the problems and algorithms in it are similar to others found in ancient Chinese mathematics. In the last part of this paper, we will further discuss these similarities, pointing out the oriental mathematical sources in *Liber Abaci*, especially those from India and the Islamic world. Throughout this analysis we hope to demonstrate the significance of Fibonacci's *Liber Abaci* as a key document for understanding the influences of oriental mathematical knowledge, much of which had its origins in China, on the further development of mathematics in the Latin West.

Fibonacci and his *Liber Abaci*

Leonardo Fibonacci was born around 1170. His father was a Pisan merchant who had extensive commercial dealings in Bugia on the North African coast (now Bejaia, in Algeria). ([Vogel 1970], 604-613) Leonardo spent much of his early life there learning Arabic and studying mathematics under Moslem teachers. Later he traveled throughout the Mediterranean. In these places he visited Islamic scholars and absorbed the mathematical knowledge of the Islamic world. After his return to Pisa in about 1200, he spent the next 25 years writing works in which he incorporated what he had learned. The ones that have been preserved include *Liber Abaci* (1202, published in a revised and expanded edition in 1228), *Practica geometriae* (1220), and *Liber quadratorum* (1225). During the period 1857-1862, the Italian bibliophile and medieval mathematical historian, Baldassarre Boncompagni, compiled a two-volume edition of Leonardo's work written in modern Latin ([Boncompagni 1857, 1862]). Still, it was not until 2002 that *Liber Abaci* was translated into modern English by Laurence Sigler ([Sigler 2002]).[3]

We should point out here that the word *abacus* in the title does not refer to the traditional *abacus* but rather means, in fact, calculation without the *abacus*, the traditional instrument for basic arithmetic. In the brief preface to *Liber Abaci*, Fibonacci describes how during his travels and studies he learned to calculate "with the new Indian numerals," and that he eventually found this method to be superior to all others. His motivation, then, for writing *Liber Abaci* was to bring this technique to the Italian people, because they are a people "who up to now are found without a minimum" ([Sigler 2002], 16). Fibonacci divided his extensive work into fifteen chapters, which we will discuss in six parts in order to to give a brief overview of its contents.

Part One, Chapters 1-7. In Chapter 1, Fibonacci introduces the nine Indian figures and the symbol for zero. By means of the decimal place value system, he gives various examples to show how numbers are to be written. In Chapters 2 to 5, Fibonacci introduces the four arithmetical operations in this system, thus multiplication, addition, subtraction, and division of whole numbers. In Chapter 6 he treats mixed numbers, that is whole numbers with simple fractions, and in Chapter 7 he explains the arithmetical operations with fractions.

Part Two, Chapter 8-11. Chapter 8 deals with finding the value of merchandise by the principal method, now called "the Rule of Three" . Chapters 9 to

[3]Laurence E. Sigler was a Professor of Mathematics at Bucknell University and an amateur Latinist whose previous translation of Leonardo's *Liber quadratorum* [Book of Squares] (1987) was well received by historians of mathematics.

Chapter 11 deal with various problems in commercial arithmetic involving business negotiations such as the barter of merchandise, financial matters between companies and their members, and calculations of the value of currencies contain metal alloys. The "Rule of Three" is developed further into a systematic account of proportions. Chapter 11 contains the well-known problem of "buying 30 birds for 30 denari" .

Part Three, Chapter 12. This chapter is the most extensive in the entire book and is particularly rich in problems, many of which are solved by the method of single false position. The very famous Fibonacci sequence appears in this chapter ([Sigler 2002], 404-405) as well as problems of indeterminate analysis.

Part Four, Chapter 13. In this chapter, Fibonacci introduces the Regola Elchatym, that is the method of double false position by which nearly all problems involving false position can be solved.

Part Five, Chapter 14. This chapter deals with methods for calculating square and cube roots. Fibonacci also discusses the multiplication, division, and subtraction of such irrationals, including the types known as binomials and apotomes.

Part Six, Chapter 15. In this chapter, Fibonacci deals with the pertinent geometric rules connected with problems of "Algebra and Almuchabala". This method is traced back to the Arabic mathematician Al-Khwarizmi, whose name, spelled "Maumeht", appears in the margin ([Sigler 2002], 554).

So we see that the *Liber Abaci* not only introduced the rules for computing with the new Hindu-Arabic numerals. It also contained numerous problems that dealt with such practical topics as calculation of profits, currency conversions, measurements, motion problems, container problems, etc. The sources for *Liber Abaci* were widespread throughout the Islamic world, however, many of these mathematical methods and problems traveled along the Silk Road that connected these sources with China and India. Among the items on Needham's list above, two are directly related to Fibonacci:

> (10) The Rule of False Position, found in the *Chiu-Chang Suan Shu* from ther period of the Han dynasty, appears in 13th-century Italy under such names as 'Regola Elchataym', indicating that the Arabs had transmitted it. They probably derived it from India, but it may well have its origin in China ([Needham 1959], 147).
> (13) In the Thang period (7th century), Wang Hisao-Thung successfully solved cubic numerical equations. In the later Sung era (12th and 13th centuries) the Chinese algebraists were particularly skilled in the management of higher numerical equations. In Europe, Fibonacci was the first to solve a problem of Wang's type, and there is reason to think that he may have been influenced by Eastern Asian sources ([Needham 1959], 147).

Fibonacci clearly played a special role in transmitting oriental mathematical knowledge to the Latin West, so we may expect that it will be fruitful to undertake a closer examination of the similarities between his *Liber Abaci* and older Chinese mathematicsl.

Similarities between *Liber Abaci* and Ancient Chinese Mathematics

Although Fibonacci was familiar to Chinese mathematicians, few Chinese scholars had the opportunity to read his *Liber Abaci* because it was written in Latin. Thanks are thus due to L.E. Sigler, who first translated the Latin text into English, and based on this version *Liber Abaci* was translated into Chinese ([Ji 2008]). After close study, we find that many problems and algorithms in *Liber Abaci* are strikingly similar to others found in ancient Chinese mathematics. We cite some of them in the following list.

Table 2: Problems in Liber Abaci and in Ancient Chinese Texts

Liber Abaci	Ancient Chinese Texts
1. He divides the chosen number by 3, and by 5, and by 7, and always you ask what are the remainders from each division. You truly for each unit of the remainder upon division by 3 keep 70, and for each unit of the remainder upon division by 5 you keep 21, and for each unit of the remainder upon division by 7 you keep 15. And whenever the total exceeds 105, you throw away 105, and that which remains for you after all the 105 are thrown away will be the chosen number.([Sigler 2002], 428)	Now there are an unknown number of things. If we count by threes, there is a remainder 2; if we count by fives, there is a remainder 3; if we count by sevens, there is a remainder 2. Find the number of things. Answer: 23. Method: If we count by threes and there is a remainder 2, put down 140. If we count by fives and there is a remainder 3, put down 63. If we count by sevens and there is a remainder 2, put down 30. Add them to obtain 233 and subtract 210 to get the answer. If we count by threes and there is a remainder 1, put down 70. If we count by fives and there is a remainder 1, put down 21. If we count by sevens and there is a remainder 1, put down 15. When [a number] exceeds 106, the result is obtained by subtracting 105.[4]([Dauben 2007], 299)

[4]《孫子算經 • 卷下》: 今有物不知其數，三三數之剩二，五五數之剩三，七七數之剩二。問物幾何？答曰：二十三。術曰：三三數之剩二，則置一百四十；五五數之剩三，置六十三；七七數之剩二，置三十。並之，得二百三十三。以二百一十減之，即得。凡三三數之剩一，則置七十；五五數之剩一，則置二十一；七七數之剩一，則置十五。一百六以上，以一百五減之，即得。[Qian 1963], 318.

2. Seven old men go to Rome; each of them has 7 mules, and on each mule there are 7 sacks; and in each sack there are 7 loaves of bread, and for each loaf of bread there are 7 knives, and each knife has seven scabbards. The sum of all the aforesaid is sought.([Sigler 2002], 439)	Suppose [a person], walking out of the gate, sees nine dams. In every dam there are nine trees, and in every tree there are nine branches, and in every branch there are nine nets, and in every net there are nine birds, and each bird has nine chicks, in every chick there are nine feathers, for each feather there are nine colors. The question: How many are there of each?[5]
3. A partridge is worth 3 denari, a pigeon 2, a turtledove 1/2 denaro, a sparrow 1/4 denaro, and I wish from them 30 birds for 30 denari. ([Sigler 2002], 257)	Now one cock is worth 5 qian, one hen 3 qian, and 3 chicks 1 qian. It is required to buy 100 fowls with 100 qian. In each case, find the number of cocks, hens and chicks bought.[6]([Dauben 2007], 307)
4. Also a fleeing fox that is 50 paces ahead of a dog goes 6 paces for every 9 of the following dog. [How many paces will it take until the dog catches up with the fox?([Sigler 2002], 276)	Suppose there is a fast walker who travels 100 bu, and a slower walker who travels 60 bu. Now the slower walker starts off 100 bu ahead, and the fast walker chases after the slower one. The question: How many bu until [the fast walker] catches up [with the slower walker]?[7]([Guo, Dauben & Xu 2013], 721)
5 Two ships are some distance apart, which journey the first can complete in 5days, the other in 7 days; it is sought in how many days they will meet if they begin the journey at the same time. ([Sigler 2002], 280)	Suppose there is a wild duck that sets out from Nan-hai, and in 7 days it reaches Bei-hai; Suppose a wild goose leaves Bei-hai, and in 9 days it reaches Nan-hai. Now, suppose the duck and goose both set out simultaneously. The question: In how many days will they meet?[8]([Guo, Dauben & Xu 2013], 759)

[5]《孫子算經 • 卷下》：今有出門望見九堤，堤有九木，木有九枝，枝有九巢，巢有九禽，禽有九雛，雛有九毛，毛有九色。問各幾何?[Qian 1963], 321.

[6]《張邱建算經 • 卷下》：今有雞翁一，值錢五；雞母一，值錢三；雞雛三，值錢一。凡百錢，買雞百隻。問：雞翁、母、雛各幾何?[Qian 1963], 402-403.

[7]《九章算術 • 均輸》：今有善行者行一百步，不善行者行六十步。今不善行者先行一百步，善行者追之。問：幾何步及之?[Qian 1963], 191.

[8]《九章算術 • 均輸》：今有鳧起南海，七日至北海；雁起北海，九日至南海。今鳧雁俱起。問：何日相逢?[Qian 1963], 198.

6. There is a vat that has four holes, and by the first hole the vat can be emptied in 1day, by the third in 3, and by the fourth in 4; it is sought in how many hours the vat will be emptied if it the said four holes are opened together. ([Sigler 2002], 281)

Suppose there is a pond, and 5 streams pour into it. If the first stream starts to flow, in 1/3 of a day [the pond] is completely full; The next, in 1day it is completely full; The next, in 2 1/2 days it is completely full; The next, in 3 days it is completely full; and the next, in 5 days it is completely full. Now, suppose all are running into the pond. The question: In how many days will the pond be full?[9] ([Guo, Dauben & Xu 2013], 781)

7. Two men had denari; the first of said to the second, if you will give to me a third part of your denari, then I shall have 14denari. The second responded to him, and he said: and if you will give to me a fourth part of your denari, then I shall have 17 denari. It is sought how many each one had.([Sigler 2002], 469)

Suppose there are 2 people, (A and B). Each of whom has an unknown amount of money. If A receives half of B's money, then he has 50 qian [altogether]; and if B receives 2/3 of A's money, he also has 50 qian[altogether]. The question: How much money does each of them, A and B, have?[10] ([Guo, Dauben & Xu 2013], 965)

8. Five men having bezants wished to buy a horse; the first man takes from the second half of his bezants, and the second takes from the third a third, and the third takes from the fourth one fourth, and the fourth takes from the fifth one fifth, and the fifth similarly takes from the first one sixth of his bezants. And each of them proposes to buy the horse. It is sought how many bezants each of them had, and what was the price of the horse.([Sigler 2002], 458)

Suppose five families share a well. 2 of family A's ropes are insufficient [to reach the bottom of the well], but adding 1 of family B's ropes [is just enough]; 3 of family B's ropes are insufficient, but using 1of family C's ropes [is just enough]; 4 of family C's ropes are insufficient, but using 1 of family D's ropes [is just enough]; 5 of family D's ropes are insufficient, but using 1of family E's ropes [is just enough], 6 of family E's ropes are insufficient, but using of 1of family A's ropes [is just enough]. If each [family] gets the 1 rope by which it is deficient, then they all can reach the [bottom of the well]. The question: What is the depth of the well, and what is the length of each of the ropes?[11]([Guo, Dauben & Xu 2013], 975)

[9] 《九章算術 • 均輸》: 今有池，五渠注之。其一渠開之，少半日一滿；次、一日一滿；次、二日半一滿；次、三日一滿；次、五日一滿。今皆決之，問：幾何日滿池?[Qian 1963], 201.

[10] 《九章算術 • 方程》: 今有甲乙二人持錢，不知其數。甲得乙半，而錢五十，乙得甲太半，而亦錢五十。問：甲、乙持錢各幾何？[Qian 1963], 230.

Besides the above list of similar problems, we also find that some algorithms employed in *Liber Abaci* closely resemble algorithms found in the *Nine Chapters on the Art of Mathematics*. We give four examples illustrating this similarity in the subsections that follow.

The Addition of Fractions

In Chapter 7 of *Liber Abaci*, Fibonacci introduced the calculation of fractions; the first example is to add 1/3 + 1/4. He gives two methods, the second one is:

> Also you otherwise write down 1/4 1/3 in this way; and you multiply the 1 which is over the 3 by the 4, there will be 4 that you write above the 1/3; and the 1 which is over the 4, you multiply by 3; there will be 3 that you write above the 1/4, and you add them together; there will be 7 that you divide by the 3 and the 4 that are under the fraction, that is by 12; the quotient similarly will be 7/12 for the addition.([Sigler 2002], 100)

This second method is exactly the same as that given in the *Nine Chapters*, where the method for adding fractions (“合分” He-Fen) is:

> Cross-multiply the numerators and denominators, combining them as the dividend; mutually multiply all of the denominators together as the divisor. Divide the dividend by the divisor; if it does not divide completely, then use the divisor to name the fraction.[12]([Guo, Dauben & Xu 2013], 39-43)

Expressing this in modern symbols, the result would be:

$$\frac{b}{a} + \frac{d}{c} = \frac{bc + ad}{ac}.$$

The Rule of Three

At the beginning of Chapter 8 in *Liber Abaci*, Fibonacci discusses how to find the value of merchandise by the principal method. This method involves finding the ‘four-proportional-number’ as follows:

> Therefore four proportional numbers are always found in all negotiations of which three are known and one is left truly unknown; the first indeed of these three known numbers is the number of the sale of nay

[11]《九章算術 • 方程》：今有五家共井，甲二綆不足，如乙一綆；乙三綆不足，如丙一綆；丙四綆不足，如丁一綆；丁五綆不足，如戊一綆；戊六綆不足，如甲一綆。如各得所不足一綆，皆逮。問：井深、綆長各幾何?[Qian 1963], 231.

[12]《九章算術 • 方田》“合分術”：母互乘子，並以為實，母相乘為法，實如法而一。不滿法者，實如法而一。[Qian 1963], 95-96.

> merchandise, or fixed number, or weight, or measure. [···]The second moreover is the price of the sale that is the first number, [···] Often the third truly will be some of the same sale of a merchandise, namely the fourth number is unknown. [···]therefore if the fourth quantity is the only unknown, indeed the multiplication of the second quantity by the third, you divide by the first, then certainly the fourth quantity results from the division.([Sigler 2002], 127-128)

The corresponding method in the *Nine Chapters* is called "Jin-You" ("今有" Jin-You), which is explained as follows:

> Use the product of the given amount multiplied by the rate of that which is sought as the dividend. [Use the given rate as divisor]. Divide the dividend by the divisor.[13]([Guo, Dauben & Xu 2013], 177-179)

In fact, following the above method, the given amount ("所有數") is the first number, the sought rate ("所求率") is the second number, the given rate ("所有率") is the third number, and we apply the 'Jin-You' Method to find the sought amount ("所求數"), or the fourth unknown:

$$\text{the sought amount} = \frac{\text{the given amount} \times \text{the sought rate}}{\text{the given rate}}$$

This, of course, is the famous method of 'the Rule of Three' ([Needham 1959], 26).

The Method of Proportional Distribution

Chapter 10 of *Liber Abaci* deals with companies and their members who wish to divide the profit when their shares are unequal. Fibonacci says:

> You write the number of shares of the first member at the head of a table in the right part, next in the same line in order towards the left you strive to write the number of shares of the other members, and the profit which they will make you put at head of the table in the same line, namely in the left part. Then you add the number shares of all the members, and you keep the total. You divide the each of the member's share by the total number of shares, and then you multiply each by the total profit, and thus you will have each member's share of the profit. ([Sigler 2002], 213)

Comparing this with the method of proportional distribution ("衰分" Cui-Fen) in Chapter 3 of the Nine Chapters, we find:

[13]《九章算術 • 粟米》"今有術"：以所有數乘所求率為實，以所有率為法，實如法而一。[Qian 1963], 114.

> Put down [on the counting board] each of the given proportional values, set to one side [the proportional values] and add [them together] as the divisor. Taking that which is to be divided, multiply that by each [of the given proportional values] not yet combined; Each respectively becomes a dividend. Divide the dividends by the divisor. If the divisions are not even, use the divisor to determine [the fraction].[14]([Guo, Dauben & Xu 2013], 259-263)

Regola Elchataym: the Rule of Double False Position

At the outset of Chapter 13 Fibonacci writes: "Here begins . . . the Method Elchataym and how with it nearly all problems of mathematics are solved." Near the beginning of this chapter, he says:

> Indeed the Arabic elchataym by which the solutions to nearly all problems are found is translated as the method of double false position. [...]we wish moreover to demonstrate how they and many other problems can be solved by elchataym. ([Sigler 2002], 447)

To explain how this method works, Fibonacci gives these details:

> [...] and the true argument is found according to the proportion of difference of one position to the other which is what occurs in the method of four proportions in which three numbers are known and the fourth unknown, namely the true argument, is found; the first number is the difference between one false position and the other. The second is the difference between the approximations to the true value. The third is the difference which is between the second approximation and the true value. ([Sigler 2002], 447)

Writing this in modern notation, we have the equation $f(x) = ax + b = c$, where c is the true value (a given number), and letting x_1 and x_2 be two false positions, we obtain the corresponding approximations: $ax_1 + b = c_1$ and $ax_2 + b = c_2$. So we now have the three difference-numbers:

$$x_1 - x_2, \qquad (c - c_2) - (c - c_1), \quad c - c_2.$$

Let x be the true argument which is sought, then according to the method of four number proportions:

$$\frac{c - c_2}{x - x_2} = \frac{(c - c_2) - (c - c_1)}{x_1 - x_2},$$

and from this it follows that

$$x = x_2 + \frac{(c - c_2)(x_1 - x_2)}{(c - c_2) - (c - c_1)}.$$

---

[14]《九章算術 • 衰分》"衰分術"：各置列衰，副並為法，以所分乘未並者各自為實，實如法而一。不滿法者，以法命之。[Qian 1963], 131.

Yet the most surprising thing is that Fibonacci goes on to give a Second Elchataym:

> There is indeed another elchataym method which is called the augmented and diminished method in which the errors are put below their positions; the first error is multiplied by the second position, and the second error is multiplied by the first position. And if both the errors are minus, or both are plus, the lesser product is subtracted from the aforesaid greater product, and the difference is divided by the difference of the errors, and thus the solution of the problem is found; and if one of the errors is plus, and the other is minus, then their products are added together, the sum is divided by the sum of their errors. ([Sigler 2002], 449)

This second elchataym is, in fact, very similar to the method of excess and deficiency ("盈不足" Yin-Buzu) which was presented in the Nine Chapters as follows:

> Put down [on the counting board] the payments rates, and the excess and the deficiency, each in their respective positions under them. Cross-Multiply these [the excess and deficiency] with the payment rates, and combine them as the dividend; add the excess and deficiency as the divisor. Divide the dividend by the divisor [to obtain the price of the goods or items in question].[15]([Guo, Dauben & Xu 2013], 815-817)

Comments and Conclusion

Reflection on Similarities in *Liber Abaci* and Ancient Chinese Sources

The exchanges and transmissions of mathematical knowledge between the West and the East have long been a topic of great interest. In 1927, Qian Baocong published a pioneering paper in which he took the method of Excess and Deficiency as an example to show how Chinese mathematics was spread in the West. ([Qian 1927]). Later Du Shiran gave various arguments that point to the direct transmission of mathematical knowledge from China to the Islamic world ([Du 1966], [Du 1984]). Recently Karine Chemla has instigated further detailed discussions regarding the world-wide dissemination of the rule of false double position. She has pointed out that "since rules of false double position have not yet been found in Sanskrit texts, here we might have a direct transmission from China to the Arabic-speaking world." ([Chemla 1997]). In consideration of the fact that the hundred fowls problem, which first appeared in *Zhang Qiujian's Mathematical Classic*, was also known in the West, Chemla has even asked "Does this indicate that this text, or a text close to it, played a specific role in the direct transmission of knowledge from China to the West?" Of course, to "leave this question open" is a good way to draw a preliminary conclusion ([Chemla 1997]).

[15]《九章算術 • 盈不足》"盈不足術"：置所出率，盈、不足各居其下。令維乘所出率，並以為實。並盈、不足為法。實如法而一。[Qian 1963], 206.

Speaking of the West, we should now return to our starting point: Needham 19(j). We already noted that in Needham's list only two items seem related to Fibonacci. Indeed, for a long time only a few fragmentary pieces of *Liber Abaci* appeared in the general historical works and source books available to Chinese historians of mathematics. In order to promote research on mathematical and astronomical exchanges between China and Middle Asia, Professor Wu Wen-Tsun, the famous Chinese mathematician, set up the Silk Road Foundation ([Li 2009],[Guo 2004]). As part of the Wu Wen-Tsun Silk Road Program, the *Liber Abaci* has been translated into Chinese ([Ji 2008]). Now, after careful study of this classic work, we find that many problems and algorithms in it are very similar to those that appear in ancient Chinese mathematics. We therefore believe that all these similarities will help us to answer how ancient Chinese mathematical problems and methods came to influence Islamic cultures and eventually made their impact felt on the Latin West.

While undertaking this study with regard to the the special role played by Fibonacci in connecting China with the West, we discovered that this had already been noted by Louis C. Karpinski. In his book *The History of Arithmetic*, Karpinski points out:

> In any event, however, the Chinese had a real gift for numerical problem quite analogous to that displayed by the Hindus and Arabs. The oriental source of many problems which appeared in Europe in 1202 in Leonard of Pisa's voluminous work on arithmetic is not to be denied. Not only the same type of problems as in the early Chinese and Hindu works are given by Leonardo, but frequently precisely the same series of numbers, so that the oriental origin is evident. These problems were taken over by Italian arithmeticians and then from them by other Europeans. By this route the problems of ancient India and China found their way into American textbooks. ([Karpinski 1925], 30)

Certainly we have to be very cautious before arriving at any conclusions about the transmission of certain mathematical contents or methods from China to the West. That is to say, all these striking similarities require careful detailed analysis with close attention to the heterogeneities involved. This makes it essential for historians to investigate a good deal of primary source material that might further document relationships between ancient China and Renaissance Europe.

Oriental Mathematical Sources in *Liber Abaci*, especially those from India and the Islamic World

In *Liber Abaci*, Fibonacci introduced the Indian nine numerals along with their arithmetical methods. We also find that some of the problems he gives have an Indian or Islamic background. For example, the tree problem, the finding of a purse (similar to Mahavira), problems involving single false position (similar to Bhaskara II); especially noteworthy are the methods in Chapter 15, "on problems of algebra and almuchabala" where he discusses the six standard equation forms of al-Khwarzimi. So the oriental mathematical sources in *Liber Abaci* indicate its special significance as a source for understanding mathematical knowledge exchanges.

As mentioned above, these topics also require careful comparison. Additionally, as long as the transmission of at least some of them can be identified, we should be able to move on to another important issue, namely the influence of these sources on the Latin West.

The Influences of Oriental Mathematical Knowledge through Fibonacci into the Latin West

It is clear that Fibonacci mastered a great deal of oriental mathematical knowledge and then passed his learning on to his European successors. Therefore, traces of Fibonacci's influence can be found in many mathematical books from the Renaissance, such as Niccolas Chuquet' *Triparty* (1484), Pacioli's *Summa de arithmetica, geometrica, proportioni et proportionalita* (1494), Cardano's *Ars Magna* (1545), Chirstopher Clavius' *Epitome Practica Arithmaticae* (1578), and so on. For example, in the collection of Johannes Regiomontanus (1436-1476), one finds problems such as the following:

> Three people buy a horse. None of them can pay for the horse by himself, but each can do so with help from one of the others.

The way they can accomplish this may be expressed in the equation:

$$x + \frac{y}{2} = 100, \quad y + \frac{z}{3} = 100, \quad z + \frac{x}{4} = 00.$$

So, "these problems had been, since the time of Leonardo Fibonacci, a standard item in recreational mathematics" ([Folkerts 2002], 416). Folkerts also mentioned an indeterminate problem such as ([Folkerts 2002], 420):

$$x + 13y + 14z = 391,$$
$$x + y + z = 40$$

and the following remainder problem ([Folkerts 2002], 421):

$$N \equiv 2 \bmod 3 \equiv 4 \bmod 5 \equiv 1 \bmod 7.$$

It is obvious that the former is related to the "Hundred Fowls Problem," which appears in Zhang Qiujian's Mathematics Classic, whereas the latter is a problem from the "Unknown Number Problem" in Master Sun's Mathematical Classic. Certainly "the writings of Leonardo Fibonacci, [...], testify to the surge of mathematical invention" ([Rose 1975], 76), so that "the mathematical renaissance in the West may be dated from him" ([Sarton 1931], 611). Indeed, through

Fibonacci's writings the Italian mathematical Renaissance already reflects a colorful picture from the viewpoint of cultural diversity. So the *Liber Abaci* will help us not only to find out which mathematical ideas and practices were actually transmitted from the East to the West, but also to draw a clear picture of this overall process, including what was made of this knowledge by those who received it. This should even help us to explore what part these influences might have played in the flowering of the Italian Mathematical Renaissance.

Acknowledgements

This paper is part of the Programe "the Communication and Transmission of Mathematical Knowledge along the Silk Road" , supported by the Fund Program of Humanities and Social Sciences of Ministry of Education(12YJAZH037)

Bibliography

[Boncompagni 1857, 1862] Boncompagni, Baldassarre, *Liber abaci, in Scritti di Leonardo Pisano*, 2 vols., Rome.

[Chemla 1997] Chemla, Karine, Reflections on the World-wide History of the Rule of False Double Position, or: How a Loop Was Closed, *Centaurus*, 39(2): 97-120.

[Cullen 2005] Cullen, Christopher, Shifting Tectonic Plates in the History of Science: Reflections on the Life and Work of Joseph Needham, Plenary Lecture, in *Book of Abstracts of the XXII International Congress of History of Science. Beijing 24-30 July, 2005.*

[Dauben 2007] Dauben, Joseph, Chinese Mathematics, in *The Mathematics of Egypt, Mesopotamia, China, India, and Islam: A Sourcebook*, Victor Katz, ed., Princeton: Princeton University Press, pp.187-384.

[Dold-Samplonius 2002] Dold-Samplonius, Yvonne, Dauben, Joseph, Folkerts, Menso, van Dalen, Benno, eds., *From China to Paris: 2000 Years Transmission of Mathematical Ideas*, Stuttgart: Franz Steiner Verlag.

[Du 1966] Du Shiran, Shilun Song Yuan shiqi Zhongguo he Isilan Guojia jian de shuxue jiaoliu, in *Song Yuan Shuxueshi Lunwenji* (On the Exchange of Mathematics between China and Islamic Countries from the Song to the Yuan Dynasty), Qian Baocong, ed., Beijing: Science Press, pp. 241-265.

[Du 1984] Du Shiran, Zai lun Zhongguo he Alabo Guojia jian de shuxue jiaoliu (Further Discussion on the Exchange of Mathematics between China and Arabic Countries), *Studies in the History of Natural Sciences*, 3 (4)(1984): 299-303.

[Folkerts 2002] Folkerts, Menso, Regiomontanus' Role in the Transmission of Mathematical Problems, in [Dold-Samplonius 2002], pp. 410-428.

[Guo 2004] Guo Shirong, Wu Wenjun: Mathematics, Astronomy & Silk Road Fund for Study on the History of Mathematics, *Journal of Guangxi University For Nationalities* (Natural Science Edition),10 (4): 6-11.

[Guo, Dauben & Xu 2013] Guo Shuchun, Joseph W. Dauben & Xu Yibao, eds., *Nine Chapters on the Art of Mathematics in Chinese-English Critical Edition and Translation* 漢英對照《九章算術》, 3 vols., Shenyang: Liaoning Education Press.

[Ji 1999] Ji Zhigang, *Sunzi suanjing, Zhang Quijian Suanjing, Xia Houyang Suanjing Daodu* 《孫子算經》《張丘建算經》《夏侯陽算經》導讀 *(Introduction to Master Su's Mathematical Classic, Zhang Qiujian's Mathematical Classic and Xia Houyang's Classic)*, Wuhan: Hubei Education Press.

[Ji 2008] Ji Zhigang, Wang Xiaoqin, Ma Dingling & Zheng Fanglei, *Fibonacci's Liber Abaci: A Translation into Modern Chinese of Leonardo Pisano's Book of Calculation*, Beijing: Science Press.

[Ji & Ma 2004] Ji Zhigang & Ma Dingling, Some Mathematical problems and Algorithms in Fibonacci's Liber Abaci Similar to Those in Ancient China, *Journal of Guangxi University for Nationalities* (Natural Science Edition), 10(4): 18-21.

[Ji & Zheng 2006] Ji Zhigang & Zheng Fanglei, Mathesis Perennis, The Characteristics and Trends of the History of Mathematics Judged from the 22nd International Congress of the History of Science, *Studies in the History of Natural Sciences*, 24(3): 269-275.

[Karpinski 1925] Karpinski, Louis Charles: *The History of Arithmetic*, Chicago: Rand McNally.

[Li 2009] Li Wenlin, In the Spirit of the Silk Road: An Introduction of Wu Wen-Tsun's Silk Road Program on the Mathematical and Astronomical Transmission in History, 數理解析研究所講究録. 第 1648 巻, 207-215.

[Liu 2002] Liu Dun, A Homecoming Stranger: Transmission of the Method of Double False Position and the Story of Hiero's Crown, in [Dold-Samplonius 2002], pp. 157-166.

[Ma & Ji 2008] Ma Dingling & Ji Zhigang, On Fibonacci's Historical Significance and Research Value in the Exchange of Mathematics between China and the West, *Journal of Shanghai Jiao Tong University* (Philosophy and Social Sciences), 16(2): 56-60.

[Martzloff 1997] Martzloff, Jean-Claude, *A History of Chinese Mathematics*, New York: Springer.

[Needham 1959] Needham, Joseph, *Science and Civilisation in China*, vol.3, *Mathematics and the Sciences of the Heavens and the Earth*, Cambridge: Cambridge University Press.

[Qian 1927] Qian Baocong, Jiu Zhang Suan Shu Yingbuzu Shu Liuchuan Ouzhou Kao (Research on the Transmission of the Double False Position Method from China to Europe), *Sciences*, 12(6): 701-714; reprinted in *Qian Baocong Kexueshi lunwen Xuanji* (Collection of Qian Baocong's Papers on the History of Science) (1927), pp. 83-96.

[Qian 1963] Qian Baocong, *Suanjing Shishu* 算經十書 (The Classics in Methematics), 2 vols., Beijing: Zhonghua Book Company 中華書局.

[Rose 1975] Rose, Paul Lawrence, *The Italian Renaissance of Mathematics: Studies on Humanists and Mathematicians from Petrarch to Galileo*, Geneva: Librairie Droz.

[Sarton 1931] Sarton, George, *Introduction to the History of Science*, Vol.2, Baltimore: Williams & Wilkins.

[Sigler 1987] Sigler, Laurence E., *Leonardo Pisano Fibonacci:The Book of Squares*, Boston: Academic Press.

[Sigler 2002] Sigler, Laurence E., *Fibonacci's Liber Abaci: A Translation into Modern English of Leonardo Pisano's Book of Calculation*, New York: Springer.

[Vogel 1970] Vogel, Kurt, Fibonacci, Leonardo or Leonardo of Pisa, *Dictionary of Scientific Biography*, Vol.4, New York: Charles Scribner's Sons, pp. 604-613.

New Perspectives on Commercial Arithmetic in Renaissance Spain

Elena Ausejo

To Joe Dauben, bridge-builder of intercultural understanding and friendship

Introduction

Commercial arithmetics in Spain date back to the fourteenth century, though they were first printed only in the following century.[1] Actually, the first printed book on mathematics in Spain was probably the second on the subject to appear in the entire Western world: Francesc Santcliment' s *Summa de l'art d'Aritmètica*, first published in Catalan (Barcelona, 1482), and later in Spanish (Zaragoza, ca. 1487) [Malet 1998]. Six more mercantile arithmetics were published by Spanish authors before the first text that included algebra appeared in Valencia in 1552: this was Marco Aurel's *Libro primero de Arithmetica Algebratica* [Aurel 1552].[2] Arithmetic thus preceded algebra in early modern Spain, a topic that has drawn the interest of a number of historians in recent decades. Still, only some of these earliest commercial arithmetics have been studied and, in a few cases, only partially.[3]

This paper approaches commercial arithmetics in the Spanish Renaissance by contextualizing the different types of people and professions involved in the production of such books. The emphasis here falls on works and authors that have previously either been overlooked or else insufficiently studied. Biographical elements, motivations, and lines of influence will be woven into the social history. We begin with one of the earliest authors, priest Juan Andrés, a renegade Muslim whose work shows that the Arabic mathematical background still played a role in early sixteenth century Spain. Andrés stopped writing about religion and started writing about mathematics, probably for personal reasons connected with the career of his protector and a shift in the policy taken toward the Moors. However,

[1]See the following recent studies: [Caunedo / Córdoba 2000], [Caunedo 2007], [Docampo Rey 2004], [Docampo Rey 2009].

[2]These six works on arithmetic are: [Ortega 1512], [Andrés 1515a], [Ventallol 1521], [Gutiérrez 1539], [Tejeda 1546], [Icíar 1549]. They coexisted with speculative arithmetics in Latin published by Spanish academic authors such as Pedro Ciruelo (Tractatus arithmeticae practice, Paris, 1495; Cursus quatttuor mathematicarum artium liberalium, Alcalá, 1516), Gaspar Lax de Sariñena (Arithmetica speculativa, Paris, 1515), and Cardinal Juan Martínez Silíceo (Ars Arithmetica in Theoricem et Praxim scissa: omni hominum conditioni superque utilis et necessaria, Paris, 1514, 1518, 1519, 1526; Valencia, 1544) [Sánchez Salor / Cobos Bueno 1996], and also with commodity trading manuals such as the Books of Almutaçafes (inspectors of weights and measures) by Adrián de Aynsa (1510, 1577, 1595) and Pascual de Abensalero (1609) [Ausejo / Hormigón / Usón 2001].

[3]See the following: [Malet / Paradís 1984] for [Ortega 1512], [Caunedo 2009] for [Andrés 1515a], [Mesquida Cantallops 1996] for [Ventallol 1521], [González Ferrando 1956] for [Tejeda 1546].

D.E. Rowe and W.-S. Horng (eds.), *A Delicate Balance: Global Perspectives on Innovation and Tradition in the History of Mathematics*,
Trends in the History of Science, DOI 10.1007/978-3-319-12030-0_9

in his *Sumario breue d'la pratica dela arithmetica* [Andrés 1515b], he carefully emphasized that his main reference, the Franciscan friar Luca Pacioli, was a master in sacred theology. Although it is well-known that Andrés' arithmetic includes the first printed references to the *Summa de arithmetica* in Spain, this paper considers how Pacioli's work may have come to Andrés' attention. At the same time, it considers why he nevertheless decided to bypass algebra and to deal only with arithmetic in his book.

Our focus next turns to Juan de Icíar, the most important calligrapher of the Spanish Renaissance, whose *Practical Arithmetic* [Icíar 1549] closed the series of non-algebraic Spanish arithmetics published during the first half of the sixteenth century. Here we discuss Icíar's choice of contents for his *Practical Arithmetic* as well as his use of fractions in order to operate with any combination of units, the main difficulty in mercantile arithmetic before the adoption of decimal systems of measurement. As a matter of fact, Icíar's detailed explanations of calculation processes with submultiples connected, step by step, contemporary mercantile practice with the new arithmetic. His books thus reflect commercial arithmetic at work during this era of incipient capitalism, whose practitioners had to acquire a full understanding of these mathematical concepts and properties. Moreover, this proves that the assimilation of modern arithmetic was not so much a problem due to the introduction of Arabic numerals and new algorithms, but rather it stemmed from practical difficulties that arose from the need to handle different systems of units.

Icíar's works on writing and reckoning [Ausejo 2012] are also considered in order to approach the publication of arithmetics as a business. Between 1559 and 1566, the bookseller Miguel de Suelves published three editions of Icíar's *Subtlest Book that teaches how to write and reckon perfectly in the same order as a teacher does with his student*[4], the first printed work in Spanish devoted to the teaching of both writing and arithmetic. This book combined Icíar's *Arte subtilissima por la qual se enseña a escreuir perfectamente* [Icíar 1555] with a reworking of a book by another author. This was *Arte breue y muy prouechoso de cuẽta castellana y arismetica*, an elementary treatise published in Toledo by Father Juan Gutiérrez de Gualda [Gutiérrez 1539], a priest from the small Castilian town of Villarejo de Fuentes. In all these works, Icíar followed in the footsteps of the Italian calligrapher Giovanni Antonio Tagliente, who also published works on arithmetic jointly with a relative, Girolamo Tagliente, in Venice. Since Icíar's biographers consider that he acquired his abilities as a calligrapher in Italy, where he drew on the works by Vicentino, Tagliente and Palatino [Three Classics 1953], this paper considers the influence of Giovanni Antonio Tagliente on Icíar's mathematical works.[5]

[4] [Icíar 1559a], [Icíar 1564], [Icíar 1566].

[5] Giovanni Antonio Tagliente is known not only as a calligrapher, but also as the first successor of Luca Pacioli in bookkeeping [Doni 2007, 78-80]; [Smith 1908, 141]. He is also listed -as Giannantonio- as co-author of printed books on arithmetic together with his relative Girolamo -Hieronymo- [Smith 1908, 114-116], according to Boncompagni exhaustive investigations [Boncompagni 1863, 140-141], although the main author seems to be Girolamo. Anyway, thirty-four

Gutiérrez's and Icíar's arithmetics have been largely overlooked, the first being rather elementary, the second because it was a truely rare book until the beginning of this century.[6] Furthermore, Icíar's *Practical Arithmetic* has been repeatedly confused with Gutiérrez's elementary treatise.[7] Here we will discuss both Icíar's and Gutiérrez's works as typical educational tools for the incipient bourgeoisie in Spain, a country in which the Italian style of *botteghe dell'abaco* has yet to be documented. The Catholic Church was by no means hostile to this new type of non-university education: in fact, three of the seven Spanish non-algebraic printed arithmetics were authored by clerics.[8] In all three cases, however, only scant biographical data are available about these authors.

Juan Andrés' *rara arithmetica*

Juan Andrés has been traditionally considered a clergyman from Zaragoza because he introduces himself as such in the foreword and again in the colophon of his book [Andrés 1515a, 2r, cxliiij]. There he also informs the reader that it was written in Zaragoza in 1514 and printed in Valencia by Juan Joffre, one of the most active printers in this town at the beginning of the sixteenth century.[9] Two variants of this edition are known, the first dedicated to the Count of Oliva, the second to Bishop Martín García. Both are very rare: eight copies of the first variant and only two of the second are listed in [Wilkinson 2010, 25].[10] This source refers to the author as *Andrés de Zaragoza, Juan* (entries 657-658) in order to distinguish him from entries 651 and 653 (*Andrés, Juan*), and entries 655-666 (*Andrés de Játiva, Juan*). But, as will be shown below, Juan Andrés was the author of all six.

The first known reference to Juan Andrés as a faqîh (an Islamic jurist) who converted to Catholicism and who wrote against Muslim errors appears rather vaguely in Nicolas Antonio's *Bibliotheca Hispana vetus* [Antonio 1696, 2, 266-267]. More precise information can be found in the enlarged and revised edition that was published almost a century later [Antonio 1788a, 2, xiii, 324-325, 369-370, 443], where he is cited as the author of a book on the Christian-Muslim religious controversy [Andrés 1515c]. One also finds mention of an author named Juan An-

editions of Tagliente's works on arithmetic were published between 1515 and 1596, which shows their wide audience among merchants. In these circumstances, it is no wonder that Icíar got to know some edition.

[6] Only two copies were known until 2005 (Columbia University Libraries Shelf mark PLIMPTON 511 1549 Ic4 [Smith 1908, 247-249]; British Library Shelf mark General Reference Collection C.62.h.5.) [Wilkinson 2010, 415, No. 10502]. I am most grateful to my colleague Manuel José Pedraza, who gave me a copy of Practical Arithmetic, which he located at the University of Salamanca. In 2006, a new copy entered the library of the Complutense University of Madrid (from the library of Francisco Guerra); this copy is available electronically since 2007.

[7] For instance: [Smith 1908, 249], [Cotarelo y Mori 2004, 366], [Mayordomo 2002, 94-95], [Martínez Pereira / Torne 2007, 60].

[8] [Ortega 1512], [Andrés 1515a], and [Gutiérrez 1539].

[9] The book was published on August 30, 1515.

[10] There is a facsimile edition of the first variant prepared by the University of Salamanca, 1999. Both variants are now electronically available.

drés who wrote about arithmetic in various places in Antonio's register of Spanish 16th-century authors [Antonio 1788b, 1, 633; 2, 413, 470, 522, 609]. Although the 19th-century Valencian bibliographer Juan Pastor Fuster [Pastor 1827, 1, 63-64] referred to Juan Andrés as the author of both books—[Andrés 1515a] and [Andrés 1515c]—most libraries and scholars continued referring to Juan Andrés—or Juan Andrés, de Zaragoza [Norton 1997, 138, No. 1192]—as the arithmetician, and to Juan Andrés de Játiva as the renegade preacher. The high prestige of Antonio's indispensable, still not superseded works, and the difficulty in comparing the two variants of the arithmetic with the first edition of the religious controversy[11] account for the persistent error. But the authorship question was raised in 2003 when the book on religion was again republished [Ruiz García 2003, 40]. It now emerged that, like the text on arithmetic, it was printed by Juan Joffre in Valencia in 1515;[12] moreover, both books share the same xylography representing Saint Mary of the Pillar with praying pilgrims on the left[13] and courtiers on the right.[14] What is more, one of the variants of the book on arithmetic was dedicated to Bishop Martín García, who is referred to as "patron and lord" of the author in the book on religion [Andrés 1515c, 3v].

The definitive answer to this authorship question can be found in the foreword to the edition of the arithmetic dedicated to the Bishop, where the author refers to himself as a "servant" to "my principal lord".[15] He writes further that he could not submit any work he might produce, *either on arithmetic or in favor of the Catholic faith*, to any worthier person in all of Aragon, Catalonia or Valencia, referring to him as an expert in liberal arts and mathematics. He explains that his book had been approved in Valencia by intelligent and reliable persons; and that it has been composed according to the doctrine, rules, and chapters of Lucas de Burgo, better known as Luca Pacioli, master in sacred theology [Andrés 1515b 2r-2v].[16]

[11]Only two copies are quoted by [Norton 1978, 431-432]; a third one is quoted by [Wilkinson 2010, 25]. None of them is in a Spanish-speaking country.

[12][Andrés 1515c] was published on November 13, but the colophon does not say it was written in Zaragoza.

[13]St. James in the foreground: according to the tradition, the first Marian apparition in history appeared to Saint James the Apostle, the brother of Saint John the Evangelist, on the bank of the river Ebro in Zaragoza, as the Virgin Mary had promised Saint James that when he needed it most in his difficult mission to the pagans in Spain, she would appear to him to encourage him. This intends to show the connection of the author with Zaragoza, and with preaching.

[14]King Ferdinand of Aragon and Queen Isabella of Castile appear in the foreground. This intends to show the connection of the author with both of them.

[15]In addressing the Count of Oliva in the other variant of the edition, he refers to himself as a clergyman. Changes from one variant to the other affect not only the dedications, but also much else in the two forewords. The title page of the Count of Oliva variant includes his coat of arms, but the title page of the edition, dedicated to Bishop Martín García, is missing in the two copies cited by [Wilkinson 2010, 25]. Following the references given by [Norton 1978, 431], who was unable to locate the latter variant and described it according to secondary sources, I found a reduced facsimile reproduction of this title page (fig. 1) [Vindel 1903, 256, No. 2839]: the ornamental separation between the coat of arms and the title was clearly designed in order to print two variants. A more elaborate printed version of Bishop García's coat of arms can be seen from the title page of [García 1520].

Figure 1: Title page of [Andrés 1515b]

Some hints of how Juan Andrés came to know Luca Pacioli's *Summa* emerge from his autobiographical notes in the foreword to his book on religion [Andrés 1515c, 2v-4r]. There we learn that he was born in Játiva (Valencia) and was educated in Islamic theology and jurisprudence by his father, faqîh Abdalla, whom he succeeded upon the latter's death.[17] According to his own testimony, "the darkness of his understanding was removed and clarified and his soul's eyes were opened" on Assumption Day (August 15) 1487.[18] He asked to be christened as Juan Andrés (referring to the two apostles), and started to preach and convert Moors in Valencia as soon as he was ordained. He was next called by King Ferdinand and Queen Isabella to preach in the conquered territory of Granada, where he was promoted to Canon. Later, he was sent by Queen Isabella to Aragon to continue preaching for the conversion of Aragon's Muslims, but this ended with her death in November 1504.[19] He then translated the Koran and the Sunnah into Aragonese by order of the royal preacher, Martín García, Inquisitor of Aragon, and after that he wrote

[16]In so doing, he supported his work with the prestige of a Catholic friar. No Arabic source is mentioned, except when dealing with multiplication [Andrés 1515a, XXXIr]. However, the reference to Luca Pacioli does not appear in the foreword of the variant dedicated to the Count of Oliva.

[17]The education of a faqîh would normally include mathematics to some extent, since inheritance law was part of this vocation [Ausejo / Benito 2006].

[18]This happened while he was listening to Juan Marqués' sermon in the cathedral of Valencia. The Dominican friar Juan Marqués was King Ferdinand of Aragon's confessor; he later became Bishop of Pati (Sicily) and died in Syracuse in 1499.

[19]As a matter of fact, King Ferdinand of Aragon followed a milder policy towards Muslims in his territories. Although the Moors were expelled from Castile by Isabella and Ferdinand in 1502, it was their great-great-grandson, King Philip III, who expelled them from Aragon a century later, in 1609.

his book on the Christian–Muslim religious controversy [Andrés 1515c]. The latter was approved by five clergymen[20] and published on November 13, 1515. Shortly before, on August 30, his book on arithmetic—written in Zaragoza in 1514–-had already been published.

These circumstances reveal how Andrés' life converged with the biography of Martín García [Montoza 2013, 7-10], a humble shepherd from Caspe (Aragon) who ended his life as a cardinal [Hebrera 1700, 114-115]. García took his Master in Theology in 1480 at the exclusive *College of Saint Clement of the Spaniards* in Bologna. He was promoted to canon of the cathedral in Zaragoza in 1481, and was later appointed Inquisitor in 1486. After delivering a sermon in the presence of Ferdinand and Isabella on November 9, 1487, he was appointed royal preacher by the king, while becoming the queen's personal confessor. In 1493 he was also appointed royal reformer of nuns, a position that enabled him to make the acquaintance of the new confessor of the queen, the Franciscan friar Jiménez de Cisneros, who was responsible for the mandatory conversions (in order to avoid expulsion) in Granada during the period 1499-1502.[21] On April 1500, Martín García was called by the queen to preach in Granada, due to his knowledge of Arabic, along with Juan Andrés. García was appointed bishop of Barcelona in 1512 and served in the Roman Curia in connection with a special mission [Hebrera 1700, 113]. He returned to Barcelona in April 1515, retired from office in 1519, and died, aged eighty, in 1521.

These biographical data show the convergence in the lives of these three clergymen, Juan Andrés, Martín García, and Jiménez de Cisneros, who were active in Granada ca. 1500. Juan Andrés made contact with the Franciscans, who comprised a potential, though non-exclusive network for the international circulation of Luca Pacioli's *Summa*.[22] He was also in close contact with Martín García, probably until the latter left for Rome. García may also be considered as a possible source for Juan Andrés' access to Pacioli's *Summa*; García might have acquired the necessary mathematical skills at the University of Bologna, and he was also in Rome when Pacioli lectured there in 1514.[23]

[20]Among them was Luis Mercader, Bishop of Tortosa since January 1514, which means that the book was only approved after this date.

[21]Cisneros had worked in Rome as a consistorial advocate between 1459 and 1465; he was a Franciscan friar since 1484 and the Minister Provincial of this Order for Castile since 1494. In 1499 he founded the Complutense University in Alcalá, now in Madrid. He was afterward elevated to higher office, becoming Cardinal Cisneros, Regent of Spain for Charles I (afterward Emperor Charles V). He died in 1517.

[22]As is well known, sometime between 1470 and 1477 Pacioli was ordained as a friar in the Franciscan order. In 1504 he was elected superior of his order for the province of Romagna, and in 1505 he was accepted as a member of the Franciscan monastery of Santa Croce in Florence ["Pacioli, Luca.", *Complete Dictionary of Scientific Biography*, 2008. *Encyclopedia.com*. 19 Feb. 2014 <http://www.encyclopedia.com>].

[23]In all events, the arrival of Pacioli's *Summa* in Spain can be documented as being no later than 1512. It was purchased in that year by Ferdinand Columbus, a Spanish bibliographer and cosmographer who was the second son of Christopher Columbus [Seville, Biblioteca Capitular y Colombina, Shelf mark 15-6-31]. Later influences of Pacioli's *Summa*, including algebra, are

As noted above, Juan Andrés claimed that his book was composed according to the doctrine, rules, and chapters of Lucas de Burgo, but this assertion refers only to arithmetic.[24] According to its title, he was only writing a brief summary that would avoid any complexities. Thus, he merely referred to Pacioli in order to justify the rules of false position based on proportion and proportionality, a subject he explicitly considered superfluous for this brief treatise [Caunedo 2009, 93]. Furthermore, he provided solutions to avoid both the rule of double false position [Andrés 1515a, CXLv-CXXXXJv], and the rule of algebra [Andrés 1515a, CXXXXJv- CXLIJr].[25] However, he did plan to write a book on algebra later [Andrés 1515a, LJv], which suggests that this commitment to mathematics was not accidental. It may well have been prompted, in fact, by a need to adapt to new circumstances. In 1514, still in Zaragoza, he surely knew that his patron, Martín García, now in Rome, would not be returning to Zaragoza, but rather was on his way to Barcelona where he would become bishop. This circumstance seems to have led Juan Andrés back to his homeland, Valencia, where he could seek new patronage. This would account for the variant edition of his book on arithmetic, dedicated to the Count of Oliva, Serafín de Centelles y Urrea (1460-1536), a member of the new nobility.[26] His new patron was an aristocrat engaged in the new world of commerce. He not only successfully traded the products of his dominion but also devoted himself to literary patronage. At the same time, he served as accountant (1514-1517) and deputy (1530-1532) of the *Generalitat*, the commission in charge of collecting taxes on trade goods.

Juan de Icíar's works on arithmetic

Little is known about the life of Juan de Icíar. He was born in 1522 or 1523 in Durango, Biscay (Basque Country, which belonged to Castile at that time).[27] By the end of 1546 he moved to Zaragoza, it would seem because of some kind of personal disgrace, this according to the dedication from his *Practical Arithmetic* [Icíar 1549, ii]. There he settled as a *scriptor de libros* and teacher. We know of

studied in [Docampo Rey 2006]: the Catalan manuscript Ms. 71 of Sant Cugat (Arxiu de la Corona d'Aragó, Barcelona), containing a very interesting set of notes on algebra mainly based on Pacioli's *Summa*, was probably written by the Majorcan Joan Ventallol, who published a commercial arithmetic [Ventallol 1521].

[24]It contained nothing on algebra, geometry, or bookkeeping. The first arithmetic containing bookkeeping printed in Spain was published by Gaspar de Texeda [Tejeda 1546], a scrivener from Zaragoza [Protocolo de Bartolomé Malo, 1547, ff. 325/326v. In: *Archivo Histórico de Protocolos Notariales de Zaragoza*, by courtesy of Prof. Manuel José Pedraza].

[25]He uses the terms rule of *algibra* or rule of *algo* and *cosa*, meaning equations of the type $ax = b$.

[26]His father had been the first Count of Oliva since 1449. According to the foreword of [Andrés 1515a, 2v], Juan Andrés was older than the Count, as he claims to know him since the latter's childhood. Thus, in 1515 Juan Andrés was probably in his seventies.

[27]This dating is based on the engraving of his portrait, first published in his *Orthographia pratica* (1548), which says he was 25 years old at the time the woodblocks for this edition were being made in 1547 [Pedraza 2007, 20-25].

several of his apprenticeship contracts and of one contract for employment as a private tutor for training in reading and writing. In 1554 he married one Catalina Carrión [Pedraza 2007, 17-20, 23], and there is documentary evidence of his activity in Zaragoza until 1559 [San Vicente 1969]; [Icíar 1559a].[28] In 1773 he retired to Logroño after being ordained [Cotarelo y Mori 2004, 351].

Figure 2: Portrait of Juan de Icíar in *Arithmetica Practica* [Icíar 1549]

Once in Zaragoza, Icíar collaborated with the engraver Juan de Vingles, to prepare woodblocks for the edition of his *Orthographia pratica* (1548). This work evinces Icíar's deep knowledge (probably acquired in Italy) of the works of Vicentino, Tagliente, and Palatino [Cotarelo y Mori 2004, 351]. For this purpose, on 5 June 1547 he and de Vingles formed a business partnership with the tallow-chandler Alonso Fraylla, who joined the company with 100 pounds to cover publishing costs [Pedraza 2007, 23-25]. Following the Italian masters, the book was printed in quarto by Bartolomé de Nájera, one of the two successors of the German Jorge Coci, a flagship printer in 16th-century Spain. This joint venture was thus made possible by combining Fraylla's innovative entrepreneurship with Icíar's and Vingles's expertise as artisans, the former as calligrapher, the latter as engraver.

Soon after the production of *Orthographia pratica*, Icíar published his *Practical Arithmetic* [Icíar 1549], though not with the same company. For this purpose, he partnered with the bookseller Miguel de Suelves, an *infanzón* (member of the lower–ranking gentry).[29] From 1553 until the end of Icíar's life they worked closely together. The book was printed in folio by Pedro Bernuz, who at the time was reputed to be the best printer in Aragon.[30] Eleven days after the publication of *Practical Arithmetic*, Icíar sold his share of the edition to the merchants Juan and

[28]This book ends with the following colophon: "Ivan de Yziar wrote it in Zaragoza in 1559." According to the colophon, the book was completed in Zaragoza on 27 June 1559, and was printed by the widow of Esteban de Nágera.

[29]An *infanzón* belonged to the untitled, lower strata of the nobility who were exempt from paying taxes.

[30]Pedro Bernuz, a former notary, was married to Jorge Coci's niece and inherited Coci's printing house.

Antón Alberite for 100 gold ducats (110 pounds). This was to be paid again should the book be reprinted under the same title anywhere in Spain or abroad [Pedraza 2007, 30-32], which may well account for why it was never reissued. This was a healthy sum of money for Icíar, especially when compared with the 15 gold ducats he charged students, his annual fee for tutoring them in reading and writing [Pedraza 2007, 20]. Moreover, the Alberite brothers were keen to circulate his work, which helped to promote his reputation.

At this point, Icíar seemed to follow in the footsteps of the Italian calligrapher Giovanni Antonio Tagliente, who also published works on arithmetic jointly with his relative Girolamo Tagliente in Venice [Tagliente 1520], [Tagliente 1525], [Tagliente 1540?]. However, Icíar's partner, Juan de Vingles, might also have inspired work on the *Practical Arithmetic*. His father and brother had been printers in Lyon,[31] where Ortega's and Ventallol's arithmetics were printed [Ortega 1512]; [Ventallol 1521].[32] At any rate, Icíar's business relationships show that he made a smart choice when he settled in Zaragoza, a place with a thriving publishing industry.

The success of *Orthographia pratica* led to the publication of a second enlarged and improved edition, entitled *Arte subtilissima por la qual se enseña a escreuir perfectamente* (1550). This was printed by Pedro Bernuz, who also played a key part in the success of this project, a typical Renaissance venture. This time the edition was sold to booksellers Suelves and Francisco García for 100 pounds, to be paid in May 1551, but under the condition that no new edition of the book could be published during the following two years. In September 1551, engraver Vingles sold his share of the company to Icíar and Fraylla [Pedraza 2007, pp. 25-34]. As a result, Icíar gained his reputation as a master among calligraphers in Spain. He was even called by King Philip II to work at El Escorial as scriptor, and also served as private tutor to the crown prince during the period 1550-1551 [Alonso García 1953, 21-22]; [Cotarelo y Mori 2004, 351].

Suelves published the third edition of *Arte subtilissima* in 1553, and the fourth in 1555.[33] The 1555 edition must have been judged positively, since in 1559 Suelves launched a new edition with a new title, *Subtlest Book that teaches how to write and reckon in the same order as a teacher does with his student* [Icíar 1559a]. This edition contains *Brief and Useful Art of Castilian reckoning and arithmetic,*

[31] His father was active there between 1495 and 1511, his brother Pierre between 1525 and 1532 [Pedraza 2007, 23].

[32] Actually, *Practical Arithmetic* contains precise references to Ortega's work as well as to that of the French author Pellos [Icíar 1549, LXIII]. However, Ortega's book was the most popular mercantile arithmetic of the time. It was already published in Italian (Rome, 1515; Messina, 1522), and then in Spain under the title *Tratado subtilissimo de arismetica y de geometria* (Sevilla, 1534, 1537, 1542).

[33] The latter [Icíar 1555] included an anonymous *Arte breue y prouechoso de cuẽta castellana y arithmetica, donde se muestrã las cinco reglas de guarismo por la cuẽta castellana, y reglas de memoria,* a work that Suelves also published separately that very year as [Arte breue y prouechoso de cuẽta castellana y arithmetica 1555]. There is also an edition of *Arte subtilissima* without arithmetic [Martínez Pereira 2006, 343-355].

where the five rules of figure on Castilian reckoning are shown, and memory rules. And now in this last edition very amusing and useful calculations, from Fray Iuan de Ortega's book, have been added, and at the end an abridged account in maravedis [Icíar 1559b]. What this long title portends will be considered briefly below. Here, we merely note that *Subtlest Book* was the first printed work in Spanish aimed at teaching both writing and reckoning. New editions were launched in 1564 [Icíar 1564] and 1566 [Icíar 1566], some including [Gutiérrez 1564a], others not.

Suelves published Gutiérrez's book a third and final time in 1569. It was licensed by the Archbishop of Zaragoza and the Inquisitors, and printed by Miguel de Güesa [Gutiérrez 1569].[34] In the meantime, Andrés de Angulo had also published Gutiérrez's work in Alcalá (Castile) in 1564 with His Majesty's privilege [Gutiérrez 1564b]. It included the table of conversion of *reales* and crowns into *maravedis* of the 1559 edition. This is the only edition where the author is identified as Reverend Juan Gutiérrez de Gualda, priest of Villarejo de Fuentes.

Altogether, Suelves published eight books of arithmetic in twenty years (1549-1569), an average rate of one edition every two years and a half. Leaving aside *Practical Arithmetic* [Icíar 1549], the remaining seven editions were published in a fourteen-year interval. These figures can only be understood by considering printing as a business within the framework of the non-university education of the incipient bourgeoisie in Spain. Not only merchants, but also the liberal professions were progressively able to appreciate the advantages of the new arithmetic based on Arabic numerals, and the importance of assimilating it as part of their education. Furthermore, those editions prove that there was a public demand for this type of elementary book during the period of transition from traditional to modern arithmetic.[35]

As to the respective responsibilities of Icíar and Suelves in this venture, it should be noticed that Icíar gave no indication he intended to combine *Ortographia Pratica* with *Practical Arithmetic*, when they were first presented to print the former in quarto and the latter in folio. Moreover, *Arte subtilissima* was marketed together with a treatise on arithmetic only after Suelves became Icíar's editor and launched his second edition [Icíar 1555]. There was no possibility of using Icíar's *Practical Arithmetic* as such, since the formats were different and the copyright had already been sold. Furthermore, *Practical Arithmetic* "intended to teach any merchant and trader" [Icíar 1549, XXVIIIv.], whereas Gutiérrez referred to "those willing to be liberal [independent] accountants" [Gutiérrez 1539, 40]; the new foreword to *Arte subtilissima* [Icíar 1555] presented it as a brief and useful book for the self-training of accountants. This is consistent with an audience of young students who aimed to earn their living as civil servants or estate managers. The

[34]The mention of this license proves that the edition was published in accordance with the law [García Cuadrado 1996, 146-148].

[35]It should be noticed, though, that the most successful author in the second half of the 16th century was Bachelor Pérez de Moya, especially his *Arithmetica practica y speculatiua* (Salamanca, 1562), which went through thirty editions in two centuries [Smith 1908, 308-311].

merchant audience could still use *Practical Arithmetic* since, as far as we know, Suelves did not sell his half of the edition to the Alberite brothers. Finally, although Suelves might have been the promoter of this venture, it is hard to believe that Icíar played no part in the successive reworkings of Gutiérrez's original treatise, at least in the 1555 and 1559 editions. It is even less likely that Suelves would have used his name, his portrait, and his renown without permission, even if only given tacitly.

Comparative Textual Analysis

The anonymously printed *Arte breue y prouechoso de cuẽta castellana y arithmetica* [1555] has traditionally been considered a mere reprint of Gutiérrez's elementary treatise [Gutiérrez 1539], although the original title is slightly different (*muy prouechoso* instead of simply *prouechoso*, and *Arismetica* instead of *arithmetica*). A careful comparison of these two arithmetics reveals, however, significant differences between them. Not only does the 1555 edition have a different typography and a new foreword, it also contains no fewer than eight new chapters. Indeed, these two editions are identical in content only from the ninth chapter onwards.

The original 1539 edition begins, after a foreword in the form of a dedication, with a statement of principles regarding *quenta castellana* and *arismetica* (Roman and Arabic numerals). The author briefly explains his purpose, namely, to explain the five fundamental operations ordinarily used in Spain—addition, subtraction, multiplication, division by a single unit, and division in general— "in order to avoid cheating" by *cuenta castellana* (Roman numerals) or with *guarismo* (Arabic numerals). He thus starts by explaining Arabic numerals and place values along with their names, from the unit all the way up to *quento de quentos* (10^{12}), emphasizing that every rule will be illustrated with examples in both Roman and Arabic numerals [Gutiérrez 1539]. The "real" check (inverse operation for addition) and the checks by sevens and nines, which involve subtraction and multiplication, are explained in the second chapter. This part of the original 1539 edition was taken out completely in 1555 and replaced by a text with a new foreword and eight new chapters, which can be summarized as follows. In the foreword, the anonymous author vaguely refers to foreign books, and highlights the brevity and usefulness of his own work, which deals with the contents for coins, and weights and measures used in Spain. He announces that he will use Roman numerals in addition and subtraction, but not in multiplication and division, since they are no longer applicable. He also states that his book can be studied by anyone who wishes to become an accountant and that it requires no teacher. The contents of the eight chapters, in brief, are:

Chapter 1: Unity and number (with references to Euclid), odd and even numbers.
Chapter 2: Arabic numerals and place values.
Chapter 3: "Letters" in *cuẽnta llana* (Roman numerals) and a nearly

one-page long table to illustrate place values in *guarismo* (Arabic numerals).
Chapter 4: Coins, weights and measures in Spain: coins in Aragon, Valencia, Barcelona, Perpignan, Navarra, Castile, and Portugal; weights and measures only in Aragon, Castile, and Valencia.
Chapter 5: Addition: definition; insistence on announcing that each example will be given in Roman and Arabic numerals, *para que cualquiere mas facilmente las pueda entender y declarar* ("so that anyone can more easily understand and declare them").
Chapter 6: Addition: two examples in Arabic and Roman numerals.
Chapter 7: Addition: check of nines (not as detailed as the check of sevens in the 1539 edition), along with a statement favoring the inverse operation as the true check.
Chapter 8: Addition of *ducados, sueldos* and *dineros* in Aragon.[36]

Two additional examples come from the 1539 edition, where certain weights and measures given in Arabic numerals are added in order to illustrate how the same process is valid in such cases, as well as for any monetary system. Chapter 10 includes a final reference to the Portuguese monetary system, which was not present in the 1539 edition. The remaining six chapters in both editions are devoted to subtraction, multiplication, division by a single unit, and division in general, followed by applications to currency exchange, including addition of (geometric) series "for those willing to be liberal [independent] accountants" [Gutiérrez 1539, 40; *Arte breue y prouechoso de cuẽta castellana y arithmetica* 1555, 44]. This part of arithmetic is indirectly related to loans and interests, which were officially condemned by the Catholic Church.

As regards changes in the 1555 edition, apart from the initial cultural remarks in the first chapter, the new material tends to reinforce the practical character of the work. The new reckoning techniques are applied to monetary systems more complex than the Castilian, and the difficulties involved in the checks of sums are avoided, following the lead of Icíar's *Practical Arithmetic.* It is also worth noting that, despite what is announced in the new foreword of 1555, the constant reference to Roman numerals seems to indicate there was a need to address this in elementary books that marked the transition from medieval to modern arithmetic. Perhaps Icíar's *Practical Arithmetic* had proved too demanding. However, the new 1555 foreword contradicted Icíar's foreword to the *Practical Arithmetic* in regard to the the need for a teacher [Icíar 1549, ij]. In the absence of data to discern the responsibility of the three main actors behind the 1555 edition—Icíar, Gutiérrez, and Suelves—we should point out that the 1539 edition may well have been free of royalties since it was printed with imperial privilege for ten years.[37]

[36]Unlike Castile, where merchants worked in *maravedis*, merchants in Aragon worked with *ducados, sueldos* and *dineros* (1 *ducado* = 22 *sueldos*, 1 *sueldo* = 12 *dineros*). This required first to sum the *dineros*, carrying the whole part of the result over to *sueldos*, then adding these, and repeating the same procedure with *ducados.*

[37]On printing privileges in Castile and Aragon see [García Cuadrado 1996, 128-129].

As noted above, in 1559 Suelves launched a new edition in two parts, *Libro Subtilissimo* [Icíar 1559a]. This ends with a colophon insisting on Icíar's authorship ("Ivan de Yziar wrote it in Zaragoza in 1559"). Still, no author is mentioned on the title page of the part devoted to arithmetic [Icíar 1559b], although Icíar's portrait is printed on the back. It was clearly reprinted from the 1555 edition of *Arte breue y prouechoso de cueta castellana y arithmetica* (including the misprint giving chapter 14 as VIII that appears in Gutiérrez original treatise [Gutiérrez 1539]). However, this new edition was also enlarged with sixteen additional pages devoted to practical rules, introduced through brief recipes, examples and problems, including the use of fractions. Among its topics are:

1. "Six very brief and useful rules for calculating house rentals, servants' wages, and other daily expenses" using simplified recipes for the rule of three. Examples were given for calculating daily wages in *dineros* from the monthly salary in *sueldos* or from the annual salary in pounds; or the monthly salary in *sueldos* from the annual salary in pounds, for instance, the daily wage in *dineros* is equivalent to two fifths of the monthly salary in *sueldos*.

2. Five kinds of Boethian proportions: $ma : a$, $(m+1) : m$, $(m+n) : n$, $(mn+1) : m$, $(mn+k) : m$. These were illustrated with twelve examples where the rule of three is implicitly used and linear equations and systems in two variables (dependent or not) are solved. As advertised in the title of the book, problems number 3, 4, 7, and 11 were inspired by corresponding problems in [Ortega 1512, 180, 177, 78].

3. Four additional solved problems, the first taken from [Ortega 1512, 178], and the second a problem involving the rule of fellowship.

4. Four proposed problems concerning the rule of fellowship.

It should be added that the exercises inspired by Ortega were not solved in the same way, nor was any mention made of the rule of three or the rule of fellowship. This use of recipes and lack of explication might mean that this arithmetic was intended to essentially displace, in practical terms, Ortega's highly reputed book with 203 folio pages. Clearly it required far less theoretical effort. The reference to Boethian proportions, the sacred core of scholastic mathematics [Høyrup 2009, 107], was surely added just to give it a touch of class.[38]

[38]Finally, the 1564 edition of Gutiérrez [Gutiérrez 1564a] (eventually included in Icíar's *Libro Subtilissimo* [Icíar 1564]) corresponds to Gutiérrez's first edition [Gutiérrez 1539], except for the initial table of contents and the final inclusion of nine exercises with beautiful illustrations from Icíar's *Practical Arithmetic* [Icíar 1549, XLIII-XLV]. It also shows an improved presentation as regards print quality and organization, with new chapter headings (resulting in eleven chapters instead of eight) and subheadings.

Juan de Icíar's Practical Arithmetic

Juan de Icíar's *Practical Arithmetic* [Icíar 1549] was a mercantile arithmetic, conceived for educational purposes as a primer for those who wished to acquire essential mathematical skills. It includes most of the basic contents of printed arithmetics: Arabic numerals and place values; fundamental operations (addition, subtraction, multiplication, division and, eventually, square root) and checks; fractions; and mercantile arithmetic. This latter subject ordinarily comprised such topics as the rule of three, the rule of fellowship, rules of alloying (eventually, rules of barter, rules of testaments), and rules of (at least single) false position; optional topics included series, financial incomes, and calculations of interest [Malet 1998, 21]. Icíar's text thus offers a spotlight on the teaching of arithmetic in mid-sixteenth century Spain, where an Italian–style *botteghe dell'abaco* was lacking. There was a great difference between *Practical Arithmetic* and the various editions of Gutiérrez's book, as will be seen from the analysis that follows. Here we will deal with Icíar's choice of contents, analyse his educational goals, and discuss the way he uses fractions in order to operate with any combination of units. This latter problem was actually the main difficulty in mercantile arithmetic before the adoption of a decimal system of measurement.

The title page of Icíar's *Practical Arithmetic* is self-explanatory, announcing a book on practical[39] arithmetic addressed to anyone who wishes to learn this subject; the figures 1234567890, printed below the author's name, implicitly suggest that calculations with Arabic numerals form a central theme. The dedication bows to classical authorities and works—Pythagoras, Caelius Rhodiginus' *Antiquarium Lectionu*, Plutarch, Plato, and Jacobus Faber Stapulensis (Lefèvre d'Estaples)—in order to legitimize mathematical learning as the basis of true knowledge. Geometry and, above all, arithmetic form the two foundational disciplines of the quadrivium, since through them we learn how to count, weigh, and measure everything that can be quantified [Icíar 1549, ij].

Figure 3: Title page of *Arithmetica Practica* [Icíar 1549]

[39] *Practical* means "commercial" [Smith 1958, 492-493].

Icíar's *Practical Arithmetic* was printed in folio, which suggests this was a work to be consulted and preserved; perhaps a teacher's textbook, but not a handbook for students attending a school. Actually, Icíar even warned against self-learning: "mathematics can be neither comprehended nor understood without a very good and provident doctor"[Icíar 1549, ij]. As compared with previous Spanish mercantile arithmetics, this book was of much higher quality and, indeed, beautifully illustrated.[40] Even more remarkable is Icíar's didactical approach, which is reflected both in the structure and content of his book. The detailed explanations, the presentation of subjects in order of increasing difficulty,[41] the combination of "theory" and practice (fifteen to twenty examples in several parts), and the choice of contents suitable for the audience he intended to reach, all these point to his mastery of mercantile arithmetic and to the originality of his *Practical Arithmetic*—insofar as a textbook on practical arithmetic can be original.[42]

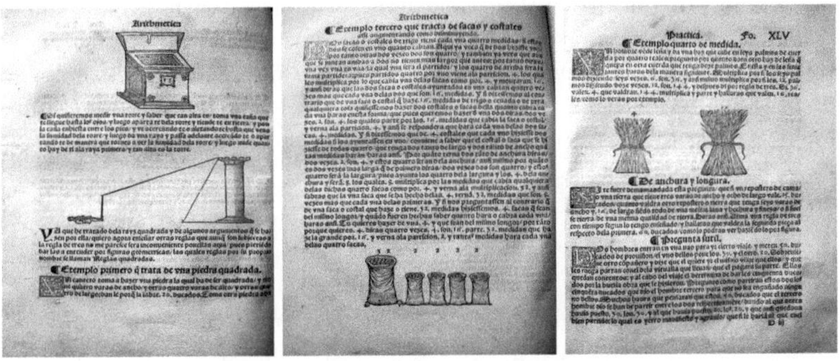

Figure 4: Illustrations in *Arithmetica Practica* [Icíar 1549, XLIIIv, XLV]

Turning to the Table of Contents, we find that *Practical Arithmetic* is divided into two parts. The first, in nine chapters [Icíar 1549, I-XI],[43] is devoted to Arabic numerals, place values, fundamental operations (addition, subtraction, multiplication, division), and checks by means of inverse operations. The second part contains 15 chapters that deal with numerous standard applications, whereas some topics are omitted. For example, Icíar dismisses the rules of barter as being too elementary; the rules of testaments, as being too similar to the rule of fellowship; and the rules of alloying (for gold and silver) as too special for a book "intended to teach any merchant and trader" [Icíar 1549, XXVIIIv], thus not merely for the

[40]As were many Italian arithmetics. This suggests an Italian inspiration that, in accordance with Icíar's primary occupation as a calligrapher, points to Tagliente [Tagliente 1520], [Tagliente 1525], [Tagliente 1540?]. However, *Practical Arithmetic* was produced for educational purposes and was printed in folio, whereas Tagliente's works were handbooks printed in quarto.

[41]Unlike Tagliente's ordering, which dealt with multiplication and division before addition and subtraction, clearly no match for Icíar's style.

[42]At the same time, he duly acknowledged the work of others, giving precise references to Pellos and Ortega [Icíar 1549, LXIII].

[43]Not fifteen, as stated in [Cotarelo y Mori 2004, 367].

education of liberal accountants.[44] Instead, he offers a series of fifteen recipes (in spite of announcing seventeen in the table of contents) for calculating the interest earned on loans over various periods of time [Icíar 1549, XXVIII-XXXII]. Icíar devoted seven chapters to such topics as: fractions [Icíar 1549, XXXII-XXXVIII]; the rule of single false position [Icíar 1549, XXXVIII-XLI]); square roots [Icíar 1549, XLIv–XLIII]; "square rules" —namely, exercises with weights, measures, widths and lengths—[Icíar 1549, XLIIII-XLVI]; lengths, weights, and conversions with coins used in several Italian, French, British, and Spanish territories [Icíar 1549, XLVI-XLVII]; fundamental operations with square roots [Icíar 1549, XLVIII-L]; and a final chapter devoted to "multiplying by minutes" (small change). Icíar was the first Spanish author to give a detailed explanation of multiplication by minutes.[45]

Icíar begins with three chapters in which he introduces Arabic numerals and place values. After presenting Castilian reckoning (with Roman numerals) as equivalent to *alguarismo* (with Arabic numerals), he decides, henceforward, to explain only the rules of *alguarismo.* As he put it, figures for reckoning are like letters in reading and writing [Icíar 1549, Iv], and he shows how to write any given quantity in Arabic numerals—from right to left—using an upper line of dots to keep track of place values:[46]

	·	·	·	·	·
2	1	0	7	4	3
	3	7	6	0	4
		4	5	0	4

In the fourth chapter Icíar commences to teach addition by using examples with Castilian coins, where most reckoning was done using one single unit (*maravedí*). He then proceeds in the fifth chapter to more complicated addition problems using coins from Aragon, Navarre and Valencia. Here the calculations involve one or two non-decimal submultiples of the currency unit (*ducado*), which implies not only summing and carrying but also conversions of units (thus, division with remainder). This shows that this book was actually addressed to an audience already familiar with the most fundamental operations, just as Molière's M. Jourdain spoke prose without knowing it.[47] For instance, in Aragon, where 1 *ducado* = 22 *sueldos*, 1 *sueldo* = 12 *dineros*, and 1 *dinero* = 2 *miajas*:

[44]This remark notwithstanding, he refers to problems of interest to such accountants elsewhere [Icíar 1549, XVI, Lv].

[45]For a discussion of this practice see [Franci/Toti Rigatelli 1982, 56-62].

[46]This was not trivial at the time, especially in multiplication: see, for instance Ortega's and Ventallol examples [Ortega 1512, 18v]; [Ventallol 1521, xxijv-XXIII].

[47]Nevertheless, by the sixth chapter, Icíar distinguished between five types of subtraction in order to explain carry overs, how to deal with zeroes, and how to subtract minutes.

4856 *ducados*	20 *sueldos*	10 *dineros*	1/2
3786 *ducados*	14 *sueldos*	9 *dineros*	1/2
4979 *ducados*	16 *sueldos*	7 *dineros*	1/2
4608 *ducados*	8 *sueldos*	8 *dineros*	
18231 *ducados*	16 *sueldos*	11 *dineros*	1/2

As for multiplication, Icíar explains this by our familiar method of multiplying—called *del ala* (wing multiplication)—which he prefers to the "prolix" column method [Icíar 1549, VIIIv]. This is the only approach he actually explains in full detail, although he displays his erudition[48] by giving examples of *multiplicar morisco* (Moresque multiplying),[49] the method of the cup [Smith 1958, 119], the multiplication by places,[50] the method of the quadrilateral [Smith 1958, 114], and two examples of the *gelosia* multiplication method.[51] According to Smith [Smith 1958, 115], the *gelosia* multiplication method might have remained popular had it not been so difficult to write and print. In this sense, Icíar's commitment to ink was not merely rhetorical, as he chooses to explain only our common method of multiplying; however, he duly acknowledges the abacus tradition when mentioning Moresque multiplying and the method of the cup as his preferred alternatives [Smith 1958, 118].[52]

For division, he gives eight examples for division by units, four examples for dividing by powers of ten, and only one for division by tens, using the galley method without cancel marks. However, an example of long division (24862662/4378) appears in the last chapter of the first part, which is devoted to checking methods. Icíar warns against the method of checks by nines and sevens, and considers only those checks that employ the inverse operation. This relative neglect of division suggests again that the choice of contents was determined by the audience he wanted to address, most of whom did not have much use for long division, except perhaps for checking multiplication.

In the second part of his book, Icíar devoted considerable attention to series. This was an optional subject in mercantile arithmetic; although commonly looked upon as one of the fundamental operations, it was rarely accorded much attention in the early printed books [Smith 1958, 497]. Like most medieval and Renaissance authors, Icíar considers only ascending series [Smith 1958, 497]. For arithmetic

[48] In extensively dealing with multiplication Icíar might have been inspired by [Tagliente 1520, 7-12]; [Tagliente 1525, 6-12]; [Tagliente 1540?, 6-12].

[49] This was a variant (with two printers' errors) of the cancellation method described in [Smith 1958, 118.]

[50] As described in [Smith 1958, 108, footnote 6].

[51] The first one, with the diagonals from the upper left-hand corner to the lower right-hand corner, places the multiplier to the right and writes the result beginning in the upper left-hand corner, as in the method of the quadrilateral, and not at the end of each diagonal.

[52] Andrés [Andrés 1515a, XXXIr] refers to Moresque multiplication as the Moorish manner and use (*modo* and *uso*), and calls it the most sound and safe (*sana* and *segura*) method. He also writes that wing multiplication is the most used method among Christian merchants because Moresque multiplication is unknown to them [Caunedo 2009, 83].

series, he employs the usual classification: natural, non-natural, continuous, and discontinuous progressions, but he does not take up even and odd progressions. He gives a single general procedure for the sum of (intertwined) arithmetic series, those of type $(2n+2)$, obtaining the formula $[(a_1+a_n)/2]n$. For geometric series, for which he uses the terms *dupla*, *tripla*, *quadrupla*, and *quintupla*, he gives the general procedure for finding the sum, namely $[(a_n - a)/(r-1)] + a_n$. As for higher series, he finds the sum of the squares of an arithmetic progression, that is, $a_n(a_n+r)(2a_n+r)/6r$.

Compared with previously printed Spanish works on practical arithmetic, this fairly detailed discussion was quite original. Thus, for instance, Santcliment [Malet 1998] devotes only a short chapter to arithmetic series; Ortega [Ortega 1512] does not include the general procedure for summing a series of squares of an arithmetic progression; Andrés [Andrés 1515a] does not give the general procedure for the sum of a geometric series; and Gutiérrez [Gutiérrez 1539] only gives the general procedure for the sum of geometric series.[53] As opposed to the profuse and detailed exposition of this subject in [Ventallol 1521], Icíar simplifies it by directly presenting general rules. However, the examples he gives unfortunately make it difficult to judge whether he was aware of the importance of geometric series for calculations of compound interest. In general terms, Icíar's explanations are more detailed than Ortega's, but not as prolix as Andrés' [Caunedo 2009, 85]. From the point of view of mathematical content, Ventallol was the most comprehensive author, whereas Ortega is generally considered the most creative, especially due to his methods for finding roots [Malet / Paradis 1984, 1984, 1, 124-127].

Unlike most earlier authors of practical arithmetics, Icíar decided to deal with fractions in the last part of his book. This topic therefore appears after he has dealt with the rule of three, the rule of fellowship and the loans, but immediately before he took up rules of false position, square roots, and multiplication by minutes.[54] As a matter of fact, Icíar's interest in fractions was related to his detailed explanation of multiplication by minutes in the last chapter of *Practical Arithmetic*. There he gives a two folio table of submultiples for all units of currency, weights, and measures in Aragon as fractions of the upper unit—for instance from 1 *dinero* to 12 *dineros* as fractions of 1 *sueldo*—to be learnt by heart so as to be able to multiply any combination of units directly without having to reduce to a common unit before and after the operation [Icíar 1549, LI-LIIv].[55]

In order to explain the procedure, Icíar proposes and solves in full detail two examples. The first asks to find the final price for 375 *cargas*, 2 *quintales*, 3

[53]Moreover, all of Icíar's predecessors in Spain—with the notable exception of Gutiérrez—included Boethian proportions in their texts.

[54]Nonetheless, in the last chapter of *Practical Arithmetic* he declares that "vocal teachers" should proceed as follows: the four fundamental operations in whole numbers, then fractions; addition, subtraction, and multiplication by minutes; rule of three in whole and fractional numbers; other rules [Icíar 1549, Lv].

[55]Tagliente's *Tariffa* [Tagliente 1525, 73-76]; [Tagliente 1540?, 73-76] was conceived as a way to convert currencies and weights while avoiding fractions; this could have inspired Icíar's innovative proposal for multiplying minutes.

arrouas, and 5 *libras* of wax at 48 *florines*, 6 *sueldos*, 5 and 1/2 *dineros* for each *carga* of wax.[56] He proceeds as follows:

1. In order to multiply 2 *quintales* by the price, he uses the previous table to find out that 2 *quintales* = 2/3 *carga*. He then multiplies (twice) 1/3 by 48 *florines*, 6 *sueldos*, 5 and 1/2 *dineros*.
2. In order to multiply 3 *arrovas* by the price, he uses the table to find out that 3 *arrovas* = 1/2 + 1/4 *quintal*, and then multiplies the previously calculated final row by 1/2, and the resulting row again by 1/2 (using that 1 *sueldo* = 12 *dineros*).
3. In order to multiply 5 *libras* by the price, he uses the table to find out that 5 *libras* = 1/6 *arrova*, and then multiplies the previously calculated final row by 1/6 (using that 1 *florín* = 16 *sueldos*, and 1 *sueldo* = 12 *dineros*).
4. Now he has to multiply 375 *cargas* by the price he starts multiplying 375 by 6 *sueldos*. He uses the table to find out that 6 *sueldos* = 1/4 + 1/8 *florín*. He first multiplies 375 by 1/4 and then the resulting row by 1/2.
5. He goes on multiplying 375 *cargas* by 5 *dineros*. He uses the table to find out that 5 *dineros* = 1/4 + 1/6 *sueldos*. He first multiplies 375 by 1/4 and then by 1/6. He also multiplies 375 *cargas* by 1/2 *dinero*.
6. He finally multiplies 375 *cargas* by 48 *florines* and adds up:

Step	Wax weight Price/carga	375 *cargas* 48 *florines*	2 *quintales* 6 *sueldos*	3 *arrovas* 5 *dineros*	5 *libras* 1/2 *dinero*
1		16 fl.	2 s.	1 din.	$\frac{2}{3}+\frac{1}{6}$
1		16 fl.	2 s.	1 din.	$\frac{2}{3}+\frac{1}{6}$
2		8 fl.	1 s.	0 din.	$\frac{1}{2}+\frac{1}{3}+\frac{1}{12}$
2		4 fl.	0 s.	6 din.	$\frac{1}{4}+\frac{1}{6}+\frac{1}{24}$
3		0 fl.	10 s.	9 din.	$\frac{1}{24}+\frac{1}{36}+\frac{1}{144}$
4		93 fl.	12 s.		
4		46 fl.	14 s.		
5-6		3000 fl.	93 s.	9 din.	
5-6		15000 fl.	62 s.	6 din.	
5				187 din.	$\frac{1}{2}$
		18196 *florines*	6 *sueldos*	6 *dineros*	$\frac{89}{144}$

Icíar's result is wrong at the level of 1/2 *dineros*: he gives 137/222 instead of 89/144, which shows that the use of fractions was neither current nor easy at the

[56]These examples seem to be artificial exercises designed to show merchants using this method how to carry out multiplication by minutes in complex cases.

time. Nonetheless, his attempt to avoid long multiplications with fractions shows a clear understanding of place values and the distributive property. According to Icíar, this method was used in Aragon, Valencia and Catalonia [Icíar 1549, Lv].[57] It was still used in the teaching of arithmetics in the late 18th century - as shown in manuscript ML-338-A (SA-M315) at the Central Military Library in Madrid - as can be seen from the following example [Navarro 1843, 65]:

24 *varas* 2 *pies* 8 *pulgadas** 8 *duros* 6 *reales* 8 *maravedises* =

= [896/(3*12)]*[5652/(20*34)] = 5064192 / 24480 = 206 *duros* 17 *reales* 14 *mrs.*

[1 *vara* = 3 *pies*, 1 *pie* = 12 *pulgadas*, 1 *duro* = 20 *reales*, 1 *real* = 34 *mrs.*]

Concluding Remarks

This paper approaches commercial arithmetic in the Spanish Renaissance by taking a largely new approach, one that seeks to contextualize the different types of people and professions involved in the production of commercial arithmetics, especially works and authors that have been either overlooked or insufficiently studied. Such an approach tries to integrate biographical circumstances, personal motivations, and cultural and political influences into the social fabric of this history. The publication of books on commercial arithmetic was evidently an attractive enteprise, and not just for publishers. Thus, it drew such diverse people as the preacher, Juan Andrés, the scrivener, Juan de Icíar, and a lower–ranking nobleman who became active as an editor and bookseller, Miguel de Suelves. These figures, and many others, were drawn to this field at a time when knowledge of the new arithmetic was becoming essential for the Renaissance world, a necessary part of the education of the incipient bourgeoisie.

Both clergymen and secular figures played a role in spreading modern arithmetic throughout Spain during the first half of the sixteenth century. None of them broadened the horizons of arithmetic by including algebra, which was only mentioned by Juan Andrés as a future project. Instead, they produced different kinds of books on arithmetic—brief summaries, handbooks, or textbooks—depending on the audience they intended to reach: accountants, scriveners, civil servants, estate managers, or merchants. The identification of Andrés—the author of the first printed references to Luca Pacioli in Spain—as a renegade Muslim reminds us of the proximity to the mathematical culture from which modern arithmetic developed. Indeed, the Arabic background still played a large role in Spain at the beginning of the sixteenth century. As suggested above, it was probably due to his religious networks that Juan Andrés introduced the arithmetical part of Pacioli's *Summa* into Spain.

By mid-sixteenth century, Icíar had published his *Ortographia pratica*, the first handbook on writing printed in Spain. His collaboration with Suelves led to numerous other publications, including the reworked publication of *Arte breue*

[57] This assertion is confirmed in [Andrés 1515a, xxxiiij] and [Ventallol 1521, XXVv.-XXVI.]

y muy prouechoso de cuẽta castellana y arismetica, the elementary treatise first published in Toledo in 1539 by Father Juan Gutiérrez de Gualda. By this time, the publication of commercial arithmetics was a flourishing business in a number of Spanish cities, where the populace was learning to make the transition from Roman to Arabic numerals. Icíar's second published work, *Practical Arithmetic* (1549), was a totally original text that was meant to address the educational needs of merchants, accountants, and others active in trade and commerce. This work provides valuable information about the teaching of mathematics in mid-sixteenth century Spain, especially with regard to then current procedures for calculations that used non-decimal measurement systems. Icíar's detailed explanation of the calculation processes with submultiples shows how he was able to connect current mercantile practices with modern arithmetic. This illustrates how this new commercial arithmetic required real understanding of the underlying mathematical concepts and properties. Moreover, it shows that the assimilation of modern arithmetic was not so much a problem due to Arabic numerals and new algorithms, but rather largely a problem that arose from having to handle different systems of units. After all, the adoption of a unified decimal system for measurements, as first proposed by Simon Stevin in 1585, only came in the aftermath of the French Revolution.

Thus, the developments described here, while in many respects idiosyncratic, due to local cultural influences, were surely quite typical for early modern Europe. This highly focused study aims, however, merely to shed light on the Spanish case. As a whole, the approach taken here to these authors and works offers a much more detailed picture and thus more accurate knowledge as to how modern arithmetic spread and was gradually assimilated in sixteenth-century Spain.

Acknowledgments

First of all, I am most grateful to my colleague Manuel José Pedraza, who gave me a copy of Icíar's *Practical Arithmetic* when he located it at the University of Salamanca. I also thank David Rowe for his useful comments, suggestions, and corrections.

This work is financed by the Research Project HAR2010-15457 (Spain's Ministry of Economy and Competitiveness, formerly MICINN)

Bibliography

[Alonso García 1953] Alonso García, Daniel, *Ioannes de Yciar, calígrafo duranqués del siglo XVI*. Bilbao: Publicaciones de la Junta de Cultura de Vizcaya, 1953.

[Andrés 1515a] Andrés, Juan, *Sumario breue d'la pratica dela arithmetica de todo el curso de larte mercãtiuol bien declarado: el qual se llama Maestro de cuento*. Valencia: Juan Joffre, 1515. (Variant edition dedicated to the Count of Oliva).

[Andrés 1515b] Andrés, Juan, *Sumario breue d'la pratica dela arithmetica de todo el curso de larte mercãtiuol bien declarado: el qual se llama Maestro de cuento.* Valencia: Juan Joffre, 1515. (Variant edition dedicated to Bishop Martín García).

[Andrés 1515c] Andrés, Juan, *Libro nuevamente imprimido que se llama confusión de la secta mahomatica y d'el alcoran.* Valencia: Juan Joffre, 1515.

[Antonio 1696] Antonio, Nicolás, *Bibliotheca hispana vetus sive Hispanorum, qui usquam unquámve scripto aliquid consignaverunt notitia.* Romae: ex Typographia Antonii de Rubeis, 1696.

[Antonio 1788a] Antonio, Nicolás, *Bibliotheca Hispana Vetus, sive hispani scriptores qui ab Octaviani Augusti aevo ad annum Christi MD floruerunt.* Matriti: Apud viduam et heredes D. Ioachimi Ibarrae, 1788.

[Antonio 1788b] Antonio, Nicolás, *Bibliotheca Hispana Nova sive Hispanorum scriptorum qui ab anno MD ad MDCLXXXIV floruere notitia.* Matriti: Apud Joachimum de Ibarra, 1788.

[Arte breue y prouechoso de cuẽta castellana y arithmetica 1555] *Arte breue y prouechoso de cuẽta castellana y arithmetica, donde se muestrã las cinco reglas de guarismo por la cuẽta castellana, y reglas de memoria.* Zaragoza: en casa de Esteuã de Nagera, a costa de Miguel de Çapila mercader de libros, 1555 [Spanish National Library (BNE) Shelf mark R/40483].

[Aurel 1552] Aurel, Marco, *Libro primero de Arithmetica Algebratica, en el qual se contiene el arte Mercantiuol con otras muchas reglas del Arte menor, y la Regla del Algebra vulgarmente llamada de Arte Mayor, o Regla de la cosa.* Valencia: en casa de Ioan de Mey, 1552.

[Ausejo 2012] Ausejo, Elena, Juan de Icíar's Practical Arithmetic (1549): Writing and reckoning in Spanish Renaissance. In: Kristín Bjarnadóttir; Fulvia Furinghetti; José Manuel Matos; Gert Schubring (Eds.), *Dig where you stand, 2.* Lisboa: Fundaçao para a Ciência e Tecnologia/Universidade Nova de Lisboa, 2012, 45-59.

[Ausejo / Benito 2006] Ausejo, Elena, and Benito, Mariano, *Le calcul des problèmes d'héritage en Al-Andalus.* In: Actes du 8e Colloque Maghrébin sur l'Histoire des Mathématiques Arabes. Tunis: Association Tunisienne des Sciences Mathématiques, 2006, 49-64.

[Ausejo / Hormigón / Usón 2001] Ausejo, Elena; Hormigón, Mariano, and Usón, Carlos, *Las aritméticas mercantiles en Zaragoza en el siglo XVI.* In: Commerce et mathématiques du Moyen Âge à la Renaissance, autour de la Méditérranée. Toulouse: Éditions du Centre International d'Histoire des Sciences Occitanes (CIHSO), Université Toulouse II, 2001, 235-270.

[Boncompagni 1863] Boncompagni, Baldassarre, *Intorno ad un trattato d'aritmetica stampato nel 1478.* Atti dell'Accademia Pont. dei Nuovi Lincei, XVI (1863), 1-64, 101, 300, 389, 503, 683, 909.

[Caunedo 2007] Caunedo, Betsabé, *La aritmética mercantil castellana en la Edad Media. Una breve aproximación.* LLULL, Revista de la Sociedad Española de Historia de las Ciencias y de las Técnicas 30 (2007), 5-20.

[Caunedo 2009] Caunedo, Betsabé, *Un manual de Aritmética mercantil de Mosén Juan de Andrés.* Pecvnia 8 (2009), 71-96.

[Caunedo / Córdoba 2000] Caunedo, Betsabé, and Córdoba, Ricardo, *El arte del alguarismo: Un libro castellano de aritmética comercial y de ensayo de moneda del siglo XIV.* Valladolid: Junta de Castilla y León, 2000.

[Cotarelo y Mori 2004] Cotarelo y Mori, Emilio, *Diccionario biográfico y bibliográfico de calígrafos españoles.* Madrid: Visor Libros, 2004 (Facsimile ed. Madrid, 1916).

[Docampo Rey 2004] Docampo Rey, Javier, *La formación matemática del mercader catalán 1380-1521. Análisis de fuentes manuscritas.* Ph.D. Dissertation. Universidade de Santiago de Compostela, 2004.

[Docampo Rey 2009] Docampo Rey, Javier, *A new source for medieval mathematics in the Iberian Peninsula: The commercial arithmetic in Ms 10106 (Biblioteca Nacional, Madrid).* Revue d'histoire des mathématiques 15 (2009), 123-177.

[Docampo Rey 2006] Docampo Rey, Javier, *Reading Luca Pacioli's in Catalonia: an early 16th-century Catalan manuscript on algebra and arithmetic.* Historia mathematica 33 (2006), 43-62.

[Doni 2007] Doni, Federica, *La teoria personalistica del conto. Aspetti evolutivi ed approfondimenti critici.* Milano: Giuffrè editore, 2007.

[Franci / Toti Rigatelli 1982] Franci, Raffaella, and Toti Rigatelli, Laura, *Introduzione all'aritmetica mercantile del medioevo e del Rinascimento.* Siena: Servizio editorale dell'Università di Siena, 1982.

[García 1520] García, Martín, *Sermones eminẽtissimi totiusq[ue] Barchinonẽsis gregis tutatoris accerrimi, necnõ ĩmarcessibilis sacre theologie paludamẽto insigniti Martini Garsie.* Impressi Cesarauguste: per Georgiũ Coci, 1520.

[García Cuadrado 1996] García Cuadrado, Amparo, *Aproximación a los criterios legales en materia de imprenta durante la Edad Moderna en España.* Revista General de Información y Documentación 6 (1996), 125-187.

[González Ferrando 1956] González Ferrando, José Mª, *Gaspar de Texeda, precursor de la teneduría de libros en España.* Técnica Económica 2 (1956), 36-43.

[Gutiérrez 1539] Gutiérrez, Juan, *Arte breue y muy prouechoso de cuẽta castellana y arismetica, dõde se muestran las cinco reglas de guarismo por la cuẽta castellana, y reglas de memoria.* Toledo: Fernando de Santa Catalina, 1539 [Spanish National Library, Shelf mark R/8172, wrongly bound].

[Gutiérrez 1564a] Gutiérrez, Juan, *Arte breve y muy prouechoso de quẽta castellana y Arithmetica, dõde se muestrã las cinco reglas de guarismo por la quenta Castellana, y reglas de memoria.* Zaragoza: a costa de Miguel de Suelues, alias çapila infançon mercader de libros, en casa de Pedro Bernuz, 1564 [Spanish National Library, Shelf mark R/5356].

[Gutiérrez 1564b] Gutiérrez, Juan, Libro de cuentas. *Arte breue y muy prouechoso de cuenta castellana y aritmética: donde se muestran las cinco reglas de guarismo por cuenta castellana y reglas de memoria, compuesto por el reuerendo Juan Gutiérrez de Gualda, cura de la villa del Villarejo de Fuentes. Dirigido al illustre señor don Juan Pacheco de Silua, señor de la dicha villa y su partido. Visto y examinado por los señores de Consejo Real y con priuilegio de su Magestad. Va añadido el repertorio de cuentas.* Alcalá: Andrés de Ángulo, 1564 [Krakow, Biblioteka Jagiellonska, Shelf mark CIM.O.846].

[Gutiérrez 1569] Gutiérrez, Juan, *Arte brebe y muy probechosa de cuenta castellana y arismetica, donde se demuestran las cinco reglas de guarismo por la cuenta castellana, y reglas de memoria.* Impresso en Caragoça: En casa de Miguel de Guessa, a costa de Miguel de Suelues infançon, 1569 [University of Sevilla, Shelf mark A Res. 13/4/21(2)].

[Hebrera 1700] Hebrera y Esmir, José Antonio de, *Vida prodigiosa del Ilustrissimo y Venerable D. Martin Garcia, Obispo de Barcelona, hijo de la fidelisima y antigua Villa de Caspe.* Zaragoza: imprenta de Domingo Gascon, por Diego de Larumbe, 1700.

[Høyrup 2009] Høyrup, Jens, *"Proportions" in and around the Italian Abbacus Tradition.* Physis, Rivista Internazionale di Storia della Scienza XLVI (2009) Nuova Serie, 5-110.

[Icíar 1549] Icíar, Juan de, *Libro intitulado arithmetica practica muy util y provechoso para toda persona que quisiere ejercitarse en aprender a contar.* Zaragoza: Pedro Bernuz, a costa del auctor y de Miguel de Çapila mercader de libros, 1549.

[Icíar 1555] Icíar, Juan de, *Arte subtilissima por la qual se enseña a escreuir perfectamente.* Zaragoza: Impresso a costa de Miguel de Çapila mercader de libros vezino de Çaragoça, 1555 [Spanish National Library, Shelf mark R/ 40714: this copy does not correspond to any of the descriptions in [Martínez Pereira 2006, 343-355]].

[Icíar 1559a] Icíar, Juan de, *Libro Subtilissimo, por el qual se enseña a escreuir y contar perfectamente el qual lleua el mesmo orden que lleua vn maestro con su discípulo.* Zaragoza: impreso a costas de miguel de çapila mercader de libros, en casa de Esteuan de Nagera que sancta gloria haya, 1559 [Portuguese National Library, Shelf mark F. 4982].

[Icíar 1559b] Icíar, Juan de, *Arte breue y prouechoso de cuenta Castellana y Arithmetica, donde se muestran las cinco reglas de guarismo por la cuenta castellana, y reglas de memoria. Y agora nueuamẽte en esta postrera impression se han añadido vnas cuentas muy graciosas y prouechosas, sacadas del libro de Fray Iuan de Ortega: y mas al cabo va añadida vna cuenta abreuiada de marauedis.* En çaragoça: ve[n]den se en casa de Miguel de Çapila, 1559 (Impressa en Çaragoça: en casa de la viuda de Esteuan de Nágera, a costa de Miguel de çapila mercader de libros, 1559) [Portuguese National Library, Shelf mark F. 4982].

[Icíar 1564] Icíar, Juan de, *Libro Subtilissimo, por el qual se enseña a escreuir y contar perfectamente el qual lleua el mesmo orden que lleua vn maestro con su discípulo.* Zaragoza: en casa dela Biuda de Bartholome de Nagera, a costas de Miguel de Suelves, alias Çapila, 1564 [Portuguese National Library, Shelf mark F. 4985 with [Gutiérrez 1564a]; Spanish National Library, Shelf mark R/8599 without [Gutiérrez 1564a]].

[Icíar 1566] Icíar, Juan de, *Libro Subtilissimo, por el qual se enseña a escreuir y contar perfectamente el qual lleua el mesmo orden que lleua vn maestro con su discípulo.* Zaragoza: A costas de miguel de Suelues, alias, çapila infançon, 1566 [Spanish National Library, Shelf mark R/12767 with [Gutiérrez 1564a]; Spanish National Library, Shelf marks R/31424 and R/13716 without [Gutiérrez 1564a]].

[Malet 1998] Malet, Antoni, *Francesc Santcliment. Summa de l'art d'aritmètica. Introducció, transcripció i notes a cura d'Antoni Malet.* Vic: Eumo editorial, 1998.

[Malet / Paradís 1984] Malet, Antoni, and Paradís, Jaume, *Els orígens i l'ensenyament de l'àlgebra simbòlica (1478-1545).* Barcelona: Publicacions i edicions de la Universitat de Barcelona, 1984.

[Martínez Pereira 2006] Martínez Pereira, Ana Mª, *Manuales de escritura de los Siglos de Oro.* Mérida: Editora Regional de Extremadura, 2006.

[Martínez Pereira / Torné 2007] Martínez Pereira, Ana Mª, and Torné, Emilio, Arte subtilissima, 1550: teoría y práctica de la escritura en la edición más perfecta de la obra de Juan de Icíar. In: Icíar, Juan de, *Arte sutilísima, por la cual se enseña a escribir perfectamente: edición facsímil de la de Zaragoza, Pedro Bernuz, 1550.* Madrid: Editorial Complutense, 2007, 35-63.

[Mayordomo 2002] Mayordomo García-Chicote, Francisco, *La Taula de Canvis: aportación a la historia de la contabilidad valenciana (siglos XIII-XVII).* Valencia: Universitat de València, 2002.

[Mesquida Cantallops 1996] Mesquida Cantallops, Joan-Antoni, *La pràtica mercantívol de Joan Ventallol. Estudi lingüistic y edició d'un text matemàtic del segle XVI.* Palma de Mallorca: Universitat de les Illes Balears, 1996.

[Montoza 2013] Montoza Coca, Manuel, *Edición, traducción y comentario de los Sermones del I-V de Don Martín García.* Master Thesis. Universitat Autònoma de Barcelona, 2013.

[Navarro 1843] Navarro, P.M., *Tratado de Aritmética, para uso de las escuelas de primeras letras.* Madrid: Imprenta de la Viuda é Hijos de la Sota, 1843.

[Norton 1978] Norton, Frederick John, *A descriptive catalogue of printing in Spain and Portugal, 1501-1520.* Cambridge: Cambridge University Press, 1978.

[Norton 1997] Norton, Frederick John, *La imprenta en España, 1501-1520 / Frederick J. Norton; edición anotada, con un nuevo índice de libros impresos en España, 1501-1520 por Julián Martín Abad.* Madrid: Ollero y Ramos, D.L. 1997.

[Ortega 1512] Ortega, Juan de, *Sigue se vna conpusicion de la arte de la arismetica y juntamente de geometria / fecha y ordenada por fray Juan de Ortega de la orden de santo domingo de los predicadores.* Imprimido a Leon: en casa de maistro Nicolau de Benedictis, por Joannes trinxer librero de Barcelona, 1512 [Libraries University of Salamanca, Shelf mark BG/36568].

[Pastor 1827] Pastor Fuster, Juan, *Biblioteca valenciana de los escritores que florecieron hasta nuestros días.* Valencia: José Ximeno, 1827.

[Pedraza 2007] Pedraza, Manuel José, Los protagonistas de la edición del Arte subtilissima: Juan de Icíar, Juan de Vingles y Pedro Bernuz. In: Icíar, Juan de, *Arte sutilísima, por la cual se enseña a escribir perfectamente: edición facsímil de la de Zaragoza, Pedro Bernuz, 1550.* Madrid: Editorial Complutense, 2007, 15-34.

[Ruiz García 2003] Ruiz García, Elisa, *Confusión o confutación de la secta mahomética y del Alcorán / Juan Andrés; estudio preliminar, edición y notas Elisa Ruiz García; transcripción del texto Ma. Isabel García-Monge.* Mérida: Editora Regional de Extremadura, 2003.

[San Vicente 1969] San Vicente, Ángel, Sobre algunos calígrafos del Bajo Renacimiento en Zaragoza. In: *Suma de estudios en homenaje al Ilustrísimo Doctor Ángel Canellas López.* Zaragoza: Facultad de Filosofía y Letras, 1969, 909-951.

[Sánchez Salor / Cobos Bueno 1996] Sánchez Salor, Eustaquio, and Cobos Bueno, José M., *"Ars Arithmetica". Dividida en Teórica y Práctica. Utilísima y necesaria para hombres de toda condición / Juan Martínez "Silíceo". Introducción, traducción y notas de E. Sánchez Salor, J. Cobos Bueno.* Cáceres: Universidad de Extremadura & Editora Regional de Extremadura, 1996 (Includes facsimile reproduction of the Latin ed. 1514).

[Smith 1908] Smith, David Eugene, *Rara arithmetica; a catalogve of the arithmetics written before the year MDCI, with a description of those in the library of George Arthvr Plimpton, of New York.* Boston: Ginn, 1908.

[Smith 1958] Smith, David Eugene, *History of Mathematics. Vol. II, Special topics of elementary mathematics.* New York: Dover, 1958. Unabridged and unaltered republication of [Boston: Ginn, 1925].

[Tagliente 1520] Tagliente, Girolamo, and Tagliente, Giovanni Antonio, *Libro da abaco che insegnia a fare ogni raxone marcadantile & apertegare le terre con larte di la giometria & altre nobilissime raxone straordinarie con la tarifa come raspondeno li pexi & monete de molte terre del mondo con la inclita citta de Venetia. El qual libro se chiama Tesauro vniuersale concesso per lo Serenissimo Dominio Venetiano per anni diexe.* Impresso in Venetia, 1520 del mese di septembrio.

[Tagliente 1525] Tagliente, Girolamo, and Tagliente, Giovanni Antonio, *Opera che insegnia a fare ogni Ragione de Mercãntia et apertegare le Terre con arte giometrical Intitolata Componimẽto di arithmetica.* Impresso in Venezia, 1525.

[Tagliente 1540?] Tagliente, Girolamo, *Libro dabaco che insegna a fare ogni ragione mercadantile & a piegare le terre cõ larte di la giometria & altre nobilissime ragione straordinarie cõ la Tariffa come raspõdeno li pesi & monete de molte terre del mõdo con la inclita citta di Vinegia. El qual libro se chiama Thesauro vniuersale.* Venezia: Luca Antonio de Uberti, n.d. [Harvard Bussiness School, Goldsmiths'-Kress no. 00029.0-3].

[Three Classics 1953] *Three classics of Italian calligraphy: an unabridged reissue of the writing books of Arrighi, Tagliente, Palatino.* New York: Dover Publications, 1953.

[Tejeda 1546] Tejeda, Gaspar de, *Suma de Arithmetica pratica y de todas Mercaderías Con la horden de contadores.* Valladolid: en la officina de Francisco Fernandez de Cordoua, 1546.

[Ventallol 1521] Ventallol, Joan, *Pratica mercantiuol.* Lyo: per mestre Joan de La Place, 1521.

[Vindel 1903] Vindel, Pedro, *Catálogo ilustrado de la libreria de P. Vindel. Tomo tercero. Obras españolas de los siglos XII á XVIII.* Madrid: Imprenta de José Rueda, 1903.

[Wilkinson 2010] Wilkinson Alexander S., *Iberian Books: Books Published in Spanish or Portuguese or on the Iberian Peninsula before 1601.* Leiden: Brill, 2010.

Mathematics and its History in Modern Cultures

Historical Events in the Background of Hilbert's Seventh Paris Problem

David E. Rowe

To Joe and in memory of Yibao

Introduction

David Hilbert's lecture, "Mathematical Problems," [Hilbert 1900] delivered in Paris in 1900 at the Second International Congress of Mathematicians, has long been recognized as marking a milestone in the history of mathematics. Certainly for Hilbert himself, this marked the single greatest event and a true turning point in his storied career. When historians and mathematicians have written about the so-called Hilbert problems, they have usually looked forward into the twentieth century, sometimes by viewing their resolution as markers for mathematical progress. In Jeremy Gray's *The Hilbert Challenge* [Gray 2000] one can read about some of these larger developments as they unfolded throughout the next fifty years, whereas Ben Yandell's *The Honors Class* [Yandell 2002] provides more personal and anecdotal information about those who gained recognition for their contributions to solving one of Hilbert's twenty-three problems.[1] Such forward-looking accounts inevitably present a picture with rather smooth contours that can easily blur out some of the messier details, many of which perhaps deserve to be forgotten. If, on the other hand, we approach Hilbert's problems from a biographical perspective, this forces us to confront many less familiar features of the mathematical world in which he grew up, particularly during his formative years in Köngisberg.[2] Indeed, it is my view that a careful exploration of events and ideas connected with the prehistory of the Hilbert problems can open up a variety of new perspectives for the history of mathematics during the last two decades of the nineteenth century (see [Rowe 2004]). At the same time, by combing this terrain, the historian is very likely to encounter certain long-buried facts that clash rather badly with received myths about the sources and even the meaning of some of those problems at the time they were first presented to the mathematical world.[3] As a case in point, I take this occasion to explore certain nearly forgotten episodes connected with the background to Hilbert's seventh Paris problem.

Hilbert's seventh was the first in a group of problems (7-12) in number theory. These six cover a vast terrain of new possibilities; nevertheless, they still form a fairly coherent subset within the entire list of twenty-three. The seventh problem

[1] For a larger picture of their place in twentieth-century mathematics, see [Struik 1987]; more technical accounts by leading experts can be found in [Alexandrov 1979] and [Browder 1976].

[2] For some telling episodes in his early career, see [Rowe 2003], [Rowe 2005].

[3] For a brief account of events that immediately preceded Hilbert's Paris lecture, see [Rowe 2013].

D.E. Rowe and W.-S. Horng (eds.), *A Delicate Balance: Global Perspectives on Innovation and Tradition in the History of Mathematics*,
Trends in the History of Science, DOI 10.1007/978-3-319-12030-0_10

concerns the transcendence of certain special numbers, about which speculation was easy, proof quite another matter.[4] Hilbert noted that such problems could easily be found in geometry as well, for instance: if the ratio of the base angle to the angle at the vertex in an isosceles triangle is algebraic but not rational, then one should be able to prove that the ratio of its base to a side must be transcendental. A more general class of numbers that Hilbert presumed must be transcendental has the form α^β, for an algebraic base α and and an irrational algebraic exponent β. As specific examples he gave $2^{\sqrt{2}}$ and $e^\pi = i^{-2i}$. These, he claimed, should always be transcendental or at least irrational numbers. His aim, though, had little to do with the special cases he set forth here, but rather to use these as a means for advancing further into this difficult terrain. As he put this in his concluding statement: "It is certain that the solution of these and similar problems must lead us to entirely new methods and to a new insight into the nature of special irrational and transcendental numbers" [Hilbert 1935], 308-309.

Methodological concerns played a major role in the background story that led up to Hilbert's engagement with these matters. Here I wish to focus close attention, first of all, on events that took place in 1882, when Ferdinand Lindemann announced a proof that π is a transcendental number [Lindemann 1882a]. This achievement brought Lindemann nearly instantaneous fame, so that his name quickly entered the annals of history as the man who proved the impossibility of squaring the circle. His paper, "Über die Zahl π" [Lindemann 1882b], published that year in *Mathematische Annalen*, belongs to that distinguished genre of mathematical literature that deserves special attention on the part of historians, namely works that fall under the category "oft cited, but rarely read." Our account here will explain why that was so with Lindemann's paper. Indeed, ironic as it may seem in consideration of the fact that Hilbert took his doctoral degree under Lindemann only a short time after this famous paper appeared, it is entirely possible that even he never read it! For in recounting this story, it will emerge that Lindemann's work has relatively little to do with the background to Hilbert's seventh problem, except for the important circumstance that it showed the fertility of a line of argument that Charles Hermite introduced ten years earlier in proving the transcendence of e [Hermite 1873a]. Once Lindemann pointed this out, others were quick to seize on these new possibilites, and so it will turn out that the events of 1882 actually involved a number of significant players, especially Georg Cantor and Karl Weierstrass, but also Felix Klein and Adolf Hurwitz. Lindemann quickly gathered his laurel wreath and then vanished from the scene, leaving this new avenue of research to others.

Hilbert's early career was spent in Königsberg, and in 1883 Lindemann succeeded Heinrich Weber there, joined by Hurwitz. Documentary evidence from Hilbert's years as a student is very thin, but later accounts nevertheless provide a clear general picture of the deep impression the twenty-five-year-old Hurwitz

[4]Hilbert considered this seventh problem to be extremely difficult, even harder than the Riemann hypothesis (!), which was his eighth Paris problem ([Yandell 2002], 174).

left on both Hilbert and Hermann Minkowski in their nearly daily encounters with him ([Reid 1970], 9-14). As we shall describe below, Hurwitz's interest in transcendence results drew directly on an unpublished paper by Weierstrass that had begun to circulate in the immediate wake of Lindemann's announcement, but which was first published three years later as [Weierstrass 1885]. This thread thus brings Hilbert indirectly into the picture. For in his Paris address from 1900, he cites Hurwitz's work from the 1880s as opening the way to the larger field of research on transcendental numbers outlined in the seventh problem ([Hilbert 1935], 308). What he did not say on that occasion, however, though many would have known this at the time, was that he himself had made a significant contribution to the simplification of the transcendence arguments given earlier by Hermite and Weierstrass. This took place in the year 1893, when Hilbert's elegant argument prompted a resurgence of interest in methodological approaches to simplifying the proofs for the transcendence of e and π [Hilbert 1893].

What happened after 1900 has been fairly well documented in the literature on the Hilbert Problems ([Gray 2000], [Yandell 2002]). The story of the events that preceded the seventh problem, on the other hand, has largely gone untold, even though it is closely linked with a very famous turning point in the history of mathematics. Part of the inspiration behind it came, of course, from a familiar chain of prior accomplishments: the breakthrough achieved in 1873 by Hermite, whose methods laid the groundwork for Lindemann's proof for π in 1882, a result that Hilbert was able to prove far more directly ten years later. Klein was an eye witness to all of this and much more, and he did his part to spread the good news. Lindemann and Hilbert were both close protégés, and so their triumph was, at least indirectly, a personal victory for him as well. Thus, when Lindemann was called to Königsberg soon after his work was published, Klein congratulated him by writing that this was a victory for "our principles" ([Lindemann 1971], 86). What he meant by this was simply that Lindemann's appointment had broken the monopoly on mathematics professorships in Prussia till then in the hands of the dominant clique in Berlin (see [Rowe 1989]). A decade later, Hilbert was already on Klein's "short list" of candidates for Göttingen; thus, his new, highly condensed proof for the transcendence of π only strengthened Klein's hand when he pushed Hilbert's appointment through in December 1894.

Around this time, Klein entered into the discourse concerning methods for proving the transcendence of π, which culminated with the publication of his popular booklet [Klein 1895], perhaps best known from its English translation under the title *Famous Problems in Elementary Geometry* [Klein 1897]. This text was based on a lecture course Klein had offered during the summer semester of 1894, the main part of which dealt with recent progress made by Hilbert, Hurwitz, and Paul Gordan in answering the oldest and most famous of all geometrical problems in the history of mathematics, the quadrature of the circle. Klein's interest in these matters was primarily didactical, as can already be seen from his remarks in his seventh Evanston Colloquium lecture, devoted to this very topic. There he expressed the opinion that this progress in dealing with the transcendence of e

and π had reached the point where these simplified proofs "should henceforth be introduced into university teaching everywhere" ([Klein 1894], 53). Klein's booklet thus showed how this topic could be taught in advanced mathematics courses. In retrospect, one can see how he himself became increasingly interested in a didactical approach that drew on similar references to recent (and occasionally more remote) historical developments. This leads us to some final reflections about the kinds of distortions that often can occur when mathematicians write about historical matters. In the case of Klein's booklet, the American historian of mathematics, Raymond Clare Archibald, took the lead in clarifying several of the slips Klein had made, some of them bearing on weighty issues [Archibald 1914].

When one hears the name Georg Cantor in the context of Hilbert's lecture [Hilbert 1900], one invariably thinks of his first Paris problem—on Cantor's continuum hypothesis—and not the seventh. Looking backward, however, it will clearly emerge that Cantor's ideas had a direct relevance for those who wished to explore the deeper implications of Hermite's work for future research. This, of course, was part of a broader set of interests that eventually brought Cantor's *Mengenlehre* into the mainstream of modern analysis and topology. During the last two decades of the nineteenth century, many mathematicians came to attach great importance to the new insights and tools provided by set theory for research in all those realms where the role of the infinite had long seemed beyond their grasp. Within this context, the story told here sheds new light on how Cantor's fundamental insight into the distinction between denumerable sets—in the present case the algebraic numbers—and the unfathomably larger sets of higher cardinality, in particular the transcendental numbers, played a major role in the background leading up to Hilbert's seventh Paris problem. Following this carefully, in fact, will reveal that Cantor himself was the first to arrive at the insight that the exponential curve contains no non-trivial algebraic points.[5]

This exciting discovery, first described by Cantor in a letter to Felix Klein from 10 June, 1882, was made just before Lindemann's announcement was published. Cantor's role as the referee for Lindemann's original submission to *Mathematische Annalen*, a no longer extant manuscript, has been noted before in the historical literature. The present account, however, provides the first full picture of what took place behind the scenes. Furthermore, it reveals that Felix Klein was the one key actor from that time who would go on to tell this story in the 1890s. Indeed, it was Klein who not only popularized this work on transcendence problems but also connected this theme directly with Cantor's new theory, albeit without acknowledging Cantor's role in recognizing the set-theoretical implications of Lindemann's work for the exponential curve. Thus, whereas Hilbert's first Paris problem surely served as a wonderful advertisement for *Mengenlehre*, Klein had long before played a major part in promoting Cantor's work, much of which

[5]This assertion may be considered as a kind of transcendental analogue of the Fermat Theorem, which can be formulated as asserting that for $n \geq 3$ the Fermat curve $x^n + y^n = 1$ contains no nontrivial rational points.

had been published in the journal he co–edited, *Mathematische Annalen.*[6] Readers familiar with Herbert Mehrtens' *Moderne–Sprache–Mathematik* [Mehrtens 1990] will recognize that the present account clashes sharply with his contention that Hilbert and Klein stood on opposite sides of the modern/ counter-modern divide with respect to Cantor's theory.[7]

Simplicity vs. Complexity

Simplicity was also one of the general themes that loomed large in Hilbert's Paris address devoted to "Mathematical Problems." Clearly, he thought mathematicians ought to be guided by the idea that a problem was never truly solved until such time as the underlying principles that lead to its solution had been fully clarified. Invoking the opinion of an elderly French mathematician (perhaps Hermite?), he explained what he meant by clarity in "pedestrian" terms: "a mathematical theory is not to be considered complete until you have made it so clear that you can explain it to the first man whom you meet on the street." Surely with a wry smile on his face, he went on to say that if this is true "for a mathematical theory, I should demand it still more for a mathematical problem if it is to be perfect; for what is clear and easily comprehended attracts, the complicated repels us" [Hilbert 1935], 291.

Hilbert's rhetorical style was simplicity itself, but a closer look at his text reveals several underlying ambiguities, particularly when we turn to his discussion of the individual problems, which varied a great deal. In fact, in several instances one might question whether these deserve to be called well-formulated problems at all, especially in cases when it is difficult to say exactly what he had in mind. Some "Hilbert problems" were already well known (like problem 1, Cantor's continuum hypothesis, or number 8, Riemann's conjecture concerning the zeroes of the zeta-function). Others were altogether vague and programmatic (like the sixth problem, dealing with the axiomatization of certain branches of physics, or problem twenty-three calling for a fresh attack on the calculus of variations). In not a few cases, however, the problems attained new meanings, and some were soon shorn from the mathematical settings in which Hilbert had originally placed them (like the fifth problem, which eventually molted into the theory of topological groups, or the tenth on the decidability of Diophantine equations, which required a sharp definition of algorithms before researchers could begin to tackle the question). Hilbert's problems thus took on lives of their own, and as noted at the outset, these new results were often taken as markers for mathematical progress over the course of the twentieth century.

On the other hand, had Hilbert been more precise, his famous set of 23 would surely not have proved so fruitful for future research. Part of what made this bold venture so successful, it seems, stemmed from the way he mixed older,

[6]For details on this, see [Purkert and Ilgauds 1987].

[7]With respect to Hilbert, I have pointed out other disparities in a lengthy essay review [Rowe 1997].

well-understood problems with hints of new arenas in which the mathematicians of the twentieth century might look for, and even answer, similar problems of comparable depth and difficulty. In short, he offered a dynamic picture of mathematical research, linking the present and future with the immediate past. Thus, in formulating the seventh problem, he began by remarking that: "Hermite's arithmetical theorems on the exponential function and their extension by Lindemann are certain of the admiration of all generations of mathematicians. Thus the task at once presents itself to push ahead further along the path here entered, as A. Hurwitz has already done in two interesting papers" ([Hilbert 1935], 308). This last remark, referring to [Hurwitz 1883] and [Hurwitz 1883], has rarely, if ever, drawn comment, and yet its importance for the story told here will soon become apparent.[8]

There are other subtle clues in Hilbert's text that tell us about how he saw this newly charted terrain. His expectations in this case were clearly stated—α^β will be transcendental, or at least irrational, whenever the base is algebraic and the exponent irrational—including the conjecture that proving this would require new, as yet unforseen methods that would involve great difficulties. This suggests that Hilbert had thought about these matters hard and for a long time, surely since the 1880s. The two papers by Hurwitz alluded to here, both bearing the title "Ueber arithmetische Eigenschaften gewisser transzendenter Funktionen," were published in *Mathematische Annalen* in the years 1883 and 1888, respectively, thus during the period when Hilbert was in steady contact with Hurwitz in Königsberg. Moreover, the first was written during the immediate aftermath of Lindemann's discovery, a story worth telling with attention to the details. As we shall see, issues of mathematical simplicity and complexity were central motifs from beginning to end.

Lindemann's Breakthrough

Klein's former pupil, Ferdinand Lindemann, found the key argument that led to his breakthrough in 1882, roughly ten years after Charles Hermite first proved the transcendence of e. Lindemann was personally acquainted with Hermite; he even knew that the distinguished French analyst regarded the proof he gave in his 1873 paper in the *Comptes Rendus* as one of the best pieces of work he had ever done ([Lindemann 1971], 70). Hermite had also thought about how to prove that π was a transcendental number, but he eventually gave up after reaching the conclusion that this problem was far more difficult than the one he had solved. According to Karl Weierstrass, he and Carl Wilhelm Borchardt, the editor of Germany's oldest research journal for mathematics, urged Hermite on several occasions to tackle this problem.[9] The latter's reaction is recorded in a paper published in Borchardt's

[8]This final sentence was almost surely added to the text after Hilbert received a letter from Minkowski, urging him to mention Hurwitz's accomplishments when he made the revisions for final publication. Minkowski to Hilbert, 5 November 1900 ([Minkowski 1973], 133.

[9]Weierstrass to Lindemann, 14 June 1882, cited in ([Lindemann 1971], 85).

Journal für die reine und angewandte Mathematik, for which he wrote: "I will not venture to search for a demonstration for the transcendence of π. If others attempt that enterprise, no one will be happier than me should they succeed, but believe me, my friend, it will cost them some real effort" ([Hermite 1873b], 342).

Hermite's result says that e cannot be the root of an algebraic equation with rational coefficients. Expressed slightly differently, this asserts that for arbitrary integers $a_0, \ldots, a_n$ the equation

$$a_0 + a_1 e^1 + \cdots + a_n e^n = 0 \tag{1}$$

only holds when all the coefficients vanish. Clearly, a natural generalization here—the one Lindemann actually found—would consider the case where the exponents were algebraic numbers. Were this to be provable, then the transcendence of π would follow immediately from Euler's formula, $1 + e^{i\pi} = 0$, since the exponent $i\pi$ would have to be algebraic if π were.

According to Lindemann's later recollections of these events, the key idea for proving this theorem came to him on 12 April 1882, during a walking tour to Günterstal outside Freiburg im Breisgau ([Lindemann 1971], 84). In his excitement he broke off his excursion and hurried back home so that he could begin writing up the argument he saw before him. By that very evening, he was convinced that he had found a sound proof showing that π had to be transcendental; soon thereafter he sent his manuscript off to his former mentor for publication in *Mathematische Annalen.*

Klein, who then taught in Leipzig, was a busy man, and even more so just now. For the past year he found himself in the thick of a largely losing struggle to keep up with the latest research announcements that were pouring out of Paris, written by a young Frenchman whose name was Henri Poincaré (see [Gray 2013], chapter 3). The stakes in the competition that developed between them were very high, in part due to the general rivalry between French and German mathematicians in the wake of the Franco-Prussian War. Nationalist fervor grew thereafter in intensity and was now on full display throughout Continental Europe. Under these circumstances, Klein, and even more so Poincaré, behaved quite civilly toward one another, as both began to open up a whole new realm in complex analysis, later called the theory of automorphic functions. Poincaré had been one of Hermite's star students, whereas Klein had rather little familiarity with the latter's work, except for Hermite's contributions to the theory of quintic equations. It is unlikely that he had ever read his now famous paper on the transcendence of e [Hermite 1873a], a topic quite remote from his own research interests.

In handling a new submission, Klein would normally have consulted with others on the editorial board, typically Paul Gordan and/or Max Noether. This time, however, he took the unusual step of forwarding Lindemann's paper to Georg Cantor, perhaps because the latter had published his most important recent work in the *Annalen.*[10] On the other hand, Klein may well have thought of Cantor in

[10] John McCleary called my attention to the possible relevance of another very early publication,

view of his famous theorem on algebraic numbers from a decade earlier [Cantor 1874]. This made clear that the transcendental numbers dominate in the continuum, owing to the fact that the algebraic numbers (including those which are complex) form only a countably infinite set.[11]

Be that as it may, Cantor, in a postcard from 8 June ([Purkert and Ilgauds 1987], 186), responded with excitement, though also mixed with some caution. Although he had no doubts that Lindemann's proof was sound, he judged that it still needed to be carefully examined. Even so, he stressed that if all stood well, then this "will be of great satisfaction for Hermite," who would have the pleasure of seeing that his methods can, indeed, lead to the goal that had eluded him. By that same afternoon, Cantor wrote again, now even more assured that "Lindemann's work represents a happy extension of Hermite's methods that deepens and exploits his work on the exponential function" ([Purkert and Ilgauds 1987], 186). Two days later, Klein received a lengthier letter from Cantor, explaining how Lindemann could sharpen the formulation of his main theorem ([Purkert and Ilgauds 1987], 187–188). He also gave a short proof showing that from this new version one could prove the striking result that the exponential curve $y = e^x$ contains no algebraic points other than $(0, 1)$. He then wrote that this result "deserved to be counted among the most beautiful and interesting in arithmetical analysis," and he thought it should rightly be named the Theorem of Hermite-Lindemann. Cantor's elation comes through strongly in this letter, though he still mentions certain gaps in Lindemann's argumentation. In view of later developments, we will return to this in order to highlight the role played by Cantor in this story.

Three days afterward, Klein received still another letter from Cantor, but this time the tone and thrust of his message sounded quite different than before ([Purkert and Ilgauds 1987], 188–189). He was still convinced that Lindemann's argument was sound, but now he stressed that the gaps in his presentation were far more serious than he had at first imagined. As a product of the "Berlin school," Cantor identified with its high standards of rigor and consequent low tolerance for sloppy writing. Still, one wonders whether he would have expressed such concerns were he merely reporting on a routine submission to Klein's journal. It seems more likely that he stressed the need for revision precisely because the problem at hand had such great import; this factor surely weighed heavily in convincing him that Lindemann needed to close all the various holes in his arguments. That, to be sure, was the main message in Cantor's letter, but not the only one; he also had another important suggestion to make. Recognizing that time was now a concern, he confided that it would also be possible merely to restate the general theorems in the form he had advised in his previous letter, while at the same time announcing that the proofs would soon follow in a sequel. This would enable Lindemann to secure his priority, while giving him the time he needed to write up detailed and

namely [Cantor 1869], in which Cantor proves a general theorem from which the irrationality of e follows immediately.

[11]For historical background related to [Cantor 1874], see chapters 2 and 3 in [Dauben 1979], and [Ferreirós 1999], 171–175.

complete proofs of the main theorems. As we shall see, Lindemann, quite possibly also following Klein's advice, opted to take this second course of action.

Cantor was not quite finished, however. He also mentioned that he had in the meantime informed Weierstrass of Lindemann's submission, citing a passage from the latter's response to him: "Should this matter, which interests me in the highest degree, be in order, perhaps you could persuade Lindemann to send his work, whether *in extenso* or also only in brief form, to our academy; it would give me great pleasure to present it for publication in the proceedings, which appear regularly every eight days following each session" ([Purkert and Ilgauds 1987], 189). Cantor thought Lindemann should certainly avail himself of this opportunity, but clearly he had in mind that Lindemann would reserve this option for the ultimate version of his work, containing full proofs of all the attendant theorems. He thus seems to have misconstrued what Weierstrass more likely had in mind, namely, that the Berlin Academy might publish a short announcement within almost no time. Cantor concluded by repeating that the paper Lindemann wished to publish in the *Annalen* should contain an announcement that a presentation of the full theory would appear in a subsequent issue.[12]

From this point onward, the situation became quite complicated, even though no one expressed any real doubts about the soundness of Lindemann's argument. By contacting Weierstrass, the great sage of German mathematics, Cantor had in effect "passed the buck" to higher authority. Lindemann and Klein now had to take the interests of the Berlin school into account, a situation made all the more delicate because of the ongoing feud between Klein and a leading representative of that school, Lazarus Fuchs (see [Rowe 1992]). In the meantime, however, Lindemann was working on a revision of his original paper, presumably in order to take account of Cantor's suggestions. But instead of working on the gaps in his arguments, he chose instead to make cuts in his original manuscript, while at the same time adding a new theorem that went beyond what was needed to prove the transcendence of π. For this more general theorem, however, he gave no hints whatsoever of a proof.[13]

Virtually no documentary evidence survives from the two-month period that

[12]Lindemann's recollections blur many of the events from this time. Thus, he writes that *after publication of his paper* Weierstrass reacted with enthusiasm in a letter to Cantor from 14 June 1882: "The generalization of Hermite's methods will ocassion admiration in all mathematicians, who will recognize and appreciate the significance of such a simple and naturally appearing and yet so deep line of thought." ("Die Verallgemeinerung der von Hermite angewandten Methode wird die Bewunderung aller Mathematiker erregen, welche die Bedeutung eines an sich einfachen und natürgemäss erscheinenden und doch so tiefsinnigen Gedankens zu würdigen wissen werden." [Lindemann 1971], 85). The passage and date could well be accurate, but at this point in time Lindemann had not yet published anything relating to Hermite's work. It is conceivable, although quite unlikely, that Cantor had sent portions of Lindemann's submission to Weierstrass in Berlin. A far more plausible explanation, however, is that the latter was merely reacting to the general information Cantor had conveyed to him.

[13]Lindemann mentions that he made certain cuts in preparing the revision ([Lindemann 1971], 86-87). This would also be consistent with the advice Cantor imparted in his letter to Klein from 13 June.

followed Lindemann's discovery in mid–April, so the precise circumstances that led him to revise his paper remain quite unclear. Nor do we have any clear idea of how far this new version differed from the original. Nevertheless, the fact that Lindemann did revise his earlier submission in June is certain,[14] and this circumstance had real consequences for what would follow. For as we shall see, what happened afterward would shape some of the most significant events that form the prehistory of Hilbert's seventh Paris problem. The threads of all this can now be traced back to Berlin in mid-June of 1882.

Weierstrass enters the Picture

Cantor's final letter to Klein regarding Lindemann's submission was dated 13 June 1882. Weierstrass may have written directly to Lindemann in the meantime,[15] but whether or not he did, Cantor's letter and the offer from Weierstrass cited in it clearly led to new deliberations. Some confusion also arose regarding the nature of the offer: did this mean that Weierstrass was willing to accept responsibility for publishing a preliminary announcement of Lindemann's sensational result? When Lindemann learned of this, he apparently had misgivings, perhaps only because his paper had already been submitted to Klein some two months earlier ([Lindemann 1971], 85). Klein, at any rate, decided to write Weierstrass on 17 June. The latter responded the next day, clearly somewhat taken aback at having to explain his intentions.[16]

Judging from this reply, Weierstrass knew nothing more at this time about the content of Lindemann's original submission than the information Cantor had sent him. It should be noted, however, that whatever other correspondence, if any, these two may have exchanged regarding Lindemann's submission, no relevant documentation seems to have survived. In any event, in his letter to Klein, Weierstrass reported that Cantor had informed him that in his opinion (which Cantor assumed was shared by Klein) Lindemann's arguments were sound, but not yet complete. For that reason, he understood from Cantor that the manuscript would need to be revised in order to fill in certain gaps, and even though these involved routine matters, such a revision might require some time. Under these circumstances, Weierstrass explained to Klein, he had raised the possibility that Lindemann might submit a brief research announcement that could then appear one week later in the proceedings of the Prussian Academy. Thus, like Cantor, Weierstrass recognized the urgency of the situation, namely that Lindemann needed to secure his priority for the main result, which represented the final episode in the long history of the quadrature of the circle. To Klein, he therefore repeated his offer to present

[14]Not only does Lindemann describe this revision process in [Lindemann 1971] but the two dates, April and June 1882, are given at the end of [Lindemann 1882b].

[15]Lindemann recalled a three-way exchange between himself, Cantor, and Weierstrass ([Lindemann 1971], 85). The extant correspondence, however, including that which he later cited, suggest that he was merely informed about what the other two had written through Klein.

[16]Weierstrass to Klein, 18 June 1882, Klein Nachlass XII, NSUB Göttingen.

a short notice at a forthcoming session of the academy, which met on a weekly basis, but to avoid any possible confusion—for he had the impression Lindemann wanted to publish his paper in *Mathematische Annalen* beforehand—he pointed out that such an announcement would have to precede any other publication.

After this, the matter was quickly settled, and Lindemann agreed to send Weierstrass an extract from his paper for the *Annalen.* This led to the historic announcement [Lindemann 1882a] proclaiming that π is transcendental, which Weierstrass presented to the academy on 22 June 1882. Lindemann's note began by indicating that his results represented a direct extension of Hermite's arguments from his 1873 paper proving that e was a transcendental number. After sketching the main ideas for his first theorem, he noted that from it one could draw the immediate consequence that e^z was transcendental for all non-zero algebraic numbers z. Lindemann noted that further details would be presented in his forthcoming paper in *Mathematische Annalen.* He then added some remarks about the second generalization he had found of Hermite's main theorem, namely that a polynomial expression in e with algebraic exponents and *algebraic* coefficients can never vanish. From this, he noted the immediate consequence that the exponential curve $y = e^x$ contains no algebraic points other than $(0, 1)$. This, of course, was the observation Cantor had pointed out earlier to Klein in his letter from 10 June, a clear indication that Cantor's suggestions for revision had, indeed, influenced the content of both this announcement and the paper that would follow it [Lindemann 1882b]. This shorter note then concludes with the following remark: "I have only partially indicated the proofs of these general theorems in my paper for the Annalen; still, I nevertheless hope to come back to these."[17] Those who read this closing remark might, indeed, have wondered whether Lindemann had a solid proof for the results he now claimed in this brief announcement.

Two days later, Cantor informed Klein that all had gone smoothly: Lindemann's note would appear the following week in the academy's proceedings.[18] Cantor was sure that this notice would only help to heighten interest in the forthcoming paper that would soon appear in *Mathematische Annalen.* Although Cantor was not present on this occasion, a member of the academy who was, the distinguished physiologist Emil du Bois-Reymond, gave a firsthand impression of what had transpired. This comes from a letter to his brother Paul, a mathematician in Tübingen (Lindemann had assumed his professorship when he came to Freiburg). At this meeting, Weierstrass called Lindemann's breakthrough a "mathematical feat of the first order," and he further recalled how he and other Berlin mathematicians had tried years ago to persuade Hermite to tackle the problem. The French analyst answered by telling them he had given up: all his efforts had been to no avail, and it seemed to him that "the devil has his hand in it" ([Lindemann 1971], 85-86). Hermite had long cultivated friendly relations with German

[17]Die Beweise für diese allgemeinen Sätze habe ich in meine Arbeit in den Annalen teilweise nur angedeutet; doch hoffe ich auf dieselben noch zurückkommen zu können ([Lindemann 1882a], 682).

[18]Cantor to Klein, 24 June 1882, Klein Nachlass VIII, NSUB Göttingen.

mathematicians, including Lindemann with whom he corresponded regularly. So he reacted enthusiastically when Lindemann informed him of his breakthrough. Soon thereafter other letters came pouring in from throughout Europe congratulating Lindemann, but also Hermite, whose methods had played such a crucial role ([Lindemann 1971], 85–86).

What Klein and Lindemann did not yet know, though they would soon find out, was that Weierstrass was taking an active interest in these matters. He was especially drawn to the challenge of proving the second theorem announced in Lindemann's notice. In the process of doing so, he also began simplifying the presentation in his paper for *Mathematische Annalen*, since Lindemann's line of reasoning required prior knowledge of the arguments Hermite had published earlier. When precisely Weierstrass actually got to read Lindemann's paper—and unless Cantor sent it to him, he never did see the original submission—would seem impossible to know. But since the published version came out in the second quarterly installment of *Mathematische Annalen*, we may safely presume that it had reached Weierstrass by no later than July of 1882. On reading it, one can easily imagine his state of mind, which never rested until he had worked through all the details needed to solve a problem in a systematic fashion. A hallmark of Weierstrass' mathematical style was the self-contained quality of his work in the classical tradition from Archimedes to Gauss. So it comes as no surprise that he felt compelled to clarify all the issues here from the ground up, without any reliance on extraneous research results.

Lindemann leaves the Arena

Weierstrass found it simply impossible to resist the challenge of trying to prove the second theorem Lindemann had announced in his note to the academy. That result can be expressed as a direct generalization of the theorem Lindemann needed in order to prove the transcendence of π, since it asserts that for arbitrary *algebraic numbers* $a_0, \ldots, a_n$ and $\alpha_1, \ldots, \alpha_n$ the equation

$$a_0 + a_1 e^{\alpha_1} + \cdots + a_n e^{\alpha_n} = 0 \tag{2}$$

can never hold unless all the coefficients vanish. Just one month later, on 16 July, Weierstrass wrote Cantor that he had found an air-tight proof for this theorem, but he also noted that he would await Lindemann's final publication before going public himself.[19] Two weeks later, he wrote Cantor again, this time to send him his manuscript, in which he had essentially rewritten Lindemann's entire paper, though only to satisfy himself that the results were correct.[20]

Weierstrass clearly expected that Lindemann would soon publish a more definitive version of his paper in *Mathematische Annalen*; perhaps he thought this final version would appear later that same year. This, after all, had been the main

[19]Weierstrass to Cantor, 16 July 1882, Cod Ms. G. Cantor 13, NSUB Göttingen.
[20]Weierstrass to Cantor, 30 July 1882, Cod Ms. G. Cantor 13, NSUB Göttingen.

advice Cantor imparted to Klein, and Weierstrass had every reason to believe that this was a consensus opinion. At the same time, there was common agreement that the gaps in Lindemann's paper represented merely technical obstacles, or small lacunae that could easily be handled. At any rate, Lindemann seems to have followed Cantor's general advice: he truncated the notes he had used to elucidate the ideas needed to prove his main theorems, saving the proofs for a subsequent publication ([Lindemann 1971], 86–87). Whether he was actually prepared to deliver what Cantor thought necessary seems more than doubtful in view of Lindemann's subsequent "contributions" to number theory, about which more below. At any rate, his brief note for the Berlin Academy and longer paper in *Mathematische Annalen* were greeted with universal approval and subsequently cited as providing the first correct proof for the transcendence of π.

Cantor may well have been right when he predicted to Klein that many mathematicians would read Lindemann's paper [Lindemann 1882b], but we can be just as sure that soon after it was published those with a serious interest in the theory of transcendental numbers no longer paid close attention to it. Two reasons why this paper was often cited, but seldom read, are fairly obvious. First of all, Lindemann's arguments are very sketchy and certainly difficult to follow. Second, his general theorem, the one he added to the final proof pages and which enables one to prove other interesting transcendence results, was only stated without any proof. In fact, this paper ends with a final remark stating that a proof would appear in a subsequent paper, just as Cantor had recommended. He and Weierstrass presumably expected that this sequel would come out soon afterward, but in fact Lindemann would never publish another paper on what is today often called the Lindemann–Weierstrass Theorem. It seems he was far less interested in this general result than in proving the transcendence of π as a special case.

A third reason why Lindemann's paper quickly became obsolete stems from Weierstrass' activities at this time. For in the meantime, Weierstrass' manuscript, or some variant of the paper he had sent to Cantor on 30 July, 1882, was now beginning to circulate within elite circles of the German mathematical community. By October, at the latest, Lindemann, too, had heard about Weierstrass's interest in the topic of his *Annalen* paper. He apparently also had heard rumors to the effect that the latter had done essential repair work on certain arguments in it. This, Lindemann later claimed, took the wind out of his sails, leaving him so depressed that he lost all interest in working on a new version of his published paper.[21] Still, he felt greatly relieved by a friendly letter he received from Weierstrass, who explained his interest in the arithmetical consequences of the theorems of Hermite and Lindemann. Weierstrass asked, therefore, if Lindemann would grant

[21]In [Lindemann 1971], 87, he recalled: "Ich erfuhre, dass Nachschriften der Weierstrass'schen Vorträge zirkulierten und so das Gerede entstand, meine Arbeit über die Zahl π sei lückenhaft und Weierstrass habe die Sache erst in Ordnung gebracht. Dadurch war mir jede Lust genommen, mich mit diesem Problem noch weiter zu beschäftigen." The rumor, however, seems actually to have been well grounded, and Weierstrass patiently awaited the announced sequel to [Lindemann 1882b] that was never written.

him permission to present his personal reflections on this topic in a public lecture. In all likelihood, Weierstrass wrote that he would withhold publication of his own work until such time as Lindemann had published the sequel to his published paper. At any rate, the record shows that Weierstrass spoke about Lindemann's earlier note for the Berlin Academy at its session held on 26 October, 1882.[22] This marked an important occasion for the early proliferation of this new theory, in particular the so-called Lindemann–Weierstrass Theorem.

Soon thereafter, if not before, Hermann Amandus Schwarz had read a copy of Weierstrass' manuscript, which was also sent to Hermite. After proving the main theorem, Weierstrass wrote that this result represents the culmination of Hermite's investigations of the exponential function. He further noted two immediate consequences that belong to the most beautiful results of arithmetic": 1) that e^x is transcendental for all nonzero algebraic numbers x; 2) similarly, that the natural logarithm is never algebraic for x algebraic, except for $x = 1$. Eventually, Dedekind would read this paper, too; in fact, both he and Schwarz later sent Weierstrass suggestions for improvements.[23] Carl Runge, who was about to begin his *Habilitation* in Berlin, also saw a version of this paper, and through him a young *Privatdozent* in Göttingen got to read it, too.[24] This was Klein's former doctoral student, Adolf Hurwitz, who would a year later join Lindemann as an associate professor in Königsberg. There, of course, is where he met Hilbert, which brings us back to the seventh Paris problem. For, as noted above, Hilbert expressly cited Hurwitz's 1883 paper on new transcendence results in his Paris lecture. One might have thought by reading Hilbert's text that the intellectual line of descent ran from Hermite to Lindemann to Hurwitz, but such was not the case. A mere glance at the latter's paper reveals that Hurwitz's immediate inspiration came from Weierstrass' unpublished manuscript, which he cites at the very beginning in a footnote ([Hurwitz 1883], 212). Hurwitz finished writing this paper in March 1883, when he was on semester break at his family's home in Hildesheim, and just before learning that he was about to be appointed to a new position next to Lindemann in Königsberg.

A few years later, on 31 October 1885, a large gathering of mathematicians took place in Berlin to celebrate Weierstrass' seventieth birthday.[25] Among those present was Lindemann, who recalled with pleasure how Weierstrass took this occasion to distribute reprints to all his guests of his freshly printed paper, "Zu Lindemanns Abhandlung: Über die Ludolph'sche Zahl" [Weierstrass 1885]. Only a short time before, he finally submitted his text for publication; this was at the October 22 meeting of the Berlin Academy, exactly three years after he had

[22]Emil du Bois-Reymond presided, so quite possibly it was after hearing Weierstrass discuss the significance of the Hermite-Lindemann theorems that he wrote the letter his brother, Paul, cited above.

[23]These tips from Schwarz and Dedekind helped to streamline the arguments, as Weierstrass gratefully acknowledged when he presented the final published version ([Weierstrass 1885], 1067).

[24]Carl Runge to Adolf Hurwitz, 31 October 1882, *Mathematiker-Archiv* 79, 31, NSUB, Göttingen.

[25]For an idea of the scale of this celebration, see [Boelling 1994].

spoken about it there. In a footnote, Weierstrass commented on the circumstances that had prompted him to write this paper, making clear that he only wished to simplify or complete the proofs of the results that Lindemann had already given ([Weierstrass 1885], 1067). The latter was, of course, greatly appreciative of the way Berlin's aged master had handled this delicate matter ([Lindemann 1971], 87).

In retrospect we can see that Ferdinand Lindemann—a solid geometer, who happened to have rather little feel for number theory—was lucky. He had stumbled onto a problem that succumbed fairly easily to Hermite's methods, once he saw how to generalize the latter's main theorem. Lindemann's sudden luck and new-found fame ended, however, there. Instead of trying to secure the foundations of this theory and apply it to find new transcendence results, he let others (Weierstrass, Hurwitz, Hilbert, et al.) pursue this new lead. Had he simply returned to geometrical research, including work on the long-awaited second volume of Clebsch-Lindemann, his subsequent career might well have resembled that of any other respected mathematician of the era. This was not to be, and so perhaps Hermite, a pious Catholic, knew what he was talking about when he claimed that the devil had his hand in the old problem of squaring the circle. After Lindemann solved that riddle, he next imagined that the time was ripe to solve the Fermat conjecture, often called Fermat's Last Theorem. His numerous claims to have solved it during the remaining decades of his long life eventually made him appear like a laughing-stock to those who had some inkling of serious research in number theory.[26]

Prussian Mathematics in Transition

During the decade that Hilbert spent alongside Lindemann in Königsberg, he learned a great deal about number theory and much else, but mainly from Lindemann's younger colleague, Adolf Hurwitz, as well as Hilbert's personal friend and fellow student, Hermann Minkowski. This talented trio loved to compete in a friendly way, though Hurwitz and Minkowski were the more playful spirits, while Hilbert was full of burning ambition. After graduating, Minkowski left Königsberg to do his *Habilitation* in Bonn, leaving Hilbert behind in the city of his birth, where he habilitated in 1886. After this, the two friends often wrote about their prospects for gaining a professorship, a goal that eluded them both until 1892. That year was decisive for mathematics at the Prussian universities, since the death of Kronecker and the retirement of Weierstrass not only meant the end of a great era for mathematics in Berlin; it also set off a chain reaction of appointments elsewhere.

After H. A. Schwarz was appointed to the chair formerly occupied by Weierstrass, his position in Göttingen had to be filled, and Klein, who had been teach-

[26]Minkowski wrote to Hilbert on more than one occasion saying that he and Hurwitz were shocked to see the kinds of elementary mistakes Lindemann made in his lectures and writings on topics in number theory; see [Minkowski 1973].

ing there since 1886, now hoped he could convince the faculty to appoint his star pupil, Hurwitz, as Schwarz's successor. Klein had been passed over entirely by the Berlin faculty, somewhat predictably in view of the partisan conflicts that had long marked his career, but he now saw the opportunity to build up Göttingen by exploiting its historical importance for the mathematical sciences, which dated back to the time of Carl Friedrich Gauss [Rowe 1989].

Much to Klein's dismay, this initial plan backfired: the faculty chose Heinrich Weber rather than Hurwitz following a series of complicated negotiations. Anti-Semitism was a pervasive obstacle throughout the German universities, and Klein surmised that this may well have been decisive in killing Hurwitz's candidacy.[27] Hilbert was deeply disappointed, too. When he learned that Hurwitz had been passed over in Göttingen, he wrote Klein in April 1892 that "es wär zu schön gewesen," and for him, too, since he would have had the chance to move into Hurwitz's associate professorship ([Frei 1985], 80). But now "everything looked bleak and hopeless (*trost- und aussichtslos*)," and he predicted that the century would be over before four chairs in mathematics would become vacant in Prussia again. This prediction, luckily enough, turned out to be totally wrong; the chain reaction had just begun.

Minkowski, who like Hurwitz was Jewish, might have had even more reason to be concerned about his future career. Yet he saw the German academic world with far more sanguine eyes than Hilbert, whom he constantly praised as the coming mathematician of their generation, and not just in his letters to him. Only one week after Hilbert wrote in despair to Klein, Minkowski had a splendid chance to sing Hilbert's praises when he met with Friedrich Althoff in Berlin. Althoff was the all-powerful ministerial official responsible for academic appointments throughout the Prussian universities. An autocratic personality, he had a well-deserved reputation for promoting excellence while running roughshod over the interests of faculties and individual professors, partly by making use of informal information that he gathered through a network of contacts with his "Vertrauensleute" (whom critics viewed as Althoff's academic spies). Just after this meeting, Minkowski dashed off a postcard to his friend reporting that Althoff had informed him that a wave of new appointments in mathematics was about to take place. According to this plan, both of them would soon receive appointments as associate professors; their days of misery as *Privatdozenten* were about to end! How this would all be orchestrated remained to be seen, of course, but Althoff proved true to his word.[28]

No doubt Friedrich Althoff was well abreast of the next round of negotiations for full professorships as well, beginning with Kronecker's chair in Berlin. Georg

[27]On this theme in general, see [Rowe 1986]; for the circumstances in Hurwitz's case, [Rowe 2007].

[28]Minkowski to Hilbert, 20 April 1892 ([Minkowski 1973], 46). Althoff named two others besides Minkowski and Hilbert, Victor Eberhard and Eduard Study. All four soon were promoted, beginning with Hilbert's appointment in Königsberg, followed by Minkowski's in Bonn. Study was offered the latter's chair in 1894, after Minkowski moved into Hilbert's post. Eberhard, who had long been in Königsberg, was then appointed associate professor in Halle.

Frobenius, a Berlin product who taught at the ETH in Zurich, had been offered that position, but he dithered for a while between accepting it or perhaps assuming the chair Schwarz had just vacated. Klein welcomed him to visit Göttingen, but soon afterward Frobenius decided he would prefer to return to the Prussian capital. That meant there was now an opening to fill at the ETH, which had long served as a waiting room ("Wartesaal erster Klasse") for mathematicians from the Prussian university system.[29] When Hurwitz was chosen to succeed Frobenius, Hilbert was first in line to move up the academic ladder. Just two months after writing his pessimistic letter to Klein, Hilbert's mood had changed completely, as he now confided to Klein that Lindemann promised to recommend him to fill Hurwitz's position.[30] Minkowski was delighted, too. He had learned from Althoff that Heinrich Weber had recommended his appointment as *Extraordinarius* (associate professor) in Marburg; but then Althoff quickly insisted, despite Minkowski's positive reaction, that he should stay in Bonn, and by August that appointment came through as well. Minkowski wondered, though, in a letter to Hilbert, whether Hurwitz might angle for Hilbert to get the other post at the ETH, now that Friedrich Schottky had gone to Marburg. In posing that query, he also commented that he realized, perhaps more than others, that his own chances to join Hurwitz in Zurich were nil, given that they were both Jews ("allein durch die Gleichheit der Konfession" [Minkowski 1973], 48). Later, after Minkowski had risen to a full professorship in Königsberg, even that obstacle no longer stood in his way, and in 1896 he joined Hurwitz on the ETH faculty.

All had gone smoothly enough, but by the time Klein wrote to congratulate Hilbert in September 1892, he could also update him on another series of events. For that past July, Klein had been offered the professorship in Munich formerly held by Ludwig Seidel. Klein immediately informed Althoff of this, and the latter wasted no time in making a generous counter-offer to ensure that the Bavarian Ministry of Education would not outbid his own and thereby entice Klein to leave Göttingen. This all took place rather quickly, so what Hilbert now learned was that Klein had, indeed, declined, and furthermore that Lindemann was next in line to be offered the chair in Munich.[31] Those negotiations, in fact, would stretch into the next year, but soon afterward, Hilbert assumed Lindemann's chair and, as a final triumph, he even succeeded in having Minkowski appointed to his old position so that they were together once again, if only briefly. Althoff played along wonderfully, though obviously no one—not even Klein—would have imagined that both Hilbert and Minkowski would eventually become his colleagues in Göttingen.

So looking back, in 1892, just two years after the founding of the *Deutsche Mathematiker-Vereinigung* (DMV),[32] both Hilbert and Minkowski had managed

---

[29]Among the prominent German mathematicians who taught at the ETH before obtaining a position in Prussia were: Dedekind, Christoffel, H. A. Schwarz, H. Weber, Frobenius, Schottky, and Minkowski.

[30]Hilbert to Klein, 23 June, 1892, ([Frei 1985], 81).

[31]Klein to Hilbert, 8 September, 1892, ([Frei 1985], 82).

[32]For background on Cantor's role in its founding, see [Hashagen 2000].

Figure 1: The inaugural meeting of the German Mathematical Union (DMV), Bremen 1890. Cantor is seated fourth from the left, Klein fifth from the right in the front row; Paul Gordan in the center. Hilbert and Minkowski are standing in the back row, second and third from the left.

to attain associate professorships. Their careers had thus been launched at just the right moment for both to make their marks on the still small, but growing German mathematical community with its new national organization. In the meantime, Hurwitz stood by in the "waiting room" at the ETH in Zurich. These years of transition also had important personal implications. Before departing from Königsberg, Hurwitz and Ida Samuel, the daughter of a professor of medicine, were married in June, right around the time that Hilbert and his fiancée, Käthe Jerosch, fixed plans for their marriage in October. Minkowski later married Auguste Adler from Strassburg, and the couple then began their life together in Zurich.

Hilbert's Simplification

Shortly after his marriage, Hilbert sent off New Year's greetings to Klein and his wife, along with a short note that he wished to have published in the *Göttinger Nachrichten.*[33] In his letter, he explained that the note gave a far simpler proof of the transcendence of e and π, and did so simply by exploiting the formal properties of an integral expression that had been introduced by Hermite. He further emphasized that his proof evaded all the technicalities that made the earlier papers of Hermite (1873) and Weierstrass (1885) so difficult to read. On the other hand, he made no mention of Lindemann's famous paper, which was by now quite obsolete. Although Hilbert had studied under both Lindemann and

[33]Hilbert to Klein, 5 January 1893, ([Frei 1985], 85–86.

Hurwitz in Königsberg, beginning in 1883, he probably had little knowledge of the events from the previous year described above, whereas Klein had been directly involved in those affairs. Klein's reaction now—"das ist ja wunderschön,"[34] with the assurance that he would be honored to present Hilbert's note at the next meeting of the Göttingen Scientific Society—was surely accompanied by vivid memories of what had transpired a decade before, when Weierstrass intervened to facilitate Lindemann's famous announcement. What Hermite, Lindemann, and finally Weierstrass had managed to build up—using fairly elaborate arguments based on the properties of special integrals related to the exponential function—Hilbert could now prove in a mere four pages!

Hilbert also wrote to Hurwitz, who quickly saw how to take still another step toward further simplification [Hurwitz 1893]; his proof for the transcendence of e eliminated altogether the need for the integrals in Hilbert's argument! Thus began an interesting back and forth competition, in which Paul Gordan soon chimed in as well [Gordan 1893]; (we return to this below when discussing Klein's role in this story). From one of his letters to Hurwitz, we learn that Hilbert got the initial idea for this new proof by reading a paper on the transcendence of e published by T. J. Stieltjes in 1890.[35] Stieltjes paper was hardly longer than Hilbert's, and had been published in the widely read *Comptes Rendus* of the French Academy. Remarkably, Hilbert made no mention of this in his publication; in fact, his notice contains no references to the literature on these problems whatsoever! Was this in the name of simplicity, or perhaps just a sign of supreme intellectual arrogance?

Minkowski had not heard from his friend for a while, but he wrote Hilbert on 23 February, only shortly after reading his paper, to congratulate him once again ([Minkowski 1973], 50). He could only think of Euler's reaction, after reading a striking result published by the young Lagrange, when he cried out: "penitus obstupui, quum hoc mihi nunciaretur" (absolutely amazed when I saw that). In his customary manner, Minkowski poured out encouragement, predicting that Hilbert's other works would from now on gain even more attention. Oddly enough, he even pictured Hermite's mood when reading this paper (for even though the French master could not read German, the entire paper consisted merely of strings of mathematical equations, all very familiar to him): "and knowing the old fellow as I do, it would not surprise me if you soon heard from him about the joy he took in being able to experience this." A little later, Minkowski showed Hilbert's note to his senior colleague in Bonn, Rudolf Lipschitz, whose reaction was not unlike Minkowski's; he returned the paper with just one comment: "meisterhaft" (masterful), (Minkowski to Hilbert, 2 June 1893, [Minkowski 1973], 52).

Hilbert's closing remark in this paper contains the only even indirect reference to earlier work. There he wrote: "it is easy to see how the most general theorem of Lindemann for the exponential function can be proved by following the same path as here" ([Hilbert 1893], 219). This comment pertains, of course, to the so-

[34]Klein to Hilbert, 7 January, 1893, ([Frei 1985], 86–87.

[35][Stieltjes 1890], a paper later reviewed by Hurwitz for the *Jahrbuch über die Fortschritte der Mathematik.*

called Lindemann-Weierstrass Theorem, announced by the former and proved by the latter. From his letter to Klein, it is clear that Hilbert was familiar with Weierstrass' 1885 paper, and yet his text makes no allusions to the prior work of Hermite, Weierstrass, or Stieltjes. Klein wondered whether Hilbert might wish to publish a more extended version of this note on the exponential function in *Mathematische Annalen*, especially since he was eager to publicize Hilbert's new proof there.[36] The curt replay he received might be summed up as saying—no, not really—or words to that effect. Instead of taking the trouble to give a more complete presentation with references, Hilbert simply wrote back to say that he had nothing more to add, though he had been pleased to receive a message from his former nemesis, Paul Gordan, who had expressed his appreciation.[37]

Yet more was to follow, as the immediate reactions of both Hurwitz and Gordan led to three short notes that first appeared in the *Göttinger Nachrichten*, then afterward, this time bundled together, in *Mathematische Annalen*.[38] Thus, once again, Felix Klein stood at the center of this new round of events which took place little more than ten years after Lindemann's original breakthrough. Hilbert's simplification of the earlier arguments now signaled that the transcendence proofs for e and π were ripe for standarization and further transmission as textbook material. No one could have been in a better position than Klein to exploit this opportunity, and he wasted no time finding ways to popularize this knowledge within the larger world of educated mathematicians. In doing so, he largely had a free hand to tell his version of the history of these famous problems and their modern-day resolution.

Klein on Hilbert's Proof

Popularizing mathematics in the context of the 1890s meant something quite different than it does today. In the present instance, what Klein had in mind was a novel way to discuss recent mathematical research, while foregoing virtually all the necessary technical details. One might call this type of approach "popularization for an intellectual elite," since its aim was not to reach the proverbial "man on the street" but rather a mathematically educated audience. By this time, Klein's lecture courses had already begun to evolve in a similar direction, and while others in Germany continued to emulate Weierstrassian rigor, he settled for short-circuited arguments and vague remarks that could hardly pass muster if one were claiming a "real proof." In Berlin, some called Klein a charlatan, and even Göttingen's Richard Courant once likened him to a visionary pilot who lacked the ability to land his plane.[39] Still, he longed for a chance to speak more informally about a whole range of topics in current mathematical research that reflected his own personal interests and tastes. By looking backward, he wanted to give a broader

[36] Klein to Hilbert, 7 January, 1893, ([Frei 1985], 86).
[37] Hilbert to Klein, 28 February, 1893, ([Frei 1985], 88).
[38] [Hilbert 1893], [Hurwitz 1893], [Gordan 1893].
[39] [Courant 1925], 772; on Klein's reputation in Berlin, see [Rowe 2001].

picture, one that largely evaded the underlying problems that would need to be solved.

This, of course, was almost the very opposite of Hilbert's Paris lecture, delivered seven years later. Still, the outward circumstances were strikingly similar. Klein's chance came from abroad in 1893 when he received an invitation from the mathematicians in the Chicago metropolitan area. He was cordially invited to attend the Mathematical Congress that would convene later that summer—a small workshop by today's standards—one of many scientific congresses that were held in conjunction with the World's Columbian Exposition, an extravaganza that would attract millions ([Parshall and Rowe 1994], chapter 7). For the mathematicians from the two local universities—the newly opened University of Chicago and Northwestern in nearby Evanston—Klein's presence proved both timely and beneficial. Two of Chicago's three faculty members, Oskar Bolza and Heinrich Maschke, were former protégés of Klein, both having come to the U.S. with hopes of gaining academic employment. Moreover, Northwestern's Henry Seely White had been one of several Americans who took their doctoral degrees under Klein, either in Leipzig, or as in White's case, in Göttingen.

Klein attended the Chicago congress as an official emissary of his state government, which sponsered a major exhibition to celebrate the intellectual tradition of the Prussian universities. Klein's former student, Walther Dyck, organized the mathematics section, which was noteworthy for its large collection of mathematical models, about which Klein gave a demonstration.[40] He also brought with him a set of papers written for the occasion by some of Germany's leading mathematicians, including a survey of research results in invariant theory offered by Hilbert. Immediately after the congress adjourned, Klein took advantage of this opportunity to deliver a series of "popular lectures," in the sense described above, held over a two-week span at Northwestern University. Another German, Alexander Ziwet from the University of Michigan, wrote up these talks using Klein's notes and the ones he had taken, so that the *Evanston Colloquium Lectures* [Klein 1894] were soon thereafter in print ([Parshall and Rowe 1994], chapter 8).

It was in the seventh of these twelve lectures that Klein spoke about the recent simplifications that had been made in proving the transcendence of e and π ([Klein 1894], 51–57). Most of this talk was devoted to outlining Hilbert's proof, but he also alluded to the contributions by Hurwitz and Gordan that would soon accompany Hilbert's note in the next issue of *Mathematische Annalen.* Klein also offered a mini-history of work on transcendental numbers over the past two decades, beginning with Hermite's paper on the exponential function. Although he mentioned the contributions of several others, he passed over the fundamental paper by Weierstrass from 1885, while highlighting Lindemann's breakthrough, published in Klein's journal. Even so, the thrust of his message suggested that all this now belonged to the past and could safely be forgotten. In the meantime, Hilbert had

[40]On the mathematics exhibition, see [Parshall and Rowe 1994], 304–309; on Dyck, see [Hashagen 2003].

overcome all the former technical difficulties, so that the proofs were now far easier to grasp. In Klein's view, mathematicians at the universities should no longer have any reservations about teaching these methods of proof to their students.

In view of the letters Klein had earlier received from Georg Cantor, written when he was reporting on Lindemann's original submission, it is noteworthy that this talk began with some brief remarks about Cantor's theory of infinite sets. In particular, Klein underscored the fact that the set of algebraic numbers was only countable, in contrast to its complement in the real number continuum, the transcendental numbers. Of course, this theory only provides a general context for the issue at hand, not for deciding whether a particular irrational number may or may not be transcendental. Still, Klein would return to this Cantorian context at the conclusion of his lecture.

Turning to Hilbert's simplification of the proof that e could not be algebraic, a proof by contradiction, Klein explained that this was achieved by treating the powers of e in the equation (1) as homogeneous coordinates $1 : e : e^2 : \ldots : e^n$, and by substituting an equivalent $(n+1)$-tuple of the form $I_0+\epsilon_0 : I_1+\epsilon_1 : \ldots : I_n+\epsilon_n$, where the I_i are integers and the ϵ_i are small. As Hilbert noted in his message to Klein, this involves finding suitable new coordinates, substituting these in (1), and then splitting the equation into two parts, $P_1 + P_2 = 0$, consisting of

$$(a_0 I_0 + a_1 I_1 + \ldots + a_n I_n) + (a_0\epsilon_0 + a_1\epsilon_1 + \ldots + a_n\epsilon_n) = 0 \tag{3}$$

in order to show this equation is impossible. The key trick is to ensure that the integer P_1 will be nonzero, while at the same time controlling that $|P_2| < 1$. This was Hilbert's new strategy, which he pushed through by using an integral identity:

$$\int_0^\infty z^\rho e^{-z} dz = \rho! \tag{4}$$

that had already played a central role in Hermite's work. Klein's account uses almost exactly the same notation as in Hilbert's note, but his text is far more discursive, making it easier to understand the thrust of the argument for the transcendence of e. When it comes to proving the transcendence of π, he merely refers the reader to Hilbert for the more technical details.

Hermite had also introduced the definite integral:

$$J = \int_0^\infty z^\rho[(z-1)(z-2)\ldots(z-n)]^{\rho+1} e^{-z} dz, \tag{5}$$

but Hilbert's idea was to use the constant ρ as a parameter to be determined after the substitution is carried out. To obtain the new equation $P_1 + P_2 = 0$, one first multiplies (1) with J, and then splits the last n integrals so that the powers of e accord with the upper limits of integration:

$$(a_0\int_0^\infty + a_1 e\int_1^\infty + \ldots + a_n e^n\int_n^\infty) + (a_1 e\int_0^1 + a_2 e^2\int_0^2 + \ldots + a_n e^n\int_0^n).$$

Finally, to get $P_1 + P_2 = 0$, the entire expression is divided by $\rho!$. One nexts observes that the integral J is divisible by $\rho!$ by virtue of (4), whereas by means of the simple substitutions $z = z' + 1, z = z' + 2, \ldots, z = z' + n$, it can be shown that the quantities $e\int_1^\infty, e^2\int_2^\infty, \ldots, e^n\int_n^\infty$ in the remaining terms of P_1 are all integers divisible by $(\rho + 1)!$. From this follows the general congruence:

$$P_1 \equiv \pm a_0(n!)^{\rho+1}[mod(\rho + 1)],$$

which holds for all possible ρ. Since a_0 and n are fixed, one can find a ρ such that the right side of the congruence is not divisible by $\rho + 1$, which then implies that the integer P_1 cannot be zero. Clearly, this argument holds for any larger value of ρ, so it involves no constraints for the second part of the argument. This merely involves showing that P_2 can be made as small as we like by choosing sufficiently large values of ρ, a straightforward matter using the mean value theorem.

At the end of his talk, Klein referred to Lindemann's paper in emphasizing that the exponential function passes continuously through the plane while missing the dense set of points with algebraic coordinates, except of course for the single point $(0, 1)$. The proof Klein gave for this closely resembles the one Cantor sketched out in his letter to Klein from 10 June, 1882. Surely, Cantor never would have written about this had Lindemann's original submission contained such an argument; but, in fact, not even in his published paper does he discuss this in any detail. In other words, what struck Cantor (but also Klein) as one of the most remarkable implications of Lindemann's theorem seems to have left little impression on Lindemann himself.

Klein's Historical and Didactical Interests

From this time forward, Klein began to take a growing interest in secondary-level mathematics education, which he often coupled with historical themes. Thus, after touring the eastern United States, he returned to Göttingen with thoughts in mind about teaching a special course for Gymnasium teachers. This took place during the Easter vacation, and the experiment encouraged him to offer a more extensive two-hour lecture course the following summer semester on the same subject, "Selected Problems in Elementary Geometry." One of the gymnasium teachers who had attended the shorter session, Friedrich Tägert, gathered up his and Klein's notes to produce a booklet, *Vorträge über ausgewählte Fragen der Elementargeometrie* [Klein 1895], published by Teubner the following year. It soon became Klein's best-known and most successful popularization of recent research in mathematics, with translations into French and Italian in 1896, and then, one year later, an English translation appeared bearing the title *Famous Problems of Elementary Geometry* [Klein 1897].[41] This booklet particularly interested American educators, including Columbia's David Eugene Smith, who helped prepare

[41] As Umberto Bottazzini has pointed out, Klein's booklet was quite influential in Italy, particularly on Federigo Enriques ([Bottazzini 2002], 83–84).

the translation. Smith and Raymond Clare Archibald clearly marveled at the way Klein approached the three famous "unsolved" problems of Greek geometry—trisection of the angle, duplication of the cube, and quadrature of the circle—all in the light of modern research. Archibald oversaw a new printing in 1930, for which he added numerous notes correcting faulty information in the original.[42]

Klein's motivation back in 1895 can easily be surmised from his opening remarks in the preface to this booklet: "The more precise definitions and more rigorous methods of demonstration developed by modern mathematicians are looked upon by the mass of gymnasium professors as abstruse and excessively abstract, and accordingly as of importance only for the small circle of specialists. With a view to countering this tendency it gave me pleasure to set forth last summer . . . what modern science has to say regarding the possibility of elementary geometric constructions" ([Klein 1897], iv). His immediate objective, in fact, was to draw the attention of those teachers who were about to convene in Göttingen for the inaugural meeting of the Association for the Promotion of Teaching of Mathematics and Natural Sciences (often simply called the *Förderverein*). So to mark this occasion, Klein presented this group with his recent historical–didactical reflections, and thus ended his preface by expressing his hope that "this unpretending little book may contribute to promote the useful work of the association" (*ibid.*).

Klein's historical interests were nearly always tied to his broader didactical concerns and teaching activities. Regarding historical matters, he rarely lingered over details, and the same can be said for this undertaking. He did, however, give references to relevant historical literature, for example the works of Moritz Cantor and H. G. Zeuthen, which had only recently appeared. Otherwise, however, his booklet concentrates entirely on the achievements of the last century, work that had finally led to the resolution of the three classical construction problems from antiquity. Two of these, angle trisection and cube duplication, Klein dispensed with in only a few pages. About these, he commented vaguely that their solution can be found "implicitly . . . in the Galois theory [presented] in treatises on higher algebra" ([Klein 1897], 2). For an explicit solution, he referred readers to the works of the Danish geometer, Julius Petersen [Petersen 1878], but he nowhere pointed out that both problems, in fact, had been resolved by Pierre Wantzel in [Wantzel 1837], and without recourse to full-blown Galois-theoretic arguments.[43]

Klein, therefore, showed little interest in these two famous problems; instead he focused solely on the third, the quadrature of the circle. So it was misleading when the American translators changed the title of the booklet to *Famous Problems of Elementary Geometry*, and then highlighted these three most famous problems as the primary subject matter ([Klein 1897], v). A better title, if one wanted to convey its contents more accurately, would have been "Famous Problems in Geometry as Solved by Gauss and Hilbert." Indeed, when read alongside Klein's *Evanston Colloquium Lectures*, these popularizations both reveal the same

[42] Archibald had already published these corrections in [Archibald 1914].

[43] Why Wantzel's short paper was so quickly forgotten has been discussed in detail by Jesper Lützen in [Luetzen 2009].

theme of glorification of the Göttingen mathematical tradition. For some, like Eduard Study, who heard the lectures in Evanston, Klein's performances at this time were merely a sign that he had given up doing serious mathematics, but could not resist taking curtain calls on third–rate stages.[44] It was during Klein's trip to the U.S. that Sophus Lie broke with him publicly, belittling Klein's mathematical competence in the preface to the third and final volume of his *Theorie der Transformationsgruppen*; his remarks sent shock waves through the German mathematical community.[45] Such personality clashes played a major role at this time, fed by massive egos that sought recognition and fame; these factors usually remained in the background, but not far from view. What Klein's critics seem to have overlooked was that, while the stages he played on may have looked unimpressive, Klein was still an actor who could take on many roles.

Having followed with keen interest the recent work on transcendence problems, beginning with Hilbert's note from January 1893, Klein was in a perfect position to expound on this for a wide mathematical public. In *Famous Problems of Elementary Geometry*, he chose to package his account with front– and back–end stories, thereby saving the quadrature of the circle and the proof for transcendence of π for the second half of his booklet. The main problem discussed in the first half of his text was a classical question made famous by the young Gauss, namely which regular n-gons are constructible by means of straightedge and compass? Since Klein merely wanted to give his readers the flavor of Gauss's argument, he restricted himself to the first non-classical case, namely $n = 17$, referring to the *Disquisitiones arithmeticae* and Bachmann's monograph [Bachmann 1872] for the general theory of cyclotomy.[46]

Perhaps not coincidentally, soon after publication of Klein's booklet Gauss's proof received much closer attention in a paper written by James Pierpont, an American who had studied under Kronecker in Berlin during the 1880s, before joining the Yale faculty. Pierpont had clearly absorbed a taste for Kronecker's style and Berlin standards of exactness. His paper [Pierpont 1895], entitled "On an Undemonstrated Theorem of the Disquisitiones Arithmeticae," also refuted, albeit implicitly, what Klein had claimed regarding Gauss's accomplishment. According to Klein, Gauss had proved necessary and sufficient conditions for the constructibility of a regular polygon, thereby showing that the number of sides n must have a prime factorization whose factors are either of the form $2^{2^{\mu}} + 1$ or else arbitrary powers of 2. Pierpont, however, clarified what Gauss had actually achieved:

> That this is a sufficient condition follows at once from an easy extension of Gauss's method as developed when n is a prime. It is, however, vastly more important to know that only these polygons can be geometrically

[44]Study and Klein collided many times before, in contrast to Hilbert, who in his younger years built a strong alliance with him. These differences were highly in [Reid 1970], 20–27.

[45]For example, Minkowski to Hilbert, 29 November, 1893, ([Minkowski 1973], 55–56).

[46]On the role of Gauss's *Disquisitiones arithmeticae* in nineteenth-century research for number theory, see [Goldstein/Schappacher/Schwermer 2007].

> constructed as thereby the theory of regular polygons, as far as their construction by ruler and compasses is concerned, is made complete. That is, in a given case we can decide whether the polygon is constructible, and in case that it is, Gauss's theory gives us the necessary directions to construct it ([Pierpont 1895], 79).

Before proceeding to prove that Gauss's criterion, indeed, constituted a necessary condition for constructibility, Pierpont cited what Gauss himself had written about this:

> we can prove with all rigor that these higher equations cannot be avoided or made to depend upon equations of lower degree; and although the limits of this work do not permit us to give the demonstration here, we still thought it necessary to signal this fact in order that one should not seek to construct other polygons than those given by our theory, as, for example, polygons of 7, 11, 13, 19 sides, and so employ his time in vain (*ibid.*).

This was only one of many historical errors that R. C. Archibald later identified in Klein's booklet, which he had used at Brown University in teaching a geometry class for teachers just before the outbreak of the First World War. Archibald published corrections in [Archibald 1914], and later added essentially the same information to the new edition of Klein's *Famous Problems of Elementary Geometry* when it was reissued in 1930.

For the second part of his booklet, Klein again offered a kind of mini–history on the arithmetical properties of e and π. Before doing so, however, he began with an introductory chapter on the fundamental results of Cantor's theory of the real number continuum, especially those needed to appreciate the vast difference between the subsets of algebraic and transcendental numbers contained within it. He had done the same in his seventh Evanston lecture, but this time he gave full proofs of these theorems. By so doing, he clearly wanted to emphasize that Cantor's theory provided a kind of intuitive explanation for the remarkable fact that the exponential curve manages to pass continuously through the plane without intersecting the dense set of points whose coordinates are both algebraic numbers. As we noted earlier, Cantor was apparently the first to take notice of this startling result and to impart to Klein that this was a consequence of Lindemann's work. All traces of his role, however, afterward disappeared, although the sources needed in order to place this story in a proper perspective have long been available.

Needless to say, Klein's booklet is of little help when it comes to reconstructing this prehistory; his interest was only in recounting the most recent events from the year 1893. After crediting Hermite and Lindemann as the two figures who ultimately resolved the issue of circle quadrature, he quickly turned to his central concern, namely simplification. Noting merely that this earlier work was very complicated, he added that the first step in simplifying it was taken by Weierstrass in his 1885 paper. As explained above, however, Lindemann's published paper was very far from complete, and once the manuscript prepared by Weierstrass began to circulate, Lindemann decided to abandon work on this topic. Yet Klein, who knew that Weierstrass was the first to prove the main theorem connected with Lindemann's results, referred to this later as "Lindemann's Corollary." How, then, should one weigh their relative achievements? Perhaps by calling in an expert on number theory?

In his introductory discussion, Klein alludes to the recently published monograph by Paul Bachmann, *Vorlesungen über die Natur der Irrationalzahlen* [Bachmann 1892], writing that this was based on the papers by Hermite, Lindemann, and Weierstrass. One might wonder, though, whether he had bothered to look at this book carefully. If so, then he apparently saw no reason to point out that Bachmann's book was based on Weierstrass's proof for the transcendence of π, not Lindemann's. Nor was this merely because the former contained a more complete and streamlined presentation than the latter. Bachmann's eighth lecture gave a fairly detailed account of Lindemann's paper, but he broke this off at the end due to technical problems he would have had to deal with in order to complete the proof. He then noted that since Weierstrass had found a way around such difficulties, he would turn instead in his next lecture to this alternative method of proof ([Bachmann 1892], 110). This, again, supports my claim that few probably read Lindemann's paper: they certainly had no reason to do so after 1885 when Weierstrass' manuscript was finally printed.

In view of the events described here, we can see how Klein's very brief remarks put this prehistory in a faulty light, though, given the brevity of his discussion, a reader would hardly register much more than names and dates. Immediately after this, he writes:

> But the spring of 1893 brought new and very important simplifications. In the first rank should be named the memoir of Hilbert in the *Göttinger Nachrichten.* Still Hilbert's proof is not asbolutely elementary: there remain traces of Hermite's reasoning in the use of the integral [(4)]. But Hurwitz and Gordan soon showed that this transcendental formula could be done away with....([Klein 1897], 60).

Here Klein makes a total departure from his Evanston lecture, in which he barely mentioned the contributions of Hurwitz and Gordan. His entire presentation back then focused, as we have seen, on Hilbert's brilliant simplification. Yet a short proof is not necessarily a simple one, and here we enter onto another rocky road involving different attitudes and opinions regarding methodological issues. Klein had only recently adjudicated a similar running debate between Gordan and Hilbert in connection with the latter's radical new approach to finiteness theorems in the theory of algebraic invariants. Here, matters were by no means so contentious, but Hilbert simply was not persuaded that the "improvements" Hurwitz and Gordan had come up with were deserving of that label. Did they really clarify anything in his argument? Suffice it to say that Klein was eventually swayed over to the algebraists' side in this particular case; hence, in his booklet he followed Gordan's purely algebraic proof, not Hilbert's, even though Hilbert's general strategy, as he had outlined it in his Evanston lecture, remained central. Leopold Kronecker surely would have been delighted; he had even once challenged Lindemann to produce just such a proof.[47]

Figure 2: Max Liebermann's portrait of Felix Klein from August 1912, shortly before his "retirement." It presently hangs in the Mathematisches Institut in Göttingen.

Glancing ahead in time, it would seem appropriate to comment briefly on Klein's well-known wartime lectures, which dealt with various developments in mathematics over the course of the nineteenth century. The circumstances involved have been described before in the historical literature,[48] but this is another one of those stories that deserves to be told in detail. Here I can only sketch a few circumstances within the general context of Klein's biography. During 1911-12 he was suffering from serious health problems that eventually forced him to give up teaching his regular lecture courses. For over a year he lived outside Göttingen in Hahnenklee, a village in the Harz mountains, where he was treated at the sanatorium owned by a Dr. Klaus. It was here that Max Liebermann came in August 1912 to paint his portrait, reproduced above. By the following October,

[47]Lindemann had an interesting meeting with Kronecker in April 1883, when Lindemann visited Berlin to discuss with Althoff the conditions for his future employment in Königsberg. Lindemann spoke first with Weierstrass and afterward with Kronecker, quickly learning from these two encounters that Kronecker rejected the theory of irrational numbers due to Weierstrass, which the latter had long taught. Kronecker's views on this score were by this time no secret, but Lindemann learned firsthand how he thought about these issues when Kronecker told him that it would be interesting to see if the transcendence of π could be proved by methods based on the rational numbers alone ([Lindemann 1971], 88). It would also have been interesting to learn Kronecker's opinion of Gordan's proof, but since he died already in 1891 we shall never know.

[48]His involvement in the large-scale Teubner encyclopedia, *Kultur der Gegenwart* is recounted in [Tobies 2008]; for an overview of the entire project, see [Stoeltzner 2008].

Klein was still too weak to teach, and so he in December of that year he resigned his professorship.

During his convalescence in Hahnenklee, Klein was visited by David Eugene Smith, who consulted with him often regarding ongoing international efforts to modernize mathematics education both in Europe and the United States. Both men also shared an interest in making topics in the history of mathematics relevant for training future teachers. For this reason, Smith was also an obvious choice as reviewer for the first volume of Klein's posthumously published *Vorlesungen über die Entwicklung der Mathematik im 19. Jahrhundert* [Klein 1926] when these were published after his death in 1925. In his review, Smith began by recalling some of his conversations with Klein from nearly twenty years earlier. Among the topics they had discussed was the need for a comprehensive historical study of the mathematics of the past century.

When Smith suggested that Klein was just the person to write such a book, he was surprised that this idea seemed to fall on deaf ears. Yet Smith persisted, urging Klein to consider that he had witnessed these developments like few others who were still living. He remembered quite vividly the reply he received, with words to this effect: "No, all that I could do would be to give a few lectures on the great events, but I am too much occupied to prepare even these" ([Smith 1928], 521). Although Smith related this anecdote without further comment, the implication he surely drew was that something had caused Klein to change his mind, perhaps the onset of the Great War?

In fact, the idea behind Klein's wartime lectures, which he himself downplayed, calling them a "pia fraus" (pious fraud) ([Klein 1926], 2), was hardly planted in his mind during Smith's visit. The actual circumstances, however, were complicated, part of a story I hope to tell on another occasion. Here, I only wish to point out that Klein had a good idea of what it meant to be a historian and to write about the history of mathematics. He was a friend of Zeuthen and knew his work, and he supervised the dissertation of Conrad Müller, who wrote a history of Göttingen mathematics during the eighteenth century. Klein was familiar with historical scholarship, but had no interest in emulating it. Like his Evanston lectures and the booklet *Famous Problems of Elementary Geometry*, his lectures on the mathematics of the nineteenth century represent a popularized account, full of anecdotes and opinions that would be out of place in a scholarly work. He surely would have been surprised to learn that these wartime lectures would later be viewed by so many readers as a serious contribution to the history of mathematics.[49]

[49]For example, [Klein 1926] receives a prominent place in *Writing the History of Mathematics: its Historical Development* [Dauben and Scriba 2002] in the chapter on contributions from Germany, ([Folkerts, Scriba, Wussing 2002], 127-128).

Bibliography

[Alexandrov 1979] Alexandrov, Paul S., Hrsg., *Die Hilbertsche Probleme*, Ostwalds Klassiker der exakten Wissenschaften, Bd. 252, Leipzig: Teubner, 1979.

[Archibald 1914] Archibald, Raymond Clare, Remarks on Klein's "Famous Problems of Elementary Geometry," *American Mathematical Monthly*, 21(8) (1914): 247-259.

[Bachmann 1872] Bachmann, Paul Gustav Heinrich, *Die Lehre von der Kreistheilung und ihre Beziehungen zur Zahlentheorie*, Leipzig: Teubner, 1872.

[Bachmann 1892] Bachmann, Paul Gustav Heinrich, *Vorlesungen über die Natur der Irrationalzahlen*, Leipzig: Teubner, 1892.

[Boelling 1994] Bölling, Reinhard, *A Photo Album for Weierstrass*, Braunschweig: Vieweg, 1994.

[Bottazzini 2002] Bottazzini, Umberto, Italy, in [Dauben and Scriba 2002], 61–96.

[Browder 1976] Browder, Felix, ed., *Mathematical Developments arising from Hilbert's Problems*, Symposia in Pure Mathematics, vol. 28, Providence: American Mathematical Society, 1976.

[Cantor 1869] Cantor, Georg, Über die einfachen Zahlensysteme. *Zeitschrift für Mathematik und Physik* 14 (1869): 121–128.

[Cantor 1874] Cantor, Georg, Über eine Eigenschaft des Inbegriffs aller reellen algebraischen Zahlen. *Journal für die reine und angewandte Mathematik* 77 (1874): 258–262.

[Courant 1925] Courant, Richard, Felix Klein, *Die Naturwissenschaften* 37 (1925): 765–772.

[Dauben 1979] Dauben, Joseph W., *Georg Cantor. His Mathematics and Philosophy of the Infinite.* Princeton University Press. 1979.

[Dauben and Scriba 2002] Dauben, Joseph and Scriba, Christoph, eds., *Writing the History of Mathematics: its Historical Development*, (Science Networks, vol. 27), Basel: Birkhäuser, 2002.

[Demidov, Folkerts, Rowe, Scriba 1992] Demidov, S., Folkerts,M., Rowe, D., and Scriba, C., Hrsg., *Amphora. Festschrift für Hans Wussing zu seinem 65. Geburtstag*, Basel: Birkhäuser, 1992.

[Ferreirós 1999] Ferreirós, José, *Labyrinths of Thought. A History of Set Theory and its Role in Modern Mathematics*, Basel: Birkhäuser, 1999.

[Folkerts, Scriba, Wussing 2002] Germany, in [Dauben and Scriba 2002], 109-150.

[Frei 1985] Frei, Günther, *Der Briefwechsel David Hilbert–Felix Klein (1886-1918)*, Arbeiten aus der Niedersächsischen Staats- und Universitätsbibliothek Göttingen, Bd. 19, Göttingen: Vandenhoeck & Ruprecht, 1985.

[Goldstein/Schappacher/Schwermer 2007] Catherine Goldstein, Norbert Schappacher, Joachim Schwermer, eds., *The Shaping of Arithmetic after C. F. Gauss's Disquisitiones Arithmeticae*, Heidelberg: Springer–Verlag, 2007.

[Gordan 1893] Gordan, Paul, Transcendenz von e und π,*Mathematische Annalen*, 43(1893): 222–224.

[Gray 2000] Gray, Jeremy, *The Hilbert Challenge*, Oxford: Oxford University Press, 2000.

[Gray 2013] Gray, Jeremy, *Henri Poincaré: A Scientific Biography*, Princeton: Princeton University Press, 2013.

[Hashagen 2000] Ulf Hashagen, Georg Cantor und die Gründung der Deutschen Mathematiker-Vereinigung, Arbeitspapier, Münchner Zentrum für Wissenschafts- und Technikgeschichte, 2000.

[Hashagen 2003] Hashagen, Ulf *Walther von Dyck (1856-1934). Mathematik, Technik und Wissenschaftsorganisation an der TH München*, (Boethius, Texte und Abhandlungen zur Geschichte der Mathematik und der Naturwissenschaften, Bd. 47), Stuttgart: Steiner 2003.

[Hermite 1873a] Hermite, Charles, Sur la fonction exponentielle, *Comptes rendus hebdomadaires des séances de l'Académie des sciences*, 77(1873): 18–24, 74–79, 226–233, 285–293.

[Hermite 1873b] Hermite, Charles, Extrait d'une lettre de Mr. Ch. Hermite à Mr. Borchardt, *Journal für die reine und angewandte Mathematik*, 76(1873): 342-344.

[Hilbert 1893] Hilbert, David, Über die Transcendenz der Zahlen e und π, *Mathematische Annalen*, 43(1893): 216–219.

[Hilbert 1900] Hilbert, David, Mathematische Probleme. Vortrag, gehalten auf dem Internationalen Mathematikerkongreß zu Paris, 1900. *Nachrichten der Königlichen Gesellschaft der Wissenschaften zu Göttingen. Math.- phys. Klasse* 1900, 253–297; reprinted in [Hilbert 1935], 339–364.

[Hilbert 1935] Hilbert, David, *Gesammelte Abhandlungen*, Band 3, Berlin: Springer-Verlag, 1935.

[Hurwitz 1883] Hurwitz, Adolf, Über arithmetische Eigenschaften gewisser transzendenter Funktionen, *Mathematische Annalen*, 22(1883): 211–229.

[Hurwitz 1888] Hurwitz, Adolf, Über arithmetische Eigenschaften gewisser transzendenter Funktionen, *Mathematische Annalen*, 32(1888): 583–588.

[Hurwitz 1893] Hurwitz, Adolf, Beweis der Transcendenz der Zahl e, *Mathematische Annalen*, 43(1893): 220–221.

[Klein 1894] Klein, Felix, *The Evanston Colloquium. Lectures on Mathematics*, New York: Macmillan, 1894.

[Klein 1995] Klein, Felix, *Vorträge über ausgewählte Fragen der Elementargeometrie,* Leipzig: Teubner,1895.

[Klein 1897] Klein, Felix, *Famous Problems of Elementary Geometry*, New York: Ginn and Co., 1897.

[Klein 1921-1923] Klein, Felix,*Gesammelte Mathematische Abhandlungen*, 3 Bde., Berlin: Julius Springer, 1921, 1922, 1923.

[Klein 1926] Klein, Felix, *Vorlesungen über die Entwicklung der Mathematik im 19. Jahrhundert*, Band I, Berlin: Julius Springer, 1926. English translation in *Development of Mathematics in the 19th Century.* trans. M. Ackerman, Lie Groups: History, Frontiers and Applications, IX. Brookline, Mass.: Math Sci Press, 1979.

[Lindemann 1882a] Lindemann, Ferdinand, Über die Ludolph'sche Zahl, *Sitzungsberichte der Königlichen Preussischen Akademie der Wissenschaften zu Berlin* Jahrgang 1882(1): 679–682.

[Lindemann 1882b] Lindemann, Ferdinand, Über die Zahl π, *Mathematische Annalen*, 20(1882): 213–225.

[Lindemann 1971] Lindemann, Ferdinand, *Lebenserinnerungen*, München: Selbstverlag, 1971.

[Luetzen 2009] Lützen, Jesper, Why was Wantzel overlooked for a century? The changing importance of an impossibility result, *Historia Mathematica*, 36(2009): 374-394.

[Minkowski 1973] Minkowski, Hermann, *Briefe an David Hilbert*, L. Rüdenberg und H. Zassenhaus, Hrsg., New York: Springer-Verlag, 1973.

[Mehrtens 1990] Mehrtens, Herbert, *Moderne — Sprache — Mathematik. Eine Geschichte des Streits um die Grandlagen der Disziplin und des Subjekts formaler Systeme*, Frankfurt: Suhrkamp, 1990.

[Parshall and Rowe 1994] Parshall, Karen Hunger and Rowe, David E. *The Emergence of the American Mathematical Research Community, 1876-1900: James Joseph Sylvester, Felix Klein, and E. H. Moore*, History of Mathematics, vol. 8. Providence: American Mathematical Society and London: London Mathematical Society, 1994.

[Petersen 1878] Petersen, Julius *Theorie der algebraischen Gleichungen*, Andr. Fred. Höst & Sohn, Kopenhagen, 1878.

[Purkert and Ilgauds 1987] Purkert, Walter, and Ilgauds, Hans Joachim, *Georg Cantor 1845–1918.* Vita Mathematica, Band I. Birkhäuser, Basel 1987.

[Reid 1970] Reid, Constance, *Hilbert*, New York: Springer Verlag, 1970.

[Rowe 1986] Rowe, David E. "Jewish Mathematics" at Göttingen in the Era of Felix Klein, *Isis*, 77(3) (1986): 422–449.

[Rowe 1989] Rowe, David E., Klein, Hilbert, and the Göttingen Mathematical Tradition, *Science in Germany: The Intersection of Institutional and Intellectual Issues*, ed. Kathryn M. Olesko, Osiris, 5(1989]: 189–213.

[Rowe 1992] Rowe, David E., Klein, Mittag–Leffler, and the Klein-Poincaré Correspondence of 1881–1882, in [Demidov, Folkerts, Rowe, Scriba 1992], 598–618.

[Rowe 1997] Rowe, David E., (Essay Review) Perspective on Hilbert: *Moderne—Sprache —Mathematik. Eine Geschichte des Streits um die Grandlagen der Disziplin und des Subjekts formaler Systeme*, by Herbert Mehrtens; *Hilbertprogramm und Kritische Philosophie*, by Volker Peckhaus; *Über die Entstehung von David Hilberts "Grundlagen der Geometrie"*, by Michael–Markus Toepell, *Perspectives on Science* 5 (1997): 533-570.

[Rowe 2001] Rowe, David E., Felix Klein as Wissenschaftspolitiker, in *Changing Images in Mathematics. From the French Revolution to the New Millennium*, ed. Umberto Bottazzini and Amy Dahan. London: Routledge, 2001, pp. 69–92.

[Rowe 2003] Rowe, David E., From Königsberg to Göttingen: A Sketch of Hilbert's Early Career, *Mathematical Intelligencer*, 25(2): 44–50.

[Rowe 2004] Rowe, David E., I Problemi di Hilbert e la Matematica del XX Secolo, *Storia della scienza*, ed., Sandro Petruccioli, vol. VIII, Rome: Istituto della Enciclopedia Italiana, pp. 104–111.

[Rowe 2005] Rowe, David E., Hilbert's Early Career: Encounters with Allies and Rivals, *Mathematical Intelligencer*, 27(1): 72–82.

[Rowe 2007] Rowe, David E., Felix Klein, Adolf Hurwitz, and the "Jewish Question" in German Academia, *Mathematical Intelligencer*, 29(2): 18–30.

[Rowe 2013] Mathematics Made in Germany: On the Background to Hilbert's Paris Lecture, *Mathematical Intelligencer*, 35(3): 9–20.

[Smith 1928] Smith, David Eugene. Klein on Nineteenth Century Mathematics, *Bulletin of the American Mathematical Society* 34(4): 521–522.

[Stieltjes 1890] Stieltjes, T. J., Sur la fonction exponentielle, *Comptes rendus hebdomadaires des séances de l'Académie des sciences*, vol. CX, 267-270.

[Stoeltzner 2008] Stöltzner, Michael, Eine Enzyklopädie für das Kaiserreich, *Berichte zur Wissenschaftsgeschichte*, 31(1)(2008): 11–28.

[Struik 1987] Struik, Dirk J. *A Concise History of Mathematics*, 4th ed., New York: Dover.

[Tobies 2008] Tobies, Renate, Mathematik, Naturwissenschaften und Technik als Bestandteile der Kultur der Gegenwart, *Berichte zur Wissenschaftsgeschichte*, 31(1)(2008): 29–43.

[Wantzel 1837] Wantzel, M. L., Recherches sur les moyens de reconnaitre si un Problème de Géométrie peut se résoudre avec la règle et le compas, *Journal de Mathématiques Pures et Appliquées* 1 (2): 366–372.

[Weierstrass 1885] Weierstrass, Karl, Zu Hrn. Lindemanns Abhandlung: "Über die Ludolph'sche Zahl," *Sitzungsberichte der Königlichen Preussischen Akademie der Wissenschaften zu Berlin* Jahrgang 1885(2): 22 October, 1885, 1067–1086.

[Yandell 2002] Yandell, Ben H. *The Honors Class. Hilbert's Problems and their Solvers*, Natick, Mass.: A K Peters, 2002.

Correspondence of 李儼 Li Yan and David Eugene Smith

XU Yibao

Introduction

During the middle of the First World War, Li Yan (1892-1963) and David Eugene Smith (1860-1944) exchanged nearly twenty letters chiefly concerning plans for a collaboration to produce a book in English on the history of Chinese mathematics. Although this idea did not come to fruition, these letters are nevertheless interesting for documenting the early stages in historical research on the history of Chinese mathematics. For this reason, this correspondence is presented here in its entirety along with some brief commentary. Zhang Dianzhou of East China Normal University first revealed the existence of these letters in his article, " A Chronicle of the Correspondence between Smith and Li Yan (1915-1917)."[1] Not only did he provide background information about these letters and their main contents, but he also translated part of the correspondence.

In the latter part of July, 2011, I took part in the 13th International Conference on the History of Science in East Asia at the University of Science and Technology of China. In Hefei, I met my good friend, Zou Dahai, who was also a participant. He informed me that a few letters had recently been found at the library of the Institute for History of Natural Sciences (IHNS), where he is a Professor of the History of Pre-Modern Chinese Science and Technology. When he went back to Beijing, he emailed me copies of these letters.[2] The IHNS part of the correspondence serves to complement the Columbia collection, making the entire correspondence a valuable source of original material for studying the history of Chinese mathematics. With permission from the Rare Book & Manuscript Library of Columbia University and the IHNS, this article publishes transcriptions of all the letters exchanged between Li Yan, D.E. Smith, Thomson Mao Yisheng, and Smith's secretary, Eleanor Mitchell.[3] To clarify some of the circumstances, additional details and a few annotations have been provided in [square brackets]. Before presenting the transcriptions, however, some biographical information about the correspondents will help orient the reader.

David Eugene Smith (1860-1944) was one of the most influential mathematicians ands mathematics educators in the first half of the 20th century. Between 1884 and 1891, he was an instructor of mathematics at the Cortland Normal School

[1]See *The Chinese Journal for The History of Science and Technology*, 12 (1) (1991): 75-83. In the David Eugene Smith Archive at the Rare Book & Manuscript Library of Columbia University, there are 4 letters from Li Yan to Smith (originals, 2 handwritten and 2 typed, 10 pages in all), and 1 postcard.

[2]Judging from these, the IHNS has five drafts of Li Yan's letters to Smith (3 handwritten, 6 pages in all; 2 typed, 7 pages in all), 3 letters from Smith to Li Yan (originals, all typed, totaling 8 pages), and 1 letter from Eleanor Mitchell to Thomson Mao (original, typed, 1 page).

[3]The transcriptions that follow are exact, so that misspellings, inconsistencies, and other inaccuracies are reproduced as in the original texts, in most all cases without remark or comment.

D.E. Rowe and W.-S. Horng (eds.), *A Delicate Balance: Global Perspectives on Innovation and Tradition in the History of Mathematics*,
Trends in the History of Science, DOI 10.1007/978-3-319-12030-0_11

in Cortland, New York; between 1891 and 1898 he was a professor of mathematics at Michigan State Normal School. In 1898, he returned to Cortland and became the principal of the New York State Normal School in Brockport, New York. In 1901, he accepted a position as professor of mathematics at Teachers College, Columbia University, where he remained until his retirement in 1926. After that, he was a professor emeritus at Teachers College until his death in 1944.[4]

Smith was also librarian (1902-1920) of the American Mathematical Society, editor of the *Bulletin of the American Mathematical Society* (1903-1909), President of the Mathematical Association of America (1920), and a founding member of the History of Science Society. He was a forerunner in using the history of mathematics as an essential part of mathematics education. A prolific author, he compiled a large number of mathematical textbooks, and published numerous articles and books on the history of mathematics. Some of his influential works include: *Rara Arithmetica: A Catalogue of the Arithmetics Written before the year MDCI with a Description of Those in the Library of George Arthur Plimpton of New York* (Boston: Ginn, 1908); *The Hindu-Arabic Numerals*, co-authored with Louis Charles Karpinski (Boston: Ginn, 1911); *A History of Japanese Mathematics*, with Yoshio Mikami (Chicago: Open Court Publishing Company, 1914); *History of Mathematics*, in 2 Volumes (Boston: Ginn, 1923 and 1925); *A History of Mathematics in America before 1900*, with Jekuthiel Ginsburg (Chicago: The Mathematical Association of America and the Open Court, 1934); and *A Source Book in Mathematics* (New York: McGraw Hill, 1929).[5]

The founder of the discipline of history of science, George Sarton (1884-1956), dedicated the inaugural issue of the journal *Osiris* to Smith, writing in the dedication: "You have been one of the first teachers of the subject (mathematics) in America ... You have written a large number of papers and textbooks devoted to our studies, textbooks which are today among the main tools for the teaching of the History of Mathematics in the English-speaking world and even outside of it."[6]

Li Yan (1892-1963) was one of the pioneers of the study of the history of Chinese mathematics. At the age of 20, he started to collect a substantial number of works and manuscripts on ancient Chinese mathematics and to devote his spare time to research on the history of Chinese mathematics. From 1917 on, he published many works on traditional Chinese mathematics. His main representative works include 中算史論叢 *Zhong suan shi luncong* (A History of Chinese Mathematics), 5 volumes, Beijing: Science Publishing House, 1954-1955; and 中國數學大綱 *Zhongguo shuxue dagang* (A General Outline of Chinese Mathematics), 2 volumes, Beijing: Science Publishing House, 1958. In 1955, after he was transferred

[4]For details of his life, refer to Lao Genevra Simons, " David Eugene Smith–In Memoriam," *Bulletin of the American Mathematical Society* 51 (1) (1945): 40-50.

[5]For a list of Smith's works, see Bertha Margaret Frick, " Bibliography of the Critical, Historical and Pedagogical Writings of David Eugene Smith," *Osiris* 1 (1936): 9-78.

[6]See George Sarton, "Dedication to David Eugene Smith," *Osiris* 1 (1936): 4-8, quoted from pp. 6-7.

to the Chinese Academy of Sciences from the Long Hai Railway Bureau, he made significant contributions to the establishment and development of the Institute for History of Natural Sciences, and to the training of historians of mathematics.[7]

Thomson Mao (茅以升 Mao Yisheng, 1896-1989) was a famous Chinese expert on bridge construction. He and Li Yan were classmates at Tangshan Engineering College. After graduating from the college, he succeeded in ap-plying for a Tsinghua University scholarship to pursue his graduate studies at Cornell University. After earning a master's degree in June, 1917, he worked as an intern in Pittsburg. In his free time, he went to night school at the Carnegie Institute of Technology for a Ph. D in engineering, which he received in 1919. After he returned to China in 1920, Mao contributed greatly to engineering education and bridge construction in China.

Of the Li-Smith correspondence preserved in the Rare Book & Manu-script Library of Columbia University and the IHNS, the earliest letter is dated January 23, 1915, and the last October 11, 1917. The main contents of the correspondence concern a collaboration to write a book in English on the history of Chinese mathematics. Unfortunately, this collaboration plan was never realized. Nevertheless, the letters are valuable original materials for studying the history of Chinese mathematics. Since portions of this correspondence published earlier are incomplete, the new finds at the IHNS shed light on the reasons for the unsuccessful cooperation. It would be very valuable if Li Yan's manuscripts on the history of Chinese mathematics were to be translated in full, and also, if possible, to determine the whereabouts of Li Maosheng's translation, as these are of interest to historians. (One handwritten draft of a letter by Li Yan, unfinished and undated, contains two pages with so many scribbles that it has not been translated here.)[8]

[7]For Li Yan's life and academic contributions, see Du Shiran, " Li Yan," in: *Biographies of Modern Chinese Scientists*, vol. 5 (Beijing: Science Publishing House, 1992): 1-5; Du Shiran, "Li Yan," in *Lives on Contemporary Chinese Scientists*, vol. 3 (Nanjing: Jiangsu Education Publishing House, 1998): 73-82; and Zou Dahai, " Li Yan," *Biographies of Scholars at the Chinese Academy of Sciences* (Beijing: Science Publishing House, 2010): 317-322.

[8]For more details about Mao's life and academic contributions, see Li Xiuzhi: " Mao Yisheng," in: *Biographies of Modern Chinese Scientists*, vol. 6 (Beijing: Science Publishing House, 1994): 715-731.

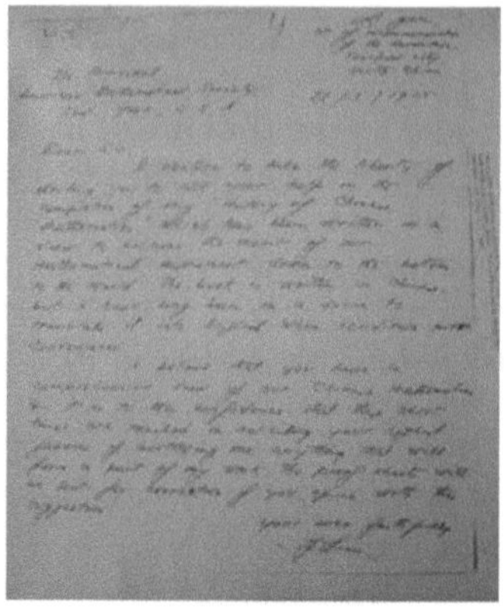

Figure 1: The first letter from Li Yan to D.E. Smith, dated January 23, 1915 (David Eugene Smith Archive, Rare Book & Manuscript Library, Columbia University)

Transcriptions of the Correspondence

(1)

Li Yen[9]
St. of Incommemoration of the Revolution[10]
Foochow City, South China

The Principal
American Mathematical Society[11]
New York, U.S.A.
23/1/1915

Dear Sir,

I venture to take the liberty of writing you to ask your help in the completion of my "History of Chinese Mathematics," which has been written in a view to expose the merit of our mathematical movement down to the bottom to the world. The book is written in Chinese, but I have long been in a desire to translate it into English when conditions prove convenience.

[9]Columbia University Archives; handwritten letter. At the top, Li Yan has used a phonetic equivalent of his name, spelling it Li Yen; at the end of the letter, he signs his name in the usual western order, as Y. Li.

[10]The Chinese return address that Li Yan gave "St. of Incommemoration of the Revolution" is 中國南方福州市革命紀念街; the street name is better rendered in English as " Commemoration of the Revolution Street." Perhaps what Li Yan meant to write was "Street in Commemoration of the Revolution."

[11]In 1915 the President of the American Mathematical Society was Ernest William Brown (1866-1938), Professor of Mathematics at Yale University; at the time, D. E. Schmit was the Society's librarian.

I believe that you have a comprehensive view of our Chinese mathematics and it is to this confidence that these short times are mailed in soliciting your highest favor of bestowing me anything that will form a part of my work.[12] The proof sheets will be sent for correction if you agree with this suggestion.

Yours ever faithfully, Y. Li

(2)

March 2, 1915[13]

Mr. Li Yen
Street of Incommemoration of the Revolution,
Foochow City, South China.

Dear Sir:

Your letter of January 23d addressed to the American Mathematical Society has been handed to me.

As perhaps you know, I got out a History of Japanese Mathematics with Mr. Mikami of Tokyo about a year ago. I have also collected a large amount of material on Chinese mathematics, particularly from the European standpoint, although I have a good many Chinese manuscripts and have most of the Chinese classics. I have been hoping to get out a history of Chinese mathematics along the same lines, and last Spring I communicated with the learned Jesuit Father, Père [Louis] Vanhee[14] of Liège, who has written so extensively upon the subject. We arranged to get out the work together, and I had already begun a little writing upon it. The war, however, has put an end to our working together for the time being, and I

[12]The word "times" is mistaken here and should have been "notes."

[13]Columbia University Archives, D.E. Smith Collection. All of the letters from Smith to Li Yan are typewritten carbon copies and do not bear Smith's name at the end. Transitions from one page to another are indicated in the transcribed letters in square brackets indicating where the break to a new page occurs, i.e. [2] denotes the beginning of the second page of the letter in question.

[14]Columbia University Archives, typewritten carbon copy. Smith writes his name Vanhée in his later correspondence with Li Yan. Louis Vanhee (1873-1951) was a Belgian Jesuit who spent nearly twenty years as a missionary working in Jiangsu Province in south-eastern China. Upon his return to Belgium in 1911, he began to publish articles about traditional Chinese mathematics in such peer-reviewed journals as *T'oung Pao* and *Isis*. Due to the experience of his missionary work in China, Vanhee was fluent in Chinese, and his articles attracted the attention of Smith and other western experts. At the time Smith received the first letter from Li Yan he had already agreed to collaborate with Vanhee in writing a history of Chinese mathematics. But due to the First World War, this project was interrupted. After the war, they very soon got in touch and again discussed the matter of their collaboration, but for some reason they still did not succeed. For information about Vanhee's life and works on the history of Chinese mathematics, see 張奠宙 Zhang Dianzhou and 王善平 Wang Shanping, "(三上義夫、赫師慎與史密斯 – 兼評本世紀初國外的中算史研究" *Sanshangyifu, Heshishen yu Shimisi-jianping ben shiji chu guowai de zhong suanshi yanjiu* (Mikami, Vanhee and Smith—their discussion of foreign research on the history of Chinese mathematics at the beginning of this century), 中國科技史料 *Zhongguo keji shiliao* (The Chinese Journal for the History of Science and Technology), 14 (4) (1993): 62-66.

am unable to communicate with him and find out if he is still living. If he is all right, we should natur- [2] ally take up the work at the conclusion of the war. It occurs to me, however, that the three of us might possibly collaborate. We could take such material as you have. I could then write up the work, combining my own material with it, and the whole thing could be submitted to Father Vanhée for final revision. I am unable to say whether this could be done, because I cannot at present communicate with him. If, however, you care to get your material into English and send it to me with this arrangement in view, I shall be glad to look it over and take it up with Father Vanhée as soon as I am able to communicate with him. As you know, it is difficult to get a work of this kind published, because there is no demand for it. Nevertheless, I think the Open Court Publishing Company might be willing to undertake it, as it did the History of Japanese Mathematics. We received no remuneration whatever for this Japanese work, nor would there be any chance of getting any for the Chinese work. The fact [3] is the publishing house loses money on books of this kind, but it has a fund enabling it to publish scientific books of merit, even at a loss, having been endowed by its wealthy founder partly for this purpose.

You might consider this matter, and if you decide to make the translation and send it to me, it will have my best attention.

Yours very truly,

(3)

St. of Incommemoration of the Revolution[15],
Foochow City, S. China, May 31, 1915.

Dr. David Eugene Smith,
Dept. of Mathematics,
Teachers College, Columbia University,
New York, U.S.A.

Dear Sir,

Your kind favor of March 2nd has been received with pleasure. In response I must express my gratitude which shall fall on the enthusiastic attention that you have taken on the history of mathematics of our country. Your task must be excellent and spiritual: I believe that it is this work that the long-buried mathematics of our country may now splash to the world with the highest speed attainable.

[15]Columbia University Archives, typewritten original. The Institute for History of Natural Sciences Archives (CSS, Beijing) has a copy of a two-page typewritten draft of this letter. Handwritten on the first page is Li Yan's address and the date "May 20, 1915." On the second page is written, in Chinese: 以上請糾正後寄出。荷京通報之譯名相似否 *Yishang qing jiuzheng hou jichu. Hejing tongbao zhi yiming xiangsi fou* (Please correct the above and then mail. Is Hejing's translation of the name similar?).

Perhaps it may be so that if you have been in acquaintance with Chinese Language the work will be greatly eased as there are numerous names and specific terms that cannot be avoided to be mistakable if is wholly relied on English. To collaborate the work, of course is the best plan: the only trouble being that so long as in accompanying with a man as awkward as I am, the work may be stained.

The general plan of my work on the History of Mathematics of China runs as follows: The manuscript consists of three volumes the first of which devoted the Ancient History and the other two, Medieval and Modern. Each volume is divided into Chapters and Sections the order being arranged chronologically. Herewith you will find a list of detailed contents of the first volume of which the chapter or the concept of Modern Analysis in Ancient China may be of some value.[16]

Such kind of work, as far as I know, cannot be complete by one hand and compelled in a short length of time. But, nevertheless, as erudite as you and Father Vanhée are the results must be invaluable. It is indeed the hardest task to translate into English the specific terms that have been extensively used in Ancient China without any loss of the true meaning. I think it may be possible-and hope it to be so, to translate the general outline of my work into English before the conclusion of the Great War. I don't think it will be appropriately entitled to be an excellent historic work, but, in any way, it may give some assistance to the works of such kind.

In our country, as you know, the books on Mathematics are too numerous to account, but as there has been little attention on these, great many of which have been lost. Although I am a native of this country yet there is no less difficulty to collect the materials than you have encountered. It is therefore the chief aim of us to devise any means in gathering the materials. By my opinion we may collect a small sum of money for the use of advertising fee (may be under the name of our three) to scrutinize the materials that are scarce and valuable. Should this be accepted by you please send me a list of books of this kind that you have already collected and I will gather the other books for you that are wanting in your bookcases. [2]

For the present as the War is still going on, it happens to me that there is a plenty of time to spend in translating my work and gathering the materials.

As I am anxious to learn any word from Father Vanhée, I hope you may communicate to him in the nearest opportunity. Have you got any news of him through the "T'oung Pao," Hague.

If it is convenient please let me have a copy of your work on the History of Mathematics of Japan.

Yours very truly,

Li Yen

[16]The draft version (in the IHNS Archives) of this sentence reads as follows: "Herewith you will find a list of detailed contents of which the first section on the concept of Differential Calculus in Ancient China may be of some value." Here the original meaning was to indicate that in ancient Chinese mathematical works there was something similar to modern Western concepts of mathematical analysis.

History of Chinese Mathematics[17]
Volume I. Ancient History
CONTENTS

[17]Columbia University Archives, typewritten original.

VII. On geometry.
Section 1. Plan.
Section 2. Solid.

VIII. Concept of Modern Analysis in ancient China.

Part III. - Biography of Ancient Mathematicians.
Chapter:

I. Sih-Sheu
Section 1. Biography and works. [2]

Chapter:

II. Chou-Kon and Shan-Koh
Section 1. Biography and works.
Section 2. On Chou Peh Soan Chin.
Section 3. Zieu Chan Soan Shue.

Chapter:

III. Sun-Sze.
Section 1. Biography and works.
Section 2. Sun-Sze Soan Chin.

IV. Conclusion.

Part IV. - Civilization of Ancient World
Chapter:

I. Brief note.

II. Babylonia.

III. Egypt.

IV. Greece and Rome.

V. India and Arabic.

Part V. - Concluded Words about Ancient Mathematics
Chapter:

I. On the method of measurement in ancient China.
Section 1. Introduction.
Section 2. Measure (length and area).
Section 3. Measure (volume).
Section 4. Weight.

II. Specific terms explained

III. Cause of regression of mathematics in Chou and Chin Dynasties.

(4)

July 7, 1915[18]

Mr. Li Yen,
Street of Incommemoration of the Revolution,
Foochow City, South China.

Dear Mr. Li Yen:

Replying to your letter of May 31st, which has just reached me, I beg to say that I am much interested in the general plan that you have suggested. One great difficulty, however, is going to arise in the plan that you have for three volumes. It will be impossible for us to get a publisher in this country who would handle this material if it exceeds in bulk the amount which we put into the History of Japanese Mathematics. I sent you a copy of this book yesterday, and I presume you will receive it at the same time that you receive this letter.

It will be absolutely necessary to bring the history of Chinese mathematics down to a single volume of about that size. The outline which you suggest here is all right, but I think you would have to treat it, or at anyrate would have to treat some of it very briefly.

I have again written to Holland to see if I can get any information concerning Father Vanhée. I shall also today write to the Vatican of Rome to see if they have any information concerning him. [2]

I suggest that you now go to work on the outline of the rest of the history of Chinese mathematics, and send the whole thing to me so that I may have that in connection with what you have sent and may go over it and make suggestions to you.

After this is done it will be in order to begin the writing and to send me the manuscript as you may complete it. I am, of course, not certain that we can get a publisher in this country, but I think that if we worked out a book on somewhat the same lines as the History of Japanese Mathematics we shall have no trouble.

With respect to Chinese books, I may say that I have all of the great Chinese classics, that I have Mei's History of Chinese Mathematics, and that that part of the great Chinese encyclopaedia that refers to mathematics is also in my library. I have also such manuscripts as the translation of Euclid made by Matthew Ricci. I, of course, have this also in printed form.[19] Any books, particularly manuscripts,

---

[18]Columbia University Archives, typewritten original.

[19]See the catalogue of the Smith collection compiled by Miwa Kai, *The David Eugene Smith Collection of Works in Chinese on Mathematics and Other Subjects*, The Rare Book & Manuscript Library, Butler Library, Columbia University. Of a total of 568 rare books in the Smith Collection, there are 88 works in Chinese, including the 算經十書 *Suanjing shishu* (Ten Books of Mathematical Classics) (Kong Jihan edition of 1890), 周髀算經 *Zhoubi suanjing* (The Arithmetic Classic of the Zhou Gnomon) (two versions, editions unclear), 算學啟蒙 *Suanxue qimeng* (Introduction to Computational Studies) (by Zhu Shijie, Liulichang (Beijing), 1871), 梅氏叢書輯要 *Mei shi congshu jiyao* (Collected Works of the Mei Family) (by Mei Wending,

which you may be able to get I should be glad to take within reasonable number and at a reasonable price. If you can get the refusal on these, and will [3] notify me I will let you know whether I will take them and in that case will send you the money.

One interesting feature of the history of Japanese mathematics was the illustration, and I think we should be very careful to get interesting illustrations from the Chinese books, especially those that have pictures in illustrating the problems.

With best regards, I remain,

Yours very sincerely,

(5)

St. of Incommemoration of the Revolution[20],
Foochow City, China, October 1st, 1915.

Dr. D.E. Smith,
Teachers College,
Columbia University,
New York, U.S.A.

1761), 曆算全書 *Lisuan quanshu* (Collected Works on Calendrical Calculations) (by Mei Wending, 1723), 數度衍 *Shudu yan* (Development of Calculations and Measures) (by Fang Zhongtong, Taiyuan Wang Family Revised Edition, 1634), 數理精蘊 *Shuli jingyun* (Selected Mathematical Works) (by Fuwen Bookstore, 1882), 句股割圜記 *Gougu geyuan ji* (Records of base-altitude and circle-division) (by Dai Zhen, Weiboxie (Ripple Pavilion) edition, 1773; also editions of 1744; 1900), 〈九章算術〉細草圖說 *Jiuzhang suanshu xicao tu shuo* (The Nine Chapters on the Art of Mathematics—Briefly Explained) (by Li Huang, Siyatang (Four Study Hall) edition, 1820), 翠薇山房數學 *Cuiwei shan-fang shuxue* (Mathematics of the Cuiwei Mountain House) (by Zhang Zuonan, 1821), 〈四元玉鑒〉細草 *Siyuan yujian xicao* (Detailed Solutions for the Precious Mirror of the Four Elements) (by Luo Shilin, 1836), 李氏遺書 *Lishi yishu* (Posthumously Published Works of Li) (by Li Rui, 1823), 中西算學大成 *Zhongxi suanxue dacheng* (Grand Compendium of Chinese and Western Mathematics) (edited by Chen Weiqi, Shanghai Hongliutang (Six Grand Halls) Publishing House, 1901), Lao Naixuan's 籌算淺釋 *Chousuan qianshi* (Brief Explanation of Computing with Counting Rods) (1897), 籌算分法淺釋 *Chousuan fenfa qianshi* (Brief Explanation of Methods for Fractions with Counting Rods) (1898), 籌算蒙課 *Chousuan mengke* (Introductory Lessons on Computing with Counting Rods) (1898), and 古籌算考釋 *Gu chousuan kaoshi* (Explanations of Ancient Computing with Counting Rods) (1898), 古籌算考釋續編 *Gu chousuan kaoshi xubian* (Continuation of Explanations of Ancient Computing with Counting Rods) (1898). Moreover, the Smith archive also has a copy of the Chinese translation of Euclid's *Elements of Geometry* published in 15 parts in 1865, as well as a handwritten copy of the Matteo Ricci and Xu Guangqi translation of the first six books of Euclid's *Elements.* Having inspected this handwritten copy, it is the conclusion of the translator (Xu Yibao) that this was a copy used by Mei Wending. As for a detailed examination of this handwritten copy and critical analysis of its contents, this will be the subject of a later study.

[20]Archives of the Institute for History of Natural Sciences, copy of 3 pages. The third page includes Li Yan's signature.

Dear Sir:

Your kind favor of July 7th and the "History of Japan Mathematics" have been duly received, for which I beg to return my sincere thanks. The plan that you have suggested is all right and I shall follow your advice in the treatment of my manuscript, as to have a thoroughly readable account of such an immense subject of thousands of years compacted into a volume not exceeding reasonable size, the ground to be covered should naturally be abridged. Accompanying this letter is enclosed a complete outline of the History, being summarized and condensed from the original work in Chinese. This may differ in some point from the one that you have received before. But the future English translation, I should say, will base on this new plan. If you have any further suggestions or even to have a better plan than this I should like to see it and I expect it to [b]e so as there may be a more concrete foundation for me to build the work upon.

Another thing I beg to write to you is the standard- [2] izations of the translated terms. Formerly, Chinese Government have appointed a Board of Standardizing Pronunciations, by which the most commonly occurred words have been put into English alphabets. In my future translations these words will be used as standard and take frequent consultation with the circulars of the Simplified Spelling Board of New York.[21] (I send you now a copy of Students' Dictionary of Chinese in which the pronunciations of the words have been standardized by the said Board).

Illustrations in the History are of course indispensable. I may say that the materials I have collected may even surpass that of the Japanese mathematics. In addition I shall also collect the portraits of modern Chinese mathematicians allied with their biographies as regarding to the date of birth and death. Toward this point I cannot deny that Chinese Historians always lost their attention. Even with Yuen Yuen, the greatest historian of Chinese mathematics, shared such common fault. It is therefore the scope of my work to have such dates as complete as possible, part of which may be gathered through newspapers. Recently, the editors of one or two papers are very enthusiastic on this line. I dare say that the advertising for [it] may be very inappreciable or even free of charge.

Valuable books or manuscripts of mathematics have always received my careful attention. The following books I [3] presume you have looked already and the Chinese editions of which were in your library[22]:

Bibliothéca Mathématica Sinensis, Pé-fou,
Li-Yé, Mathématicien Chinois,
Algébre Chinoise.

[21]The Simplified Spelling Board was founded in New York City (USA) in 1906. Its purpose was to reform English spelling, so that English would be easier to learn. The Board was funded by the steel magnate Andrew Carnegie (1835-1919); funding for the Board ended in 1920.

[22]About these works, see the articles written by the Belgian Jesuit scholar Louis Vanhee: *T'oung Pao* "Algèbre Chinoise," *Toung Pao* 13 (1912): 29-300; "Li Ye, Mathématicien Chinois du XIIIe Siècle," *T'oung Pao* 14 (1913): 537-68; "Bibliotheca Mathematica Sinensis Pè-Fou," *T'oung Pao* 15 (1914): 111-64.

At present time there are only a few men in this country who have time in devoting the Ancient Mathematics; this consequently brings down the price of the books. Should I have a chance in getting some of it I will send to you, money should not be the predetermined thing.

I have a complete set of Encyclopaedia of our ancient "Rod Computation" (22 volumes) which is written by a modern mathematician (still flourish now) and is of some value. I shall send it to you if you will be on the affirmative side.

With best regards, I remain,
Yours very sincerely
[signed] Y. Li

History of Chinese Mathematics[23]
CONTENTS

Part I. - Introductory.
Chapter:

1. The source and development of Chinese Civilization.
2. Chronological view of Chinese Mathematics.
3. Mathematics in I-Ching.
4. Relations existed between Mathematics, and Philosophy and Astronomy.

Part II. - Ancient Mathematics. I. (3000 B.C.?-200 B.C.)
Chapter:

1. On the origin of number and the method of counting.
2. On Arithmetic.
3. On methods of computation.
4. On equations.
5. On Geometry.
6. On the concept of modern Higher Mathematics.

Part III. - Ancient Mathematics. II.
Chapter:

1. Biographies of Ancient Mathematicians (3000 B.C.?-200 B.C.).
2. The system and institutions of ancient science.

[23] Archives of the Institute for History of Natural Sciences, copy of 2 pages.

3. A research of ancient Mathematical Classics.
4. On ancient Metrology.
5. Conclusion: Civilization of ancient World.

Part IV. - Medieval Mathematics. (200 B.C.-600 AD)
Chapter:

1. The period of 200 B.C.-200 A.D.
2. The period of 200 A.D.-400 A.D. (Progress is of Circle-Squaring)
3. The period of 400 A.D.-600 A.D.
4. Biographies of mathematicians. (200 B.C.-600 A.D.)
5. A reform of Metrology.
6. Conclusion: A research on the international intercourse.

Part V. - Modern Mathematics. I. (600 A.D.-1900 A.D.)
Chapter:

1. The period of 600 A.D.-1000 A.D.
2. The period of 1000 A.D.-1400 A.D.
3. The period of 1400 A.D.-1900 A.D. (Introduction of Occidental Mathematics)

Part VI. - Modern Mathematics. II.
Chapter:

1. On mechanical computations.
2. On Ta Yen Ch'in I.
3. On "T'ien Yüan" "Szu Yüan"
4. On circles and right triangles.
5. On Algebra.
6. A sequel to Euclid.
7. Magic square and other geometric figures.
8. Mathematical recreations and miscellaneous. [2]

Part VII. - Modern Mathematics. III.[24]
Chapter:

1. Biographies of Mathematicians (600 A.D.-1900 A.D.).
2. Different Schools in modern mathematics.
3. Conclusion

(6)

November 4, 1915[25]

Mr. Y. Li
St. of Incommemoration of the Revolution,
Foochow, China

My dear Mr. Li:

Your letter of October 1st has duly reached me. I am much pleased to know of your continued interest in the work. I have meantime been slowly gathering together my note and pamphlets on the whole subject, and shall be prepared to undertake the writing whenever your material reaches me.

It strikes me, however, that it would be possible to improve your general plan a little. It seems to me that it would be much better to put the Introduction by itself. Then after the Introduction is out of the way, I think it would be better to divide the rest of the work into three parts,—the first dealing with ancient mathematics, the second with mediaeval mathematics, and the third with modern mathematics. How the work should be arranged under these various parts I am not certain until I see what you have got together and what I can find in my own notes and library. We always have the difficulty in writing the history of getting the biographical material in so that it comes [2] at just the place it is wanted. In my suggestion for part of the ancient mathematics, I have put the biographical material as No. IV so as to get it before the mathematical classics. On the general topics of algebra, geometry, and so on, in Part II on mediaeval mathematics I have not bee so successful. My impression is that it would be better to put the discoveries in with the biography in Parts II and III, and possibly the same may be true for Part I. On the other hand, in Part I, it is quite desirable to have the algebra and geometry stand by themselves, and it is possible that the same may be true for Parts II, and III.

[24]Xu Yibao, working from a very blurry photograph of this page of Li Yan's outline, added a note at this point saying: "This section consists of 3 chapters, but it so illegible that the chapter headings cannot be identified." Thanks to Zou Dahai, who reexamined the original in the INHS archives, it has now been possible to identify this as Part VII, devoted to modern mathematics, with the intended Chapter headings as they appear in Li Yan's outline.

[25]Columbia University Archives; typewritten original.

Of course what I suggest is merely tentative, and you may see still further improvements. I am quite sure that when I come to work the material over I shall myself see places in which the order can be changed to advantage.

I am also somewhat uncertain as to whether it is best to number the chapters right through the book, or to drop the word "chapter" and simply use the Roman numeral heading for the sub-divisions of the various parts. [3]

I am unable yet to hear from Father Vanhee although I have taken pretty much every mean that I can think of to reach him. I am afraid that he was killed in the early days of the war.

I shall be glad to receive the copy of the Students' Dictionary of Chinese which you mention as having sent. No doubt it will come by a later steamer.

With respect to illustrations I may say that I can have the photographs made here from any of the material which I may have or which you may send on. If there is any material which you cannot send, it would be well for you to have the photographs made there, being sure that they are very distinct.

With relation to modern Chinese mathematics I am inclined to think that you may find it better not to include the living mathematicians. This was our plan in working out the history of Japanese mathematics.

The European books on Chinese mathematics which you mention are in my library, or are available in the University Library.

I should be glad to have you send the Encyclo- [4] paedia on "Rod Computations" together with the bill for the same. I shall also be glad to purchase any old Chinese manuscripts that can be got for a reasonable sum.

Yours very sincerely,

Enclosure:

HISTORY OF CHINESE MATHEMATICS
By David Eugene Smith and Y. Li
Introduction

I. Origin and Early Development of Chinese Civilization

1. Various traditions.
2. Scientific data.
3. Probable truth of the matter

II. Chronology of Chinese mathematics.

1. Extreme claims of antiquity.
2. Scientific data as to antiquity.
3. Probable truth as to antiquity of Chinese mathematics.
4. Chronological summary, arranged both by dynasties and by dates according to the Christian era.
5. Basis of the three-fold division into ancient mathematics (c. 3000-200 B.C.), medieval (200 B.C.-600 A.D.), and modern (600-1900 A.D.).

III. Mathematics in the I-Ching. Probable date. Possibility of alterations in the manuscripts.

IV. Relation of ancient mathematics to philosophy and astronomy. Astronomy in China. (Schlegel's theory of the signs of the zodiac.)

Part I. Ancient Mathematics (c. 3000-200 B.C.)

I. The origin of number. Systems of counting; of writing numbers.

II. Ancient arithmetic.
Two-fold division, corresponding to the Greek logistic (computation) and arithmetic (theory of numbers). Examples of the theory of numbers.

III. Methods of computation.
Possible antiquity of the bamboo rods.
Possible use of knotted cords, like the quipu of the Peruvians.
Uncertainty as to ancient methods. [2]

IV. Biographical sketches of ancient mathematicians.

V. The ancient mathematical classics. A Scientific study of their dates. Their contents.

VI. Early Chinese algebra.

VII. Early Chinese geometry.
Relation to mensuration.
The Pythagorean theorem.
Comparison with early Egyptian geometry.

VIII. Ancient ideas of higher mathematics.

IX. Ancient metrology.

X. The origin and early history of the magic square.

XI. Probable effects of intercourse with other countries.

XII. Conclusion as to the ancient civilization and ancient mathematics.

Part II. Mediaeval Mathematics (200 B.C.-600 A.D.)

I. From 200 B.C. to 200 A.D. Biographies. Discoveries. Treatises.

II. From 200 A.D. to 400 A.D. Biographies. Discoveries. Treatises.

III. From 400 A.D. to 600 A.D. Biographies. Discoveries. Treatises.
The development of the theory of the circle.

IV. Reforms in metrology.

V. Probable effects of intercourse with other countries.

VI. Conclusion as to the mediaeval civilization and mediaeval mathematics. [3]

Part III. Modern Mathematics (600 A.D.-1900 A.D.)

I. From 600 A.D. to 1000 A.D. Biographies, Discoveries. Treatises. Growth of algebra.
Mechanical computation. On Ta Yen Ch'in I.

II. From 1000 A.D. to 1400 A.D. Biographies, Discoveries, Treatises. Golden age of algebra.
Possible effect upon Persian and Italian algebra. On T'ien Yüan and Szu Yüan.
On the rise of trigonometry.

III. From 1400 A.D. to 1900 A.D. Biographies. Treatises. The introduction of Occidental mathematics. Influence of Matthew Ricci. Reaction against European mathematics. Magic squares and recreations. Typical problems.

IV. Different schools in modern mathematics.

V. Conclusion.

(7)

December 6, 1915[26]

Mr. Y. Li
St. of Incommemoration of the Revolution,
Foochow, China

My dear Mr. Li:

In chapters 48, 49 of the Yüan shi (completed in 1370) is a discussion of the astronomy (T'ien wen) of the Mongol dynasty. We know some of it through Wylie, but I am not sure that we know it all. In chapters 114—there is a long list of biographies of the Mongol period, and among those biographies there may possibly be some of interest to us. I presume you know of this, but it occurs to me to mention it to you.

I believe you told me that you were familiar with Mikami's book on Chinese Mathematics.[27] Although it is very poorly written, it has, of course, a good deal of information.

[26] Columbia University Archives; typewritten original.

[27] The work on history of Chinese mathematics referenced here is Yoshio Mikami's *The Development of Mathematics in China and Japan* (Leipzig: Teubner, 1913; reprinted New York: Chelsea, 1974).

It seems to me that what we need chiefly is translations of problems and solutions from the older works, so that we shall thus get the reader in touch with the original material.

Yours very sincerely,

(8)

South China, Foochow City,
St. of Incommemoration of the Revolution[28]
December 28th, 1915

Dr. D.H. Smith,
Teachers College,
Columbia University,
New York, U.S.A.

My dear Sir:

Your kind letter of November 4th has been received with pleasure. I am quite satisfied with the plan that you suggested and beg to say that the translated piece of the work being based on your arrangement is just now at its start.

I send you by mail a complete set of the Encyclopaedia on Rod Computations which I beg you to accept and trust that it will reach you safely.[29]

Pray accept my sincere felicitations on the New Year.

Yours very sincerely,
[signed] Y. Li

(9)

Henan Kwan-Yin- Tang,[30]
Republic of China postcard
April 6th, 1916.

Dr. D.E. Smith,
Teachers College,

---

[28] Columbia University Archives; typewritten original.

[29] The above note (18) referred to the Smith Archive at Columbia University possessing copies of Lao Naixuan's 籌算淺釋 *Chousuan qianshi* (Brief Explanation of Computing with Counting Rods) (1897), 籌算分法淺釋 *Chousuan fenfa qianshi* (Brief Explanation of Methods for Fractions with Counting Rods) (1898), 籌算蒙課 *Chousuan mengke* (Introductory Lessons on Computing with Counting Rods) (1898), and 籌算考釋 *Gu chousuan kaoshi* (Explanations of Ancient Computing with Counting Rods) (1898), 古籌算考釋續編 *Gu chousuan kaoshi xubian* (Continuation of Explanations of Ancient Computing with Counting Rods) (1898). The reference here to the complete set of the Encyclopaedia on Rod Computations may refer to these books.

[30] Columbia University Archives.

Columbia University,
New York, U.S.A.

My dear Sir:

I am inclined to believe that you are in receipt of a complete set of the Encyclopaedia on Rod computations which was forwarded on the 28th of December. If you are fail to receive the book, kindly let me know and the matter will be adjusted.

Please write my address in full as:

Y. Li,
Lung Hai Railway
Kwan-Yin-tang, Honan, China

when you write to me next.

I am remain
Yours very sincerely
Y. Li [signed]

(10)

Kwan-Yin-Tang,[31]
Honan, China
February 28th, 1917

Dr. D.E. Smith,
Teachers College
Columbia University,
New York, U.S.A.

Dear Sir,

Your favor dated 4th Jan. was duly received with many thanks. With regards to the translation of the History of Chinese Mathematics, my friends & I are working on it, the first three parts (Introductory, ancient & medieval) in Chinese language, have been finished already + were posted to Mr. Thomson Mao who is a post-graduate in Cornell University. He will translate them into English & forward the same to you directly.

Mr. Mao my old school-mate at Tangshan Engineering College, wishes to have the honour of being acquainted with you during his summer vacation if he got any chance, so I beg to introduce him to you now. He, Mr. Mao, had written a book on the History of Chinese "π" which has not been [2] printed yet.[32] He is

[31] Archives of the Institute for History of Natural Sciences, Beijing (CAS); handwritten draft, 3 pages.

[32] The article referred to here is: 中國圓周率略史 *Zhongguo yuanzhou luge shi* (History of Chinese rates of the circumference of the circle), 科學 *Kexue* (Science), 3 (4) (1917): 411-423.

very fond of Chinese mathematics too & therefore it is good for him to call on your.

I shall be very glad if you are going to make up a library of Chinese mathematics & I am quite willingly to help you if possible could you please send me a list of your Chinese books in order that I can collect & try to secure the rest for you.

It is rather difficult to have the manuscript of any before 1300 A.D., what I hope is the complete discovery of those since the thirteenth century, some of the books were never printed since they had been written. Although these are not easy to purchase, but still we may have them by copying. Moreover, in some cases these are impossible to get even the old rare printed books for you. I shall again let you have the copied ones either from what I have had, or from those who are interesting on this line of research, or from public library. The average expense including lodging, paper, salary and postage amounts to about [3] 0.40 gold per thousand words in Chinese by my estimation.

Some books on mathematics written by eminent writers are enveloped in the Encyclopedias, & they may be of one or two per cent in the set. These Encylclopedia are not sold separately, as it is also a matter to be considered. I would like to know the limits of the amount of fund you are going to invest for the collection, because this will serve as a correct guide to me.

LY [I] am very glad to [me] tell you that I got a copy of Siki Kawa manuscript on "Yang Hui Suan Fa" (in 3 volumes) from Mr. Y. Mikami.[33] The author of this work is a well known Chinese mathematician in Song Dynasty. This invaluable copy differs from the early printed ones in the respect that there are chapters devoted on the magic squares, circles, rings, & c which, sorry to say, disappeared in the printed book. Sike Kawa's copy, however, contains quite a bit of miscopied [4] words, so I am taking the work to revise them & have finished two volumes already. By some concrete proofs, I have reinforced in the revision that they are the real original work of Yang's. It is indeed a beautiful talking that the magic squares have been well developed by the ancient Chinese during the thirteenth century. Should you amuse with the revise copy, I will send you a copy with pleasure.[34]

[33]Mikami sent Li Yan a transcription of the *Yang Hui suanfa* (Yang Hui's Mathematical Methods), now preserved (as part of Li Yan's collection) in the Library of the Institute for History of Natural Sciences (Beijing). With 5 works which consisted of 7 volumes, and were altogether bound into 3 books, this copy was transcribed in 1661 from Seki Kowa's copy of the block-printed version (1378) made by Yu's Qingde Bookstore at Guhang. The transcription was revised in red using bronze movable type (1433) of Gyeongju in Korea, with additional revisions at the top of the margins, which must have been made by Li Yan. The transcription is reproduced in the first volume of *General Literature on Chinese Science and Technology—Mathematics*, pp. 1047-1117. This note is provided by Zheng Cheng, who also reviewed the translation of the text and made a few suggestions, and to whom I express my gratitude!

[34]The translator reviewed *The David Eugene Smith Collection of Works in Chinese on Mathematics and Other Subjects* (compiled by Miwa Kai, The Rare Book & Manuscript Library, New York, 1989) and *The David Eugene Smith Collection of Works in Japanese on Japanese Mathematics* (compiled by Miwa Kai, The Rare Book & Manuscript Library, New York, 1986). No copy revised by Li Yan of Seki Kowa's transcription of the *Yang Hui suanfa* was found.

Waiting for your kind favor & best compliments to you.
I am
Yours very sincerely
Y. Li [signed]

(11)

April 10, 1917[35]

Mr Y. Li,
Kwan - Yin - Tang
Honan, China

Dear Mr Li:

Your letter of February 28 has just reached me. I have written to Mr. Mao, as suggested. I hardly know what to say about having the Chinese books copied. I am more anxious to secure the original works for historical purposes. I think perhaps it is better to let the matter lie a little until I can see if I can purchase some of these originals. When the war is over I hope to take another trip to China, and at that time I could see you and could talk over the matter of having copies made in case the originals cannot be secured.

Yours very sincerely,
[signed] David Eugene Smith

(12)

[Cornell University
Ithaca NY]

[Mao Yisheng[36]]
319 Dryden Road, Ithaca, N.Y.
April 11th, 1917

Dr. D.E. Smith,
Teachers College,
Columbia University,
New York.

Dear Sir:

I have the honor to acknowledge receipt of your favor of the 10th inst. In availing myself of this opportunity I beg to inform you some of the relations which I was held with the translation of the materials which Mr. Li has collected for the

[35] Archives of the Institute for History of Natural Sciences, Beijing (CAS), typewritten copy.
[36] Columbia University Archives.

History of Chinese Mathematics. There are altogether three or four men undertake the same task and I am but one of them. Although I have had a profound interest in mathematics for a long time, I am not yet prepared to be a specialist on this line. I have been a student of Civil Engineering for five years and am now taking up the post-graduate work in Cornell University. It is therefore not to be expected that my translation will ever dispose any real merit of the original writing although I am endeavoring to utilize all that is within my ability. The fact that Chinese Mathematics is often [2] embodied in the sayings of ancient sages and classics makes the task rather uneasy as it involves a clear understanding of the ancient literature.

I am delighted to hear that I may have the honor of meeting you in some time in future although not in the coming summer. It will give me immeasurable pleasure when the opportunity comes.

I have the honor to be, Sir,
Your obedient servant,
Thomson Mao [Mao Yisheng]

P.S. If you are interested with the original writing of Mr. Li's work I can forward his MS to you upon notification.

(13)

[Cornell University
Ithaca NY]

[Mao Yisheng[37]]
319 Dryden Road, Ithaca, N.Y.
April 18th, 1917.

Dr. D.E. Smith,
Teachers College,
Columbia University,
New York.

Dear Sir:

I beg to acknowledge receipt of your favor of the 16th last. The translation of the material supplied by Mr. Li was begun a short time ago although the overwhelming collegiate work has prevented any perceptible progress. I am endeavoring to finish it before the commencement in time so that the MS and Mr. Li's original may reach you by the middle of that month.

Yours most respectfully,
Thomson Mao [Mao Yisheng]

[37] Columbia University Archives.

(14)

[Typewritten, no header]

[Mao Yisheng[38]]
319 Dryden Road, Ithaca,
June 16th, 1917.

Dr. D.E. Smith,
Teachers College,
Columbia University,
New York.

Dear Sir:

I take pleasure to inform you that the translation of Mr. Li's materials for the History of Chinese mathematics was just finished and is now given a thorough correction. It will be sent before the Commencement (27th inst.).

I remain
Yours most respectfully,
Thomson Mao

(15)

[Handwritten]
[Cornell University
Ithaca NY]

[Mao Yisheng[39]]
319 Dryden Road, Ithaca,
June 25, 1917

Mr. E.A. Mitchell
Columbia University,
New York.

Dear Sir,

I beg to send my insured parcel post the translations of Mr. Li's materials for Chinese Mathematics and his original.

Please forward them to Dr. Smith, with oblige,

I remain,
Yours most sincerely,
Thomson Mao.

---

[38]Columbia University Archives.
[39]Columbia University Archives.

(16)

[Teachers College
Columbia University
New York]

June 28, 1917[40]

Mr. Thomson Mao
319 Dryden Road,
Ithaca, N.Y.

My Dear Mr. Mao:

I beg to acknowledge the receipt of your letter of the 25th, and also of the insured parcel post package, containing the translations of Mr. Li for Chinese Mathematics. This I shall send to Dr. Smith, as you request.

I remain,
Yours very sincerely,
[signed] Eleanor A. Mitchell
Secretary to Dr. Smith

(17)

[Teachers College
Columbia University
New York]

July 25, 1917[41]

Mr. Y. Li,
Kwan - Yin - Tang,
Honan, China.

Dear Mr. Li,

I have just read the manuscript sent by Mr. Mao. I have been delayed in this because I am away from home on a vacation. I do not know whether this represents all the material that you have in mind to send on ancient and medieval mathematics, and I shall hold it until I hear further from you. As I wrote you on December 6, 1915, "It seems to me that what we need chiefly is translations of problems and solutions from the older works, so that we shall thus get the reader in touch with the original material." Now in your manuscript there are almost no translations from any of the ancient works mentioned. There is nothing from the Yih-King, although the text relative to the ho-t'u and the lo-shu

[40] Columbia University Archives.

[41] Archives of the Institute for History of Natural Sciences, Beijing (CAS), typewritten original; Columbia University Archives, typewritten carbon copy.

would probably be very interesting. I believe I have this in my own notes, however, so it does not matter. [2]

But when we come to the Chou-pei there is not a word of translation, although Mikami gives a number of valuable quotations, and the book must contain other interesting material. Then, too, the Chiu-chang (Kew Chang, as Mr. Mao transliterates it) contains a large amount of interesting matter, and Mikami has translated some of it; but your manuscript has little exact translation from the work. The same may be said of all the other classics and leading works on algebra and geometry,—that Western readers want the translation of certain typical parts, not merely a statement of the contents of the books. The latter is already available and is not of much significance.

A second difficulty that I find is in the fact that many names are mentioned with little or no reference to what the men did or even when they lived. For example, the following names are given with not the slightest indication of when the men worked or what the contributed: Chang Hoe, Chang Hung, Chang Wen Hu, Chien Ta Hin, Chu Kao Pao, Fang Chung Tung, Fu Shih, Ho Chun, Hsu Yao, Hua Heng Tang, Hwang Chung Shei, Lao Leh Shuen, Tien Mu, [3] Wah Heng Fang, Wang Yen Ming, and various others.[42] I may have something about these men in my own notes, but in your manuscript merely the names are given. The men may have lived anywhere between 3000 B.C. and 1000 A.D. Of course such names have no significance to a Western reader, as mere names.

A third difficulty lies in what seems to be inaccuracies. For example, as I understand the manuscript, you say that Chang Chang (whom I take to be the Chang T'sang of Mikami) lived in the East Han Dynasty, which was 35-221 A.D. (according to Williams), and was a statesman in the reign of Wan Te (c. 170 B.C.).[43] Now if it were the Han Dynasty (208 B.C.-25 A.D. according to Williams), this would be possible, but not the East Han. Mikami is quite exact, saying that Chang T'sang was made premier in 176 B.C. and died in 152. You give Chang Heng as a geometrician of about 100 A.D. (Mikami says he died in 139, and such exact dates are always desirable), but you also say that he lived in the Han Dynasty (306 B.C.-35 A.D.), which of course is impossible. [4]

You give Chun Ma Yen Men as living in the time of Chien Sze (32 B.C.) but as also living in the East Han Dynasty (25-221 A.D.), if I understand the manuscript correctly. A similar difficulty arises in connection with Kang Shou Chang), who lived under Shuen Te (60 B.C.) and also in the East Han Dynasty (25-221 A.D.) You give Lew Hui, son of Lew Heang, as living c. 10 A.D. and also as of the Tsin Dynasty (265-322).[44] Lo Heä Hung is given as of the East Han Dynasty (25-221

[42]Chung Hoe, Chang Hung, Ho Chun, Hua Heng Tang are all Chinese names that have yet to be identified.

[43]According to Han Qi, this is Samuel Wells Williams (1812-1884), who went to China in 1833 as an American missionary and diplomat, and later taught the Chinese language at Yale University. He was the author of *The Middle Kingdom* (1848) and *A History of China: Being the Historical Chapters from the Middle Kingdom* (1897).

[44]Here there is a confusion in the transliteration spellings of the names of "Liu Hui" and "Liu

A.D.) and as having calculated the calendar of Tai Chu (104 B.C.). There are various other similar difficulties which I will not take the time to mention.

A fourth difficulty lies in the matter of transliteration and capitalization. There must be some recognized standard, and Mikami seems to have tried to follow it, but I have much trouble with your manuscript. For example, one would think that Chen Long is Mikami's Ch'en Luan; but is he? You may have [5] mentioned Ching Ch'ou-ch'ang, who revised the Nine Sections in the reign of Hsüan-Ti (73-49 B.C.); but, if so, I do not recognize the name; and similarly for Li Ch'un-feng, Lü-Shih (or Lü Pu-i) of the third century B.C. Of course I recognize Tsoo Ch'ung Che as Tsu Ch'ung-shih, but there is probably some recognized law as to small letters and hyphens if not as to transliteration. Similarly, there must be some standard for Kew Chang Swan Shuh or Chiu-chang Suan-shu.

There are many other difficulties that I have met in my hurried examination of the manuscript, but the above are typical of the important ones.

I feel sure that you will see, from what I have written, that I cannot make use of your manuscript for the purposes I have in mind. What I wish is material which I do not have already, and to have statements that are so exact that I can use them with absolute confidence. My own material is much more extensive than what your manuscript contains, but it is deficient in exact translations of interesting material from the greatest Chinese works.

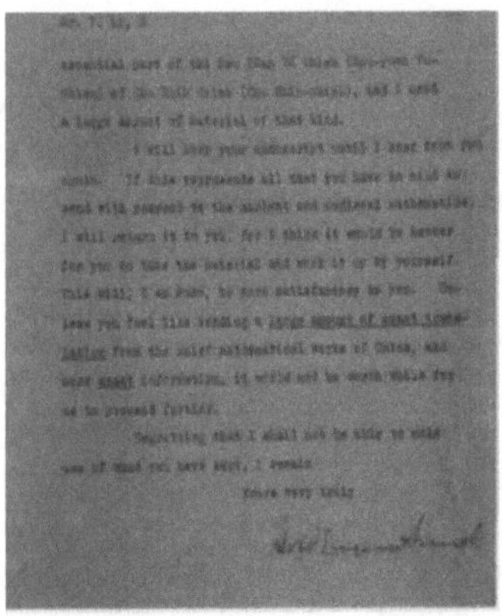

Figure 2: Page 6 of Smith's letter to Li Yan of July 25, 1917 (Archives of the Institute for History of Natural Sciences, Beijing (CAS))

I have, for example, an exact translation of the [6] essential part of the Seu Yüan Yü Chien (Szu-yuen Yu-chien) of Chu Shih Chieh (Chu Shih- chieh), and I need a large amount of material of that kind.[45]

Xin."

[45]This is a reference to Zhang Zaixin's master's thesis written in 1912, entitled "Ssu Yuan Yu Chien of Chu Shih Chieh" (*Zhu Shijie's Jade Mirror of the Four Unknowns*, Columbia

I will keep your manuscript until I hear from you again. If this represents all that you have in mind to send with respect to the ancient and medieval mathematics, I will return it to you, for I think it would be better for you to have the material and work it up by yourself. This will, I am sure, be more satisfactory to you. Unless you feel like sending a large amount of exact translation from the chief mathematical works of China, and more exact information, it would not be worth while for us to proceed further.

Regretting that I shall not be able to make use of what you have sent, I remain

Yours very truly
[signed] David Eugene Smith

(18)

Cheng Chow[46]
Sept. 4th, 1917

Dr. D.E. Smith,
Teacher College,
Columbia University,
New York, U.S.A.

Dear Sir,

Reply to your letter of July 25, which has just reached me, I beg to say that I am much agreed with the honorable proposal that you have suggested. Pray return me the manuscripts in the near future, as you desire.

Mr. Mao may set post-Han for former-Han & vise versa, as to performed the third difficulty. He may also drop the important data, as those Chang Heng died in 139 & Chang Tsuang was made premier in 176. B.C. & died in 152 that I quoted is entirely disappeared. In short, the [third are plainly] work is partly mistranslated, and [the fourth] partly mis-transliterated.

Hereafter [I am desire to] will forward to you some rare text in Chinese mathematics instead of my work, as request. Accordingly, I send you per mail three copied manuscripts of Chinese rare mathematics in the beginning of 16th century which I beg you to accept.

University, 1912). The thesis contains translations of the preface and postscript of the *Jade Mirror of the Four Unknowns*, of *Jialing Sicao*, and of several typical examples of the use of [the method of] four unknowns. Zhang Zaixin studied at Columbia University from 1910 to 1912, with David Eugene Smith as his thesis advisor. Upon his return to China in 1912, Zhang Zaixin taught at Yenching University and translated the entire *Jade Mirror of the Four Unknowns* into English. The translation had a checkered history until its eventual publication in 2006 by Liaoning Education Press. For details about the translator and his work, see Guo Jinhai: "Zhang Zaixin and his translation of the Jade Mirror of Four Unknowns," *Chinese Journal for the History of Science and Technology* 26 (2) (2005): 142-154.

[46] Archives of the Institute for History of Natural Sciences, Beijing (CAS), handwritten draft, 1 page.

(19)

October 11, 1917[47]

Mr. Y. Li,
Lung-Hai-Railway,
Chengchow, Honan, China.

My dear Mr. Li:

Your letter of September 6th has just reached me. I am returning herewith, by registered post, your manuscript and the translation made by Mr. Mao.

I am glad to know that you feel disposed to send me some translations of the rare works on Chinese mathematics. I should be very glad to have them.

With best regards, I remain,
Yours very sincerely,
[signed] David Eugene Smith

P.S. The manuscript goes to you under separate cover.

47 Archives of the Institute for History of Natural Sciences, Beijing (CAS), typed original.

"A New Era in the Development of Our Science": The American Mathematical Research Community, 1920-1950

Karen Parshall

Happy Birthday, Joe!

It was the end of August 1950 and some 2,300 mathematicians had gathered from all over the world in Cambridge, Massachusetts for the eleventh International Congress of Mathematicians (ICM). An ICM had never before been held in the United States, and the American mathematical research community had a point to make. Princeton's Oswald Veblen, President of the ICM and one of its principal organizers, put it this way: "If this Congress could have been held, as originally planned in 1940, it would have marked in rather a definite sense the coming of age of mathematics in the United States."[1] By 1950, however, the mathematical landscape had changed even more dramatically. North America had absorbed some of Europe's best mathematical minds as political refugees had sought new homes on the other side of the Atlantic. That brain-drain had been a blow for mathematical Europe but a boon for the United States. In Veblen's view, in 1950, it was more than "possible to hope that ...the American gains will be permanent."

Veblen was speaking for the entire American mathematical research community from that podium in Cambridge. He and his colleagues sensed that a definite shift had occurred over the course of the three decades from 1920 to 1950. As one mathematician expressed it, the interwar years marked "a new era in the development of our science."[2] What had changed in the mathematical landscape to give them this sense?

Building the Financial Infrastructure for Mathematical Research: 1920–1929

World War I had served as a kind of wake-up call for America's scientists, who had realized that their technical expertise could and should play a key role in the defense of the nation if only it could be effectively mobilized. In 1916, the National Research Council (NRC) had been founded under the aegis of the National Academy of Sciences (NAS) to coordinate the nation's non-governmental scientific and technical resources with those of the military. Two years later, U. S. President Woodrow Wilson issued an Executive Order that charged the NAS to make the NRC a permanent organization. After a year of planning, the new NRC was launched in 1919 "to promote research in the mathematical, physical, and biological sciences, and in the application of these sciences to engineering, agricul-

[1] [Veblen 1952], p. 124. The quotation that follows in this paragraph may also be found here.
[2] Roland Richardson to Oswald Veblen, 19 December, 1923, Box 10, Folder: "Richardson, R. G. D. 1923" in [Veblen].

D.E. Rowe and W.-S. Horng (eds.), *A Delicate Balance: Global Perspectives on Innovation and Tradition in the History of Mathematics*,
Trends in the History of Science, DOI 10.1007/978-3-319-12030-0_12

ture, medicine, and other useful arts, with the object of increasing knowledge, of strengthening the national defense, and of contributing in other ways to the public welfare" ([Anonymous 1919], p. 458 and [Barrows 1933], p. 7). Funded initially through the income from an endowment of $5,000,000 provided by the Carnegie Corporation of New York,[3] the NRC also received significant support from the Rockefeller Foundation for a post-doctoral fellowship program that aimed at providing "the more able young scientists with opportunities to acquire 'momentum' in their research, before settling down to permanent positions" ([Richtmyer 1919], p. 15). The close of World War I thus marked at least the symbolic beginning of what would increasingly become the symbiotic relationship between private and governmental funding sources, American science, and American institutions of higher education.[4]

Although mathematics had been included among the sciences that the NRC would ostensibly support, the post-doctoral fellowship program was initially open only to candidates in physics and chemistry with the stated purpose of "promoting fundamental research" in those fields "primarily in the educational institutions of the United States" ([Anonymous 1919], p. 462). In 1919, those were the two sciences widely deemed to have been the most relevant to the war effort, but they were also the two sciences believed by influential forces within the Rockefeller philanthropic organization to hold the greatest promise for further advances in the medical sciences, a key interest of the Rockefeller Institute of Medical Research ([Kohler 1991], p. 138 and [Siegmund-Schultze 2001], p. 28). What kind of brave new world was evolving for American science? Who knew? One in which patrons would significantly affect the direction of scientific research? And/or one in which negotiation and persuasive skills would be as important to scientific production as talent in basic scientific research? And/or ...?

In 1919, these questions—much less their answers—were only dimly perceived, at least by the mathematicians. They had largely not been part of the belated U. S. scientific mobilization for World War I, and, when they had participated, it had been to use their most basic skills in the service of ballistics.[5] World War I had been the chemists' and—to a lesser extent—the physicists' war, with the development especially of chemical weapons. The chemist Arthur Noyes of MIT, the physicist Robert Millikan of the University of Chicago, the Caltech astrophysi-

[3]See [Barrows 1933], p. 9 and [Siegmund-Schultze 2001], p. 29. In 2013 U. S. dollars, this was equivalent in "economic status" (considering the money as income or wealth) or in "income value" (considering the money as a commodity) to an endowment of some $353,000,000 based on the nominal gross domestic product per capita. All currency conversions to follow will use this measure. For more, see www.measuringworth.com.

[4]The Carnegie Institution of Washington ostensibly included mathematics among the subjects it supported from its founding in 1902, but, effectively, its support of the field was spotty at best and came primarily in the form of underwriting specific publications. See [Fenster 2003]. The John Simon Guggenheim Foundation, founded in 1925, also provided funds for the support of individual researchers in all academic fields—mathematics among them—beginning in the 1926–1927 academic year. See [Anonymous 1925], p. 471.

[5]On the involvement of American mathematicians in World War I, see [Archibald et al., to appear].

cist and president of the National Academy of Sciences, George Ellery Hale, these had been the major players in the new game of foundation and governmental linkage in a time of war. They had quickly become Rockefeller Foundation insiders. They had begun to capture foundation dollars for research in their fields. They had begun to define the twentieth-century category of the "scientific statesman." For the mathematicians, these were uncharted waters.

Enter Princeton's Oswald Veblen, that same Veblen who would serve as President of the 1950 ICM. Veblen became a member of the National Academy of Sciences in 1919 just as the NRC was being formed. The thirty-nine-year-old mathematician had earned his doctorate from the University of Chicago in 1903, had taken a position at Princeton in 1905, and had quickly established himself as one of America's leading research mathematicians for his work on the foundations of geometry, projective geometry, and analysis situs or what would today be termed topology.[6] As early as 1910, in fact, he had earned a star in the second edition of James McKeen Cattell's *American Men of Science*. In establishing their disciplinary hierarchy, Veblen's peers had not only placed him near the top but also pronounced him one of America's elite scientists.[7] Veblen would also quickly prove himself to be one of the first of a new breed of mathematician, the *mathematician* as scientific statesman, that is, the first-rate researcher who also recognized the need to use and had the skills to adapt mathematics to the changing environment of American philanthropy.

After just one year as a member of the Academy, the American Mathematical Society (AMS) appointed Veblen as its member of the NRC's Division of Physical Sciences, the division of which mathematics was a part and the division that administered the post-doctoral fellowships. As a savvy NAS and NRC insider, Veblen quickly came to realize the critical roles that the NRC fellowships were playing in physics and chemistry. Not only were they providing newly minted Ph.D.s a leg up in research, but they were also serving to enhance the standing of physics and chemistry within the scientific and, more broadly, the academic community in the United States. When he became the seventeenth AMS President for a two-calendar-year term in 1923, Veblen cannily adopted an agenda that aimed to situate mathematics squarely within this new philanthropic environment for science. The fact that he also served as chair of the NRC's Division of Physical Sciences during the 1923–1924 fiscal year and on the post-doctoral fellowship board beginning in 1923 even better positioned him to try to make things happen for

[6]For a brief overview of Veblen's work, see [Parshall and Rowe 1994], pp. 383-384 and 448-449.

[7]As explained in the preface to the first edition of 1906, "a star is prefixed to the subject of research in the case of about a thousand of the biographical entries. ...In each of the twelve principal sciences the names were arranged in the order of merit by ten leading students of the science. ...The thousand are distributed among the sciences as follows: Chemistry, 175; physics, 150; zoology, 150; botany, 100; geology, 100; mathematics, 80; pathology, 60; astronomy, 50; psychology, 50; physiology, 40; anatomy, 25; anthropology, 20. The star means that the subject of the biographical sketch is probably among the leading thousand students of science in the United States ...." See [Cattell 1906], pp. vi-vii. For more on starring and, in particular, on mathematicians starred, see [Visher 1947].

mathematics.

On 11 October, 1923, Veblen addressed a letter to Simon Flexner, chair of the NRC's fellowship board, President (1901–1935) of the Rockefeller Institute of Medical Research, and close adviser to the Rockefellers. Since Flexner had made the case for expanding Rockefeller support into the sciences of physics and chemistry because of their potential to aid the medical sciences, it would be key to convince him of the need for further expansion of the program to include mathematics. Veblen crafted a syllogistic argument that emphasized Flexner's main interests at the same time that it stressed the interconnectedness of the sciences. It went like this: chemistry and physics should be funded because medicine depends critically on them; chemistry and physics depend critically on mathematics; therefore, mathematics should also be funded. As a mathematician, he might well have added Q.E.D.[8] Aligning itself in this way with physics and chemistry would thus allow mathematics, Veblen and his colleagues in the AMS hoped, to establish the kind of bonds needed to bring it solidly into the fold of grant-supported sciences.

Veblen, the pure mathematician, was not being disingenuous in calling for this alignment. He had, in fact, become deeply interested in the applications of his own field of geometry to electrodynamics and general relativity and would soon use the platform of his retiring AMS presidential address in December 1924 to try to convince his colleagues of the importance of such applications. "The foundations of geometry," he asserted, "must be studied both as a branch of physics and as a branch of mathematics." "To understand the exact way in which the euclidean geometry and the Newtonian mechanics can be regarded as a first order approximation to the Einstein theory of gravitation," he continued, "seems to me to be of great importance in understanding the physical significance of geometry altogether."[9] Despite these pronouncements, Veblen equally recognized the importance of maintaining the disciplinary autonomy necessary for America's mathematicians to continue to pursue their primarily pure research interests.

Two months after Veblen made his pitch to Flexner, the Rockefeller Foundation delivered its verdict: the research fellowships would be extended to include mathematics. Veblen could hardly contain his excitement in writing to his right-hand man at the AMS, the Society's Secretary Roland Richardson of Brown University. "The Rockefeller Foundation has voted to extend the research fellowships in Physics and Chemistry so as to include Mathematics also," he announced, "and for this purpose they have increased their appropriation from $100,000 a year to $125,000 a year. This is to run for five years from the first of July, 1925, but in the meantime I have every reason to believe that mathematical fellowships will be granted this year if suitable candidates turn up."[10] Although on the surface,

[8]Oswald Veblen to Simon Flexner, 11 October, 1923, Box 28, Folders: "Board meetings, Correspondence, 1929–1937" in [Veblen]. See also [Feffer 1998], p. 479.

[9][Veblen 1925], p. 121. For more on Veblen's sense of the links between mathematics and physics, see [Feffer 1997], especially pp. 74-75.

[10]Oswald Veblen to Roland G. D. Richardson, 15 December, 1923, Box 10, Folder: "Richardson,

this may have sounded like mathematics received an appropriation of just $25,000, the facts were that there was no formal division of the funding among the three sciences and that the prior appointments in physics and chemistry had been so conservative that the fellowship board had failed to expend all of the funds allotted.[11] Veblen thus saw this as a real opportunity to jump-start funding for mathematical research.

Veblen's argument — reductionistically logical though it might have been—had worked. He had sensed the desire of the Rockefeller Foundation's executives to support science writ large in the 1920s, that is, science beyond the wartime sciences of chemistry and physics for which its external scientific advisers continued to advocate. He had equally appreciated that the Foundation did not want to take on something too big, again, as the chemists and physicists were urging, and that it did not want to be too hands-on. It viewed fellowships as ideal: "not too expensive, low risk, and universally popular."[12] Veblen had played his cards perfectly.

The next problem, of course, as both Veblen and Richardson realized, was to get the good news out as quickly as possible to the mathematical community in order to assure a strong showing of applications. Fortunately, the AMS was scheduled to hold its thirtieth annual meeting in New York City at the end of December, so he and Richardson had the perfect opportunity to, as Veblen put it, "give the thing as much publicity and emphasis as possible."[13] Richardson reported on what he termed "this remarkable addition to the resources for mathematical research" in his published minutes of the meeting, noting that Veblen had "urged" the AMS's membership "to give information concerning talented and well-equipped young men and women" so that "adequate use" could be made of this new funding opportunity ([Richardson 1924], p. 203).

Not only was it important to get applications from strong candidates, but it was also critical to have the "right" people on the selection committee. To that end, Veblen, with Flexner's blessing, solicited the participation of the University of Chicago's Gilbert Bliss. As immediate past-President of the AMS, Bliss, a specialist in the calculus of variations, was a recognized leader in American mathematics. Moreover, Chicago was arguably the best program in mathematics in the country, even though it was increasingly clear by the 1920s that it was being eclipsed by both Harvard and Princeton.

An especially revealing manifestation of this increasing sense of pecking order was the (non-scientific) 1924 survey conducted by Raymond Hughes, a trained

R. G. D. 1923" in [Veblen]. In 2013 U. S. dollars, $125,000 was equivalent to $8,610,000.

[11]Oswald Veblen to Gilbert Bliss, 15 January, 1924, Box 3, Folder: "G. A. Bliss 1923–40" in [Veblen].

[12][Kohler 1991], p. 107. After three years of negotiations, biology was also included in the postdoctoral fellowship program in 1923. Together with mathematics, astrophysics and geophysics were added in 1923 as well. As Robert Kohler has put it, the latter three additions were made "with an ease that astonished some NRC regulars." See [Kohler 1991], p. 106.

[13]Oswald Veblen to Roland Richardson, 15 December, 1923, Box 10, Folder: "Richardson, R. G. D. 1923" in [Veblen].

chemist and the President of Miami University of Ohio. Motivated by the need to hire the best people for his faculty—a concept that by the 1920s increasingly meant those with doctorates from the strongest programs nationally—Hughes aimed at determining a ranking of American graduate schools by field and, to this end, asked experts in the various areas to rank-order the programs in their disciplines. Nineteen mathematicians responded (seventeen from outside, or a return of about 50%, and two, presumably, from Miami of Ohio). Hughes then tabulated the rank-orderings. In mathematics, the University of Chicago was the unequivocal leader, with Harvard a close but distinct second and Princeton a more distant but still clearly delineated third. These top three were followed by the University of Illinois, Columbia, Cornell, Yale, Wisconsin, the Johns Hopkins, Michigan, the University of California at Berkeley, the University of Pennsylvania, and the University of Minnesota.[14]

Despite possible flaws in the survey that had generated them, Hughes's rankings nevertheless roughly mirrored data that Roland Richardson later compiled on American Ph.D. production in mathematics, with Chicago having produced by far the most mathematics Ph.D.s (177) by 1924, followed by the Johns Hopkins (77), Yale (63), Harvard (59), Cornell and Columbia (45 each), the University of Pennsylvania (32), Princeton (26), Illinois (25), the University of California at Berkeley (23), and Wisconsin (14).[15] Given these data on Ph.D. production as well as this overall sense of the hierarchy within American mathematics in 1924, it should perhaps come as no great surprise that of the twelve mathematics fellows ultimately named in that first round of NRC funding for the 1924–1925 academic year, University of Chicago graduates led the field.[16] As the program continued into the 1930s, however, graduates from other universities were increasingly represented. This reflected the concomitant development from coast to coast of universities and their graduate programs (see the next section).

If the inclusion of mathematics in the NRC fellowship program marked one of the successes of Veblen's agenda as AMS President, another was the AMS's legal incorporation in October 1923. The new corporation had a Board of Trustees responsible for the Society's financial affairs, principal among them defraying the costs of publication of the Society's two journals, the *Bulletin* and the *Transactions*. The thirty-one mathematicians initially named to the original Board included two women, Anna Pell, a 1910 Chicago Ph.D. and professor at Bryn Mawr

[14]See Raymond Hughes to George D. Birkhoff, 29 October, 1924, HUA 4213.2, Box 4 "Correspondence 1924 'K–M' " in [Birkhoff]. The University of Chicago received seventeen first-place and two second-place ratings; Harvard got thirteen first-place and six second-place rankings; and Princeton garnered five first-place, seven second-place, five third-place, and two fourth- or fifth-place ratings.

[15]See [Richardson 1936], Table 1, p. 203 (or see p. 366 in the reprinted edition). In 1924, the Universities of Michigan (12) and Minnesota (4) had been giving the Ph.D. in mathematics for less than a decade.

[16]See [Anonymous 1924a], p. 378 and Wilbur E. Tisdale to Oliver D. Kellogg, 1 October, 1925, "Memorandum" on "Research Fellowships in Physics, Chemistry and Mathematics," Box 14, Folder: "Tisdale, W. E. 1923–24" in [Veblen].

College, and Clara Smith a 1904 Yale Ph.D. who took the position at Wellesley in 1908 that she would hold until her retirement in 1934. Perhaps not surprisingly, these Trustees elected Veblen as their Chair and Richardson as their Secretary. Together with Harry Tyler of MIT, Herbert Hawkes, Dean of Columbia College, and Robert Henderson, actuary with the Equitable Life Assurance Society of the United States, these five men formed the Executive Committee responsible for overseeing the business of the Board.

That business included continuing Bliss's initiatives as AMS President to increase the Society's membership, and thereby its financial base, as well as to raise a $100,000 endowment to help offset the ever-increasing cost of the Society's publications.[17] Late in 1922, Bliss had appointed Julian Coolidge of Harvard as chair of the endowment campaign, and, thanks to his efforts, those within the Society had contributed a remarkable $25,000 in the campaign's "silent" phase that ended at the close of 1923.

Not surprisingly, this initiative had largely tapped out the Society's membership. Other potential donors would need to be identified in order for the drive to be a success, but success meant more than money to the mathematicians. In the words of AMS historian, Raymond C. Archibald, "the campaign was to be more than an attempt to put the finances of the Society on a firm basis; it was to be also a campaign of education of the public concerning the basic character of mathematics *in our present civilization* and the importance of mathematical research in *advancing that civilization*."[18] Veblen had seemingly succeeded in so educating the powers-that-be at the NRC; perhaps the AMS could have similar success with the broader public. In the view of Coolidge, Veblen, and others, that public should include the other foundations as well as industries dependent on mathematics ([Archibald 1938], p. 31). In 1924, Veblen and his team turned in earnest to the problem of marshalling support from those potential funding sources.

Some of the groundwork had already been laid. In September of 1923, the AMS Council had voted to create a series of more popular lectures "on topics in mathematics or its applications" that would be held periodically and that would be named in honor of the distinguished, nineteenth-century Yale mathematical physicist and physical chemist, Josiah Willard Gibbs [Richardson 1923]. The idea was that by highlighting applications of specific kinds, the AMS could more meaningfully appeal to various industrial concerns. Two Gibbs lectures were given in 1924. The first—by Michael Pupin, a professor of electromechanics at Columbia —attempted to convey a sense of the aesthetics of then-modern physics, while the second—by Robert Henderson—treated "life insurance as a social science and as a mathematical problem."[19]

The Gibbs lectures represented just one strategy that the AMS leadership

---

[17]See [Archibald 1938], p. 30. In 2013 U. S. dollars, this was a $7,880,000 million endowment campaign.

[18]For the quote, see [Archibald 1938], p. 31 (my emphasis). See also [Feffer 1998], pp. 480-481.

[19]For the talks, see [Pupin 1924] and [Henderson 1925]. On the Gibbs lectures in general, compare [Feffer 1998], pp. 481-482.

explored in 1924 for interesting business and industry in the affairs of research mathematicians in the United States. In March of that year, for example, Richardson wrote provocatively to the members of the AMS Council, proposing that the AMS "enter into contract with firms to provide a clearing house service for general mathematical information."[20] This scheme modeled one that MIT had then recently included in its endowment campaign, but Richardson was quick to add that he was not advocating "that the Society ...undertake to give detailed advice, but rather to indicate where such advice can be obtained in return for a compensation agreeable to both parties." He freely admitted that his was "a very radical proposal."

The Council ultimately opted not to move in this direction, but it did support Coolidge's initiative of soliciting sustaining memberships from a variety of business concerns. That initiative had, in fact, already been paying off. By May of 1924, the publishers Allyn & Bacon and Ginn & Company in addition to General Electric had taken out such memberships, and by the end of the year, eleven more concerns had been added to these ranks, among them, two more electric companies, four insurance companies, and two institutions of higher education. By 1926, these numbers had almost tripled to a total of thirty-eight.[21] As Richardson reported to the AMS Council in 1926, with the $25,000 raised in the silent phase of the endowment campaign and with the additional $30,000 raised in the campaign's second phase through non-corporate sources, if sustaining "membership can be kept near its present level, the Society is in a financial condition even more favorable than if the sum [of $100,000] originally named had been procured for the treasury." "[T]he endowment campaign," he concluded, "may be considered a complete success in attaining its goal" ([Richardson 1926], p. 205).

As noted, this fund-raising activity aimed to address the ever-increasing publication costs of the AMS's two journals. In September of 1923, when the endowment campaign was in full swing, the AMS learned that the Lancaster Press in Pennsylvania, which had been serving as the Society's printer, would likely increase its charge per page for printing the *Transactions* to $15.00, whereas the AMS could secure the services of a press in Hamburg, Germany for the equivalent of $3.00 per page.[22] It certainly made sense for the AMS to deal with just one printer, so if the *Transactions* moved abroad, then the *Bulletin* should as well. The problem was the timing. As Harvard's George Birkhoff scrawled in a P.S. to

[20]Roland G. D. Richardson to the Members of the Council, American Mathematical Society, 29 March, 1924, Box 10, Folder: "Richardson, R. G. D. 1924" in [Veblen]. The quotes that follow in this paragraph are also from this letter.

[21]Julian Coolidge to Roland G. D. Richardson, 24 May, 1924, Box 10, Folder: "Richardson, R. G. D. 1924" in [Veblen]. See also [Anonymous 1924b] and [Anonymous 1926]. Owing to the Depression as well as to other factors, the income from sustaining memberships dropped from approximately $4,800 (or $312,000 in 2013 U. S. dollars) in 1928 to $1,800 (or $210,000 in 2013 U. S. dollars) by 1933. See [Archibald 1938], p. 34.

[22]Oswald Veblen to George Birkhoff, 11 September, 1923, Box 2, Folder: "Birkhoff, George D." in [Veblen]. For more on the issue of mathematics publishing in the United States in the interwar period, see [Siegmund-Schultze 1997].

a letter to Veblen a month later, "*[e]verybody* here seems to favor publication of the Transactions in Germany, at least for the further volume, *and perhaps further*. But they think it would be bad policy to *adopt* any such policy—it might hurt the endowment drive seriously."[23] Immediate financial concerns ultimately trumped other considerations, however, and the AMS contracted with a German printer beginning with the first numbers of 1924.

Still, as AMS President and chair of the AMS's Board of Trustees, Veblen continued to search for a solution to the problem of publication costs at home. By October of 1924, he had laid the groundwork for discussions with the General Education Board (GEB) of the Rockefeller Foundation about defraying the costs of the Society's publications. The GEB's Wycliffe Rose informed Veblen that the biologists and others had already approached the GEB with a similar proposal and that the GEB was seriously considering allocating funds to be administered through the National Academy of Sciences for the purpose. As Veblen described the meeting to Richardson, he asked Rose point blank "whether it would in the least jeopardize our request if we enlarged it so as to include the American Journal of Mathematics, The Annals of Mathematics, and perhaps a prospective Journal of Applied Mathematics," that is, the other publications—and even a journal not yet born—that served the American mathematical research community.[24] "He reported that it would not," Veblen explained to Richardson. "[T]hey could just as well do the whole thing as a part of it, and would rather not do it piecemeal." By June of 1925, the GEB seemed close to allocating some $10,000 (or $672,000 in 2013 U. S. dollars) a year to the National Academy of Sciences, and Veblen was meeting with the NAS's committee on publication funds to pitch the case for mathematics.

His argument was fundamentally nationalistic in tone. Is it not a disgrace that the AMS has been forced, on account of cost, to publish its journals in Germany? Writing to his successor as AMS President, George Birkhoff, Veblen recounted that "[t]he members of the Committee were so impressed with the importance of bringing the Bulletin back to this country that they assured me informally that it is perfectly safe to go ahead and make the arrangements ...on the assumption that this would require the $1600 which I estimated."[25] In fact, through the National Academy of Sciences, the GEB initially allocated funds in excess of $3,000 (or $202,000 in 2013 U. S. dollars) annually for the publication of research-level mathematics, a commitment soon extended at least through 1931. As Richardson put it to Birkhoff, "The Society owes very much to Veblen for having carried through this matter with the General Education Board. The financial problems of the Society are solved for the next five years. And the Secretary will breathe more

[23]George Birkhoff to Oswald Veblen, 5 October, 1923, Box 2, Folder: "Birkhoff, George D." in [Veblen] (Birkhoff's emphases).

[24]Oswald Veblen to Roland G. D. Richardson, 3 October, 1924, Box 10, Folder: "Richardson, R. G. D. 1924" in [Veblen]. The next quote is also from this letter.

[25]Oswald Veblen to George Birkhoff, 3 June, 1925, Box 2, Folder: "Birkhoff, George D." in [Veblen].

freely."[26] In Veblen, the American mathematical research community had come to recognize one of its first bona fide "scientific statesmen" over the course of the 1920s. Another was his successor as AMS President, George Birkhoff.

As AMS President in 1925 and 1926, Birkhoff actively continued the initiatives that Veblen had begun. Four years Veblen's junior, Birkhoff, too, had earned his doctorate at the University of Chicago; he, too, had been "starred" in the second edition of Cattell's *American Men of Mathematics*; he, too, had been named to the National Academy of Sciences (in fact, one year earlier than Veblen); he, too, was a recognized leader in American mathematics.[27] In the 1920s, Veblen at Princeton and Birkhoff at Harvard fundamentally shaped the American mathematical landscape through their efforts within the AMS, through their activities in the NRC and the NAS, through their work with the Rockefeller Foundation, and through the program-building that they oversaw at their respective institutions. Birkhoff put it crisply in a letter to Veblen on 17 February, 1927. "[Y]ou and I," he said, "have rather worked together to 'boost' mathematics into the position it deserves to have."[28] That "boosting," as well as the example set by the program at their doctoral *alma mater* and at a handful of other schools, served as the model for the ethos of research and its support that continued to spread in American mathematics to state-supported and other institutions in the decade of the 1930s.

An Evolving Mathematical Landscape: 1929–1939

The wave of financial optimism on which Veblen, Birkhoff, and their colleagues had been riding high in the mid to late 1920s crashed with the stock market in the final days of October 1929. There continued, however, to be some tangible manifestations of the development of research-level mathematics in the United States. Private benefactions made before the market crash had led to the completion and dedication of Eckhart Hall at the University of Chicago in 1930 and of Fine Hall at Princeton a year later in 1931, both of which were state-of-the-art homes for mathematics and symbols of support for the mathematical endeavor ([Aspray 1988], pp. 353-356 and [Feffer 1998], pp. 492-493).

With similarly propitious timing, Louis Bamberger and his sister, Caroline Bamberger Fuld, had taken their family fortune out of the market just before the crash and were ultimately persuaded by Abraham Flexner, younger brother of Simon Flexner, to endow a new sort of institution for the support of research. It

[26] Roland G. D. Richardson to George D. Birkhoff, 29 September, 1925, HUA 4213.2, Box 5 "Correspondence 1924/1925 'A–Z'," Folder: "R" in [Birkhoff]. The AMS received publication subventions through the GEB until 1936 but instituted a program of sustaining and contributing institutional memberships in 1934 when the GEB made it clear that its funding would soon come to an end. This served to bridge the gap and to make the AMS's publications self-sustaining. See [Archibald 1938], pp. 34-36 and compare [Feffer 1998], pp. 482-484.

[27] For a brief overview of Birkhoff's work, see [Parshall and Rowe 1994], pp. 390-393 and 446-447.

[28] George Birkhoff to Oswald Veblen, 17 February, 1927, Box 2, Folder: "Birkhoff, George D." in [Veblen].

would *not* be a graduate school. *That* model was already in place. In the junior Flexner's conception, it would be what he liked to style "a paradise for scholars," but it would be even more. It would be a model of philanthropic support that served to "elevat[e] research to a full-time endeavor" as well as to secure the "recognition of the high status of researchers and, consequently, of the deference due them in a proper national culture" ([Reingold and Reingold 1981], p. 434). Although formally launched in 1930, the Institute for Advanced Study actually only opened its School of Mathematics, the first of several schools planned, in 1933 in space shared with Princeton's Department of Mathematics in Fine Hall. Perhaps not surprisingly, Flexner, as the Institute's first director, ultimately secured Oswald Veblen as his primary mathematical adviser. Working together in close concert with Flexner, Veblen thus helped shape the institution as a whole.[29]

Despite these manifestations of support for research-level mathematics in the early 1930s, those years are much more accurately characterized by "financial difficulties," a sentiment that ran through the correspondence of mathematicians nationwide.[30] At Michigan, for example, topologist Raymond Wilder wrote despairingly to fellow topologist Gordon Whyburn, who was then at the Johns Hopkins, about prospects for tiding over a promising recent graduate who had been unable to afford to proceed to graduate school for the 1932–1933 academic year. "I am going to see if I can get him a job as an assistant or something," Wilder wrote, "not at Michigan, for we are broke (and when I say broke, I mean *broke*)."[31] "Have your faculty been cut yet—either in numbers o[r] in salary?," Wilder asked Whyburn. "We've been cut both ways," he volunteered, "and it looks like we are due for a beautiful cut next year. Beginning with our April salary, we are going to get only half pay for the rest of the year (this means for four of our ten months' salary), the rest to be paid maybe, when, and if the state is willing. The proposal in the legislature, now taking up the appropriation for next year, is to cut us 50%. In addition there is a joker in the bill making our allowances payable only on prop[o]rtion as the taxes are paid. If only 50% of the taxes are taken in that would mean a 75% cut. Well," Wilder closed, "enough of such mournful stuff."

The NRC fellowship program took a hit as well. By 1935, the Depression had so severely affected the Rockefeller Foundation's ability to fund the program that, while it was not canceled, it was significantly curtailed. As topologist Leo Zippin reported to Wilder during the course of their effort to secure an NRC fellowship to Michigan for a young up-and-comer, "Veblen says the appropriation has been cut one third ...and there will be so few appointments that no one has very much

---

[29]For more on the history of the Institute, see [Batterson 2006].

[30]See, for just one example, Gilbert Bliss to Oswald Veblen, 27 January, 1933, Box 3, Folder: "G. A. Bliss (1923–40)" in [Veblen].

[31]Raymond Wilder to Gordon Whyburn, 18 April, 1933, Box 86-36/14, Folder 5: "General Correspondence Whyburn, G. T." in [Wilder] (his emphasis). The quotation that follows is also from this letter. The student in question was Norman Steenrod, who went on to make seminal contributions to algebraic topology after earning his Ph.D. from Princeton in 1936 under the direction of Solomon Lefschetz.

of a chance."[32]

While financial stringencies were a harsh reality, the 1930s also—and paradoxically—witnessed the targeted building of programs in mathematics at a number of institutions across the nation as forward-thinking university administrators, especially in the Midwest, West, and South, worked with members of their faculties to transform their institutions into bona fide *research* universities. Consider, for example, the case of the state-supported University of Michigan. As indicated by Raymond Hughes's unscientific poll, it was already perceived by some in the mathematics community as in the top ten. Still, it lagged far behind the "big three" of Chicago, Harvard, and Princeton in the 1930s. As a state-supported, as opposed to a privately endowed, institution, its efforts to build programs were much more susceptible to the vagaries of public opinion and the state legislature than were those of its elite competitors. Moreover, as Wilder had lamented to Whyburn, the Depression had hit the school hard in the early 1930s, making that seem hardly the best of times for a shift toward greater support for pure research.

At first, indications that institutional change was afoot were subtle. In the spring of 1933, for example, Wilder was presented with the opportunity to visit the Institute for Advanced Study during its first year of operation in 1933–1934. He was not sanguine about his chances for securing the necessary sabbatical at half-pay. His argument to Michigan's Board of Regents was purely research-oriented. "[T]his offer," he explained, "1) is intended virtually as a fellowship, 2) is probably inspired by a desire on the part of the Institute to enable me to give practically all my time for a year to furthering a program of research outlined in my recent Chicago symposium lecture ..., and 3) will provide contacts (with such men as Einstein, Weyl & Veblen) and opportunities which will, I believe, be of benefit in my future work in the University of Michigan."[33] Although his salary, like everyone else's on the faculty, was further cut, the Regents granted him the leave at half pay, and the Institute not only matched the half-salary but also supplemented it to defray the cut, at least partially, and thereby to make it feasible for the family of six to spend the year in Princeton.[34] By the time Wilder resumed his work in Ann Arbor in the 1934–1935 academic year, even more change was in the air.

Turmoil over the resignation of the department chair in the fall of 1934 had resulted in a political clash between the majority—the non-research-oriented full

[32]Leo Zippin to Raymond L. Wilder, undated but around 11 March, 1935, Box 86-36/14, Folder 5: "General Correspondence Zippin, L." in [Wilder]. The student, William Claytor, was an African-American who had earned his Ph.D. under John Kline at the University of Pennsylvania in 1933. The results that he proved in his dissertation and published in the *Annals of Mathematics* sparked much interest in the international community of topologists, and a concerted effort was made in the 1930s to secure a position for him at a research-oriented university. For more on his story, see [Parshall, in preparation]. A version of this study in preparation was presented at the October 2013 meeting of the American Mathematical Society held in Philadelphia, Pennsylvania.

[33]Draft "For communication to the Board of Regents," Box 86-36/42, Folder 4: "Faculty and Administrative Activities, University of Michigan, Sabbatical Leaves, 1933-1934, 1956–1957" in [Wilder]. For the content of Wilder's symposium, see [Wilder 1932].

[34]See Raymond Wilder to Robert L. Moore, 3 September, 1934, Box 86-36/8, Folder 6: "General Correspondence R. L. Moore" in [Wilder].

professors—and the minority—the younger researchers like Wilder. "It does not seem to count with the full professors that is it vastly more important to have a real mathematician who will do the dep[artmen]t some good," Wilder wrote in disgust to a friend. "[T]hey act as if they were electing the president of a country club. No doubt the chairman will ultimately be selected by the President."[35] When announced, that choice was Theophil Hildebrandt, and, in Wilder's words, "the dope is that he is supposed to build up the research side of the dep[artmen]t."[36]

Hildebrandt had joined the Michigan faculty in 1909, a year before officially earning his Ph.D. at the University of Chicago, and had continued to do respected research in functional analysis. One of the first issues that confronted him on becoming chair was an outside offer to Wilder from Hopkins. Although Hildebrandt could not ultimately match the salary, he succeeded in retaining Wilder by securing his promotion to full professor. In the five years that followed, Hildebrandt—with the help of Wilder in topology and another of the younger turks, George Rainich, in the theory of relativity—attracted seven new research-oriented faculty members to Ann Arbor and further developed the program's graduate offerings.[37]

Just behind Michigan in Hughes's rankings, the University of California at Berkeley on the West Coast also made major strides in mathematics in the 1930s. California's land-grant university, Berkeley had opened in 1868 and, despite Daniel Coit Gilman's early efforts as President to orient the university more toward research, had remained an institution strongly under the control of a state legislature unsympathetic to that ethos.[38] Spurred by the opening of nearby Stanford University in 1891, however, Berkeley began to strengthen its programs in the sciences, first in chemistry and then in physics. Mathematics, according to one administrator, however, had "not been maintained at the proper standard" and badly needed an infusion of new, younger faculty members in order both to establish some sort of critical mass and to provide support for the other, rapidly developing departments in science and engineering ([Rider 1989], p. 286). In fact, those other departments argued, and in a way that supported Veblen's claims for mathematics to the NRC in the mid-1920s, that a strong mathematics department was needed better to support their own work.

When he assumed the presidency at Berkeley in 1930, Robert Sproul took those calls to heart. On the advice of a hand-picked and high-powered committee,

[35]Raymond Wilder to Leo Zippin, 15 October, 1934, Box 86-36/14, Folder 5: "General Correspondence Zippin, L." in [Wilder].

[36]Raymond Wilder to Leo Zippin, 20 January, 1935, Box 86-36/14, Folder 5: "General Correspondence Zippin, L." in [Wilder].

[37]Compare [Kaplan 1989] and [Wilder 1989]. George Yuri Rainich (born Rabinowitsch) was a Russian Jew who had emigrated to the United States in the 1920s, settling first at the Johns Hopkins University before moving to Michigan in 1926. His emigration to the United States, like that of Solomon Lefschetz and Oscar Zariski, thus preceded the more conspicuous influx of Jewish mathematicians that occurred in the 1930s.

[38]"Land-grant" institutions were created in each state by Federal law and with Federal funds in 1862 for the promotion primarily of the utilitarian sciences of agriculture, mining, and engineering.

Sproul marshalled some of his scarce, Depression-era resources to hire Griffith Evans from Rice University in Texas to lead an initiative in mathematics. Sproul specifically gave Evans the task of turning Berkeley's department of mathematics into a major center for mathematical research ([Rider 1989], pp. 287-292).

Evans, a proven and respected researcher who had done seminal work on the theory of integral equations as well as in applied mathematics and mathematical economics, not only had a broad vision of mathematics both pure and applied but also worked hard to make that vision a reality at Berkeley. Even before he arrived to assume his new post in 1934, he encouraged the addition to the department of two researchers active in mathematical physics and analysis, areas likely to serve as bridges to the other sciences and especially physics.[39] Working with a supportive administration, Evans continued to strengthen the department through the 1930s and into the 1940s (see below).

While Berkeley was the only university in the West that had risen high enough in the collective mathematical consciousness by the mid-1920s to rank in Hughes's survey of graduate programs, the South had not been represented in that survey at all. In 1934, the year in which Gordon Whyburn was hired to head the Department of Mathematics at the University of Virginia, the South accounted for only 13% of the membership of the AMS. Only Maryland (36), Texas (49), and Virginia (32) had more than twenty AMS members resident in the entire state, while Arkansas (7), Mississippi (6), South Carolina (6), and Tennessee (8) had fewer than ten.[40]

Whyburn had earned his Ph.D. in topology under Robert L. Moore at the University of Texas in 1927. A gifted student, he had thoroughly embraced the sense of the primacy of research that Moore had fostered since his own student days at the University of Chicago in the earliest years of the twentieth century and that he had instilled in the many students with whom he had worked first at the University of Pennsylvania and then at Texas.[41] In his successful application to the Guggenheim Foundation in 1929, Whyburn gave eloquent expression to that ideal when he wrote that "[m]y ultimate purpose as a student is to locate as a professor of mathematics in some American university of high rank, where the spirit of research is fostered and respected, and where I would be able to teach and direct research in my chosen field of work and to develop students of ability into research workers of high standing."[42] He would have seemed to have landed

---

[39] Alfred Foster specialized in the former, Charles B. Morrey in the latter. See [Rider 1989], pp. 288-289.

[40] See [Anonymous 1935] for the geographic distribution of AMS members in 1934 and [Kline 1935], p. 153 for the total membership figure.

[41] In fact, Robert L. Moore animated what many consider to have been a "school" of mathematics based on his so-called "Moore Method" of instruction, a socratic method that hinged on students' personal discovery of mathematical facts. For one of many discussions of Moore, his school, and his method, see [Parker 2005].

[42] Typescript of "Some interesting data as to Whyburn's plans for study [quoted from his application to the Guggenheim Foundation]," Box 86-36/14: Folder 5: "General Correspondence Whyburn, G. T." in [Wilder].

such a position on his return from his Guggenheim year in Vienna when he was appointed to an associate professorship at the Johns Hopkins.

Although Hopkins had led the way in graduate-level mathematics teaching and research from its founding in 1876 into the 1880s, it had been successively eclipsed by programs at the "big three" and had been unable to prevent its slide to the ninth-place ranking it had garnered in Hughes's survey.[43] In the late 1920s and early 1930s, moreover, the Hopkins department was top-heavy with senior professors who effectively called the shots. That left little opportunity for the younger members of the department to exert much influence. As Whyburn explained in a letter to Moore on the day on which he accepted the call to Virginia, "this will give me an opportunity of the type which would not be available for me at Hopkins for many years, if ever."[44] That opportunity afforded, in Whyburn's words, "a great deal of freedom in developing the Mathematics Department along lines which seem best to me."[45]

With the blessing and encouragement of the university's second President, John Newcomb, Whyburn set to work. In addition to drafting both a "plan embodying the future aims and purposes of the department and a tentative curriculum," he solicited recommendations for possible hires from, among others, Veblen and Moore. Since the number of graduate students would be small for the foreseeable future, it would be important to find someone whose research interests complemented but did not duplicate Whyburn's own. It would ideally be "a man whose research has been somewhat between the fields of topology and the theory of functions and if possible one who is interested in applying the results and methods of topology to problems in function theory and conversely."

By the end of his first year at Virginia, Whyburn had hired a man who filled that bill, Edward J. ("Jimmy") McShane, a Chicago Ph.D. specializing in the calculus of variations whom he coaxed from an assistant professorship at Princeton. By 1939, Harvard-trained Gustav Hedlund had also joined the Virginia faculty, rounding out the cluster of offerings to topological dynamics. So successful, in fact, were Whyburn's efforts at Virginia perceived in the wider mathematical community that he was actively courted in the early 1940s for headships at Ohio State, Purdue, and Washington University of St. Louis and even for the deanship of the Graduate School at the University of Iowa.[46] As this list suggests, state-supported, land-grant, and other schools in the Midwest actively worked to build and strengthen their research profiles in mathematics in the late 1930s and into the 1940s. Exactly why this was the case is not yet clear, but it *is* clear that the

[43] On the early Hopkins program under the English mathematician, James Joseph Sylvester, see [Parshall 1988] and [Parshall and Rowe 1994], pp. 53-146.

[44] Gordon Whyburn to Robert L. Moore, 10 May, 1934, Box 4RM76, Folder: "Whyburn, Gordon Thomas (1927-1968)" in [Moore].

[45] Gordon Whyburn to Robert L. Moore, 18 November, 1934, Box 4RM76, Folder: "Whyburn, Gordon Thomas (1927-1968)" in [Moore]. The quotes in the next paragraph are also from this letter.

[46] See Gordon Whyburn to Robert Moore, 15 July, 1944, Box 4RM76, Folder: "Whyburn, Gordon Thomas (1927-1968)" in [Moore].

emphasis in American higher education was increasingly on the development of *research* universities in the 1930s and 1940s, in which graduate training in general and graduate training in mathematics in particular were deemed necessary components.

This growing emphasis on research and graduate training—coming from university administrators above as well as from within the faculties below—increasingly clashed with the employment realities of most with higher mathematical training, however. They were employed as teachers of mathematics in junior colleges, four-year colleges, normal schools, and universities. They earned their livings as teachers, and, to the extent to which they could do it, research was done on the side.

The implications of this were apparent in a statistical study that Roland Richardson did on "The Ph.D. Degree and Mathematical Research" in 1935. Of the 4,444 teachers of mathematics he documented, just under 30% had earned a doctoral degree, although that figure rose to 44% when the teaching institutions were limited to the 285 colleges and universities approved by the Association of American Universities.[47] The proportion of Ph.D.s actually making "substantial contributions to research," which Richardson defined modestly as "the publication of three or more research articles," was not more than one third. He gauged, moreover, that not more than one fifth had "really been consistently productive in their contributions" and that "[a]bout 60 (or 5%) of the doctors are responsible for half the published pages of research." With teaching loads at most universities of twelve or more instructional hours a week, and with severely limited resources for research at all but those institutions most involved in graduate training, it was decidedly difficult for the majority of America's Ph.D.s in mathematics to maintain an active research program.

This employment profile presented problems in the 1930s as the Depression stretched on and as pressure increased in the United States to find positions for academics and other intellectuals who were affected by events in Europe. Some Europeans, like the Hungarian mathematical prodigy John von Neumann, had come to the United States from Germany before April 1933 when non-Aryans were expelled from the German universities. First as a visiting lecturer at Princeton for the 1930–1931 academic year, then as a regular faculty member there from 1931 to 1933, and finally as one of the first professors at the Institute for Advanced Study from its opening in 1933, von Neumann landed in the United States before the escalation of the crisis in Europe and in the best research environments for mathematics that the nation had to offer.

Those who tried to relocate beginning in 1933 faced much greater difficulties, despite the organization in 1933 of the Emergency Committee for Displaced German (later "European") Scholars and of financial support made available by the Rockefeller Foundation. The Emergency Committee served as a kind of scholarly

[47]See [Richardson 1936], pp. 205 and 207 (or pp. 368 and 370 in the reprinted edition). For quotations that follow in this paragraph, see *ibid.*, p. 210 (or p. 373).

clearinghouse, working with institutions to place available scholars. The idea was that it would supply a grant, often for two years, for a particular individual. That grant would then be matched by the institution or by the Rockefeller Foundation. These individuals, moreover, would be senior, distinguished researchers, and the positions would be research- as opposed to teaching-oriented. The Committee's hopes were that 1) the institution would absorb the individual into its faculty at the end of the two-year period and 2) by scattering the displaced scholars to as many different institutions as possible, the scholars would both be better assimilated into American culture and raise the research profile of a maximal number of American institutions. Oswald Veblen, as a member of the committee, firmly believed in these ideals. Together with Richardson as AMS Secretary, he worked through the committee as well as through the AMS to bring mathematicians to American shores and to place them in suitable academic positions.

One of the earliest mathematical emigrées to secure a position under this system was the German algebraist, Emmy Noether.[48] Widely heralded as one of the creators of so-called modern algebra, Noether had earned a doctoral degree from the University of Erlangen in 1908, but only after a law had been enacted in 1904 that finally permitted women to matriculate and sit for examinations there with the same rights and privileges as male students. In order to teach at the German universities, however, candidates required the *Habilitation* and when Noether applied for that in Göttingen in 1915, it was denied her on account of her sex.[49] Noether had come to Göttingen at the invitation of Felix Klein and David Hilbert, the former the doyen of late-nineteenth-century German mathematics and the latter widely regarded as the leading mathematician of the early twentieth century. Both men, recognizing her talent, worked, in vain until 1923, to secure for her an actual salaried position.

In Göttingen, Noether inspired a loyal following of students through her courses and seminars on her evolving ideas, but that ended in April of 1933 when, as a Jew, her right to teach in Göttingen—or, indeed, anywhere in Germany—was withdrawn. Through the agency of the Emergency Committee and the Rockefeller Foundation, Noether was appointed at all women's Bryn Mawr College effective 2 October, 1933 and arrived in the United States to take up her position shortly thereafter. With only advanced students to teach and with both Princeton University and the newly opened School of Mathematics at the Institute for Advanced Study also in nearby Princeton, she was able to benefit from the stimulation both of bright students at Bryn Mawr and of the joint research seminars held in Princeton by the faculties of the University and the Institute. The spring of 1935 would have found her at the side of her first American Ph.D., Ruth Stauffer, as she graduated from Bryn Mawr had fate not intervened. Noether died unexpectedly from complications after surgery to remove an ovarian cyst in April of 1935. One can only imagine the implications for the graduate training in mathematics of women

[48]On Noether's life and career, see, for example, [Brewer and Smith 1981] and [Dick 1981].

[49]Noether finally received the *Habilitation* along with the *venia legendi*, or official right to lecture, in 1919.

on American shores had Noether's life not been cut short, but it is clear that in her short time at Bryn Mawr, she was well on her way to realizing the Emergency Committee's noble aims.

While, as Noether's example suggests, it may have been possible to realize such idealistic goals, the financial realities of higher education in the 1930s and the paucity of actual research positions sometimes resulted in feelings of resentment—of a protectionist or even an anti-Semitic tinge—by Americans saddled with heavy teaching loads and little to no support for research.[50] Griffith Evans expressed the sentiments of many when he wrote to Veblen just months before taking on his new post as chair of the Department of Mathematics at Berkeley.

Veblen and Richardson had been working hard to relocate Richard Courant, a talented applied mathematician, an even more gifted academic administrator, and the then head of the Rockefeller Foundation-funded Mathematics Institute at Göttingen. Well aware of Berkeley's move to build its program in mathematics, Veblen naturally thought of the West Coast school as a soft landing place for Courant and approached Evans with the idea. Evans's response was categorical. "My own desire," he wrote, "would be that Professor C should not be appointed permanently at Berkeley, this without any reference to ability as a mathematician or personality. ...To say that there are too many foreigners in American universities is not chauvinism, but merely, that the careers of promising students in America are being cut off at the top. I do not see how this can be anything but an unfavorable situation in which to develop intellectual life. A generation ago we were in need of direct stimulation and there was plenty of room; now we could well interchange."[51]

Evans may have been right about the American mathematical research community's maturity in the mid 1930s in a number of pure sub-specialties, but he was wrong relative to the more applications-oriented type of mathematics done by Courant and a number of other leading European refugees. Although Courant was ultimately offered a temporary position at Berkeley, he declined it when the temporary position he already held at New York University evolved into a permanent one. Thanks to the strong recommendations of Veblen and others associated with both the Emergency Committee and the Rockefeller Foundation, NYU's Chancellor was sold on Courant's ability to build up a strong graduate program in mathematics there. Courant set to work immediately, hiring James Stoker, then an assistant professor of mechanics at the Carnegie Institute of Technology, and Kurt Friedrichs, a former student of Courant's at Göttingen, a refugee from Nazi Germany, and another mathematician of an applied bent.[52] By 1941 and the United

[50]Much has been written, both in general and relative to mathematics in particular, about the so-called "intellectual migration" from Europe to the United States and elsewhere in the 1930s and 1940s. The account here draws primarily from [Fleming and Bailyn 1968], [Fermi 1971], [Reingold 1981], and [Siegmund-Schultze 2009].

[51]Griffith Evans to Oswald Veblen, 16 January, 1934, Box 4, Folder: "Courant, Richard 1923–38" in [Veblen].

[52]Although Friedrichs was not Jewish, his fiancée and later wife was. It was thus impossible

States's entry into World War II, Courant was well on the way to establishing a major graduate center for applied mathematics in New York City.[53]

Back in Berkeley, Evans had recognized the short-sightedness of his earlier position on the influx of mathematicians from Europe. By 1942, he had, in fact, hired four Europeans who had been affected in one way or another by political events abroad. Hans Lewy, a student of Courant's, came in 1935. He was followed by the Polish statistician, Jerzy Neyman, in 1938; the Polish logician and mathematician, Alfred Tarski, in 1942; and, also in 1942, Frantisek Wolf, a complex analyst from what later became the Czech Republic. These four were among the sixteen mathematicians whom Evans brought to Berkeley during the course of a fourteen-year chairmanship that ended in 1949. In all, they directed some fifty-five students successfully to the Ph.D., an 83% increase over the previous fifteen-year period ([Rider 1989], pp. 291-295).

By capitalizing on a supportive administration, on world events, and on his own sense of both mathematical talent and the future directions of the field, Evans had "realized the ambition of a major center of mathematical research and teaching at Berkeley" ([Rider 1989], p. 296). In so doing, he had proven the wisdom of the strategy employed by the Emergency Committee and the Rockefeller Foundation in placing displaced scholars. In all, just over eighty German-speaking mathematicians, among them at least five women, emigrated to the United States during the Nazi period ([Siegmund-Schultze 2009], pp. 343-357) and contributed to varying degrees to the American mathematical research endeavor.

At least symbolically, that research endeavor was fifty years old at the time these European emigrés were resettling in the United States. An American mathematical research community centered on the activities of the AMS had, in fact, been founded in 1888 and celebrated its semicentennial in 1938 at a four-day-long gala held in September at Columbia University. Ten years (!) in the planning, it was a time of great fanfare and self-congratulation.[54]

Over 400 AMS members and some 700 people in all attended the celebration. They came from all over the country—204 from the North Atlantic, seventy-eight from the Midwest, sixty-five from New England, thirty-nine from the South, thirteen from the Rockies and West, sixteen from Canada, and three from abroad—to fête fifty years of American mathematical research. There were toasts, a lavish dinner, excursions up the Hudson River and to Long Island, over a hundred research talks by mathematicians from coast to coast, and ten plenary lectures that aimed to sketch developments in a variety of mathematical areas. As President of the Society, the Texas topologist, Robert L. Moore, set the tone for the festivities

for them to remain in Germany.

[53]For a non-technical biography of Courant, see [Reid 1976]. Another early American center of applied mathematics was animated by Roland Richardson at Brown University. For a first-person account of it by mathematician and history of mathematics, Clifford Truesdell, see [Rowe 2012], especially pp. 31-37. For a comparative study of the situation of applied mathematics in Germany and the United States in World War II, see [Siegmund-Schultze 2004], pp. 803-807.

[54]For accounts of the meeting, see [Hollcroft 1938] and [Richardson et al. 1939]. The description that follows draws primarily from these sources.

in opening remarks that highlighted what he saw as America's mathematical "rise" over the preceding fifty years ([Richardson et al. 1939], p. 4).

To a certain extent, contextualizing the advances that American mathematicians had been making and were continuing to make to (primarily) pure mathematical research had been the charge of the plenary speakers, who had been chosen from North America's best mathematicians, and the goal of the research presentations that augmented each of the plenary sessions. Those sessions highlighted developments in, among other areas, algebra, differential equations, geometry, harmonic analysis, topology, and even the more applied subfield of hydrodynamic stability. Almost all of them had boosteristic overtones, but no two more than Eric Temple Bell's "Fifty Years of Algebra in America, 1888–1938" and George Birkhoff's "Fifty Years of American Mathematics."[55] Both told largely triumphal stories—explicitly citing the work of their fellow Americans—of the research productivity and world-class accomplishments of the community defined by the AMS.

If these talks represented the intellectual capstone of the festivities, the social capstone was what was described as the "gala dinner" in the "beautiful banquet room of the Hotel Astor" ([Richardson et al. 1939], p. 7). There, congratulatory letters were read from U. S. President Franklin Delano Roosevelt and Canadian Prime Minister W. L. MacKenzie King as well as from the London Mathematical Society and other learned groups both at home and abroad. Roosevelt's letter must have been particularly gratifying, since it reflected an appreciation of the message that the leadership of the AMS—men like Veblen, Birkhoff, Richardson, and others—had been trying to get across since at least the 1920s. "It is sometimes difficult," Roosevelt admitted, "to comprehend the values accruing to society from mathematics." Yet, he acknowledged, "[i]t is deeply rooted in social progress as a large part of the technological advance made in recent centuries would have been impossible had it not been for the constant refinement of the essential tool of technology—mathematics. Social sciences also owe a large debt to mathematics. Future advances in the social sciences will be largely dependent upon mathematical treatment of their data" ([Richardson et al. 1939], p. 10).

In point of fact, neither Roosevelt nor his ghost writer understood mathematical research. It was too esoteric. It was beyond the ken of those without the necessary training and expertise. The general public could, however, accept that mathematics of some sort underlay such spectacular interwar, technological and engineering feats as the Empire State Building (1929–1931), the Hoover Dam (1931–1936), and the Golden Gate Bridge (1933–1937). And, Roosevelt could appreciate that the mathematical sciences could help to tame the deluge of data generated in attempts to understand economic and social trends like the study, *Recent Social Trends in the United States*, commissioned by U. S. President Herbert Hoover in 1929 and published in two volumes in 1933 ([President's Research Committee 1933]).

[55]For the lectures themselves, see [Semicentennial Addresses 1938]. See [Parshall 2000], pp. 390-395 for the lectures in historical overview. Raymond C. Archibald also gave a plenary address on the history of the AMS, but his findings were published as the free-standing book [Archibald 1938].

It was in these ways that mathematics had been incorporated into the popular perception by the 1930s. And, it may well be that this addresses at least one aspect of the seeming paradox of the Depression-era building of research departments of mathematics in the United States.

Guiding "the greatest group of mathematicians in the world": 1939–1950

The American mathematical research community entered its second half-century in a buoyant mood. "The center of gravity of mathematics has moved more definitely toward America," Roland Richardson stated with apparent satisfaction in 1939.[56] With a strong and vibrant community as reflected in the AMS, with the continuing absorption of first-rate mathematicians from Europe, and with the first International Congress of Mathematicians to be held in the United States just months off, the United States seemed poised, at long last, to wrest mathematical hegemony from Germany. Attention was fundamentally redirected, however, by the German invasion of Poland on 1 September, 1939 and by the subsequent declarations of war on Germany by both France and the United Kingdom.

Berkeley's Griffith Evans had been in the presidency of the AMS for just nine months on 1 September, 1939. Days after the invasion, he and some 350 of his mathematical colleagues convened at the University of Wisconsin in Madison for the AMS's forty-fifth summer meeting, deeply aware of what the minutes officially described as "the distressful world situation" ([Ayers 1939], p. 804). In Madison, in fact, it was resolved that the International Congress of Mathematicians that had been scheduled to take place in Cambridge, Massachusetts in September of 1940 would have to be postponed. It was clear that it would not be business as usual. Then, the Japanese bombed Pearl Harbor in December 1941. America's mathematical community, like the country at large, went to war.

It is fair to say that World War I had witnessed little or no *concerted* mobilization of America's mathematicians. When the U. S. had entered that war in 1917, some American mathematicians enlisted and served on active duty at the front. Those who used their mathematical expertise in the war effort did so primarily in ballistics at the Aberdeen Proving Ground or as basic mathematics instructors for the military.[57] Between the two wars, work in mathematical logic like that of Princeton's Alonzo Church and his student, Alan Turing, had begun to suggest that deep, pure mathematical research might have fundamental applications to, among other things, computing and cryptanalysis. In World War I, decoding the Zimmermann telegram—that proposed a German-Mexican alliance against the United States—had required little more than literacy and the good fortune of purloining a code book from a sinking German cruiser. In a second world war,

[56]Roland G. D. Richardson to "Dear Colleagues," 25 April, 1939, Box 11, Folder: "Richardson, R.G.D. 1939" in [Veblen].

[57]See [Rothrock 1919] as well as [Archibald et al., to appear].

given the intervening advances in mathematics, science, and technology, things were not likely to be that "easy." As that second world war loomed, America's mathematicians anticipated that their expertise had the potential to contribute much more profoundly than it had in the 1910s, and they began self-consciously to mobilize.

As early as 1940, Evans and his presidential counterpart in the Mathematical Association of America (MAA), Cornell's Walter Carver, had formed what they termed a joint War Preparedness Committee (WPC). As they and their colleagues understood, the mobilization of mathematicians meant not only those actively engaged in research but also those involved in basic mathematics teaching, the fostering of the latter being the stated mission of the MAA.[58] The country might need solutions to specific mathematical problems, but it would also need armed forces with mathematical training adequate to the tasks it would be called on to perform. To these ends, the WPC, chaired by Marston Morse, Professor of Mathematics at the Institute for Advanced Study and soon to be President-elect of the AMS, made its first report to the AMS Council on 9 September, 1940. It had set three objectives for itself: 1) "[t]he solution of mathematical problems essential for military or naval science, or rearmament, 2) [t]he preparation of mathematicians for research essential for the preceding objective, and 3) "[t]he strengthening of undergraduate mathematical education to a point where it affords adequate preparation in mathematics for military and naval service of any nature."[59]

Each of these objectives translated administratively into a distinct subcommittee. The Subcommittee on Research, headed by Dunham Jackson of the University of Minnesota, was charged with receiving problems and assigning consultants to solve them either from among the members of the subcommittee or from among the mathematical community at large.[60] The Subcommittee on Preparedness for Research, led by Harvard's Marshall Stone, canvassed the nation's departments of mathematics to assess the extent to which appropriately sophisticated training in applied mathematics was available in the predominantly purist environment of mid-century American mathematics and to coordinate the teaching of required subjects.[61] While these two subcommittees largely reflected the research orientation of the AMS, the third, the Subcommittee on Education for Service, centered on the teaching focus of the MAA. Under William Hart of the University of Minnesota, this third subcommittee reviewed mathematical materials in use in the

[58]On the differentiation between and stratification of the AMS and the MAA following the MAA's founding in 1915, see [Parshall 2015].

[59]"Minutes of Council: 9 September, 1940," Ms.75.2, Box 15, Folder 29: "War Preparedness Committee" in [AMS Papers]. See also [Morse 1940], p. 711.

[60]The subcommittee's other members were E. J. McShane (University of Virginia), Marshall Stone (Harvard), John H. Van Vleck (Wisconsin), Norbert Wiener (MIT), and Samuel Wilks (Princeton). For the full list WPC members and their subcommittee or consulting assignments, see [Ayers 1940], p. 182.

[61]The subcommittee's other members were Bernard Koopman (Columbia), R. E. Langer (Wisconsin), Hans Lewy (Berkeley), Frank D. Murnaghan (Johns Hopkins), and Howard P. Robertson (Princeton).

nation's service schools and made recommendations on the mathematical backgrounds "*desirable*," as Hart put it, "for workers in government and industry, and for officers and enlisted men in the Army and Navy."[62] These subcommittees were supplemented by a distinguished corps of "chief consultants": from the Institute for Advanced Study John von Neumann for ballistics; CalTech's Harry Bateman for aeronautics; MIT's Norbert Wiener for mechanical and electrical aids to computation; Thornton Fry of Bell Labs for industry; Howard Engstrom of Yale for cryptanalysis; and Princeton's Samuel Wilks for statistics and probability. At least as the mathematical societies saw it, *these* would be the main areas in which mathematical expertise would likely be brought to bear on a war effort.

By June of 1941, an additional subcommittee on "Supply and Demand for Mathematicians" had been added to the WPC. Chaired by Tracy Thomas of UCLA, it aimed to address what was increasingly becoming the critical issue of the deferment of mathematicians.[63] Late in the fall of 1940, and with the endorsement of General Lewis Hershey, the Director of the Selective Service System, the American Council on Education had issued a bulletin that went to local Selective Service Boards and that advised them "to give special consideration to men in chemistry, physics, engineering, biology, et cetera, with reference to the desirability of granting 'individual occupational deferment' " ([Morse 1941], p. 830). Mathematicians were conspicuously absent from this list of specialists judged critical and potentially in short supply should the United States enter the war. The new subcommittee's charge was thus to gather the data and to make the case that mathematicians should be added to the list of those considered for individual occupational deferments. By June of 1942, and thanks to the efforts of Thomas's WPC subcommittee, mathematicians were "listed among those engaged in 'critical occupations' considered essential to the support of the war effort."[64]

The initial omission of mathematics had been symptomatic, in the words of Marshall Stone, of "the failure of the various governmental agencies which are handling the scientific aspects of the present emergency to make effective use of the mathematical skills which are available in the country."[65] On 27 June, 1940, the National Defense Research Committee (NDRC) had been formed by presidential decree "to conduct research for the creation and improvement of instrumentalities, methods and materials of warfare" and thereby "to aid and supplement the experimental and research activities of the War and Navy Departments" ([Stewart 1948],

[62]The subcommittee's other members were R. S. Burington (Case School of Applied Science), Haskell B. Curry (Penn State), Elmer C. Goldsworthy (Berkeley), Frank L. Griffin (Reed College), M. H. Ingraham (Wisconsin), and Ernest J. Moulton (Northwestern). For the quote, see [Hart 1941], p. 354 (his emphasis).

[63]Other committee members were Harry Bateman (CalTech), E. T. Bell (CalTech), Griffith Evans (Berkeley), and William Whyburn (UCLA).

[64]J. R. Kline to Chairs of Departments of Mathematics, "Memorandum D. Occupational Classification of Mathematicians," 30 June, 1942, Ms.75.5, Box 28, Folder 42 in [AMS Papers]. See also [Anonymous 1942].

[65]Marshall Stone to Gilbert Bliss and Griffith Evans, 15 September, 1941, Box 38, Folder 16-17 in [Stone].

p. 8). Created as a result of the direct lobbying of President Roosevelt by Vannevar Bush, the President of the Carnegie Institution, it was comprised of eight men and aimed to provide an efficient link between the civilian scientific community and the military. Bush, an electrical engineer, was the committee's chair. Together with the Commissioner of Patents and one high-ranking representative each from the Navy and the Army, he served with four of his close friends: Richard Tolman, a physicist and Professor at CalTech; Karl Compton, a physicist and the President of MIT; James Conant, a chemist and the President of Harvard; and Frank Jewett, an electrical engineer and the President of both the National Academy of Sciences and Bell Labs. Thus dominated by engineers, chemists, and physicists, the NDRC, perhaps naturally, failed to place mathematics at the forefront of its agenda. Moreover, as Bush candidly put it in a 1970 memoir, "[t]here were those who protested that the action of setting up NDRC was an end run, a grab by which a small company of scientists and engineers, acting outside established channels, got hold of the authority and money for the program of developing new weapons. That, in fact, is exactly what it was" ([Bush 1970], pp. 31-32).

By June of 1941, moreover, the NDRC, which, as its title suggested, focused on *research* into new weaponry, had, by yet another presidential decree, been subsumed by a newly created Office of Scientific Research and Development (OSRD), also under Bush and with the former NDRC members prominently placed in the chain of command. The OSRD had a much broader purview than the NDRC, encompassing not only weapons research but also military medicine and, especially, the development aspect of turning basic research into actual implementations for use in the field. The OSRD also had almost unlimited power and resources, with Bush reporting directly to Roosevelt and having what was essentially an open Federal bank account on which to draw ([Stewart 1948], pp. 35-40). This was a new playing field, one different from that on which Veblen had so successfully competed in the 1920s. This involved dealings *not* with a foundation of relatively limited scope but with nothing less than the Federal government. At stake was support *not* on the order of magnitude of tens of thousands of dollars per year but on that of the U. S. Treasury.

Now enter Marshall Stone. Stone had matriculated at Harvard in 1919 at the young age of sixteen and had earned his doctorate there seven years later under George Birkhoff. By the 1930s, he found himself back at Harvard in a professorship and, in 1938, was elected to the National Academy of Sciences at the age of only thirty-five for his seminal work in functional analysis and Boolean algebra. Son of the then Chief Justice of the U. S. Supreme Court, Harlan Stone, Marshall moved naturally through many of the corridors of influence in the capital. On leave from Harvard and based in Washington for much of the war, he used that access in the service of American mathematics and, in so doing, took mathematical statesmanship to a new level.[66]

[66]For more on Stone and his activities during and immediately after World War II, see [Parshall 2009].

Stone quickly sized things up, recognizing the implications for mathematics should it not have a place at what was a newly developing Federal table. He approached AMS President Morse about the matter of establishing lines of communication between the AMS's War Preparedness Committee and the growing Washington scientific bureaucracy. In March 1942, he and Morse met with Bush, Conant, and Jewett, and they were frustrated by the outcome.[67] By August, Morse was writing in exasperation to Stone that "[t]he situation is not yet clear with regard to whether or not [the NDRC] will use our committee. My conclusion is either that Conant is *too busy to use us*, or that his subordinates ...*don't want to use us*, or that the Army and [N]avy *don't wish to go further* with the deeper investigations."[68] Unbeknownst to Morse and Stone, Conant and Bush were, indeed, preoccupied—with the Manhattan Project—but by November 1942, an Applied Mathematics Panel *had* officially been added to the government's growing scientific bureaucracy directly to coordinate the work of mathematicians with the needs of the military. Mathematics was thus finally and officially recognized at the highest levels as relevant, if perhaps not as *critically* relevant, to the war effort.

Stone succeeded Morse in the AMS presidency on New Year's Day in 1943, just days after he had been authorized by the AMS Council to replace the joint AMS-MAA War Preparedness Committee by a War Policy Committee (WPC) to "handle all problems which concern the most effective use of mathematicians in the war effort" as well as to "study means for alleviating the serious shortage of teachers of mathematics which will confront colleges and universities" ([Kline 1943], p. 199). A month later, Stone was able to report positively on discussions he had already had with the War Manpower Commission, noting that it was likely that the manpower committee on physicists would be enlarged to a committee on physicists *and* mathematicians.[69] Partly at issue was staffing courses at the nation's colleges for the approximately 90,000 men from the Army, 40,000 from the Navy, and 70,000 from the Air Corps, who required immediate, basic mathematical training in order effectively to carry out their specialized military duties.[70]

To meet these needs, a new WPC subcommittee "on Available Teachers of Collegiate Mathematics" was formed with Oberlin's William Cairns as chair.[71]

---

[67]Larry Owens details this meeting in [Owens 1989], p. 294. Although Owens states that the memorandum was co-authored by Stone, Bliss, Evans, and Jackson, correspondence in [Stone] makes it clear that Stone was the author and that he asked for and secured the approval of the other three mathematicians. See Box 37, Folder 3 "Scientific Mobilization," in [Stone] and specifically letters exchanged in January and February of 1942.

[68]Marston Morse to Marshall Stone, 14 August, 1942, Box 37, Folder 3: "Scientific Mobilization" in [Stone] (his emphases).

[69]Although this happened two months later in April, the joint committee was soon dissolved, and deferments continued to present problems for the duration of the war.

[70]For the numbers, which came from the Army and the Navy at the inauguration of their training programs, see J. R. Kline, Minutes of the War Policy Committee, 21 February, 1943, Ms.75.5, Box 28, Folder 110 in [AMS Papers].

[71]See [Hollcroft 1943], p. 341. The other members of the committee were Arnold Dresden (Swarthmore) and AMS Secretary J. R. Kline (Pennsylvania). For the numbers, which came from the Army and the Navy at the inauguration of their training programs, see J. R. Kline,

This committee's existence represented further recognition of the role of mathematics in wartime, albeit the role of mathematics instruction as opposed to research expertise. It also highlighted a key intersection between mathematicians and the armed services.

The WPC actually only began to realize the originally desired role of mathematical coordinator—and, by extension, active scientific contributor—through the mobilization of first-rate mathematical problem-solvers toward the very end of the war. Perhaps as a result of Stone's earlier outreach to the Army and Navy, the WPC was approached by Major General Gladeon Barnes, Chief of the Research and Development Service of the Ordnance Department, about "appoint[ing] an Ordnance Advisory Committee which could be consulted by the Ordnance Department on mathematical problems ... [and which would] continue into the postwar period."[72] The committee—consisting of Marston Morse, Garrett Birkhoff (son of George), Griffith Evans, Herman Goldstine, Jimmy McShane, John von Neumann, and Charles B. Morrey—was in place before the end of April 1945. The WPC clearly could organize effectively if called upon.[73] Barnes's stipulation that this group continue into the postwar period undoubtedly had to do with the still secret, contemporaneous development of the first computer, ENIAC, partly under his aegis.

By the time Barnes made this overture, however, the war was almost over. The American mathematical community—through the Applied Mathematics Panel, through the AMS's War Policy Committee, and through the individual efforts of mathematicians at the Aberdeen Proving Ground, in the Army, in the Navy, and elsewhere—had made solid contributions to the war effort ([Rosser 1982] and [Rees 1980]), but the wartime AMS Presidents—and especially, Morse and Stone —had wanted more. They had wanted to help realize an effective mobilization of the country's *élite* mathematical talent and, in so doing, to effect solutions to problems that would contemporaneously as well as in retrospect be recognized as critical to American victories in both the European and Pacific theaters.[74] They recognized this as all the more important in 1945 as the Congress began to formulate bills for the creation of a federally funded foundation for the support of scientific research.[75]

Minutes of the War Policy Committee, 21 February, 1943, Ms.75.5, Box 28, Folder 110 in [AMS Papers].

[72]"Minutes of Council, 24 February, 1945," Ms.75.2, Box 15, Folder 83 in [AMS Papers].

[73]J. R. Kline to Major Gen. Gladeon Barnes, 23 March and 2 April, 1945, Ms.75.2, Box 15, Folder 87 in [AMS Papers].

[74]This sense of history was present early on. At the second meeting of the WPC on 11 July, 1943, Stone's call for "the desirability of gathering complete and reliable information concerning the war activities of mathematicians" was re-sounded. In his view, "[b]y keeping a careful and detailed record of his own activities and experience during the war period, each mathematician will contribute to the preservation of the data without which a record of this extraordinarily significant period in our history would be impossible." See C. C. MacDuffee, "Report of the War Policy Committee," 1 September, 1943, Ms. 75.2, Box 15, Folder 68: "War Policy Committee" in [AMS Papers].

[75]In fact, as a result of its wartime contributions to the Navy, mathematics was funded by the

The history of what finally became the National Science Foundation in 1950 is tortuous. Suffice it to say, however, that the mathematicians had finally won their battle for recognition, since the politicians drafting the various bills that would ultimately result in the NSF included the mathematical sciences among the areas to be supported from the beginning. Whereas Veblen had had to argue hard for the inclusion of mathematics in the NRC fellowship program in the 1920s, whereas the AMS had had to convince the government of the need for deferments for mathematicians as critical to the war effort in 1942, and whereas Stone had had to secure the enlargement of the War Manpower Commission to include mathematicians a year later in 1943, it was recognized by 1945 that mathematics commanded a place at the Federal table. Stone and Morse, the mathematicians as "scientific statesman" of the 1940s, may have failed to realize their most idealistic ambitions for mathematics, but at a more pragmatic level, they and their colleagues had succeeded in representing mathematics as a tenacious player in the new Federal game of science.

If that new game characterized the immediately postwar period of American mathematics, so, too, did the increasing recognition of the United States as the global center of gravity of mathematical activity. The decades of the 1920s and 1930s had witnessed a strengthening of the infrastructure for mathematics in the United States, a reorientation of a significant number of America's universities toward the active support of research, a significant amount of mostly pure mathematical results that had swollen the pages of the country's journals, and an influx of mathematicians from Europe talented in pure subfields already represented in the United States as well as in applied mathematics.

A newly re-amalgamated research community was in evidence in December of 1946 when the Princeton Department of Mathematics helped to celebrate the University's bicentennial with a conference ambitiously entitled "Problems of Mathematics." Almost a hundred mathematicians from the United States and abroad participated in the event which was intended to supersede nothing less than the famous lecture on twenty-three "Mathematical Problems" given by David Hilbert at the Paris ICM in 1900 [Hilbert 1901]. The three-day meeting was rife with symbolism. It took place in the United States. It was recognized as "the first international gathering of mathematicians in a long and terrible decade" ([Princeton 1946], p. 310). It debuted a unified American mathematical community consisting of new emigrés as well as "native" American talent. It reflected the centrality of what *was* that newly refashioned American mathematical research endeavor in its analysis of important open problems and the work suggestive of them.[76]

The Princeton conference, however, was but a small, localized event. Four years later, the International Congress of Mathematicians met in Cambridge, Massachusetts. What had been the symbols of Princeton had become the reality of

Office of Naval Research from its inception in 1946.

[76]For an overview of the mathematical content of the conference, see [Parshall 2000], pp. 396-400. David Rowe analyzed this conference as well in [Rowe 2003].

the passing of the mathematical torch from Europe to the United States. In the words of Raymond Wilder, the American mathematical research community had come to constitute by 1950 "the greatest group of mathematicians in the world."[77]

Conclusion

The story told here is ultimately a success story, but one of success in the face of many obstacles. At numerous points along the way, things could have turned out dramatically differently. In the 1920s, the Rockefeller Foundation might have decided *not* to provide significant support to the NRC; the most talented recent Ph.D.s in the sciences and in mathematics would then *not* have been given the research "leg up" provided by the NRC fellowship program that allowed them to establish active research programs. In the 1930s, in the depths of the Depression, a significant number of university administrators might *not* have recognized research as the wave of the future in higher education and *not* made the commitment to reorient their institutions in that direction; academic departments, among them mathematics, thus might *not* have had the institutional support self-consciously to hire committed research faculty. Also in the 1930s, the Rockefeller Foundation might have chosen *not* to direct significant amounts of money toward placing displaced European scholars; those eminent emigrés thus might *not* have been spread throughout the country in departments reorienting toward research. The entire thirty-year period from the 1920s through 1950 might *not* have witnessed the "capitalization" of the sciences in general and of mathematics in particular via, first, the foundations and, then, the Federal government; this would have *removed* a major impetus toward the fashioning of a new kind of leader, namely, the research mathematician as savvy statesman. Viewed in this light, a fascinatingly complex web of contingencies may be seen to characterize the history of the American mathematical research community between the two World Wars.

Acknowledgments

I would like to thank Carol Mead, archivist of the Archive of American Mathematics at the Dolph Briscoe Center for American History of the University of Texas, Austin, as well as the staffs at the Harvard University Archives, the Hay Library at Brown University, and the Library of Congress for all of their help. David Rowe provided helpful suggestions as did my colleague and economic historian, John James. Joe Kett and Albert Lewis have been constant sources of support, guidance, and enlightenment on this project.

[77]Raymond Wilder to Saunders Mac Lane, 16 February, 1946, Box 86-36/28: Folder 13: "AMS Records, 1922–1948" in [Wilder].

Archival Sources

[AMS Papers]: American Mathematical Society Papers, John Hay Library, Brown University, Providence, Rhode Island.

[Birkhoff]: George David Birkhoff Papers, Harvard University Archives, Pusey Library, Cambridge, MA.

[Moore]: R. L. Moore Papers, 1875, 1891–1975, Archives of American Mathematics, Dolph Briscoe Center for American History, The University of Texas at Austin.

[Stone]: Marshall H. Stone Papers, John Hay Library, Brown University, Providence, Rhode Island.

[Veblen]: The Papers of Oswald Veblen, Library of Congress Manuscripts Division, Washington, D.C.

[Wilder]: Raymond L. Wilder Papers, 1914–1982, Archives of American Mathematics, Dolph Briscoe Center for American History, The University of Texas at Austin.

Bibliography

[Anonymous 1919] Anonymous. "The National Research Council." *Science* 49 (16 May, 1919): 458–462.

[Anonymous 1924a] Anonymous. "Notes." *Bulletin of the American Mathematical Society* 30 (1924): 376-378.

[Anonymous 1924b] Anonymous. "Sustaining Members of the Society, December, 1924." *American Mathematical Society: List of Officers and Members, 1923–1924* 30 (1924): 7.

[Anonymous 1925] Anonymous. "Notes." *Bulletin of the American Mathematical Society* 31 (1925): 470-475.

[Anonymous 1926] Anonymous. "Sustaining Members of the Society, October, 1926." *List of Officers and Members of the American Mathematical Society* 32 (1926): 9.

[Anonymous 1935] Anonymous. "Geographical Distribution of Members." *Bulletin of the American Mathematical Society* 40 (1934): 50-64.

[Anonymous 1942] Anonymous. "Occupational Classification of Mathematicians." *Bulletin of the American Mathematical Society* 48 (1942): 792-794.

[Archibald 1938] Archibald, Raymond C. *A Semicentennial History of the American Mathematical Society, 1888–1938.* New York: American Mathematical Society, 1938; Reprint ed., New York: Arno Press, 1980.

[Archibald et al., to appear] Archibald, Thomas; Dumbaugh, Della; and Kent, Deborah. "A Mobilized Community: Mathematicians in the United States during World War I." In *The War of Guns and Mathematics.* Ed. David Aubin and Catherine Goldstein. HMATH. Providence: American Mathematical Society, to appear.

[Aspray 1988] Aspray, William. "The Emergence of Princeton as a World Center for Mathematical Research, 1896–1939." In *History and Philosophy of Modern Mathematics.* Ed. William Aspray and Philip Kitcher. Minneapolis: University of Minnesota Press, 1988, pp. 346-366.

[Ayers 1939] Ayers, William L. "The September Meeting in Madison." *Bulletin of the American Mathematical Society* 45 (1939): 801-811.

[Ayers 1940] Ayers, William L. "The Annual Meeting of the Society." *Bulletin of the American Mathematical Society* 47 (1941): 175-187.

[Barrows 1933] Barrows, Albert L. "General Organization and Activities." In *A History of the National Research Council, 1919–1933.* Washington, DC: National Research Council, 1933, pp. 7-11.

[Batterson 2006] Batterson, Steve. *Pursuit of Genius: Flexner, Einstein, and the Early Faculty at the Institute for Advanced Study.* Wellesley, MA: A K Peters, Ltd.

[Brewer and Smith 1981] Brewer, James W. and Smith, Martha K. *Emmy Noether: A Tribute to Her Life and Work.* New York: Marcel Dekker, 1981.

[Bush 1970] Bush, Vannevar. *Pieces of the Action.* New York: Morrow, 1970.

[Cattell 1906] Cattell, James McKeen, Ed. *American Men of Science: A Biographical Dictionary.* New York: The Science Press, 1906.

[Duren et al. 1988–1989] Duren, Peter L., *et al. A Century of Matheamtics in America.* 3 Vols. Providence: American Mathematical Society, 1988–1989.

[Dick 1981] Dick, Auguste. *Emmy Noether, 1881–1935.* Basel: Birkhäuser Verlag, 1981.

[Feffer 1997] Feffer, Loren Butler. "Mathematical Physics and the Planning of American Mathematics: Ideology and Institutions." *Historia Mathematica* 24 (1997): 66-85.

[Feffer 1998] Feffer, Loren Butler. "Oswald Veblen and the Capitalization of American Mathematics: Raising Money for Research, 1923–1928." *Isis* 89 (1998): 474-497.

[Fenster 2003] Fenster, Della Dumbaugh. "Funds for Mathematics: Carnegie Institution of Washington Support for Mathematics from 1902 to 1920." *Historia Mathematica* 30 (2003): 195-216.

[Fermi 1971] Fermi, Laura. *Illustrious Immigrants: The Intellectual Migration from Europe 1930/41.* Chicago: University of Chicago Press, 1971.

[Fleming and Bailyn 1968] Fleming, Donald and Bailyn, Bernard, Ed. *The Intellectual Migration: Europe and America, 1930–1960.* Vol. 2. *Perspectives in American History* (1968).

[Hart 1941] Hart, William L. "On Education for the Service." *American Mathematical Monthly* 48 (1941): 353-362.

[Henderson 1925] Henderson, Robert. "Life Insurance as a Social Science and as a Mathematical Problem." *Bulletin of the American Mathematical Society* 31 (1925): 227-252.

[Hilbert 1901] Hilbert, David. "Mathematische Probleme." *Archiv für Mathematik und Physik* 1 (1901): 44-63. Trans. Mary F. Winston. "Mathematical Problems: Lecture Delivered before the International Congress of Mathematicians at Paris in 1900." *Bulletin of the American Mathematical Society* 8 (1902): 437-479.

[Hollcroft 1938] Hollcroft, Temple R. "The Summer Meeting in New York." *Bulletin of the American Mathematical Society* 44 (1938): 745-755.

[Hollcroft 1943] Hollcroft, Temple R. "The February Meeting in New York." *Bulletin of the American Mathematical Society* 49 (1943): 339-344.

[Kaplan 1989] Kaplan, Wilfred. "Mathematics at the University of Michigan." In [Duren et al. 1988–1989], 3: 179-189.

[Kline 1935] Kline, John R. "The Annual Meeting in Pittsburgh." *Bulletin of the American Mathematical Society* 41 (1935): 150-161.

[Kline 1943] Kline, John R. "The Annual Meeting of the Society." *Bulletin of the American Mathematical Society* 49 (1943): 195-204.

[Kohler 1991] Kohler, Robert. *Partners in Science: Foundations and Natural Scientists 1900–1945.* Chicago: University of Chicago Press, 1991.

[Morse 1940] Morse, Marston. "Report of the War Preparedness Committee of the American Mathematical Society and Mathematical Association of America at the Hanover Meeting." *Bulletin of the American Mathematical Society* 46 (1940): 711-714.

[Morse 1941] Morse, Marston. "Report of the War Preparedness Committee of the American Mathematical Society and Mathematical Association of America at the Chicago Meeting." *Bulletin of the American Mathematical Society* 47 (1941): 829-831.

[Owens 1989] Owens, Larry. "Mathematicians at War: Warren Weaver and the Applied Mathematics Panel, 1942–1945." In *The History of Modern Mathematics: Vol. 2, "Institutions and Applications.* Ed. David E. Rowe and John McCleary. Boston: Academic Press, 1989, pp. 287-305.

[Parker 2005] Parker, John. *R. L. Moore: Mathematician and Teacher.* Washington, DC: The Mathematical Association of America, 2005.

[Parshall 1988] Parshall, Karen Hunger. "America's First School of Mathematical Research: James Joseph Sylvester at The Johns Hopkins University 1876–1883." *Archive for History of Exact Sciences* 38 (1988): 153-196.

[Parshall 2000] Parshall, Karen Hunger. "Perspectives on American Mathematics." *Bulletin of the American Mathematical Society* 37 (2000): 381-405.

[Parshall 2009] Parshall, Karen Hunger. "Marshall Stone and the Internationalization of the American Mathematical Research Community." *Bulletin of the American Mathematical Society* 46 (2009): 459-482.

[Parshall 2015] Parshall, Karen Hunger. "The Stratification of the American Mathematical Community: The Mathematical Association of America and the American Mathematical Society, 1915–1925." In *A Century of Advancing Mathematics.* Ed. Donald Albers, *et al.* Washington, DC: Mathematical Association of America, 2015, to appear.

[Parshall, in preparation] Parshall, Karen Hunger. "Mathematics and the Politics of Race: The Case of William Claytor (Ph.D., University of Pennsylvania, 1933)," manuscript in preparation.

[Parshall and Rowe 1994] Parshall, Karen Hunger and Rowe, David E. *The Emergence of the American Mathematical Research Community, 1876-1900: James Joseph Sylvester, Felix Klein, and E. H. Moore.* HMATH 8. Providence: American Mathematical Society and London: London Mathematical Society, 1994.

[President's Research Committee 1933] President's Research Committee on Social Trends. *Recent Social Trends in the United States: Report of the President's Research Committee on Social Trends.* 2 Vols. New York and London: MacGraw-Hill Book Company, Inc., 1933.

[Princeton 1946] "The Princeton University Bicentennial Conference on the Problems of Mathematics." In [Duren et al. 1988–1989], 2: 309-333.

[Pupin 1924] Pupin, Michael. "From Chaos to Cosmos." *Scribner's Magazine* 76 (1924): 3-10.

[Rees 1980] Rees, Mina. "The Mathematical Sciences and World War II." *American Mathematical Monthly* 87 (1980): 607-621; Reprinted in [Duren et al. 1988–1989], 1: 275-289.

[Reingold 1981] Reingold, Nathan. "Refugee Mathematicians in the U.S.A." *Annals of Science* 38 (1981): 313-338; Reprinted in [Duren et al. 1988–1989], 1: 175-200.

[Reid 1976] Reid, Constance. *Courant in Göttingen and New York: The Story of an Improbable Mathematician.* New York: Springer-Verlag, 1976.

[Reingold and Reingold 1981] Reingold, Nathan and Reingold, Ida Ed. *Science in America: A Documentary History 1900–1939.* Chicago: University of Chicago Press, 1981.

[Richardson 1923] Richardson, Roland G. D. "The Josiah Willard Gibbs Lectureship." *Bulletin of the American Mathematical Society* 29 (1923): 385.

[Richardson 1924] Richardson, Roland G. D. "The Thirtieth Annual Meeting of the Society." *Bulletin of the American Mathematical Society* 30 (1924): 199-216.

[Richardson 1926] Richardson, Roland G. D. "Report of the Secretary to the Council for the Years 1921–1925." *Bulletin of the American Mathematical Society* 32 (1926): 203-211.

[Richardson 1936] Richardson, Roland G. D. "The Ph.D. Degree and Mathematical Research." *American Mathematical Monthly* 43 (1936): 199-215; Reprinted in [Duren et al. 1988–1989], 2: 361-378.

[Richardson et al. 1939] Richardson, Roland G. D. et al. "The Semicentennial Celebration: September 6-9, 1938." *Bulletin of the American Mathematical Society* 45 (1939): 1-30.

[Richtmyer 1919] Richtmyer, F. K. "Division of Physical Sciences." In *A History of the National Research Council, 1919–1933.* Washington, DC: National Research Council, 1933, pp. 12-16.

[Rider 1989] Rider, Robin. "An Opportune Time: Griffith C. Evans and Mathematics at Berkeley." In [Duren et al. 1988–1989], 2: 282-302.

[Rosser 1982] Rosser, Barkley. "Mathematics and Mathematicians in World War II." *Notices of the American Mathematical Society* 29 (1982): 509-515; Reprinted in [Duren et al. 1988–1989], 1: 303-309.

[Rothrock 1919] Rothrock, David.

[Rowe 2003] Rowe, David E. "On Projecting the Future and Assessing the Past —the 1946 Princeton Bicentennial Conference." *The Mathematical Intelligencer* 25 (2003): 8-15.

[Rowe 2012] Rowe, David E. "Mathematics in Wartime: Private Reflections of Clifford Truesdell." *The Mathematical Intelligencer* 34 (2012): 29-39.

[Semicentennial Addresses 1938] *Semicentennial Addresses of the American Mathematical Society.* New York: American Mathematical Society, 1938.

[Siegmund-Schultze 1997] Siegmund-Schultze, Reinhard. "The Emanicipation of Mathematical Research Publishing in the United States from German Dominance (1878–1945)." *Historia Mathematica* 24 (1997): 135-166.

[Siegmund-Schultze 2001] Siegmund-Schultze, Reinhard. *Rockefeller and the Internationalization of Mathematics between the Two World Wars.* Basel: Birkhäuser Verlag, 2001.

[Siegmund-Schultze 2004] Siegmund-Schultze, Reinhard. "The Ideology of Applied Mathematics within Mathematics in Germany and the U. S. until the End of World War II." *LLULL* 27 (2004): 791-811.

[Siegmund-Schultze 2009] Siegmund-Schultze, Reinhard. *Mathematicians Fleeing from Nazi Germany: Individual Fates and Global Impact.* Princeton: Princeton University Press, 2009.

[Stewart 1948] Stewart, Irvin. *Organizing Scientific Research for War: The Administrative History of the Office of Scientific Research and Development.* Boston: Little, Brown and Company, 1948.

[Veblen 1925] Veblen, Oswald. "Remarks on the Foundations of Geometry." *Bulletin of the American Mathematical Society* 31 (1925): 121-141.

[Veblen 1952] Veblen, Oswald. "Opening Address." *Proceedings of the International Congress of Mathematicians.* 2 Vols. Providence: American Mathematical Society, 1952, 1: 124-125.

[Visher 1947] Visher, Stephen Sargent. *Scientists Starred 1903–1943 in "American Men of Science": A Study of Collegiate and Doctoral Training, Birthplace, Distribution, Backgrounds, and Developmental Influences.* Baltimore: The Johns Hopkins Press, 1947.

[Wilder 1932] Wilder, Raymond L. "Point Sets in Three and Higher Dimensions and Their Investigation by Means of a Unified Analysis Situs." *Bulletin of the American Mathematical Society* 38 (1932): 649-692.

[Wilder 1989] Wilder, Raymond L. "Reminiscences of Mathematics at Michigan." In [Duren et al. 1988–1989], 3: 191-204.

Traditional Chinese Mathematics

Joseph W. Dauben, Great Wall, Spring 1988

Some Distinctive Features of Ancient Chinese Mathematics: A Brief Introduction and Comparison with its Greek Counterpart

Liu Dun

Introduction

What distinctive features does ancient Chinese mathematics possess? To answer this question, it is helpful to make a comparison with another familiar system of knowledge from the past. The ideal candidate for this purpose is, without doubt, ancient Greek mathematics, which not only exemplifies a paradigm for mathematical works in the West but also represents the pinnacle of spiritual achievement in ancient Greece. Another reason for such a comparison, however, stems from mathematics education. For today we still encounter a standard myth: the claim that the remarkable achievements of contemporary mathematics are solely due to a certain "Royal Road" first pioneered by the ancient Greeks. Even a brief overview of the situation, however, will show that, compared with the Greeks, ancient Chinese mathematicians reached a high level of mathematical knowledge using quite different techniques. This essay describes some of these characteristic differences by highlighting some of the most striking qualities in the classical works of Chinese mathematics. These special aspects include a distinctive numeral system and calculation tools for working with it, but also a unique style of reasoning with so-called "constructed techniques and mechanized algorithms." Alongside these technical elements, Chinese mathematicians were guided by the standard set by the most famous of all Chinese works, the *Nine Chapters on the Art of Mathematics*. This celebrated classic played an exemplary role in guiding subsequent research, but also for promoting the strong utilitarian orientation within Chinese mathematics, one of the hallmarks of its longstanding tradition.

Chinese Numeral and Calculation Tools

So simple and deep-seated is the principle of the place-value system in decimal arithmetic that we often overlook the great significance this invention has had for the development of higher mathematics. Just imagine a different scenario: what if the numeral system we used today were based on the alphabetic letters introduced by Archimedes in *The Sand Reckoner* ([Heath 1912], lxix-lxxi, 221-232)? How much time and energy would a person have to devote to studying and memorizing these numbers just so they could learn to count, and this from day one of their education? It is possible, of course, to carry out simple elementary operations using the ancient Babylonian or Greek numeral systems, but, needless to say, more complicated operations would be a tremendous headache even for

D.E. Rowe and W.-S. Horng (eds.), *A Delicate Balance: Global Perspectives on Innovation and Tradition in the History of Mathematics*,
Trends in the History of Science, DOI 10.1007/978-3-319-12030-0_13

a well-trained accountant. In other words, the adoption of a reasonable numeral system was of vital importance for all mankind.

For a long time, Chinese mathematicians stood at the forefront when it came to the arithmetical and algebraic sciences. They were probably not smarter than their foreign peers, but they were surely luckier, for they had an unequalled edge because of their calculation tools based on a place-value system with base ten. Years of experience with counting-rod calculations not only gave the Chinese effective numerical technqiues but also led them quite naturally to the important mathematical method of separating coefficients. Their convenient numeral system and efficient counting tools were thus linked with some of the most universally acclaimed achievements of ancient Chinese mathematics. These include techniques not just for handling fractions, negative numbers, and decimal numbers, but also for finding ratios, common fractions, or for solving the rule of three as well as equations of higher degree, but also linear systems of equations. Further innovations were the methods of *Tianyuan* (天元術) (rule of polynomial equations) and *Siyuan* (四元術) (rule of simultaneous polynomial equations of high degree in several variables); algorithms for finding the greatest common divisor and least common multiple; and methods for determining congruence relations, to list just a few.

A culture that employs a clumsy numerical system and which lacks appropriate counting tools is unlikely to develop a sophisticated mathematical system. As for the origins of our modern number system, it was long popular to speak of the "Hindu-Arabic numeral system." However, Lam Lay Yong has pointed out that this system "could only have originated from the rod numeral system, which was developed centuries earlier. Transmission from China to India would not have been difficult, given the extensive use of rods in ancient China." She also argued that "like printing, gunpowder, and the magnet, three inventions which Francis Bacon said had 'changed the whole face and state of things throughout the world', the concept of our numeral system should rank as one of China's most significant contributions to human science and civilization" ([Lam 2004], 185).

Rich and Diverse Reasoning Methods

In contrast with the ancient Greeks, who mainly relied on deductive reasoning, ancient Chinese mathematicians followed no set patterns and employed a variety of reasoning methods.[1] Here, we should first debunk a myth that has become one of the most deeply entrenched viewpoints in contemporary philosophy of science and mathematics, namely the view that any knowledge not derivable by strict deduction cannot be attributed to exact science. This position would entail that much of physics and chemistry (not to mention biology), as we know these subjects today, do not belong to the exact sciences. Likewise, many branches of

[1] We note here, however, that the influential historian of Greek mathematics, Wilbur Knorr, underscored the importance of the older problem-solving tradition.

mathematics would also fail to qualify as exact science, surely an unreasonable conclusion. After all, it has been known since Gödel's work that even within a purely formal logical system it is impossible to deduce all of the propositions that happen to be compatible with the system.

Secondly, we should clarify the nature of ancient Chinese mathematics. Generally speaking, Chinese mathematics represents a system of knowledge that developed on the basis of practical problem solving. The methods were then analyzed, refined, and summarized as a set of theories and principles for solving a series of similar problems. We see this with Liu Hui's (劉徽, ca. 263 AD) work, which uses both the "out-in complementary principle",[2] as well as sums of infinite excisions. A similar approach can be seen in Zu Geng's (祖暅, ca. 500 AD) technique of comparing cross sections when tackling problems concerning volumes of various solids.[3] These examples reflect Chinese mathematicians' preliminary understanding, which preceded their axiom-like thinking. Liu Hui's annotation of the *Jiu zhang suan shu* (九章算術, *Nine Chapters on the Art of Mathematics*, ca. 50 AD, hereafter *Nine Chapters*) also indicates a strong inclination toward deductive reasoning [Guo 1983] and [Guo 1987], although in the constrained form of his annotations the text fails to unfurl fully its logical structure. Similarly, certain deductive features can also be found in the "Distinguishing Notes" in Li Zhi's (李冶, 1192-1279) *Ceyuan haijing* (測圓海鏡, *Sea Mirror of Circle Measurements*, 1248), but also in the "Ten Common Senses" of Li Shanlan's (李善蘭, 1811-1882) *Fangyuan changyou* (方圓闡幽, *Explanation of the Secrets in Squares and Circles*, 1845). These deductive methods are used, respectively, in the discussions of the problem of inscribing a circle in a right triangle (勾股容圓) and for finding an algorithm for the volume of a pyramid (尖錐術) [Wang 1983].

Nevertheless, in most ancient Chinese mathematics the results are presented in a non-deductive manner, using other forms of reasoning based on intuition, analogy, observation and induction, etc. Thus, for example, Liu Hui did not shy away from experience and intuition in making his arguments; his use of diagramm verification (圖驗法) and solid model verification (棊驗法) are cases in point. A glimmer of reasoning by analogy can already be seen in the chapter *Shanggong* (商功, construction consultations) of the *Nine Chapters.* Such reasoning was mastered and used dexterously by mathematicians like Sheng Kuo (沈括, 1031-1095), Yang Hui (楊輝, 13th century), Zhu Shijie (朱世傑, 1291-1314), Wu Jing (吳敬, 15th century), and Li Shanlan. Observation and induction were also often used by Chinese mathematicians in the course of their reasonings. Successful examples include Zhu Shijie's research on sums of triangles and on interpolations of high order [Du 1966], as well as Li Rui's (李銳, 1769-1817) research on the relations between the roots of an equation and its coefficients [Liu 1989]. For Chinese mathematicians, it was perfectly natural to combine forms with numbers and to employ such principles within their calculations (寓理於算): the methods for extracting

[2]The "out-in complementary" (出入相補) is a phrase from Liu Hui's annotation on *Nine Chapters*, see [Wu 1983].

[3]This is equivalent to Cavalieri's principle in geometry, see [Du 1954].

square and cube roots when solving higher-order equations were always derived from geometric models. The diagram demonstration (演段) that mathematicians of the Song and Yuan Dynasties took so much delight in elaborating was essentially no different from the ancient method of diagram verification, although these later mathematicians put more emphasis on explaining the calculations rather than emphasizing the associated geometric relations.

In general, we should not imagine that Chinese mathematics lagged behind because of a failure to produce a purely deductive system of knowledge. That fact that there was no counterpart to Euclid's *Elements* in ancient China does not justify the conclusion that ancient Chinese mathematicians lacked methods of logical reasoning or ignored proofs. On the contrary, they wielded an impressive array of techniques and reasoning methods.

Constructive Techniques and Mechanized Algorithms

If ancient Greek mathematicians took pride in discovering geometric theorems, their Chinese counterparts went about the task of constructing elaborate algorithms. Indeed, this was most characteristic feature of ancient Chinese mathematics, which developed a methodology based on grouping practical questions into different counting patterns by constructive means. These problems were then worked out using specific numerical solutions obtained by means of mechanized algorithms. No wonder the forebears of this Chinese tradition called mathematics "counting learning" (算學). These algorithms, however, varied in complexity, important examples being: *Jinyou* (今有, rule of three), *Qitong* (齊同, rule of homogenization and uniformization), *Kaifang* (開方, rule of extraction), *Fangcheng* (方程, rule of linear simultaneous equations), *Geyuan* (割圓, rule of circle division), *Gengxiang jiansun* (更相減損, rule of reciprocal subtraction, i.e. the Euclidean algorithm), *Zengcheng kaifang* (增乘開方, rule of additive-multiplicative root extraction), *Tianyuan, Siyuan, Dayan qiuyi* (大衍求一, rule of determination of unity), *Tiaori* (調日, rule of common fractions), and *Zhaocha* (招差, rule of interpolation). All of these methods can be regarded as, in essence, sets of mechanized calculation procedures. Some can even be transcribed and implemented using modern computer programming.

Chinese mathematical writings thus contained an abundance of clearly conceived mechanized procedures. This is illustrated in the *Nine Chapters*, where we find such phrases as: "find the solution by the *Yingbuzu* method" (盈不足, the rule of excess and deficit, i.e. double false position)', "use the *Zhengfu* method (正負, the rule of sign) to solve", and "use the *Fangcheng* method". These techniques are mathematically equivalent to a set of commands for running computer programs which would carry out, respectively, the rule of double false position, the rule of signs, and an algorithm for solving simultaneous linear equations. In programs that implement the *Geyuan* method and the *Dayan* method, loop statements and subprograms are required [Liu 1996]. A basic formula of *Gougu suanshu* (勾股算術, the arithmetic of right triangles) can be used to express a second-order algebraic identity, but also to make further deductions.

While learning such basic calculation methods, mathematicians came to grasp the scope of these applications, an essential step for those who would go on to acquire higher skills. Thus, for instance, the *Yingbuzu* method can be applied not only to the five basic types of data for excesses and deficits, but also to solve any algebraic problem of order one. This was done by supposing twice the *Chongcha* (重差, the rule of double differences for indirect survey), a method which stemmed from a model for measuring the height of the sun on a (hypothetical) flat earth. Quite complex algorithms are specified in *Haidao suanjing* (海島算經, *Sea Island Mathematical Manual*, 3rd century AD). The methods presented by Li Chunfeng (李淳風, 602-670 AD) in his annotation of this book (before 656 AD) show how surveying problems relating to an inclined plane or terraced field, can be reduced to a standard type. The method makes use of an "imaginary" surface, where two poles are set up with the help of properties of similar triangles; the problem can then be transformed to the standard *Chongcha* model [Fu 1988] [Liu 1993B].

As for the famous diagram drawn by Jia Xian (賈憲, ca. 1050), an equivalent of Pascal's triangle, this occupies a vital position in ancient Chinese algebra. One finds this in calculations for additive-multiplicative root extraction (*Zengcheng kaifang*), interpolations (*Zhaocha*), stack accumulations (*Duoji* 垛積), as well as in later studies involving combinatorics by Qing scholars (18th-19th centuries). All of these are related to this *Jiaxian* pattern, which embodied distinctly constructive properties.

The Exemplary Role of a Great Classic

Just as the West had its "Holy Book of Science", Euclid's *Elements*, so the Chinese had the *Nine Chapters*, which they viewed, alongside its annotations, as setting the highest standard for mathematical writing and research [Dauben 1998]. This tradition gave rise to a rich mathematical literature, which includes such works as: *Xiangjie jiuzhang suanfa* (詳解九章算法, *Detailed Explanation of the Nine Chapters*, 1261), *Shushu jiuzhang* (數書九章, *Computational Techniques in Nine Chapters*, 1247), *Jiuzhang tongming suanfa* (九章通明算法, *General Algorithms in Nine Chapters*, 1424), *Jiuzhang suanfa bilei daquan* (九章算法比類大全, *Fully Comprehensive Collection of Computational Methods in Nine Chapters*, 1450), *Jiuzhang yi* (九章翼, *The Wing of Nine Chapters*, 1898), etc. The titles of these books alone make evident that they were part of an ongoing fascination with the "Holy Book" of Chinese mathematics, the *Nine Chapters*.

For two millennia, from the Western Han Dynasty to the end of the Qing Dynasty (ca. 50-1911), the *Nine Chapters* served as the unique paradigm and principal model for the Chinese mathematical research tradition. Not only did it equip mathematicians with a unified glossary, dictionary, and writing format, it also furnished them with a diverse array of innovative conceptual models. The standard terminologies changed very little after the first century, so that this genre of literature continued to follow a tightly structured pattern. Most ancient Chinese mathematical books contain groups of topics divided into chapters and sections. A typical topic has different entries, such as "question", "answer", "method", and "note". First a specific "question" is raised. The "answer" then provides a concrete solution in numbers; then the "method" sets forth a more general solution procedure, sometimes equivalent to a formula or a theorem; finally, the "note" explains the basis for the "method", which in essence amounts to giving a proof for the given procedure. During the Song and Yuan Dynasties (960-1368), probably due to the invention and development of printing, another type of entry, called "draft", was added, in order to give details about the calculation procedure [Wu 1987]. The formation of this traditional writing style was, in fact, closely associated with the various annotations produced for the *Nine Chapters*.

This great classic also served as the original source for ancient Chinese mathematical problems and methods. Thus, the researches on the circle and sphere by Liu Hui and Zu Chongzhi (祖冲之, 429-500 AD) were refinements of the methods of *Yuantian* (圓田, area-circumference of the circle) and *Kai liyuan* (開立圓, volume of the sphere) found in the first and fourth chapters of the *Nine Chapters*. Likewise, the techniques for solving higher-degree equations by successive approximations, pursued by Wang Xiaotong (王孝通, 7th century), Liu Yi (劉益, 10th century), Jia Xian, Qin Jiuzhao (秦九韶, 1208-1261), and Li Zhi, had their origin in the *Kaifang* (開方, extraction of square root) method, which is found in the fourth chapter. Later research on the arithmetical properties of right triangles was inspired by the results presented in the final chapter, and the same holds true for the massive formula systems in *Ceyuan Haijing*, which stem from problem 15 of that chapter.[4] Again, the study of stack accumulations by Song and Yuan mathematicians involves deductive reasoning that can be traced to the fifth chapter *Shanggong* as well as the fourth *Shaoguang* (少廣, shorten width) [Fu 1988]. Even the multifaceted works on infinite power series by Qing mathematicians (18th-19th centuries) used as their main methodology either continued proportions or stack accumulations, methods essentially derived from the *Nine Chapters*.

Strong Utilitarian Tendency

In contrast to the Hellenic view of mathematics as a purely contemplative activity, ancient Chinese mathematical practice was imbued with a strong sensitivity for the needs of society and human life. Plato labeled mathematics as a subject "compelling the soul to reason about abstract number, and rebelling against the introduction of visible or tangible objects into the argument" ([Kline 1972], 43).[5] While Euclid in his *Elements* erased all traces of the practical origins of Greek geometry, Chinese mathematicians were never averse to drawing on questions taken

[4]The original text is "Suppose there is a *gou* of 8 *bu*, and a *gu* of 15 *bu*. The question: how large is the diameter of the circle inscribed inside the *gou* [-gu figur]?" ([Dauben 2013], II: 1111)

[5]According to M. Kline, this quotation is from one of Plato's *Dialogues*, refering to B. Jowett ed., *Dialogues of Plato*, Book VII, sec. 525, Clarendon Press, 1953.

from immediate social practice.

The *Nine Chapters* represented this viewpoint by offering problems of a multifaceted nature. Many had direct relevance for politics, economics, military, or cultural affairs, especially problems that reflected life in feudal China after the unification of the Qin and Han empires (221 BC-220 AD). This can be illustrated by a number of examples. Under the Western Han government, which encouraged the cultivation of new lands, a vast amount of former wasteland was reclaimed. Thus, it became an urgent matter to undertake a land survey, and the very first chapter of the *Nine Chapters*, *Fangtian*, addresses this pressing concern. The second chapter, *Sumi* (粟米, millet and rice) concerns the bartering of commodities. A list of exchange rates of various grains is given at the beginning, from which we can discern that the central government maintained control over the grain trade. The third chapter, *Cuifen* (衰分, distribution by proportion), concerns rankings for the allocation of rewards; the first and sixth problems mention the five ranks of nobility, revealing that the higher ranks were assigned more goods, whereas the reverse applied for taxation. The background of the fifth chapter, *Shanggong*, concerns a series of state projects involving civil constructions and irrigation engineering. The sixth chapter, *Junshu* (均輸, fair levies or impartial taxation), deals with official allocations of labor—in fact, the term *Junshu* refers to a financial act issued during the Wu Empire (156-87 BC) of the Western Han dynasty.[6]

The *Nine Chapters* also reflects the author's vast knowledge of astronomy, geography, meteorology, music, architecture, as well as business and military affairs. Many problems deal with measuring units and scales, reflecting those in use during the Qin-Han times. The inscription on the vessel of *Dasinong* (大司農, the highest official in charge of agriculture, trade, and finance), dated 179 AD, stipulates that all the measuring tools must be standardized in accordance with the *Nine Chapters*. Coming back to the chapter *Junshu*, it is noteworthy, that the first four problems deal with an assignment involving thousands upon thousands of laborers and vehicles from several different counties, suggesting a picture of a well-organized project led by a powerful central authority. The ninth problem from this chapter solves the problem *Taicang shusu Shanglin* (太倉輸粟上林, allocate and transport grain from imperial storehouse to capital); and the twenty-first problem, *Jia fa Chang'an, wuri zhi Qi* (甲發長安五日至齊, one left Chang'an city and arrives in Qi state in five days), deals with the current state of traffic and communication, providing information that still remains unavailable for the pre-Qin periods [Liu 1995].

This utilitarian trend lasted several centuries, up until the time of Liu Hui. As Joseph Dauben wrote in his introduction to the new English version of annotated *Nine Chapters*, "The dramatic changes in mathematics from the pre-Qing period to the time of Liu Hui may be measured in a number of ways. Not only did the unification of the formerly diverse warring states serve to provide a common language,

[6] *Junshu*, literally means "to make tax levy equally in order to transport and pay tribute effectively". Before the act was carried out under the Western Han dynasty, there was a special official in charge of *Junshu* affairs in the Qin empire (221-207 BC).

uniformize and standardize weights, measures, and currency, but great national works required the creation and maintenance of canal and roads, the construction of granaries, public buildings and palaces on unprecedented scale. This all led to the need for government administration on diverse levels." ([Dauben 2013], I:73)

This background can be seen, for example, with the *Gougu* Theorem, described in *Zhoubi suanjing* (周髀算經, *Mathematical Canon of Zhou Gnomon*, 1st century BC) and which derives from the ancients' need to measure the heaven and earth. Another example is the *Wucao suanjing* (五曹算法, *Computational Rules of the Five Offices*, 6th century), which reflects some of the technical requirements demanded of state officials following Confucian concepts of governance.[7] In his annotations of the *Nine Chapters*, Liu Hui quotes extensively from such classics as *Mozi* (墨子, Mohist), *Zuo Zhuan* (左傳, *Chronicle of Zuo*), and *Kaogong ji* (孝工記, *Book of Diverse Crafts*). He appealed in these notes to objects, such as a gold sphere with diameter of one *cun* (寸)[8] and *Lüjia lianghu* (律嘉量斛)[9] to explain the mathematics. Nearly all the outstanding achievements in ancient Chinese mathematics share this sense of purpose or practicality: the *Dayan qiuyi* method was related to computation of the initial date for a calendar. The concepts of fraction, decimal, and the conversion of different numerical systems originated from a problem in musicology, namely to produce standardized pitch pipes.

In addition, the social standing of mathematicians within the state bureaucracy reinforced the importance of mathematics within the educational system. This institutional grounding, toegether with the strong interactions between mathematical knowledge and Chinese societal needs gave rise to a unity and cohesion that was reflected in academic trends spanning several dynasties. These features all help to account for the longevity and stability of the Chinese tradition. They also serve to strengthen our conclusion that ancient Chinese mathematics was imbued with strong interests in human affairs while reflecting the distinctive social characteristics of its age ([Liu 1993A], 432-443).

Bibliography

[Dauben 1998] Dauben, J., Ancient Chinese Mathematics: The Nine Chapters of the Art of Mathematics versus Euclid's Elements, Aspects of Proof and the Linguistic Limits of Knowledge, in Chandrasekhar Memorial Issue, *International Journal of Engineering Science*, **36**: 1339-1359.

[7] *Wucao* refers to five major functional departments in local government; they were charged, respectively, with agriculture, warfare, trade, store, and finance.

[8] *Cun* is a length unit and 1 Han *cun* roughly equals to 2.31 cm.

[9] This bronze *hu* mentioned here is an especially prized artifact that survives from the Han dynasty and is today preserved in the Palace Museum in Taipei. It is mentioned by Liu Hui in chapter one following Problem 1.32, in a controversial comment about computing the area of the circle as described in the course of an inscription on the standard measure indicating its dimensions. ([Dauben 2013], I:74)

[Dauben 2013] Dauben, J. (with Guo Shuchun and Xu Yibao), *Nine Chapters on the Art of Mathematics, English Critical Edition and Translation with Notes*, Liaoning Education Press, Shenyang.

[Du 1954] Du, S.R., Zu Geng's Axiom, *Bulletin des Sciences Mathématique*, 3, 9-11.

[Du 1966] Du, S.R., Research into Zhu Shijie, in Qian Baocong ed., *Collected Papers of the History of Mathematics during the Song-Yuan Period*, Science Press, Beijing

[Fu 1988] Fu, D.W., On the Development and Transition in the History of Zhoubi Research Tradition, *Tsing Hua Journal of Chinese Studies*, New **18**:1, 1-41.

[Fu 1992] Fu, D.W., A New Research of Duoji in the History of Chinese Mathematics, Historical Development of the Two Lines Shanggong and Shaoguang, in *Yishikong li de zhishi zhuizhu*, Dongda Book Co., Taipei, 69-113.

[Guo 1983] Guo, S.C., A Tentative Analysis of the Definition and Deductive Logic in Liu Hui's Annotations to the Nine Chapters of the Mathematical Art, *Studies in the History of Natural Sciences*, **2**(3): 193-203.

[Guo 1987] Guo, S.C., On Liu Hui's Theoretic System of Mathematics, *Journal of Dialectics of Nature*, **9**(2): 42-48.

[Heath 1912] Heath, T. L., *The Works of Archimedes with the Method of Archimedes*, Dover Publications, New York.

[Kline 1972] Kline, M., *Mathematical Thought from Ancient to Modern Times*, Oxford University Press, New York.

[Lam 2004] Lam, L.Y., *Fleeting Footsteps, Tracing the Conception of Arithmetic and Algebra in Ancient China*, Revised Edition, World Scientific Co., Singapore.

[Liu 1989] Liu, D., Li Rui and Descartes' Sign Rule, *Studies in the History of Natural Sciences*, **8** (2): 127-137.

[Liu 1993A] Liu, D., *Daz yanshu*, Liaoning Education Press, Shenyang.

[Liu 1993B] Liu, D., On Li Chunfeng's Method of Double Differences Applied to an Inclined Plane, *Studies in the History of Natural Sciences*, **12** (2): 101-111.

[Liu 1995] Liu, D., The Triumph of Utilitarian Mathematics, in K. Hashimoto *et. al.* eds., *East Asian Science: Tradition and Beyond*, Kansai University Press, Osaka, 457-460.

[Liu 1996] Liu, D., A Comparison of Archimedes' and Liu Hui's Studies of Circle, in Fan, D.N. & Cohen, R. eds., *Chinese Studies in the History and Philosophy of Science and Technology* (Boston Studies in the Philosophy of Science, vol. 179), Kluwer Academic Publishers, Dordrecht, 279-287.

[Liu 2002] Liu, D., A Homecoming Stranger: Transmission of the Method of Double False Position and the Story of Hiero's Crown, in Dold-Samplonius, Y. *et. al.* eds., *From China to Paris: 2000 Years Transmission of Mathematical Ideas*, Steiner, Stuttgart, 157-166.

[Wang 1983] Wang, Y.S., The Sharp Pyramids Algorithm of Li Shanlan, *Studies in the History of Natural Sciences*, **2**(3): 266-288.

[Wu 1983] Wu, W.J., The Out-In Complementary Principle, in *Ancient China's Technology and Science*, Foreign Languages Press, Beijing, 66-89.

[Wu 1987] Wu, W.J., Recent Studies of the History of Chinese Mathematics, in *Proceedings of the International Congress on the History of Science and Technology, Berkeley, 1986*, 1657-1667.

On the Algorithmic Tradition in the History of Mathematics

LI Wenlin

Introduction

There have been two major tendencies in the history of mathematics—theorem-proving and algorithm-creating. While theorem-proving, which originated in ancient Greece, has been the backbone of the deductive tradition in the history of mathematics, the algorithm-creating style flourished in the ancient and medieval Orient, where a strong algorithmic tendency shaped the evolution of mathematics in the East. In comparison with the deductive tradition, this algorithmic direction has only recently become the focus of historical studies aimed at showing its function and influence. The development of modern computer techniques, however, has helped draw the attention of mathematicians and historians to the importance of this algorithmic tradition in the history of mathematics. Thus, for example, D. E. Knuth has investigated certain algorithms employed by ancient Babylonian mathematicians, and he has concluded that "much more can be told" for the cases of both Indian and Chinese mathematics.[1] The present article aims to shed further light on this theme. It emphasizes the algorithmic spirit of oriental mathematics by describing some representative Chinese, Indian, and *Wasan* algorithms in ancient and medieval times. At the same time, it examines the historical roots of modern algorithmic languages by analyzing the structures of several older algorithms. Finally, it argues that the role played by this algorithmic tradition in the evolution of mathematics should be seen as comparable to that of deductive methods in the Western tradition.

Chinese Algorithms in Ancient and Medieval Times

Like the Babylonians, Chinese and Indian mathematicians were accustomed to deriving formulas following a step-by-step list of computational rules; they did this, however, in a more general way than the Babylonians. In ancient Chinese mathematical works, one always finds a standardized paragraph called "shu" (術, method). This amounts to a program for computing a formula in a step-by-step manner, or likewise for solving a certain class of problems. In subsequent problems of the same class, the text of "shu" is then either repeated or this repetition is simply indicated by the phrase "as above." Here is an example from the *Fangcheng* (equation) chapter of the *Jiuzhang suanshu* (*Nine Chapters on the Mathematical Art*, 1st cent. A.D.):[2]

[1]Knuth, D. E., Ancient Babylonian Algorithms, *Communications of the Association for Computing Machinery*, 15(7), 1972, 671-677.

[2]For the original text see Guo Shuchun ed., *Nine Chapters on the Mathematical Art* (Liaoning Education Press 1990) (郭書春:《九章算術》匯校, 遼寧教育出版社), ; the English translation given

D.E. Rowe and W.-S. Horng (eds.), *A Delicate Balance: Global Perspectives on Innovation and Tradition in the History of Mathematics*,
Trends in the History of Science, DOI 10.1007/978-3-319-12030-0_14

Problem: From 3 *bing* (秉 sheaves) of *shanghe* (上禾 first class corn), 2 *bing* of *zhonghe* (中禾 second class corn), and 1 *bing* of *xiahe* (下禾 third class corn), you get 39 *dou* (斗 Chinese bushel) of *shi* (實 grain); from 2 *bing* of *shanghe*, 3 *bing* of *zhonghe* and 1 *bing* of *xiahe*, you get 34 *dou* of shi; from 1 *bing* of *shanghe*, 2 *bing* of *zhonghe* and 3 *bing* of *xiahe*, you get 26 *dou* of *shi*. How many dou of *shi* can you get from each *bing* of *shanghe*, *zhonghe* and *xiahe* respectively?
Method (At first, the data were arranged as follows):

Left	Middle	Right	
1	2	3	*Shanghe*
2	3	2	*Zhonghe*
3	1	1	*Xiahe*
26	34	39	*Shi*

Multiply each term of the middle column by *shanghe* of the top right (3), then repeatedly subtract each term of the right column from the resultant terms (until a zero is attained in the top place). Multiply each term of the left column by *shanghe* of the top right (3), then repeatedly subtract each term of the right column from the resultant terms. Multiply each term of the left column by the *shanghe* remainder of the middle column (5), then repeatedly subtract each term of the middle column from the resultant terms of the right column. Divide the *shi* remainder of the left column (99) by the *xiahe* remainder of the left column (36), and you will obtain the answer for the number of *dou* of *shi* which you will get from each *bing* of *xiahe*.

What comes next is a similar program for obtaining the numbers of *dou* of *shi* one will get from each *bing* of *zhonghe* or each *bing* of *shanghe*. The fundamental operation here is known as *biancheng zhichu* (遍乘直除), i.e. each term of the second equation is multiplied by the coefficient of a certain unknown in the first equation, then each term of the first equation is repeatedly subtracted from the resultant terms of the second equation. The author of *Jiuzhang Suanshu* called this procedure for solving simultaneous linear equations the *Fangcheng shu* (the square-array method). In subsequent problems of this class, it was always said that "the method is as the *Fangcheng*," followed by an arrangement of new data in the corresponding places. In modern algorithmic language, this amounts to "specifying" a standard procedure and "assigning" new values to the parameters.

Thinking about such algorithmic structures became more and more sophisticated among mathematicians down to the Song dynasty. A famous example can be seen from the method for solving systems of linear congruences in Qin Jiushao's *Shushu jiuzhang* (數書九章, *Mathematical Treatise in Nine Sections*, 1247 A.D.). In the beginning of the "Dayan" (大衍 Great Extension) chapter, Qin designed a general program—*dayan zongshu shu* (大衍總數術: the general method of *Dayan*)

above is the author's own.

—for solving all problems of the Dayan class, which are systems of such linear congruences.[3]

A flow chart for this program is shown in Fig.1.

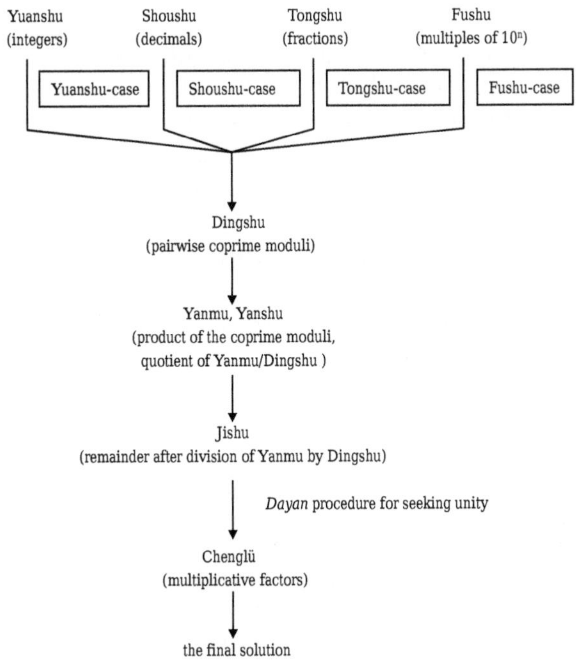

Figure 1

This was a really large program for that time, and it is noticeable that Qin's procedure consisted of several parts which could be compiled in various combi-

[3]In modern algebraic language, the outline of Qin's method is as follows:
A system of first order congruences

$$X \equiv R_i (\text{mod}\, A_i) \quad i = 1, 2, \ldots, n \tag{1}$$

is equivalent to the system

$$X \equiv R_i (\text{mod}\, a_i) \quad i = 1, 2, \ldots, n$$

where "a_i" is a factor of "A_i", L.C.M. $(a_1, a_2, \ldots, a_n)$ = L.C.M. $(A_1, A_2, \ldots, A_n)$, and the $a_1, a_2, \ldots, a_n$ are pairwise prime. Let $M = a_1 \times a_2 \times \ldots \times a_n$, then if we can find numbers k_i, satisfying $k_i g_i \equiv 1 (\text{mod}\, a_i)$, $i = 1, 2, \ldots, n$, where $g_i = M/a_i (\text{mod}\, a_i)$, then $X = \sum_{i=1}^{n} k_i R_i M/a_i (\text{mod}\, M)$ is a solution of the system (1).
Qin denominated A_i as *wenshu* (問數 problem numbers) ; a_i as *dingshu* (定數 definite numbers); M as *yanmu* (衍母 multiple denominator); M/a_i as *yanshu* (衍數 expansion numbers); g_i as *jishu* (奇數 remainders); k_i as *chenglü* (乘率 multiplying factors).

nations, i.e. it included several sub-programs within the general program.[4] We mainly refer here to Qin's procedures for determining the *dingshu* (which is distinguished by four kinds of modules: *yuanshu* 元數, *shoushu* 收數, *tongshu* 通數, and *fushu* 複數)[5] and his *dayan qiuyi shu* (大衍求一術: *Dayan* Procedure for Seeking Unity) for determining the Chenglü.

We can diagram the structure of Qin's program (see Fig. 2). Precisely this type of form, in which a large block program includes smaller ones, is a fundamental structural feature in modern algorithmic languages. The sub-program can

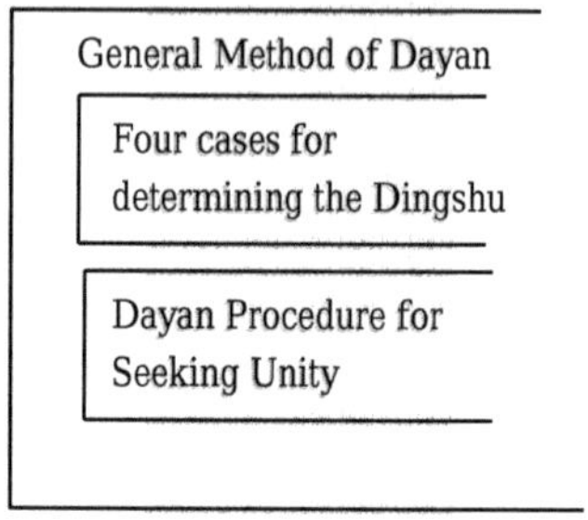

Figure 2

be considered a complete program in itself. Qin also often "called" directly the four cases and he used these cases in the *dayan qiuyi shu* when computing problems of the *dayan* class. In applying this procedure he always used expressions such as "use ×× case" or "use *dayan qiuyi shu*". His *dayan zongshu shu* represents the strong tendency among Song mathematicians to create elaborate algorithmic structures. Other important algorithmic phenomena can be excavated in it as well.

Iterations and conditionals are two basic components for constructing algorithms of a non-trivial type. D. E. Knuth has noted that "there is very little evidence of this in Babylonian texts,"[6] however, it is never difficult to find these in the traditional Chinese and Indian mathematical literatures. The famous *geyuan shu* (割圓術: method for determining segment areas) is a typical iterative procedure that appeared even earlier. Liu Hui (3. A.D.) gave the following program for calculating the length of each side of a regular polygon of $2n$ ($n = 3, 4, 5, \ldots$) sides,

[4]A detailed explanation of Qin's method can be found in Qian Baocong: "*Collection of Essays on the Song-Yuan Mathematics*" (宋元數學史論文集), Beijing, 1966 and U. Libbrecht: *Chinese Mathematics in the Thirteenth century, The Shu-Shu Chiu-chang of Chin Chiu-Shao*, Cambridge, Massachusetts and London, England, 1973. Qin's method for determining the *Dingshu* was also discussed by Li Wenlin and Yuan Xiangdong in "*On some problems about the Indeterminate Analysis in the History of Chinese Mathematics*" (中國古代不定分析若干問題探討), 科技史文集 —數學史專輯, 上海科學技術出版社, 1982 (Collection of Essays on the History of Science and Technology–Special Issue on Mathematics, Shanghai, 1982).

[5]Qin wrote four special paragraphs for computing the Dingshu, and he used a new term, "Ge" (格 -case), for these sub-programs.

[6]Knuth, D. E., Ancient Babylonian Algorithms, *Communications of the Association for Computing Machinery*, 15(7), 1972, 671-677.

successively inscribed in a given circle:[7]

> Bisect the segment of a 6-*gua* (觚, regular polygon inscribed in a circle; 6-gua means a regular hexagon), so that you get a 12-gua (regular dodecagon);
> Let the diameter be 2 *chi* (尺 Chinese feet);
> Take half of it, and you get the *gua-mian* (觚面 the length of each side of the regular polygon, here the hexagon), i.e. 1 *chi*;
> Let the radius i.e. 1 chi be the *Xian* (弦 hypotenuse of a right triangle); and let the half-mian (of the 6-*gua*), i.e. 5 *cun* (寸, 1 *chi*=10 *cun*) be the *gou* (勾 short leg of a right triangle);
> (Use the *gougu shu* [勾股術 Pythagorean theorem] to) calculate the *gu* (股 the long leg);
> Subtract the square of the short leg (i.e. 25 square cun) from the square of the hypotenuse, thus obtaining 75 square *cun*;
> Extract its root, and you get the long leg (i.e. 86625 2/5 *hu*=8.66254 *cun*);
> Subtract it from the radius, and you get the short leg of the smaller triangle (i.e. 133974 3/5 *hu*);
> Let the half-mian be the long leg;
> (Use the *gougu shu* to) calculate the hypotenuse and you get its square 267949173445 square *hu*;
> Extract its root, thus obtaining the mian of 12-*gua*.

The following instructions use exactly the same procedure for computing the "mian" of polygons with 24, 48, and 96 sides:

> Bisect the segments of 12-*gua*, and then you get 24-*gua*;
> Let the radius be the hypotenuse, and let the half-mian (of 12-*gua*) be the short leg;
> ...
> Thus obtaining the mian of 24-*gua*;
> Bisect the segment of 24-*gua*, and you get 48-*gua*;
> Let the radius be the hypotenuse, and let the half-*mian* (of 24-*gua*) be the short leg;
> ...
> Thus obtaining the mian of 48-*gua*;
> Bisect the segment of 48-*gua*, and you get 96-*gua*;
> Let the radius be the hypotenuse, and let the half-mian (of 48-*gua*) be the short leg;
> ...
> Thus obtaining the mian of 96-*gua*.

[7]For the original text see Guo Shuchun ed., *Nine Chapters on the Mathematical Art*, Liaoning Education Press 1990 (郭書春: 《九章算術》 匯校, 遼寧教育出版社), the English translation here was given by Li Wenlin.

The only difference between the above cases is the new values assigned to the "mian" (as a matter of fact, Liu "assigned" the result from each stage to the parameter "mian" in the next stage). So we have here an example of a "DO I=1 to N" (here N=4, but Liu had also gone on to N=8).

As to the conditionals, we look again at the *dayan qiuyi shu* (*Dayan* Procedure for Seeking Unity) in more detail. We can list Qin's program by modern symbols in this way:

STEP 1:	*divide* Dingshu	Quotient	Remainder	Calculate c_i
	by *jishu*, i.e. A/G	q_1	r_1	$c_1 = q_1$
STEP 2:	G/r_1	q_2	r_2	$c_2 = c_1 q_2 + 1$
STEP 3:	r_1/r_2	q_3	r_3	$c_3 = c_2 q_3 + c_1$
...				
STEP n:	r_{n-2}/r_{n-1}	q_n	r_n	$c_n = q_n c_{n-1} + c_{n-2}$

If $r_n \neq 1$, go on, otherwise do

STEP $n+1$: If n is even, then k *(Chenglü)* $= c_n$,
otherwise do
$q_{n+1} = r_{n-1} - 1$, $c_{n+1} = c_n q_{n+1} + c_{n-1}$, then $k = c_{n+1}$.

It is clear that this program includes a complex conditional of the type "IF S_1 GO TO S_2, IF S_3 GO TO S_4, etc."

More evidence of such sophisticated algorithmic structures can be cited if we were to examine other procedures, such as Zhu Shijie's *zhaocha shu* (招差術 -interpolation method), Qin Jiushao's *zhengfu kaifang shu* (正負開方術 -method for solving "positive-negative" equations of high degree), etc. Taking one more example, we analyze here the *Zhengfu Kaifang Shu*, which was in fact a method for numerical solving higher equations and appeared also in Qin's work *Shushu jiuzhang*.

Qin's method was illustrated in detail in volume 5 of *Shushu jiuzhang*; it represents an essential improvement over the earlier techniques for calculating the roots of an equation as developed by a series of mathematicians prior to Qin. Given such an equation

$$f(x) = a_0 x^n + a_1 x^{n-1} + \ldots\ldots + a_{n-1} x + a_n = 0,$$

where we assume $a_0 \neq 0$, $a_n < 0$. Qin first obtained an initial approximation c for a root (in Qin's term, *shou shang*, i.e. initial quotient) by trial and error. He then set $x = c + h$ to obtain a new equation for h:

$$\bar{f}(h) = \bar{a}_0 h^n + \bar{a}_1 h^{n-1} + \ldots\ldots + \bar{a}_{n-1} h + \bar{a}_n = 0$$

from which Qin approximated the next value of the root (*xu shang*, i.e. successive quotient) and he then repeats the operations all over again. Qin gave a mechanical algorithm for calculating the coefficients of the reduced equations, which can be tabularized as in appendix I.

This is an effective iterative process which enabled Qin to approximate to arbitrary precision a real solution of algebraic equations.

Qin's method itself was based on a simpler iterative algorithm, called the "method for calculating coefficients by additions and multiplications." This was first created by Jia Xian (ca. 1050) to produce a so-called "diagram for solving equations" (i.e. the numerical triangle of binomial coefficients):

1	$1+5=6$				
1	$1+4=5$	$5+10=15$			
1	$1+3=4$	$4+6=10$	$10+10=20$		
1	$1+2=3$	$3+3=6$	$6+4=10$	$10+5=15$	
1	$1+1=2$	$2+1=3$	$3+1=4$	$4+1=5$	$5+1=6$
1	1	1	1	1	1

Indian Cases

In many Indian mathematical works, one often finds a "rule"-paragraph called "paricarma" (algorithm, or logistics) used for solving certain classes of problem: This is quite similar to the "shu" in Chinese cases. To begin with a simple example for the summation of a geometric progression from Chapter V §127 of the *Lilavati* of Bhaskara:[8]

> **Rule:** If n, the number of terms in a geometric progression, is odd, then $(n-1)$ is called a "multiplier" (M), and if it is even, $n/2$ is called a "square" (S). Now beginning with n, continue this process $(n-1)$ for odd and $n/2$ for even until 0 is reached. Then keep the common ratio (C.R.) r against 0 and start writing M or S in the opposite direction. Carry out the operations and then subtract 1 from the final result and divide it by $(r-1)$. The result is the required sum.

This is a step-by-step program to compute the sum of a geometric progression equivalent to the formula:

$$S = a + ar + ar^2 + ar^3 + \ldots + ar^{n-1} = \frac{a(r^n - 1)}{r - 1}.$$

This uses a sub-program to compute the powers r^n at even numbers n (multiplying by r at odd ones), which produces:

$$\begin{aligned} &\mathrm{P}_0 = \mathrm{r}^0 \times \mathrm{r} \\ &\mathrm{P}_1 = (\mathrm{P}_0)^2 \times \mathrm{r} \\ &\mathrm{P}_2 = (\mathrm{P}_1)^2 \times \mathrm{r} \\ &\ldots\ldots\ldots\ldots \\ &\mathrm{P}_{i+1} = (\mathrm{P}_i)^2 \times \mathrm{r} \end{aligned}$$

[8] See K. S. Patwardhan, S. A. Naimpally and S. Lal Singh, *Līlavātī of Bhāskarācaya, A Treatise of Mathematics of Vedic Tradition*, Jainendra Prakash Jain at Shri Jainendra Press, India, Delhi, 2001.

This example illustrates the interest of ancient Indian mathematicians in such algorithms.

In fact, one can also find many examples of iterations and conditionals in ancient Indian algorithms. From early times onward, Indian scholars had been adopting iterative formulas to produce tables for their astronomical calculations. For instance, the construction of the difference table of sine in Aryabhata I's *Aryabhatiya* (499AD)[9] depended on an iterative procedure given in rule 1 of Chapt. 2 of this book:

> By what number the second sine [difference] is less than the first sine, and by the quotient obtained by dividing the sum of the preceding sine [difference] by the first sine, by the sum of these two quantities the following [difference] are less than the first one.

This procedure amounts to the following recursion formula:

$$K_{i+1} = K_i - \frac{(K_1 - K_2)J_i}{J_1}$$

For a more sophisticated example in this regard, however, let us consider the Indian method for determining solutions of the indeterminate quadratic equations $\mathrm{Nx}^2 + 1 = \mathrm{y}^2$. It is well-known that Brahmagupta (598-665) found a rule for obtaining solutions of the equation $\mathrm{Nx}^2 + 1 = \mathrm{y}^2$ by making use of auxiliary equations of the form:

$$\mathrm{N}\alpha^2 + \mathrm{k} = \beta^2, \quad \mathrm{k} = \pm 1;\ \pm 2;\ \text{or}\ \pm 4. \qquad (*)$$

Brahmagupta Principle of Composition (BPC)

If $\mathrm{N}\alpha^2 + \mathrm{k} = \beta^2$ *and* $\mathrm{N}\alpha'^2 + \mathrm{k}' = \beta'^2$,
then $\mathrm{N}(\alpha\beta' \pm \alpha'\beta)^2 + \mathrm{kk}' = (\beta\beta' \pm \mathrm{N}\alpha\alpha')^2$.

A special case of the BPC is:
If $\mathrm{N}\alpha^2 + \mathrm{k} = \beta^2$, then $\mathrm{N}(2\alpha\beta)^2 + \mathrm{k}^2 = (\beta^2 \pm \mathrm{N}\alpha^2)^2$.

To obtain the auxiliary equations Brahmagupta mainly relied on trial and error.

The problem was treated systematically by Bhaskara II(1114-1185?), who gave a beautiful mechanical method to arrive at auxiliary equations of the form (*) beginning with an initial equation: $\mathrm{Na}_0^2 + k_0 = b_0^2$. Bhaskara called his approach the "cyclic method," which was described by H. T. Colebrooke as follows:[10]

> **Rule for the cyclic method** Making the "lesser" and "greater"" roots and additive, a dividend, addend and divisor, let the multiplier be thence found. The square of that multiplier being subtracted from the

[9]Aryabhata I, *Aryabhatiya*, ed. by K. S. Shukla, New Delhi, Indian National Science Academy, 1976

[10]Colebrooke, H. T., trans., *Algebra with Arithmetic and Mensuration from Sanscrit of Brahmegupta and Bhaskara*, John Murray, 1817.

given coefficient, or this coefficient being subtracted from that square, (so as the remainder be small;) the remainder, divided by the original additive, is a new additive, which is reversed if the subtraction be (of square) from the coefficient. The quotient corresponding to the multiplier (and found with it) will be the "lesser" root, whence the "greater" root may be deduced. With these, the operation is repeated, setting aside the former roots and additive. This method mathematicians call that of the circle. Thus are integral roots found with four, two, or one(or other number, for) additive, ...

The method, as here described, is highly iterative. Together with Brahmagupta's principle of composition, it constitutes an ingenious program for producing solutions of the famous equation $Nx^2 + 1 = y^2$. The role of iterations and conditionals in this procedure can be seen in the flow chart shown below in Fig. 3.

This program was based on a theorem known as Bhaskara's Lemma, which states:

Bhaskara's Lemma (BL) If $Na^2 + k = b^2$, where a, b, k are integers, k being positive or negative, then for any arbitrary whole number m:

$$\mathrm{N}(\frac{\mathrm{am}+\mathrm{b}}{\mathrm{k}})^2 + \frac{\mathrm{m}^2-\mathrm{N}}{\mathrm{k}} = (\frac{\mathrm{bm}+\mathrm{Na}}{\mathrm{k}})^2.$$

We should emphasize that Bhaskara here uses Aryabhata's *Kuttaka* method as a sub-program, starting with the initial values $\mathrm{a_0}, \mathrm{b_0}, \mathrm{k_0}$. The *Kuttaka* method is applied to the data a, b, k (given, respectively, as dividend, addend and divisor) in order to find integers m so that $(\frac{am+b}{k})$ will be integral. Among these m, the algorithm chooses the one (denoted m_0) for which $|m^2 - \mathrm{N}|$ has the smallest value possible. It then makes a new array $\mathrm{a_1}, \mathrm{b_1}, \mathrm{k_1}$:

$$\mathrm{a_1} = \frac{\mathrm{a_0 m_0} + \mathrm{b_0}}{\mathrm{k_0}}, \quad \mathrm{b_1} = \frac{\mathrm{b_0 m_0} + \mathrm{N a_0}}{\mathrm{k_0}}, \quad \mathrm{k_1} = \frac{\mathrm{m_0^2} - \mathrm{N}}{\mathrm{k_0}}.$$

BL ensures the cycling process:

$$\{\mathrm{a_1}, \mathrm{b_1}, \mathrm{k_1}\} \to \{\mathrm{a_2}, \mathrm{b_2}, \mathrm{k_2}\} \to \ldots \{\mathrm{a_i}, \mathrm{b_i}, \mathrm{k_i}\},$$

and **BPC** serves to deduce the final solutions for the original equation (see Fig. 4 below).

The programs of Chinese and Indian mathematicians were, of course, a kind of "software" designed for their own computing tools. Nevertheless, they can easily be translated into a contemporary algorithmic language for execution by modern computers. (We have done this for the case of the *Kuttaka* algorithm and also for the procedures in Qin Jiushao's *zheng fu kaifang shu.* In appendices II and III,

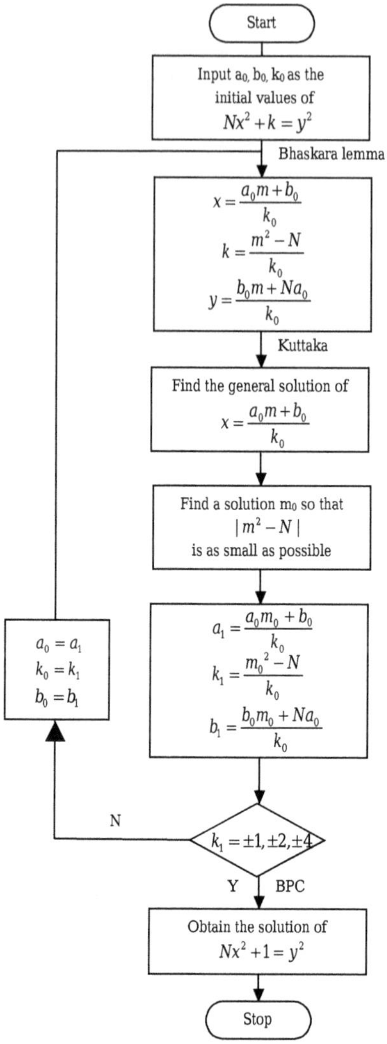

Figure 3

we have given the corresponding flow chart and program in JAVA to show how to solve indeterminate equations in the *Aryabhatiya* of Aryabhata. Here we have made only a minute modification of the Kuttaka algorithm in order to mechanize its only trial step, which gives a solution to the last in the chain of equations.)

These features show that both Indian and Chinese mathematicians paid great attention to the requirements for a scientific algorithm, which may be summarized

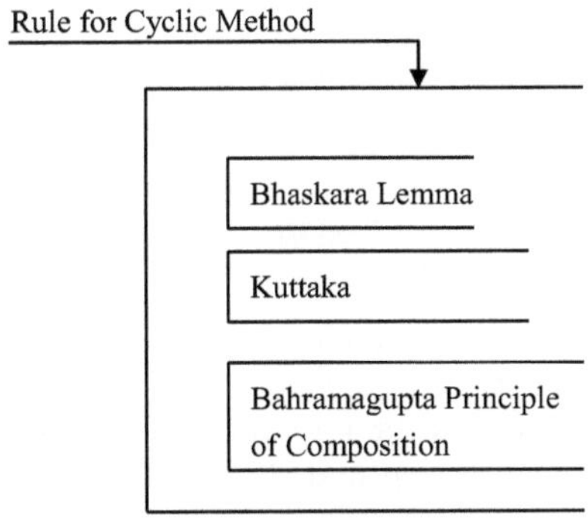

Figure 4

as follows:[11]

(a) *Mechanical character*, it should contain a sequence of instructions which are unambiguous, standardized, and easily iterated, so that anyone following these instructions in a step-by-step manner will always arrive at the same result;

(b) *Generality*, it should be applicable to a whole class of homogenous problems;

(c) *Effectiveness*, it should solve a given problem with certainty in a finite series of steps beginning with the initial data.

Like many other fields of human knowledge, modern algorithmic languages have very ancient roots.

Algorithmic Tendency in *Wasan*

Until the end of the sixteenth century, Japanese mathematics had developed as a transplantation of the traditional methods that came from China. This situation changed in the seventeenth century when Japanese mathematicians, represented by Seki Takakazu (1642-1708) and Takebe Katahiro (1664-1739), founded Wasan. Wasan mathematicians succeeded in greatly promoting the algorithmic tradition of Chinese mathematics. It seems that one of the original characteristics of Wasan mathematicians was their synthetic use of methods they learnt from Chinese mathematical texts, which inspired them to create even more advanced methods. Thus, we note that:

- They extended Liu Hui's *geyuan shu* (割圓術 —method for determining segment areas) to cases of solid figures while developing a naïve integral calculus—圓理 (*enri*, circle theory).
- They integrated *fangcheng shu* (方程術 –method for solving simultaneous linear equations by elimination) and *tianyuan shu* (天元術 -celestial element

[11]See Rybnikov, K. A., On the Role of Algorithms in the History of Mathematical Analysis, *Actes du VIII Congress International d'Historie des Sciences*, vol. I. Firenze & Paris, 1958.

method, a method of listing and solving higher–degree equations in one variable by setting a "celestial element") into the elimination theory for higher–degree equations with several unknowns along with the introduction of determinants.

- They expanded *zhaocha shu* (招差術, interpolation method) into their own approach to dealing with infinite series, ...*etc.*[12]

To see how strongly the algorithmic tendency was embodied in Wasan and to illustrate how its practitioners carefully worked out elaborate algorithms, we take up the example of Takebe's algorithm for approximating the value of π by use of 累遍増約術 (*ruihen zōyaku jutsu* — method of successive accelerated approximation). Takebe's method can be described as follows:

Let the diameter of the circle $d = 1$, and let a_n be the side length of a regular n–gon inscribed in the circle, and let p_n be its perimeter. Then the square of the perimeter $p_n^2 = T_n^{(0)}$, which expresses the initial sequence for approximating the value of π^2. Taking p as the perimeter of the circle, obviously $T_n^{(0)} = p_n^2 \to p^2 = \pi^2$. Takebe obtained the sequence $\{T_i^{(0)}\}$ $(i = 1, 2, 3, \ldots, n)$ by using Liu Hui's *geyuan shu*, as described above. But he then went on to calculate the ratios of first differences and used these to form a geometric series with common ratio of $\frac{1}{4}$:

$T_{i+1}^{(0)} - T_i^{(0)} = \Delta_i^{(1)}$ (First differences),

$\dfrac{\Delta_{i+1}^{(1)}}{\Delta_i^{(1)}} = r_1 \approx \dfrac{1}{4}$ (First common ratio),

$T_i^{(1)} = T_i^{(0)} + (1 + r_1 + r_1^2 + r_1^3 + \cdots)\Delta_i^{(1)} = T_i^{(0)} + \dfrac{4\Delta_i^{(1)}}{3}$ (first approximating value for the square of the perimeter).

Takebe called this procedure *jōyaku* (増約). The first *jōyaku*, as described above, produces a sequence of the second approximating values $\{T_i^{(1)}\}$ $(i = 1, 2, 3, \ldots, n-1)$. Taking second differences to form a geometric sequence with common ratio of $\frac{1}{4^2}$ is the second *jōyaku*:

$T_{i+1}^{(1)} - T_i^{(1)} = \Delta_i^{(2)}$ (Second differences),

$\dfrac{\Delta_{i+1}^{(2)}}{\Delta_i^{(2)}} = r_2 \approx \dfrac{1}{4^2}$ (Second common ratio),

$T_i^{(2)} = T_i^{(1)} + (1 + r_2 + r_2^2 + r_2^3 + \cdots)\Delta_i^{(2)} = T_i^{(1)} + \dfrac{4^2\Delta_i^{(2)}}{4^2 - 1}$ (second approximating value for the square of the perimeter).

---

[12] For details of such development see LI Wenlin, Some Reflections on Main Lines of Mathematical Development, in E. Knobloch, H. Komatsu, D. Liu. (eds.): *Seki: Founder of Modern Mathematics in Japan, A Commemoration on his Tercentenary*, Springer, 2013, pp.21-30.

One proceeds similarly to produce a sequence for the third approximating values $\{T_i^{(2)}\}$ $(i = 1, 2, 3, \ldots, n-2)$, taking third differences to form a geometric series with common ratio $\frac{1}{4^3}$, carrying out the third *jōyaku*, and so on. In general, the operations are these:

$$\Delta_i^{(m)} = T_{i+1}^{(m-1)} - T_i^{(m-1)} \ (m^{th} \text{ differences}),$$

$$r_m = \frac{\Delta_{i+1}^{(m)}}{\Delta_i^{(m)}} \approx \frac{1}{4^m} \ (m^{th} \text{ common ratio}),$$

$$T_i^{(m)} = T_i^{(m-1)} + (1 + r_m + r_m^2 + r_m^3 + \cdots)\Delta_i^{(m)} = T_i^{(m-1)} + \frac{4^m \Delta_i^{(m)}}{4^m - 1} (m^{th}$$
approximating value of square of the perimeter)

Takebe stopped at $m = 8$ (ie $i = 1, 2, 3, \ldots, 9$), and took $T_9^{(m)}$ as the resultant approximating value of π^2, he thus obtained

$$\pi = 3.1415\ 9265\ 3589\ 7932\ 3846\ 2643\ 3832\ 7950\ 2884\ 1971\ 2$$

an approximation of π that was far more accurate than those given earlier by Liu Hui and Zu Chongzhi.

In fact, Takabe's method can be summarized by a recusion formula:

$$T_i^{(m)} = T_i^{(m-1)} + \frac{4^m \Delta_i^{(m)}}{4^m - 1} =$$

$$T_i^{(m-1)} + \frac{4^m (T_{i+1}^{(m-1)} - T_i^{(m-1)})}{4^m - 1} = \frac{4^m T_{i+1}^{(m-1)} - T_i^{(m-1)}}{4^m - 1}.$$

It can be proved,[13] in fact, that Takebe's *ruihen jōyaku* is actually equivalent to a modern algorithm—*Romberg's method* for numerical integration—established in 1955 by the German mathematician Werner Romberg (1909–2003), who taught for many years in Norway after the Nazis rose to power in 1933.[14]

Some Concluding Remarks

The design of the structure of an algorithm is closely connected with the creation of the algorithm itself. Chinese and Indian mathematicians in ancient and medieval times as well as Wasan scholars created a great number of important algorithms, and here we have analyzed only a few representative cases. The algorithms created by these mathematicians, as illustrated above, were sophisticated, even when seen from a modern point of view. This kind of algorithmic thinking was

[13] For the proof and details of Takebe's method, see Xu Zelin: *On the Chinese Source of Wasan —Wasan Algorithms and Their Origins in Chinese Mathematics*, pp. 270-278, Shanhai Jiaotong University Press, Shanghai, 2012

[14] Werner Romberg, Vereinfachte numerische Integration, *Det Kongelige Norske Videnskabers Selskab Forhandlinger* (Trondheim) 28 (7) 1955: 30–36.

closely connected with, or later bred, a series of important mathematical concepts and ideas such as:

The place–valued decimal system, which provided the whole algorithmic tradition in China and India with a firm foundation;

Introduction of negative numbers, which were recognized in both China and India at an early stage due to their utility for an algorithmic approach to solving equations;

Algebraic discoveries, such as the triangle of binomial coefficients, determinants, etc.;

Symbolic expressions, especially for equation solving, which formed a central part in mathematical activities of an algorithmic character, for which Chinese and Indian mathematicians established their own respective symbolic systems for expressing equations;

Algebraization of geometric problems, Asian mathematicians, in contrast with the purely synthetic approach to geometry typical in the Euclidean tradition, transformed geometric problems into algebraic equations. This approach accorded well with the direction advocated by Descartes when modern mathematics was first beginning to arise in Europe.

All these aspects show the dynamic role of algorithmic thinking in advancing the development of mathematics, a tradition that actually rivaled, but also complemented, the Greek approach based on deduction. The discovery of great truths in mathematics has often resulted from the search for new algorithms. The emergence of both analytic geometry and calculus are perhaps among the most convincing examples of this. Calculus, as Rybnikov has pointed out,[15] was, indeed, a triumph of the algorithmic trend. As regards Descartes' analytic geometry, it is well known that analytic geometry represents an algebraization of geometry that already stands poles apart from the Euclidean tradition. By studying Descartes' sources, including his methodological works, we find clear evidence that his deeper motivation for developing a new approach to geometry grew out of a strong dissatisfaction with traditional deductive methods.

In his unfinished work, entitled *Rules for the Direction of the Mind* (1619-1628?) (abbreviated hereafter as *Rules*),[16] Descartes strongly criticized the "superficial demonstrations" of the Greeks, which did not seem to teach people "why things are so and how they discovered them." Being convinced that the Greeks had made their discoveries "more frequently by chance," Descartes felt the need to uncover a general method for finding new truths. Thus, the major objective of his *Rules* was to search for such a general method, which Descartes called a *universal mathematics*. This was, in fact, nothing less than a universal problem-solving program that can be summarized by the following series of reductive steps:

---

[15]Rybnikov, K. A., On the Role of Algorithms in the History of Mathematical Analysis, *Actes du VIII Congress International d'Historie des Sciences*, vol. I. Firenze & Paris, 1958.

[16]René Descartes, *Regulae ad Directionem Ingenii*, in C. Adam & P. Tannery, eds., *Oeuvres de Descartes*, vol. X, Paris, 1974, p. 375; English translation in E. S. Haldane & G. R. T. Ross, eds., *The Philosophical Works of Descartes* vol. I. Cambridge, 1973.

(1) Reduce any kind of problem to a mathematical problem;
(2) Reduce any kind of mathematical problem to a problem of algebra;
(3) Reduce any problem of algebra to a system of algebraic equations;
(4) Reduce the system of algebraic equations to a single equation.

The extant version of the *Rules* does not offer readers any details about how these steps are to be carried out, nor does it say how one should proceed once a problem has been reduced to a single algebraic equation. Nevertheless, a comparative reading of the *Rules* and Descartes' *Geometry* (1637) shows that in that classic work he definitely realized his program of *universal mathematics* within the realm of geometry.[17] The *Geometry* starts with a method for translating geometrical problems into a system of algebraic equations, and then proceeds to reduce the latter to a single equation in just one unknown. Descartes wrote these equations in descending order of the powers of the unknown and classified them by their degree. The third book of the *Geometry* was then devoted to constructing the roots of the equations of each class by a standard procedure. Thus the need to carry out these constructions for equations of the third degree and higher was what led Descartes to investigate the nature and classification of the curves appropriate for a given equation, and this eventually led to his coordinate geometry, which firmly established Descartes' position as one of the founders of modern mathematics. For Descartes himself, however, coordinate geometry was little more than a means to an end: his major concern was to show how all geometric problems can be solved by standard constructions of the roots of the corresponding algebraic equations. He was able to algebraicize these construction procedures, i.e. to base them on step-by-step algebraic calculations, precisely because his methods reflected algorithmic principles, as described above. Descartes thus freed himself from the Greek tradition by moving along the algorithmic road.[18]

The above argument does not mean, of course, that Descartes was aware of works written by ancient Chinese, Indian or Japanese mathematicians. The transmission of mathematical knowledge falls, at any rate, outside the range of this article, whose main purpose has been to focus attention on the algorithmic tendency in the development of mathematics, a trend more apparent today due to fast-breaking developments in computer science that are opening up a new era in mathematics. From this vantage point, it has become easier to recognize the larger role of the algorithmic tradition in the history of mathematics. Indeed, algorithmic construction and deductive reasoning are two different, but quite complementary components in mathematical intelligence, and the parts they played in the history of mathematics cannot be considered interchangeable. In fact, it appears that the development of mathematics is characterized by an alternation between the algorithmic and deductive approaches as dominant trends. Thus, for example, the mathematical knowledge of the ancient Babylonian and Egyptian type was fol-

[17]See Henk Bos, *Redefining Geometrical Exactness: Descartes' Transformation of the Early Modern Concept of Construction*, New York, Springer 2001.

[18]See LI Wenlin, On the Mechanical Character of Descartes' Geometry, *Studies in the History of Natural Science* 12(3) 1993, Science Press, 1993.

lowed by an era in which the methods of Greek geometry dominated, and these, in turn, gave way to the style of the Chinese, Indian, and Arabic mathematicians during medieval times. Then came the seventeenth and eighteenth centuries, which oversaw the breakthrough to a new era of modern mathematics. This was an heroic age full of searching for new infinitesimal algorithms; but from the 19th century onward the deductive method again thrived, though at a much higher level. This overview thus suggests a spiral development, in which two main lines of deveop-ment interacted: a deductive line—dominated by theorem-proving activities—and an algorithmic line, in which creative problem solving was central to mathematical praxis.

Acknowledgment

The author would like to thank Prof. Xu Zelin for his offer of the examples of Wasan algorithms, and Dr. Yang Jing for her help in producing the flow chart and program in JAVA for the Kuttaka algorithm in the appendices.

APPENDIX I *Qin's method*

(shi) a_n	$r^1_{n-1}c + a_n = r^1_{n-1}(= \bar{a}_n)$		......		
(fang) a_{n-1}	$r^1_{n-2}c + a_{n-1} = r^1_{n-1}$	$r^2_{n-2}c + r^1_{n-1} = r^2_{n-1}(= \bar{a}_{n-1})$	......		
(shanglian)	$r^1_{n-3}c + a_{n-2} = r^1_{n-2}$	$r^2_{n-3}c + r^1_{n-2} = r^2_{n-1}$	......		
a_{n-2}	$\vdots$	$\vdots$	......		
(lian)			......		
$\vdots$	$\vdots$	$\vdots$			
$\vdots$					
a_3	$r^1_2 c + a_3 = r^1_3$	$r^2_2 c + r^1_3 = r^2_3$			
a_2	$r^1_1 c + a_2 = r^1_2$	$r^2_1 c + r^1_2 = r^2_2$			
(xialian) a_1	$a_0 c + a_1 = r^1_1$	$a_0 c + r^1_1 = r^2_1$		$a_0 c + r_1^{n-1} = r_1^n(= \bar{a}_1)$	
(yu) a_0	a_0	a_0		a_0	$a_0(\bar{a}_0)$

APPENDIX II *A Flow Chart for the Kuttaka Algorithm*

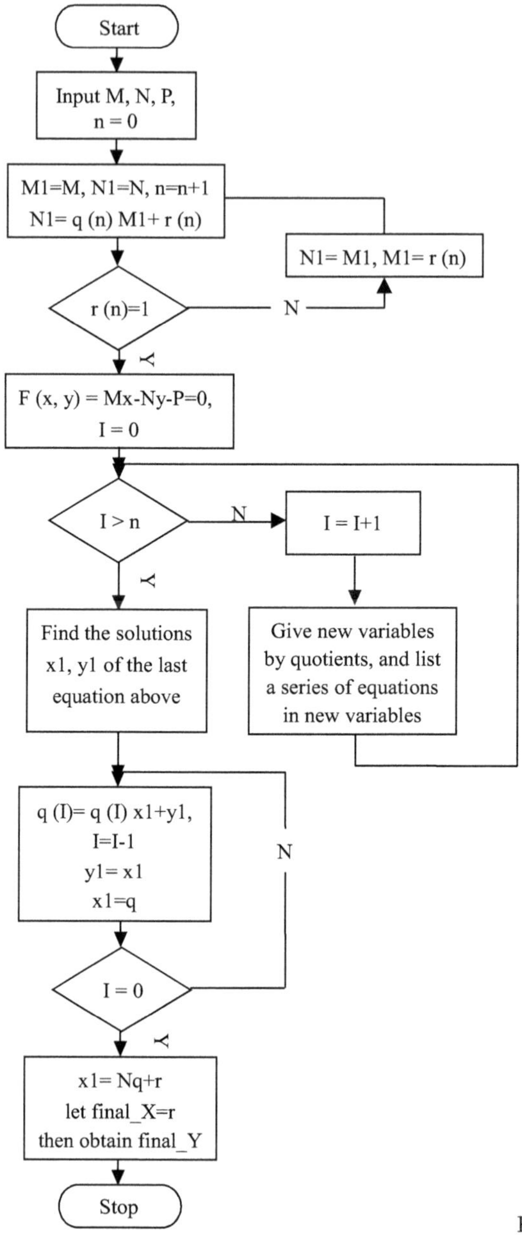

Figure 5

APPENDIX III *The program in JAVA for Kuttaka*

```
void jButton1_actionPerformed(ActionEvent e) {
try{
  int final_X,final_Y;
  M = new Integer(jTextFieldM.getText()).intValue();
  N = new Integer(jTextFieldN.getText()).intValue();
  P = new Integer(jTextFieldP.getText()).intValue();
  if(P!=0){
    int M1 = M, N1 = N, r;
    Vector v = new Vector();
    do {
      r = N1 % M1;
      v.add(new Integer( (N1 - r) / M1));
      N1 = M1;
      M1 = r;
    }
    while (r != 1);
    //while (r != 1 && r != 0);
    for (int i = 0; i < v.size(); i++) {
      System.out.println("q:" + v.elementAt(i).toString());
    }
    int args[] = new int[v.size() + 2];
    args[0] = M;
    args[1] = N;
    for (int i = 0; i < v.size(); i++) {
      args[i + 1] = args[i + 1] -
          new Integer(v.elementAt(i).toString()).intValue() *
          args[i];
      args[i + 2] = args[i];
      for (int j = 0; j < args.length; j++) {
        System.out.println(args[j]);
      }
      System.out.println();
    }
    int x = -1, y = -1;
    if (args.length % 2 == 0) {
      System.out.println(new Integer(args[args.length - 2]).
                toString() +
                "X" + "-" +
                new Integer(args[args.length - 1]).
                toString() +
                "Y-" + new Integer(P) + "=0");
```

```
    if (new Integer(args[args.length - 2]).intValue() == 1) {
      y = P / new Integer(args[args.length - 1]).intValue() +
        1;
      x = new Integer(args[args.length - 1]).intValue() * y +
        P;
    }
    if (new Integer(args[args.length - 1]).intValue() == 1) {
      x = P / new Integer(args[args.length - 2]).intValue() +
        1;
      y = new Integer(args[args.length - 2]).intValue() * x +
        P;
    }
  }
  else {
    System.out.println(new Integer(args[args.length - 1]).
               toString() +
               "X" + "-" +
               new Integer(args[args.length - 2]).
               toString() +
               "Y-" + new Integer(P) + "=0");
    if (new Integer(args[args.length - 2]).intValue() == 1) {
      x = P / new Integer(args[args.length - 1]).intValue() +
        1;
      y = new Integer(args[args.length - 1]).intValue() * x -
        P;
    }
    if (new Integer(args[args.length - 1]).intValue() == 1) {
      y = P / new Integer(args[args.length - 2]).intValue() +
        1;
      x = new Integer(args[args.length - 2]).intValue() * y -
        P;
    }
  }
  System.out.println("X:" + x + " Y:" + y);
  int middle[] = new int[v.size() + 2];
  for (int i = 0; i < v.size(); i++) {
    middle[i] = new Integer(v.elementAt(i).toString()).intValue();
  }
  if (middle.length % 2 == 1) {
    middle[middle.length - 2] = y;
    middle[middle.length - 1] = x;
  }
  else {
    middle[middle.length - 2] = x;
```

```
          middle[middle.length - 1] = y;
        }

        for (int i = 0; i < v.size(); i++) {
          middle[middle.length - i - 3] = middle[middle.length -
            i - 3] * middle[middle.length - i - 2] +
            middle[middle.length -
            i -
            1];
        }
        int X = middle[0];
        final_X = X % N;
        System.out.println("Final_X:" + final_X);
        final_Y = (M * final_X - P) / N;
        System.out.println("Final_Y:" + final_Y);
      }
      else{
        int mn = minMN(M,N);
        final_X = mn/M;
        final_Y = mn/N;
      }
      jLabel5.setText("FINAL_X:"+final_X+" "+"FINAL_Y:"+final_Y);
    }
    catch(Exception newe){
      jLabel5.setText("ERROR: "+newe.getMessage());
    }
  }
  int minMN(int m,int n){
    int mn = m;
    while(!(mn%n==0)){
      mn = mn+m;
    }
    return mn;
```

Gu Yingxiang's Method of Solving Numerical Equations with the Abacus

Guo Shirong (郭世榮)

Introduction

Gu Yingxiang (顧應祥) (1483-1565) was a major Chinese mathematician in the sixteenth century. Since the end of the eighteenth century, however, he has usually been regarded as a mediocre figure, one who merely represents the decline of mathematics during the Ming Dynasty (1368-1644). This negative assessment is mainly due to the fact that he openly acknowledged not understanding the *tianyuan shu* (天元術)(the celestial element method), one of the most important developments in Chinese mathematics during the twelfth and thirteenth centuries. Nevertheless, a proper and detailed investigation of the four mathematical treatises that Gu Yingxiang published—in particular both the *Ceyuan haijing fenlei shishu* (測圓海鏡分類釋術) (*Categorizing and Annotating on the Sea Mirror of Circle Measurement*, 1550) and the *Ceyuan suanshu* (測圓算術) (*Calculation Procedure for Circle Measurement*, 1553)—reveals that this usual view of his work is unjust. In fact, the main purpose of these two treatises was not to popularize the *Ceyuan haijing* (測圓海鏡) (*Sea Mirror of Circle Measurement*), a major work on the method of celestial element written in 1248 by Li Ye (李冶), but rather to demonstrate how to solve numerical equations by means of the Chinese abacus. More importantly, it should be noted that Gu Yingxiang's treatises were among the first published texts which explain the methods for solving numerical equations with an abacus, a simple and yet powerful tool that afterward gained widespread poularity among Chinese mathematicians. The present brief account of Gu Yingxiang's work should help rectify the established, but faulty view of Gu's accomplishments, and also to restore his reputation as a leading mathematician in sixteenth-century Ming China.

Extracting square or cube roots of a number, and the closely related, but more general problem of solving quadratic or cubic numerical equations, have a long history in China. Indeed, these problems can be traced back to the *Jiuzhang suanshu* (九章算術) (Nine Chapters on the Arts of Mathematics) of the first century BC. These early calculation procedures for extracting roots were later adapted into a general algorithm for solving numerical equations during the Song (960-1279) and Yuan (1279-1368) dynasties. Traditional procedures for root-extraction or equation-solving were carried out by manipulating counting rods on a counting board within the context of a numerical-rod system ([Qian 1966]). This system, however, was replaced by the abacus during the middle of the Ming dynasty (1368-1644). Both Gu Yingxiang and his contemporary Wang Wensu (王文素) witnessed the beginnings of the transformation process that led to the eclipse of the counting-rod techniques by the new approach based on the Chinese abacus.

D.E. Rowe and W.-S. Horng (eds.), *A Delicate Balance: Global Perspectives on Innovation and Tradition in the History of Mathematics*,
Trends in the History of Science, DOI 10.1007/978-3-319-12030-0_15

Nevertheless, historians have paid far more attention to Wang Wensu's role in this process, while largely overlooking the far more important contributions of Gu Yingxiang. Another work, Yan Gong's (嚴恭) *Tongyuan suanfa* (通原算法) (1372) ([Hua 1987], 379]), has sometimes been portrayed by historians of the Chinese abacus as an early example in which this device was used to carry out root extractions. However, in [[Guo 2010], 574-575], I showed that this was not the case, since Yan Gong's work still reflects the methods of the older tradition of counting rod calculations. Concerning the early influence of Gu Yingxiang's texts, it should be emphasized that his treatises from the period 1550-1553 represent the first significant published works on the use of the abacus for extracting roots and solving numerical equations in the history of mathematics in China. For while it is true that Wang Wensu's *Suanxue baojian* (算學寶鑒) from 1524 dealt with these topics and was written earlier, his manuscript was never published. Therefore, this work had very limited circulation and exerted scarcely any influence on contemporary Chinese mathematicians, unlike the works of his contemporary, Gu Yingxiang.

In this paper, I will deal mainly with Gu Yingxiang's extraction procedure in terms of the abacus. For this purpose, I will begin with a brief account of his mathematical treatises in Section 2. Since his *Ceyuan haijing fenlei shishu* was devoted to a discussion of the methods used in Li Ye's *Ceyuan haijing*, in Section 3 I will give a short description of the content and format of this mathematical classic from the thirteenth century. Drawing on the *Ceyuan haijing*, I will then in Section 4 show how Gu Yingxiang demonstrated his "*fenlei shishu*." In Section 5, I give two examples to illustrate how he solved equations using the abacus. Finally, my primary concern in Section 6 will be to compare this approach based on the abacus with the alternative methodology that used counting rods.

Gu Yingxiang's Mathematical Treatises

As a mathematician and astronomer who lived during the middle of the Ming Dynasty (1368-1644), Gu Yingxiang had contacts with two other contemporary mathematicians, Tang Shunzhi (唐順之) (1507-1560) and Zhou Shuxue (周述學).[1] Although he was one of the three top mathematicians and astronomers of his time, Gu Yingxiang has been held in rather low esteem for quite a long time, in fact, ever since Ruan Yuan (阮元) (1764-1849) and Li Rui (李銳) (1769-1817) published their *Chouren zhuan* (疇人傳) (Biographies of Mathematicians and Astronomers) in 1799.[2] As noted above, this negative assessment was comparative in nature: he was criticized because he had been unable to follow up on the great mathematical accomplishments attained by the mathematicians of the Song and Yuan dynasties,

[1] Tang Shunzhi styled himself Yingde (字應德), or Yixiu (一字義修)，alias Jingchuan (號荆川), and was, therefore, usually called Master Jingchuan (荆川先生). Zhou Shuxue (active in the mid-sixteenth century), called himself Jizhi (字繼志)，alias Yunyuanzhi (號雲淵子).

[2] The editors of the *Siku Quanshu* (四庫全書) (The Complete Collection of the Four Treasuries, 1772-1778) already expressed a similarly negative view of his work, but their opinion was less influential than that of Ruan Yuan and Li Rui.

most notably Li Ye (李冶) (1192-1279) and Zhu Shijie (朱世杰). This verdict, in fact, was largely based on his own frank acknowledgement that he could not understand Li Ye's *tianyuan shu* (天元術) (the celestial element method):

> Although the detailed demonstrations attached to each problem [in Li Ye's work] are checked again and again by setting up the *tianyuan* (celestial element), no method has yet been found to deal with them; therefore scholars are puzzled about what to do. I have to omit [these] ...and add another exposition.[3]
>
> But, the detailed demonstration in the solution of each problem is only dealt with by means of setting up the *tianyuan* [in Li's work] so that it is very difficult to get in hand.[4]

Tianyuan shu was an algebraic method invented by mathematicians during the Song dynasty. Li Ye gives a general explanation of this method in his works, but it seems that by the time of the Ming dynasty knowledge of *tianyuan shu* was largely lost ([Guo 2001a, Guo 2002]), while mainstream mathematical interests shifted during this period. In fact, the research of mathematicians during the Ming dynasty moved in entirely new directions. Whereas the mathematicians of the Song and Yuan dynasties had focused on "theoretical" concerns, mathematical research in the Ming era gave more weight to applications and popularizations of mathematics, or what might be characterized as "practical mathematics" ([Guo 2001b]). Nevertheless, in a recent study, I have shown that the three Ming–era mathematicians mentioned above—Tang Shunzhi, Gu Yingxiang, and Zhou Shuxue—were all deeply concerned with theoretical matters involving both content and format ([Guo 2010]). In fact, all three mainly worked on problems in pure mathematics rather than applications.

In the case of Gu Yingxiang, this can be seen merely by noting the topics in his four treatises on mathematics, which were written in the following chronological order:

(1) *Gougu suanshu* (勾股算術) (Calculation Procedures for Right-Angled Triangles), 1533;

(2) *Ceyuan haijing fenlei shishu* (測圓海鏡分類釋術) (Categorizing and Annotating the *Sea Mirror of Circle Measurement*), 1550;

(3) *Hushi suanshu* (弧矢算術) (Calculation Procedures for Arcs and *Sagitta*), 1552;

(4) *Ceyuan suanshu* (測圓算術) (Calculation Procedures of Circle Measurement), 1553.

---

[3]See ([Gu 1550/1993], Preface). This passage reads as follows: "每條下細草，雖經立天元一反復合之，而無下手之術，使後學之士茫然而無門路可入。輒不自揆，每章去其細草，立一算術."

[4]See ([Gu 1553/1993], 1109). The cited original reads: "但每章細草止以天元一互算，而漫無下手之處."

All four of these works reflect striking theoretical features. Among them, the treatises (1) and (3) were investigated by Li Rui and Luo Shilin (羅士琳) (1789-1860) in the eighteenth and nineteenth centuries, respectively. The subject of (1) is concerned with methods for determining the three sides of a right-angled triangle based on knowing the sums or differences of the three sides. Treatise (3), on the other hand, deals with the relationship between arcs and their *sagitta* in a circle. Both of these studies are the first monographs in China devoted to these subjects, which explains why historians have long been interested in them.[5] Gu Yingxiang's other two works, namely the treatises (2) and (4), have received far less historical attention. Even worse, historians of Chinese mathematics have very often cited his confession regarding his difficulties in understanding the *Tianyuan shu* as evidence that Gu was an inferior mathematician. In what follows, I will deal with Gu Yingxiang's treatises (2) and (4), mainly in order to discuss his methods for solving equations using the traditional Chinese abacus. This aspect of his work has never been dealt with in studies of the history of Chinese mathematics, nor does one find this topic in the historical literature on the abacus.

Li Ye's *Ceyuan Haijing* (測圓海鏡)

Before turning to these methods, however, we should first make some brief comments in order to understand the source of Gu Yingxiang's equations. Both of his works, the *Ceyuan haijing fenlei shishu* and his *Ceyuan Suanshu*, are based on Li Ye's classic text, *Ceyuan haijing* (測圓海鏡) (Sea Mirror of Circle Measurement, 1248). The subject of the *Ceyuan haijing* concerns an elaborate set of problems that use given data for a set of right-angled triangles in order to determine the diameter of a circle inscribed in a larger right-angled triangle.[6] All the problems have the same setting: a circular city outside of which two people walk along different paths, reaching points where they can either just see each other, or else a landmark, such as a tree. From any two given lengths in the diagram (see Fig. 1 below), the problem is to determine the diameter or radius of the circle. The diagram is the same for every problem, and the diameter of the circular city is set as 240 *bu*, so this turns out to be the solution for virtually every one of the 170 problems in the book. Indeed, its entire purpose is to construct a model that illustrates the celestial element method.

Li Ye named this figure the *yuancheng tushi* (圓城圖式) (illustration of a circle town); it has the following features:

(1) A right-angled triangle $\triangle ABC$ and inscribed circle with center O.

(2) DE and FG are two perpendicular diameters parallel to the legs of $\triangle ABC$. The circle is also inscribed in the square $JKLB$.

[5] For examples see [Li 1954] and [Wang 2002].
[6] See [Dauben 2007], 324-327.

(3) The figure is completed by drawing lines IP, NS, MQ, HR from the intersection points I, N, M and H on the hypotenuse AC and parallel to the legs of triangle $\triangle ABC$.

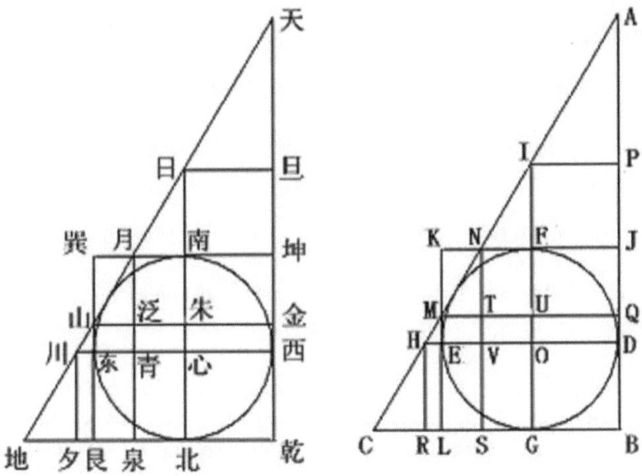

Figure 1: Illustration of A Circle Town [yuancheng tushi]

Li Ye's *yuancheng tushi* thus has one circle and 15 right-angled triangles. There are, however, two congruent pairs among the 15 triangles, so only 13 of them are quantitatively different. Although Li Ye wrote that this scheme was inspired by nine formulae he had learned from Master Dong Yuan (洞淵), the geometric figure originally derives from similar ones found in the *Nine Chapters on the Art of Mathematics* (九章算術). These relate to the problems of finding a right triangle that contains a circle (勾股容圓) (*gougu rongyuan*) or a square (勾股容方) (*gougu rongfang*), or the problem of the square town (邑方) (*yifang*). The circle in the diagram represents a town with gates located at D, E, F and G. In ([Guo 2013]) I have pointed out that Li Ye actually constructed an important mathematical model through which he could display his concepts and methods as well as his skill in carrying out these mathematical researches.

Using this mathematical model, the *yuancheng tushi*, Li Ye derived 682 identities, which he referred to as the *Shibie zaji* (識別雜記). They demonstrate the relations among the 15 right-angled triangles in the figure in Book 1 of the *Ceyuan haijing* ([Chemla 1993, Kong 1988]). Li Ye went on to enunciate 170 problems that illustrate, in every conceivable case, how one can always determine the diameter of the inscribed circle,[7] if just two data relating to the sides of the 13 right triangles

[7]This is why the word *ceyuan*, or circle measurement, appears in the title of the work *Ceyuan haijing*.

are given.[8]

Here is an example of one such problem:

> It is asked: Person β walks out of the south gate and proceeds 135 *bu* before stopping. Person α walks 600 *bu* southward from the corner B, and he then observes that he, β and the circular town all lie on a straight line. Question and answer are the same as before.[9]

As shown in Figure 1, β goes from the south gate F to point I, and α from point B to A, respectively.[10] Thus, FI and BA are the two given conditions from which Li Ye derives the following equation by means of his *tianyuan shu*, giving details step by step:

$$-4x^2 + 4AB - 2IFx - [2(AB - IF) - AB]AB = 0.$$

At this point, he merely noted that the diameter x of the circle could be obtained by solving this quadratic equation, but without mentioning the method for solving the equation.

In general, Li Ye's text explains how to construct the polynomial equations associated with each problem. He describes in detail exactly where the counting rods must be placed on the counting board to set up the equation, and then how the manipulations of the counting rods must be carried out in order to solve the problem. This is all designed to illustrate the new computational technique with which the treatise is concerned, the *tianyuan* or celestial element method, the celestial element being the unknown or x of the equations. Each of the powers of $x, (x^2, x^3)$, etc., has a particular position on the counting board, which designates where the coefficients for each power must be placed. From there, Li Ye gives step-by-step algorithms to work out the solutions to the given problems, which involve polynomials of degree at most 6. Applying his identities and *tianyuan shu*, Li Ye establishes one numerical equation for each problem, except for fourteen basic problems.

Gu Yingxiang's Arrangements of the Problems

Gu Yingxiang was interested in Li Ye's mathematical model of the circle measurement as well as this general subject matter. He thus studied Li Ye's classification of problems and their underlying structure, about which he commented that

[8]To each of the 13 right triangles, Li Ye considered 13 possible quantities. Thus, for example, if the three sides of one such triangle are a, b, c with $a < b < c$, then the 13 quantities are $a + b, a + c, b + c, c + (b - a), a + b + c, b - a, c - a, c - b, a + b - c, c - (b - a), a, b, c$. The first five are called *wuhe* (五和, literally five sums), the next five are called *wujiao* (五較, literally five differences), followed by the three sides of the triangle.

[9][*Ceyuan haijing* Book 5, Problem 1]. The original Chinese text reads as follows: "或問：乙出南門直行一百三十五步而立，甲從乾隅南行六百步，望乙與城參相直，問答同前".

[10]Notice that when reading an ancient Chinese map, the southern direction points upward, whereas the northern points downward.

Li Ye's system was "based on the *tong gou-gu* ($\triangle ABC$), *bian gou-gu* ($\triangle ADH$), *si gou-gu* ($\triangle IGC$) and so on, and combined other sides to find the diameter. It is very clear and complete."[11] Nevertheless, he was not completely satisfied by this system, and so he offered two new arrangements of the problems that followed different principles.

First, though, we must describe the original arrangement of problems in Li Ye's *Ceyuan haijing*, which consists of twelve books. Book 1 introduces the basic mathematical model and gives the identities that can be deduced from the geometric relations in it. Book 2 is entitled *zhenlu* (正率), or ordinary problems, and deals with the most basic problems in the subject, for instance, to find the diameter from known values of AB and CB. The next four books, 3–6, are each based on knowing, respectively, one of the following four primary sides: *bian-gu* (邊股) AD, *di-gou* (底勾) CG, *da-gu* (大股) AB, and *da-gou* (大勾) CB. In Books 7 and 8 we are given, respectively, one leg of the right triangle $\triangle IFN$ or of $\triangle MEH$. Book 9 is based on knowing either AC (4 problems) or $AB + BC$ (8 problems), whereas in Book 10 we are given the sum of three sides of a triangle. Finally, Books 11 and 12 concern problems involving remainders and fractions.

In his *Ceyuan hajing fenlei shishu*, Gu Yingxiang re-arranged the order of Li Ye's problems by classifying them into new categories. The 10 books of the *Ceyuan hajing fenlei shishu* present the original problems based on the following given elements:

Book 1: Two legs of one right triangle;

Book 2: Two *gou* (short legs), or two *gu* (longer legs), or two *xian* (hypotenuses);

Book 3: One *gou* and one *gu* in different triangles;

Book 4: One *gou* and one *xian* in different triangles;

Book 5: One *gu* and one *xian* in different triangles;

Book 6: One of *gou*, *gu* or *xian*, and the sum or difference of two legs;

Book 7: Sum of *gou* and *gu* of the largest triangle, and one sum, difference, or *xian*;

Book 8: Two sums of *gou*, *gu* or *xian*;

Book 9: Two differences of *gou*, *gu* or *xian*;

Book 10: Sum or difference with fractional relations.

Comparing the above list with the contents of Li Ye's *Ceyuan haijing*, we can easily distinguish the different ideas behind the classifications these two mathematicians used. Li Ye's approach fixes one particular line segment or combination

[11] Gu Yingxiang's foreword to *Ceyuan haijing feilei shishu*. The Chinese text reads as follows: "以通勾股，邊勾股，底勾股等錯綜而求之，極為明備."

of segments to categorize the problems, whereas Gu Yingxiang employed relationships between the three sides—*gou*, *gu* and *xian*—independent of the particular right triangles involved. He therefore introduced a new principle for categorizing the problems.

Gu Yingxiang's other treatise on this same subject, *Ceyuan suanshu*, has usually been regarded as a simplified or popularized version of his *Ceyuan haijing fenlei shishu*. Thus, for example, Ma Xiang writes that: "the *Ceyuan suanshu* is only a simplified version of the *Ceyuan suanshu fenlei shishu* ... we can also say that the *Ceyuan suanshu* is a popularized version." ([Ma 1993]) Wang Lianfa also writes that: "This treatise could be regarded as a simplified and popularized version of the *Ceyuan haijing fenlei shishu*, and as a rudimentary textbook for the study of problems of the inscribed circle."([Wang 2002]) These comments are in some ways correct, but still it should be noticed that Gu Yingxiang had another purpose in mind, namely to highlight the way relations between the parallels in the mathematical model lead to another way to classify the problems associated with it. He explains this point very clearly in his foreword to the *Ceyuan suanshu*:

> Inside and outside the circle [in the figure], the horizontal lines are short legs and the vertical lines longer legs [of right triangles]. There are beautiful relations of natural parallelism between the horizontal lines and vertical ones, or between the line segments in two horizontal lines and those in two vertical lines, or [those] between horizontal lines with hypotenuses and vertical lines with hypotenuses. Through studies and comparisons, the relations and results can be obtained. I, therefore, try to summarize my thinking and compile this work. The calculation methods are attached for those too difficult to be understood [in the work]. Although the old academic terms [from previous work] are still engaged, the explanations are as simple and clear as possible so that scholars can gather the essentials [of the theory] by reading this book. The contents about the combinations of sum and difference of lines seem too redundant to be discussed in this work, because the circle measurement methods are complete even without them.[12]

So again in this work, Gu Yingxiang rearranged the problems, introducing new categories in accordance with the parallel lines and their relations with the known elements. Here he considered not only the parallels in relation to the shorter and longer legs of the various right triangles but also those involving the hypotenuses as well, thereby taking up a structural feature of the entire mathematical model. The *Ceyuan suanshu* should, therefore, not be regarded merely as a simplified version of the *Ceyuan haijing fenlei shishu*. Indeed, this aspect had actually been exploited by Li Ye when he deduced the identities in Book 1 of his

[12]The Chinese text cited reads as follows: "蓋圓之內外，其横為勾，其直者為股。一横一直，或兩横兩直夾，或一横一斜、一直一斜，自有天然對待之妙。比而合之，皆可推類而知者。於是別出己見，復為編次。其難曉者，附以布算之法。名號雖因其舊，而詞則務簡而明，庶使學者一覽而可得其要領焉耳。若諸和諸較雜揉之分，似涉繁冗，故俱不錄。非略之也，測圓之法，止於是足矣".

Ceyuan haijing ([Chemla 1993]), and Gu Yingxiang was clearly well aware of this from his study of the text. He did not, however, discuss Li Ye's *shibie zaji*, nor did he address the identities Li Ye had derived from the geometric relationships. What is more, he left out the *tianyuan shu* method for finding equations that correspond to the various problems, frankly admitting that he could not understand these. However, his main interest was in solving those equations, a matter that was regarded as elementary and which was thus not explained in Li Ye's work. Accordingly, he was later criticized by mathematicians from the period of the Qing dynasty, especially since they had eventually come to understand Li Ye's theory of *tianyuan shu*. Ruan Yuan, for instance, expressed this opinion in a review:

> When Gu Yingxiang wrote his *Ceyuan haijing fenlei shishu* and *Ceyuan suanshu*, he left out the detailed explanations in each chapter because he could not manage to deal with the *tianyuan shu*. He only paid attention to solving the equations. He therefore attended to trifles and neglected the essentials. For his faults in writing this commentary on a subject he did not understand, Yingxiang would have no words to exculpate himself.[13]

Gu Yingxiang's Methods for Solving Equations

Methods for extracting roots or solving equations form a very important and time-honored subject in the history of Chinese mathematics. For this reason, historians of Chinese mathematics have paid much attention to them. For example, Qian Baocong has carefully discussed all kinds of traditional methods for finding roots of equations using the counting rods ([Qian 1966]). Joseph Needham and Wang Ling made a comparison of the Chinese techniques, especially those found in *Jiuzhang suanshu*, with Horner's method ([Needham and Wang 1955]). Hua Yinchun gave a survey of the history of root-extraction methods in his history of the abacus in China ([Hua 1987]), and there are other studies on this subject as well. Nevertheless, historians of mathematics have overlooked Gu Yingxiang's contributions to this field.

Before describing the techniques Gu Yingxiang introduced, it is necessary to explain briefly some standard Chinese terms used in discussing equations and their roots, called *shang* (商) (quotients). For an equation in x:

$$A_n x^n + A_{n-1} x^{n-1} + \ldots + A_1 x = N,$$

the constant N is called *shi* (實), and the other coefficients have special names: A_n is *yu* (隅); A_1 is *fang* (方) or *zong* (從); and $A_2, \ldots, A_{n-1}$ are the first *lian* (廉),..., the $(n-1)$th *lian*, respectively. A_2 is also called the upper *lian*, and A_{n-1} the lower

[13] Ruan Yuan, Preface to the New Edition of *Ceyuan haijing fenlei shishu*. The Chinese text reads as follows: "顧若溪應祥作《測圓海鏡分類釋術》、《測圓算術》等書，以立天元一無下手之處，每章輒刪去細草，而但演開帶諸方法，捨其本而求其末，不知妄作之罪，應祥實無可辭焉。(阮元：重刻測圓海鏡細草序)."

lian. Any positive coefficient of $A_1, A_2, \ldots, A_{n-1}$ is called *zong*, for instance, A_2 is call the first *zong lian* when it is positive, while a negative coefficient is called *yi* (益). If not all $A_1, A_2, \ldots, A_{n-1}$ are 0, the equation is with *zong* (*dai-zong*) (帶從). The roots of an equation are the first quotient (初商), second quotient (次商), and so on, from the highest to the lowest value.

Let us now consider two examples from Gu Yingxiang's *Ceyuan haijing fenlei shishu* in order to illustrate his methods.

Example. $x^2 + 720x = 230400$

The procedure for finding a root of a quadratic equation with *zong:* Put the *shi* on the left. (Double the difference, we get 720, and) it is *zong*. [Put it on the right of *shi*]. Estimating the first figure of the quotient, we get 200. Put it on the left place of *shi*, it is upper *fa* (法). Also put it [on the most right place] as *yu-fa*(隅法). *Yu-fa* together with *zong* becomes 920 as lower *fa* (下法). Subtracting 184000 [which is the product of *zong* and lower *fa*] from the *shi*, the difference is 46400. Doubling the *yu-fa*, we get 400 as *lian-fa* (廉法) . Estimate the second figure of the quotient, we get 40. Put it next to the left figure as upper *fa* (上法) and also put it as *yu-fa*. The sum of *zong* and *lian-fa* is 1160. Multiplying it by upper *fa*, and subtracting the product from the *shi* then the *shi* becomes 0.[14]

[14] Book 1 of the *Ceyuan haijing fenlei shishu*. ([Gu 1550/1993], 999) The original text reads: "帶從開平方術曰：列實於左。倍較得七百二十為從，約初商，得二百。置一於左上爲法，置一為隅法，帶從方，共九百二十為下法。除實一十八萬四千，餘實四萬六千四百。倍隅法得四百，為廉法，約次商，得四十，置一于左次，為上法，置一為隅法，並從方、廉法共一千一百六十，為下法，與上次法相乘，除實盡。"

This procedure can be summarized as shown here:

Table 3: Procedure to find a root of $x^2 + 720x = 230400$

Step	*fa* (upper *fa*)	*Shi*	*Zong* or *zong fang*	*yu-fa* or *lian-fa*	lower *fa*
1		230400	720		
2	200	230400	720	200	$720 + 200 = 920$
3	200	$230400 - 920 \times 200 = 46400$	720	$200 \times 2 = 400$	920
4	240	46400	720	440	$720 + 440 = 1160$
5	240	$46400 - 1160 \times 40 = 0$			

Example. $x^4 + 1406x^3 + 511907x^2 + 4730640x = 10576065600$

The procedure for finding a root of a 4th–degree equation with *zong*. Put the fourth power product [15076065600] as *shi*. Estimate the first quotient by dividing *shi* with the sum of *zong*, *fang* and *yu* to obtain 100. Put the first quotient on the upper left as upper *fa*. Also put it [on the right of *shi* and] multiply the first *lian*, it is [the new first *lian*] 51190700. Put it next, and multiply it by itself, then multiply by the second *lian*, the product is [the new second *lian*] 14060000. Put it next, multiply itself twice, it is *yu* 1000000. Adding together the *zong-fang*, *lian*[s], and *yu*, the sum is lower *fa* 70981340. Multiply the upper *fa*, then subtract the product from *shi* 10576065600, the difference is 3477931600. It is the *shi* of the next quotient. Doubling the first *lian* gives 102381400, multiplying the second *lian* by 3 is 42180000, and multiplying the second *lian* by 4 is 4000000. Multiplying the first quotient by itself and then multiplying by 6 is 60000; tripling the first quotient and multiplying the lower*lian* is 421800; their sum plus the first *lian* [511907] is the [new] upper *lian* 993707. Multiplying the first quotient by 4 plus the second *lian* [1406] is the lower *lian* 1806. The next quotient is 20. Put one of it as upper *fa*; also put one of it multiplying by the upper *lian* [993707] gives 19874140; multiplying itself and then multiplying by *yu*, it is 8000.[15] Adding together the *fang*, *lian*[s], and *yu*, the sum is lower *fa* 173896580. Multiplying the lower *fa* by the upper *fa*, and then subtracting the product from shi, the difference is 0.[16]

[15] The author has added 12 missing figures based on calculation.

[16] Book 3 of the *Ceyuan haijing fenlei shishu*. The Chinese text reads as follows: "帶從方廉開三乘方曰：置所得三乘方積為實，以從方廉約之，初商得一百。置一於左上爲法；置一乘從一廉得五

The calculation in this procedure is summarized in:

Table 4: Procedure for solving
$x^4 + 1406x^3 + 511907x^2 + 4730640x = 10576065600$

Step	up *fa*	*shi*	*fang*	first *lian*	second *lian*	*yu-fa*	lower *fa*
1		10576065600	4730640	511907	1406		
2	100	10576065600 − 70981340 × 100 = 3477931600	4730640	100 × 511907 = 51190700	100^2 × 1406 = 14060000	100^3 × 1 = 1000000	4730640 + 51190700 + 14060000 + 1000000 = 70981340
3	100			51190700 × 2 = 102381400	14060000 × 3 = 42180000	1000000 × 4 = 4000000	A, B
4	120	3477931600	4730640	102381400	42180000	4000000	C, D
5	120	3477931600− $E = 0$					lower *fa*= E

Note: There should be two other columns in the table for the so-called upper *lian* (上廉) and lower *lian* (下廉). To save space, these are abbreviated here as A, C and B, D, and placed in the lower *fa* column instead of listing two new columns.
$A = 1002 \times 6 + 100 \times 3 \times 1406 + 511907 = 481800 + 511907 = 993707$
$B = 100 \times 4 + 1406 = 1806, C = 20 \times 993707 = 19874140, D = 202 \times 1806 = 722400$
$E = 4730640 + 102381400 + 42180000 + 4000000 + C + D + 8000 = 173896580$

Generally speaking, given an equation $ax + bx^2 + cx^3 + dx^4 = F$ and let $x = x_1 + x_2$, then

$$\begin{aligned} F = ax + bx^2 + cx^3 + dx^4 &= a(x_1 + x_2) + b(x_1 + x_2)^2 + c(x_1 + x_2)^3 + d(x_1 + x_2)^4 \\ &= a(x_1 + x_2) + b(x_1^2 + 2x_1x_2 + x_2^2) + c(x_1^3 + 3x_1^2x_2 + 3x_1x_2^2 + x_2^3) \\ &\quad + d(x_1^4 + 4x_1^3x_2 + 6x_1^2x_2^2 + 4x_1x_2^3 + x_2^4) \\ &= x_1(a + bx_1 + cx_1^2 + dx_1^3) + \\ &\quad x_2[(a + 2bx_1 + 3cx_1^2 + 4dx_1^3) + x_2(b + 3cx_1 + 6dx_1^2) + x_2^2 \cdot 4dx_1 + dx_2^3] \end{aligned}$$

The solution procedure is then based on the last expression in the above formula. By setting $x_1 = c$, a first approximation of the root, one then gets a quartic equation for x_2, so solving this gives the sought for value of the root as $x_1 + x_2$.

千一百一十九萬〇七百；置一自之，以乘從二廉，得一千四百〇六萬；置一自乘再乘，得一百萬為隅法。併從方廉隅共七千〇九十八萬一千三百四十爲下法，與上法相乘，除實七十〇億九千八百一十三萬四千，餘積三十四億七千七百九十三萬一千六百爲次商之實。倍從一廉得一億〇二百三十八萬一千四百，三因從二廉得四千二百一十八萬，四因隅法得四百萬。初商自之，又六因，得六萬；初商三之，以乘下廉得四十二萬一千八百; 相併，加入從一廉，得九十九萬三千七百〇七，為上廉。初商四之，帶從二廉，得一千八百〇六，為下廉。次商二十，置一爲法，置一乘上廉得一千九百八十七萬四千一百四十，置一自之，以乘下廉得七十二萬二千四百，[置一自乘再乘，以乘隅，得八千。] 併方廉隅共一億七千三百八十九萬六千五百八十爲下法，與上法相乘，除實盡."

Gu Yingxiang also solved the equation in another way by using the formula:

$$\begin{aligned}F &= ax + bx^2 + cx^3 + dx^4 \\ &= a(x_1 + x_2) + b(x_1 + x_2)^2 + c(x_1 + x_2)^3 + d(x_1 + x_2)^4 \\ &= x_1(a + bx_1 + cx_1^2 + dx_1^3) + \\ &\quad x_2\{a + b(2x_1 + x_2) + c[3(x_1 + x_2)x_1 + x_2^2] + 4dx_1^3 + 6x_1^2x_2 + 4x_1x_2^2 + dx_2^3\}\end{aligned}$$

This form, however, is not relevant for the theme of the present paper. Here, we only point out that Gu Yingxiang solved equations from degree two to six in his works. He gave dozens examples with detailed explanations for different symbols for the coefficients of the equations.

Counting Rods vs. Abacus

The key question for our present discussion concerns the calculation tools Gu Yingxiang used to solve these equations. As I now will show, he did not employ traditional counting rods for this purpose; he worked instead with an abacus. Historian Hua Yinchun has discussed very briefly the similarities and differences between these two tools and their use in carrying out algorithms for solving equations ([Hua 1987], 379, 396]). Here we will give a more detailed explanation before discussing Gu Yingxiang's computational procedures.

Counting rods served as the main calculation tool in traditional Chinese mathematics. From the thirteenth century onwards, mathematicians also completed their calculations in written form on paper; nevertheless, their calculations were still carried out within the framework of the counting–rod system ([Li 1991]). This means that their mathematical expressions, terms and symbols remained unchanged. The only innovation was that the final results, obtained by the manipulation procedures of counting rods on the counting board, were recorded on paper.

A much more significant shift took place during the fifteenth and sixteenth centuries when Chinese mathematicians shifted away from the old techniques based on counting-rods to a new system that used the abacus. There are major differences between these two systems when it comes to handling numerical expressions and operating on them. In the counting-rod system a numerical equation $A_nx^n + A_{n-1}x^{n-1} + \ldots + A_1x = N$ is expressed by writing the coefficients in a vertical column (see Fig.2). All procedures that lead to the solution of the equation are carried out by operations on the different levels of the vertical columns, where the sought for root appears at the highest level.

There are several standard steps in the procedures for solving a numerical equation. One of these involves estimating the number of digits required to express that part of the root which is a whole number. This is done by means of "pacing" (subtracting successively) the constant N by using the coefficient A_n. When this cannot be done, then the coefficients A_m $(1 \leq m \leq n)$ are each moved m places to the left, sometimes one rod is borrowed (*jieyisuan*) (借一算) as an indication rod of pacing and the pacing begins again with A_{n-1}. The pace value gives the number of digits for the root. Another key step is formula transformation. Here, the number at the level of A_m is moved backward to the right.

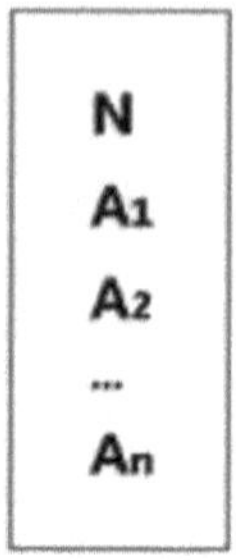

The procedures for extracting roots or solving equations in the abacus system are quite different from the counting-rod methods. Many of these differences can be understood simply because the representation of numbers with an abacus requires placements along a horizontal line. This affects the techniques used to carry out an algorithm. Thus, one of the distinguishing features in calculations using the Chinese abacus is the lack of shifting of the columns. This contrasts sharply with calculations in the counting–rod system, where a digit can express different values depending on its placement. Thus, for instance, the digit 1 can stand for 10 when situated in the tens–place or 100 in the hundreds–place. In the abacus system, however, the digits cannot be freely moved: their value is defined and fixed by naming the place value of the digit.

This accords, for example, with Gu Yingxian's procedure when making the calculations for A, B, C, and D in Table 2 above. If we consider, in general, the manner in which he carries out his algorithms, it becomes evident that he made his calculations using an abacus. As illustrated in the examples above, Gu Yingxiang's methods for extracting roots and solving numerical equations can be distinguished from the counting-rod methods due to the fact that:

(1) All his calculations are carried out along one horizontal line, which means that his numbers are arranged horizontally. Some of the technical terms he uses indicate placements in a horizontal direction, such as "left" and "right" in locating the terms.

(2) He makes no use of standard techniques employed with traditional counting rods, such as "pacing N" or "borrowing one rod" to determine the number of digits of the root. Thus, the procedures involving movements of the coefficients A_m are absent in his algorithms.

(3) Every digit of a number is listed in his calculations, including 0 when it appears at the end of a number. This is also a feature consistent with recording abacus calculations.

(4) All root-extraction methods are based on transformations of polynomials instead of using the iterated multiplication method (*zengchen kaifangfa*) (增乘開方法).

Gu Yingxiang provided numerous detailed explanations for the various equations he studied, which covered examples ranging from quadratics up to sixth-degree polynomials. For each equation, which represented a typical case for a certain type, he explained the calculation procedure required. Altogether, one finds more than 60 different root-extraction procedures in his works, and all of them exhibit the features (1)-(3) indicated above, namely the characteristic methodology employed when using the abacus to solve the equations.

Concluding Remarks

In summary, we noted at the outset how Gu Yingxiang's studies of Li Ye's *Ceyuan haijing* have been evaluated from a perspective that took little account of the motivations behind this work. For the mathematicians in the Qing dynasty as well as later historians of mathematics, Gu's works simply reflected a poor understanding of the *tianyuan shu.* This view, however, reflects a very limited understanding of what he accomplished. Indeed, in our view Gu Yingxiang's four treatises should be regarded as among the most important works on mathematics undertaken during the Ming dynasty. Moreover, like those of Tang Shunzhi and Zhou Shuxue, these were "theoretical" studies, in contrast with other "application-oriented" mathematical work from this period [Guo 2001b].

With regard to Gu Yingxiang's two studies devoted to circle measurement —his *Ceyuan haijing fenlei shishu* and *Ceyuan suanshu*—one should not simply fault him for his failure to understand the *tianyuan shu.* His two proposals for the classification of the problems in Li Ye's *Ceyuan haijing* clearly reflect real understanding and originality, particularly his second system based on parallelism; for this alone, his work deserves more detailed investigation. Most significant of all, however, were Gu Yingxiang's methods for extracting roots and solving numerical equations by means of the abacus, an aspect of his studies that has received almost no attention by historians of mathematics. Historians of the Chinese abacus, on the other hand, have mistakenly regarded Yan Gong's *Tongyuan suanfa* (通原算法) (1372) ([Hua 1987], 379]) as a pioneering work for the methods of root extraction with the abacus. This claim can hardly be upheld, however, as pointed out in [[Guo 2010], 574-575]. Yan Gong's approach shares only certain characteristics with the standard techniques introduced by Gu Yingxiang, whereas he still employs the operation of "borrowing one rod," a procedure that played a major part in calculations with counting rods. Quite clearly, Yan Gong's use of the abacus appeared at a time when counting rod calculations still dominated Chinese mathematics.[17]

Finally, in considering the influence of Gu Yingxiang's two texts based on

[17]This period also saw the introduction of written records for the calculations [Li 1991].

calculations with the abacus, we have stressed that they were far more important than Wang Wensu's *Suanxue baojian* (算學寶鑒) (1524), a work that was original, but which had little historical significance. Since it remained unpublished, very few mathematicians had a chance to study it, unlike the two treatises of Gu Yingxiang, which were widely known after their publication in the middle of the sixteenth century.

Acknowledgment

I would like to take this opportunity to thank Jia-Ming Ying for helpful comments on an earlier version of this article. My thanks also go to the editors, Wann-Sheng Horng and David Rowe, for their help in preparing the present version.

Bibliography

[Chemla 1993] Chemla, Karine (1993). Textual Structure of Li Ye's *Sea Mirror of Circle Measurement* and Expression of Mathematical Knowledge, in Li Di ed., *Shuxue Shi Yanjiu Wenji* Vol. V. Huhhot: Inner Mongolia University Press / Taipei: Jiu Zhang Press, 1993, 122-140.

[Dauben 2007] Dauben, Joseph, Chinese Mathematics, in *The Mathematics of Egypt, Mesopotamia, China, India, and Islam: A Sourcebook*, Victor Katz, ed., Princeton: Princeton University Press, pp.187-384.

[Gu 1550/1993] Gu Yingxiang (1550/1993). *Ceyuan haijing fenlei shishu* (*Categorizing and Annotating on the Sea Mirror of Circle Measurement*), in Guo Shuchun ed., *Zhongguo Kexue Jishu Dianji Tonghui* (Shuxue juan) (Selected works of Chinese Science and Technology (mathematics section). Zhengzhou (鄭卅): Henan Education Press, 1993, pp. 995-1078.

[Gu 1553/1993] Gu Yingxiang (1553/1993). Ceyuan Suanshu (*Calculation Procedure Circle Measurement*), in Guo Shuchun ed., *Zhongguo Kexue Jishu Dianji Tonghui* (Shuxue juan) (Selected works of Chinese Science and Technology (mathematics section). Zhengzhou: Henan Education Press, 1993, pp. 1109-1140.

[Guo 2001a] Guo Shirong (2001a). La perdita della conoscenze matematiche e astronomiche, Sandro Petruccioli ed., *Storia della scienza*, Roma, Istituto della Enciclopdia Italiana, v. II.

[Guo 2001b] Guo Shirong (2001b). The Influence of Yang Hui' s Works on the Mathematical Mainstream in the Ming Dynasty, in Alan K. L. Chan, Gregory K. Clancey & Hui-Chieh Loy eds., *Historical Perspectives on East Asian Science, Technology and Medicine*, Singapore: World Scientific, 2001, pp. 358-367.

[Guo 2002] Guo Shirong (2002). On the Problem of the Lost of Mathematical and Astronomical Knowledge, *France Sinology* (法國漢學), vol. 6, 2002, 320-335.

[Guo 2010] Guo Shirong (2010). The Transformation of the Mathematical Mainstream and the Development of Abacus, in Guo Shuchun ed., *History of Science and Technology in China (Mathematics)*, Beijing: Science Press, 2010, pp. 511-610.

[Guo 2013] Guo Shirong (2013). The Axes of Mathematical Methodology in the Song and Yuan Dynasties: the Construction of Mathematical Models , in Eberhard Knobloch, Hikosaburo Komatsu, and Liu Dun eds., *Seki, Founder of Modern Mathematics in Japan, A Commemoration on His tercentenary*, Springer 2013, pp.105-118.

[Hua 1987] Hua Yinchun (1987). A History of Chinese Abacus (in Chinese). Beijng: China Financial Press, 1987.

[Kong 1988] Kong Guoping (1988). Biography of Li Ye (in Chinese). Hebei Education Press, 1988.

[Li 1991] Li Di (1991). On the Changes of Mathematics in Song and Yuan dynasties (in Chinese), in*Selected Papers on the History of Science and Technology in China* (1). Huhhot: Inner Mongolia Education Publishing House, 1991, pp. 219-233.

[Li 1954] Li Yan (1954). Chinese Mathematicians' Theories for the Arc and Sagitta (in Chinese), in Li Yan, *Zhongsuanshi luncong* (中算史論叢) Vol. 2 Beijing: Science Press, 1954.

[Ma 1993] Ma Xiang (1993). Summary to the *Ceyuan suanshu*, in Guo Shuchun ed., *Zhongguo Kexue Jishu Dianji Tonghui* (*Shuxue juan*) (Selected works of Chinese Science and Technology (mathematics volume II). Zhengzhou: Henan Education Press, 1993. Vol. 2, pp. 2-1107.

[Needham and Wang 1955] Needham, Joseph and Wang Ling (1955). Horner's Method in Chinese Mathematics: Its Origins in Root-Extraction Procedures of the Han Dynasty, *T'oung Pao* XLIII, 345 (1955).

[Ruan 1798] Ruan Yuan (1798). Preface to the Reprinted *Cenyuan Haijing*, collected in *Zhibuzu Zhai Congshu* (知不足齋叢書), 1798.

[Qian 1966] Qian Baocong (1966). *Zengcheng kaifangfa de lishi fazhan* (The Historical Development of Root-extraction by Iterated Multiplicatoin, in *Song-Yuan Shuxue Shi Lunwen Ji* (Papers on the History of Song-Yuan Mathematics in China). Beijing: Science Press, 1966.

[Wang 2002] Wang Lianfa (2002). Gu Yingxian: A High Officer and Mathematician, *HPM Tongxun* 5(5) (2002), 10-15.

Chinese Mathematics: Transmissions and Transformations

History of Korean Mathematics, 1657-1868: An Overview

Wann-Sheng Horng

Introduction

The history of Korean mathematics, just as that of any other civilization, deserves due attention for its own sake. However, this subject drew hardly any serious scholarly attention until the last decade of the twentieth century.[1] This situation has in the meantime changed, mainly due to several investigations involving the critical analysis of key mathematical texts and their relationship with those texts that were introduced into Korea from China.[2]

A large number of these contributions are devoted to analyzing the mathematical texts collected in Kim Yong-Woon's edition, *Hanguk Kwahak Kisul Sa Charyo Taekye: Suhak Pyon* (Collection of Source Materials of the Korean History of Science and Technology: Mathematics Section, 1985). By joining efforts to explore this subject, a number of scholars have published articles on Korean mathematics. These intend not only to show how *chungsan* (中算) (Chinese mathematics) was transmitted to Korea but also, and even more significantly as Korean historians of science have proclaimed, how *chungsan* was adapted and transformed as a key ingredient in what came to be understood by *tongsan* (東算) (Korean mathematics).[3]

[1]Cf. [Kim and Kim 1978]; [Kim 1998]; [Song 1986]; [Park 1990]. As one of the leaders in the community of historians of Korean science and technology, Kim Yung Sik has encouraged his graduate students to take up studies on the history of Korean mathematics. This work complements the efforts of Kim Yong-Woon's to produce new editions of Korean mathematical texts (Cf. [Kim 1985]). In addition to his articles on the subject ([Kim 1973a]; [Kim 1973b]; [Kim 1973c]; [Kim 1986]), Kim Yong Woon published, in collaboration with Kim Yong Guk, *Kankoku Sugakushi*, a work which provides a general overview in Japanese of the history of Korean mathematics ([Kim and Kim 1978]). Hedeki Kawahara has also made substantial contributions to the history of Korean mathematics in the *Choson* period from the middle 17th century to the early 18th century ([Kawahara 1998]).

[2]I have been undertaking a comprehensive study of the subject since the year 2000. Around 1989, when I was teaching at the Institute of History, Tsing Hua University, Taiwan, I came across Kim's edition in the library, from which I copied a few pages from the Nam Pyong-gil texts. I did not read these texts carefully, however, until early 2000. Just before hosting the HPM 2000 in Taipei, I published my first article on the history of Korean mathematics ([Horng 2000a]). This compares *Jiegen fang* (the Western method, based on Cardano's work, for setting up equations, which was transmitted to China in the early eighteenth century) and *tianyuan shu* (the traditional Chinese method as presented by the Song-Yuan mathematician Li Ye) in Nam's *Vu Yi He* (無異解) (*Muihae*, 1855). Some years later, Guo Shirong published his *Zhongguo Shuxue dianji zai chaoxian bandao de liuchuan yu yingxiang* (中國數學典籍在朝鮮半島的流傳與影響, Circulation and Influence of Rare Chinese Mathematical Texts in Korea), which represents a substantial contribution to the study of the transmission and influence of Chinese mathematical texts in Korea ([Guo 2009]).

[3]In this article, Korean names and terms are spelled in accordance with the McCune-Reischauer system, whereas for Chinese the pinyin system is employed. However, for some transmitted Chinese texts, such as the *Tongwen suanzhi* (同文算指), the pinyin system will be used when there is no contemporary Korean spelling available.

D.E. Rowe and W.-S. Horng (eds.), *A Delicate Balance: Global Perspectives on Innovation and Tradition in the History of Mathematics*,
Trends in the History of Science, DOI 10.1007/978-3-319-12030-0_16

At first glance, Korean *Choson* mathematics strongly resembles Chinese work from the same period. The major texts of both cultures were written in Chinese. In addition, the educational and recruiting system for Korean mathematical practitioners—including textbooks, teaching requirements, and the form of the examinations—closely followed the models established in Ming and Qing China. Even the rhetoric of *Choson* Confucians appealed strongly to *sirhak* (實學) (practical learning) concerns, and thus echoed similar views expressed by their Chinese counterparts. Little wonder, then, that some historians maintain that Korean mathematics should be regarded as a particular branch of Chinese mathematics ([Guo 2009], 335).

Most studies of the history of Korean mathematics in the last two decades, whether they aim to address this issue or not, have mainly been concerned with textual explication, though often taking into account larger contextual issues. In particular, Guo Shirong's comprehensive investigation of Sino-Korean mathematical transmission has made a substantial contribution toward resolving several larger issues ([Guo 2009]). Hideki Kawahara has studied especially how *Choson* mathematics developed in the context of *songnihak* (性理學) (neo-Confucianism), demonstrating that this was a major factor in shaping the character of Korean mathematics ([Kawahara 2010]). By now an overall picture, if only sketchy, of the history of Korean mathematics during the *Choson* period has begun to emerge.

In this article, besides referring to Hideki Kawahara and Guo Shirong's contributions to this theme ([Kawahara 1998]; [Kawahara 2010]; [Guo 2009]), I will basically synthesize the conclusions drawn from careful readings of Chinese and Korean texts made by me and my former students.[4] My aim is to give an overview of Korean mathematics during the *Choson* (朝鮮) period of 1657-1868. This was an era in which Korean mathematical texts, written or printed in Chinese, provide easy access for such exploratory studies. In particular, I will focus on how the Korean mathematicians came to be aware of what the term *tongsan* meant in the *Choson* context. In this connection, arguments about how the Korean Confucians explained mathematics in terms of *songnihak* are also given in order to explain the evolutionary aspects of the status of mathematical knowledge (See Sections 3 & 4). I will also comment on some relevant historiographical issues, such as the "Problem of China" and certain "Why not" questions (as paraphrased by Kim Yung Sik), with the hope that this discussion will lead to a better understanding of Korean mathematics in its socio-cultural context (See Section 6). Concerning contextual issues that surround the practical applications of mathematics, I will in Section 2 discuss the intellectual tradition of *sirhak*, which the *yangban* (兩班) (ruling class) adopted in an effort to reform the state. It is interesting to

[4]Aside from my own publications, seventeen master's theses were written under my supervision. In addition, one doctoral thesis dealt with Nam Pyong-gil's reinterpretation of classical mathematics. What follows is the list of the master's and doctoral theses: [Chang 2005]; [Chen 2003]; [Chen 2007]; [Chou 2003]; [Hsieh 2003]; [Hung 2003]; [Kuo 2007]; [Li 2003]; [Li 2010]; [Lin 2003]; [Sun 2003]; [Wang 2011]; [Wen 2009]; [Wu 2003]; [Wu 2007]; [Xia 2003]; [Yeh 2002]; [Ying 2010].

note the distinctive features of Korean mathematical knowledge, as compared to the Chinese, especially when comparing Chinese translations of Western works on mathematics and science. As for social background, I will explore how the role of *chungin* (中人) (secondary status group) mathematicians came to be regarded as significant in *Choson* history.[5] Indeed, their dialogue with the mathematical *sadaebu* (士大夫) (literati) eventually made the *tongsan* more meaningful than ever and in due course helped to raise the epistemological status of mathematical knowledge in the period (see in particular Section 5).

Intellectual Concerns and Social Background

Our story begins in the year 1657 when Kyong Song-jin (慶善徵) (1616-?) gave Kim Si-jin (金始振) a copy of *Sanhak gemong* (算學啟蒙) (Introduction to Mathematical Study, by Zhu Shijie (朱世傑)).[6] As far as we know, Kyong Song-jin was the first Korean mathematician from whom a mathematical manuscript has survived to the present.[7] His *Muksajip sanbop* (默思集算法) (Calculation Method of Writings of Contemplation) not only followed the format of the *Sanhak gemong* but also synthesized the *Yanghwi sanbop* (楊輝算法) (Yang Hui's Method of Computation, by Yang Hui (楊輝)) and *Sangmyong sanbop* (詳明算法) (Explanations of Arithmetic, by an anonymous author) and *Sanbop tongjong* (算法統宗) (Systematic Treatise on Arithmetic, by Cheng Dawei (程大位)).[8] These first three texts on mathematics introduced from China played, on the one hand, a significant role in the transmission of knowledge into Korea before the second half of the 18th century. On the other hand, among the three the *Sanhak gemong* was particularly important in helping to shape the *tongsan*.[9] Some brief explanation of this transmission process is needed in order to do justice to this early stage in the history of Korean mathematics.

Going back to the seventh century, the Korean people learned mathematics mainly by means of Chinese texts introduced during the early Tang. In the course of the Tang dynasty (618-907) they also duplicated its bureaucratic systems, which included training and recruiting of mathematical practitioners ([Yeh 2002],

[5]Literally, the word *chungin*, which is commonly used in Korean literature, would be rendered in English as "middle people." Here I instead translate it as "secondary status group" following the lead of Kyung Moon Hwang in his *Beyong Birth: Social Status in the Emergence of Modern Korea* (Harvard University Press, 2004). I am grateful to Professor Joseph Dauben for bringing this work to my attention.

[6]In the Chinese pinyin system, *Sanhak gemong* is written *Suanxue qimeng*. In this article I will use both the Korean and Chinese terms, depending on the context.

[7]Kyong Song-jin's *Muksajip sanbop* is the first text collected in Kim Yoon Woon's *Hanguk Kwahak Kisul Sa Charyo Taekye: Suhak Pyon* (韓國科學技術史資料大系・數學篇).

[8]For a detailed textual analysis of the *Muksajip sanbop*, see [Li 2003]. It is noted there that Kyong Song-jin did not make use of the *tiayuan shu* (天元術) in this text, though he was apparently quite familiar with the method.

[9]In fact, the *Sanhak gemong* also played an indispensible role in shaping *wasan* (Japanese mathematics). This story is discussed in [Horng 2002b].

[Jun 2006]). In order to appreciate how *tongsan*[10] arose from this context, it is important to take account of the historical circumstances that shaped the *Choson* or *Yi* dynasty (李朝) (1392-1910), especially in relation to its foreign policy regarding contemporary China.

Due to his non-aristocratic background, the first *Choson* King Yi Song-gye (李成桂) (reigned 1392-1398) had to struggle in order to win the confidence of the traditional *Koryo* (高麗) aristocracy.[11] He therefore cultivated the support of the Ming dynasty, adopting the term *sadae* (事大) ("serving the great") to describe his foreign policy toward Ming China ([Lee 1984], 189). This may explain why the Yi government dispatched three regular delegations to China each year. "The purpose of all these missions was chiefly political, but they served also as the medium for cultural borrowing and economic exchange" ([Lee 1984], 184).

Among the articles imported from China, Korea obtained silk fabrics, medicines, porcelain wares, and books. Already during the late *Koryo* period (918-1392), the above mentioned mathematical texts—*Sanhak gemong, Yanghwi sanbop*, and *Sangmyong sansop*—had already been transmitted to Korea ([Yeh 2002]; [Yeh 2004]). Thanks to King Sejong's promotion of mathematical studies ([Guo 2009], 22-26), these texts exerted a significant influence on Korean mathematics by no later than the fifteenth century. King Sejong (世宗大王) (reigned 1418-1450) set an example for royal patronage that proved important for the history of mathematics and astronomy in Korea. I will return to this story in the next section.

After Manchuria conquered Ming China and established the Qing dynasty, the *Choson* kingdom maintained a similar foreign policy towards China. This made it possible for Western mathematical knowledge, which first entered China at the beginning of the seventeenth century, to be introduced into Korea by means of Chinese translations of Western works. Korea's regular envoys to Qing China provided access to these books. For example, the *Chongzhen lishu* (崇禎曆書) (*Chongzhen* Reign Treatise on Astronomy and Calendrical Science), a comprehensive translation of Western astronomy first edited by Xu Guangqi (徐光啟) and then by Li Tianjing (李天經) in the late Ming period, may have been transmitted into Korea by 1644. It was in that year that Kim Yuk (金堉), then the Chief of the Astronomical Bureau, led a mission to Qing China so that he and his two associates could learn the Qing's *Shixian* Calendar (時憲曆) ([Chu 2009]; [Guo 2009]). Moreover, beginning in 1741 the Korean envoys always included an astronomer whose duty was to study under the Jesuit astronomers in Beijing. This official practice was due to the need to amend the Korean calendar, which was based on the *Shixian* Calendar. In fact, this 1741 envoy brought back the *Shuli jingyun* (數理精蘊) (Collected Basic Principles of Mathematics), *Lixiang kaocheng* (曆象考

[10]Some Korean scholars prefer the term *Hansan* (Korean mathematics (韓算)) rather than *tongsan* (literally Eastern mathematics). The latter term was associated with the *Tongguk* (literally Eastern state (東國), with reference to China as the central state), a term used by the *Choson* Koreans for their own country.

[11]See [Lee 1984] for more details on the history of Korea.

成) (Compendium of Calendrical Science and Astronomy), as well as other Chinese versions of Western treatises on astronomy, trigonometry, and logarithmic tables. Fifty years later, the *Lixiang kaocheng* and the *Shuli jingyun* (in Korean, *Suri chongon*) were assigned as official textbooks for educating and recruiting astronomers. This initiative in education and in reforming the examination system proved very successful ([Guo 2009], 298-299). The *Suri chongon* would become influential from the early nineteenth century onwards.

Some of the Chinese books collected during these missions may have been gifts from the Qing emperor. For example, in 1783 Emperor Qianlong showed his hospitality by giving the Korean visitors the *Siqu quanshu* (四庫全書) (Complete Library of the four Branches of Literature) ([Wen 2009], 14). Thanks to this voluminous edition, several classic mathematical texts found their way to Korea, including the *Suanjing shishu* (算經十書) (Ten Mathematical Manuals) and of course *Jiuzhang suanshu* (九章算術), but also *Qigu suanjing* (輯古算經) (Continuation of Ancient Mathematics) by Wang Xiaotong (王孝通) as well as Song-Yuan Chinese mathematical texts like *Shushu jiuzhang* (數書九章) (Mathematical Treatise of Nine Sections) by Qin Jiushao (秦九韶), *Ceyuan haijing* (測圓海鏡) (Sea Mirror of Circle Measurement), and *Yigu yanduan* (益古演段) (New Steps in Computations) by Li Ye (李冶).[12] Still, another very significant factor affected this transmission process, namely, the circumstance that these Korean envoys included a number of *yangban* mathematical practitioners, like Choe Sok-chong (崔錫鼎), Cho Tae-koo (趙泰耉), So Ho-su (徐浩修), and Hong Tae-yong (洪大容), all of whom played important roles in *Choson* mathematics during the seventeenth and eighteenth centuries ([Lee 2001]).

Even so, the Korean tradition of *tongsan* could only emerge once its practitioners began to develop the discipline on their own. This, however, became possible through interactions between two distinct social classes, the *yangban* and the *chungin*. The *yangban* rose to exert influence in Korean society beginning around the second half of the fifteenth century, after which they attained high social status due to their role in public service ([Yi 1996], 50). Historian Lee Ki-baik contrasts this class with the *chungin*, noting that:

> The sole profession of the *yangban* was the holding of public office. Yet they did not serve in the technical posts that alike were components of the bureaucracy. Positions as medical officers, translator-interpreters, technicians in the astronomy and meteorology office, accountants, statute law clerks, scribes, and government artists all were the hereditary preserve of the so-called middle people (*chungin*). Nor did the *yangban* perform the routine duties of petty clerks and local civil functionaries (*sori* (胥吏) and *ajon* (衙前) or of military cadre members (*kungyo* (軍校)). These too, on their lower level, were the preserve of hereditary classes of functionaries. To be sure, in the broad sense that

[12]See [Guo 2010], 678-679 for the mathematics texts included in the *Siku quanshu*, for which Dai Zhen (戴震) served as editor of the section on astronomy and mathematics.

> they too were assured of substantive positions within the administrative apparatus, these specialists, clerks, and military distinguished from the *yangban*. The *yangban* scholars also were not interested in working in agriculture, manufacture, or commerce, for these were but the occupations of farmers, artisans, and merchants. The *yangban* scholars might well dream of achieving an ideal polity through the moral cultivation of *Choson* people, but this did not mean in fact that they were prepared to do away with distinctions of social status. ([Lee 1984], 174-175).

In fact, *yangban* scholars were primarily Confucians,[13] and like their Chinese counterparts they regarded mathematics as one of the six inferior arts; they thus downgraded its epistemological status as a discipline. This remained true, even though they tried to make sense of mathematical knowledge in terms of the *songnihak* (性理學) (neo-Confucianism), which they learned from Chinese Song scholars like Zhu Xi (朱熹) (1130-1200) ([Kim 2003]). Nevertheless, by no later than the mid-seventeenth century some *yangban* scholars began to pay close attention to mathematical studies. One reason for this came from urgent social problems that could only be met by the use of mathematical methods. In the wake of the devastation caused by invasions from Japan (1592-1593, 1597-1598) and Manchu China (1627, 1636-1637), the Koreans faced serious challenges in reconstructing and reforming their society. Farmland ruined by these events had to be surveyed and relocated. In this connection, techniques of mathematical measurement became urgent for the bureaucratic system. This may well explain why Kim Si-jin, who endeavored to restore and reprint *Sanhak gemong*, had been appointed to head the national surveying team.[14]

The privileged social status of *yangban* scholars and their efforts to promote mathematical knowledge by means of institutional resources were major factors that led to a gradual improvement in status for the discipline. It was within this context that issues like *tongsan* versus *sirhak* (實學) (practical learning) took their place on the intellectual stage. Meanwhile, a parallel exchange took place regarding mathematical practices that led to a crossing of borders that formerly stood between the *yangban* Confucians and the *chungin* mathematicians. Yet, why did some of these *yangban* scholars become so seriously interested in mathematical studies? One of the major reasons was closely connected with the development of the *sirhak* in the *Choson* period, a theme to which we now must turn.

According to historian Lee Ki-baik, "the birth of *Sirhak*" was part of an effort to censure "those who held political power and was intent to bring about changes in the political and social order" ([Lee 1984], 232). This came about in reaction to the political power of a small number of *yangban* families who exploited their position and failed to deal with the social and economic problems caused by this

[13]During the period of the Three Kingdoms in Korean history—the *Koguryo* (37 BC-668AD), *Paekche* (18 BC-660AD) and *Silla* (1 BC-668 AD)—the Confucian ethos inculcated teachings that helped maintain the traditional aristocratic social orders ([Lee 1984], 58).

[14]*Sillokchong* (實錄廳) (Bureau of the *Sillok*) 36 (1978), 343, record of 6 August 1662, cited in [Jun 2006].

system. Agriculture in the countryside was controlled by rich farmers, whereas economic life in the urban areas was dominated by the wealthy merchants who controlled trade and handcraft production. "As the numerous social ills attendant upon these phenomena became more grave, the problems *Yi* dynasty society now faced demanded serious reflection on the part of the members of its educated class. Their response is embodied in the scholarship and thought known today as 'Practical Learning,' or *Sirhak*" ([Lee 1984], 232). The scholars of the *sirhak*, active in the *Choson* period since the seventeenth century, can be divided into three groups: "Those who advocated the renovation of the farmland system and administration machinery, those who were interested in the orderly distribution of commodities and the innovation of production techniques, and those who sought truth by ascertaining historical evidence in the classics and ancient monumental inscriptions" ([Kim 1986]).

The first group was led by Yu Hyong-won (柳馨遠) (1622-1673) and afterward by Yi Ik (李瀷) (1681-1763). Yi Ik's major work, *Songho sasol* (星湖僿說) (Essays of Songho), reflects the diversity of his scholarship, including his broad interests in science and technology. He even referred to the remarks of the Ming Chinese authority, Xu Guangqi, regarding the didactical importance of the *Jihe yuanben* (幾何原本), the Chinese translation of Euclid's *Elements*, which was deemed useful for training students' mental reasoning powers. On the other hand, he also criticized the disregard of mathematics in contemporary Korean society ([Horng 2008]). His attitude towards mathematics would prove very significant in the period that followed, since he "attracted many disciples, and thus *sirhak* gradually emerged as the dynasty's dominant school of thought" ([Lee 1984], 234).

The second *sirhak* school was concerned with the "economic enrichment" of the state. Among its leaders, Hong Tae-yong (1731-1783) deserves special attention here because he devoted himself to mathematical studies and their relevance for *songnihak* ([Hung 2003]). Like his predecessor, Yu Su-won (柳壽垣) (1695-1755), Hong Tae-yong also wrote an account of his travels to Beijing. In both accounts, these visitors recognized the superiority of the Chinese civilization in which they "attached great value to commercial and manufacturing activity and to agricultural work" ([Lee 1984], 236). This may explain why Hong Tae-yong was enthusiastic about learning mathematics and science from the Qing Chinese, including the Western knowledge that had been introduced into China earlier. I will discuss Hong Tae-yong's mathematics in Section 3.

Among the leaders of the third group, Kim Chong-hui (金正喜) (1786-1856) "adopted the approach of the Ch'ing (Qing) empirical school and is known in particular for his deeply learned studies of historical inscriptions" ([Lee 1984], 256). The goal was to "find solutions to the pressing problems of the day by searching for their roots in the Korean historical process" ([Lee 1984], 255). Another scholar from this group, Choe Han-gi (崔漢綺) (1803-1875), came from a "fallen" *yangban* background.[15] By committing himself to the new *yangban* ideals, he developed a

[15]Literally, the fallen *yangban* were "those whose capacity to actually exercise their claim to

progressive view of history. This may explain why in his political treatise, *Injong* (人政) (On Personnel Administration), he argued that "the way to restore good government was through proper recruitment of officials, and he proposed that men of talent be educated and employed from *yangban*, peasant, craftsman, or merchant backgrounds, without discrimination" ([Lee 1984], 256). I will later describe his ambitious plans for integrating mathematics into *songnihak*. In any event, *yangban sadaebu* and *chungin* mathematicians now made serious efforts to work together and to show respect for each other. The close collaboration of *yangban* Nam Pyong-gil (南秉吉) and *chungin* Yi Sang-hyok (李尚爀) marks an important turning point for the elevation of the social status of Korean mathematicians in the second half of the nineteenth century, as we will show in Section 4.

Tongsan in Context: Kyong Song-Jin and Hong Chong-Ha

The term *tongsan* made its first appearance in the manuscript *Tongsan cho* (東算抄) (Manuscript of Eastern mathematics), which was written by an anonymous author.[16] This text may have provided source material for Hong Chong-ha (洪正夏) when he wrote his *Kuil chip* (九一集) (Writings on Nine and One) ([Horng 2002a]; [Chen 2003]). On the other hand, Hong Chong-ha had family connections with Kyong Song-jin—Hong was a maternal grandson of Kyong's nephew—so Kyong's *Muksajip sanbop* was perhaps an even more likely influence on the *Kuil chip*.[17] Hong Chong-ha and Kyong Song-jin, who are regarded as the most important mathematicians in Korea during the early eighteenth century, were the two figures mainly responsible for the formation of the *tongsan*.[18]

In order to assess their contributions, one should not ignore their social status as *chungin* mathematicians. As technocrats in the nine-class system, the *chungin* could only be promoted to rank six, whereas the *yangban* could be promoted to the first class, the highest in the state bureaucratic system.[19] There was thus a

yangban status had been eroded." Due to the dominance and political power of a few families, "many *yangban* thus excluded from government buried themselves in the countryside and fell to the status of a kind of local gentry." In addition, "within *yangban* lineages unable to obtain a government appointment for any of their members for several generations, some had been forced by circumstances to become small-scale farmers. These were the so-called ruined *yangban* (*chanban* (殘班)), who had sunk to the point where they no longer could maintain the dignity and authority that marked traditional *yangban* class status" [Lee 1984], 250.

[16]The *Tongsan cho* (Manuscript of Eastern mathematics) was probably written in 1718 ([Chen 2003], 7). According to Guo Shirong, the author was Hong Chong-ha ([Guo 2009], 38-42).

[17]This accords with what Kim Yun Sik argues about the social characteristics of the *chungin* class, namely that it formed a closed system through intermarriage in *Choson* society ([Kim 1998]). See also [Kawahara 1998], in which the author points to the genealogies of Kyong and Hong, respectively, in order to show the *chungin* people indeed formed a tightly knit group.

[18]Kawahara also refers to Hong Chong-ha's and Kyong Song-ji's genealogies in order to demonstrate that Kyong's contributions to mathematics helped to shape the *tongsan* ([Kawahara 1998]).

[19]By adopting the Chinese system of bureaucratic ranking, *Choson* Korea had nine classes to which officials could be promoted. Nevertheless, the mathematics candidates who completed their training courses and passed the examination could only serve in the four lowest classes. By

striking contrast between the high–ranking *yangban* and these *chungin* mathematicians. How the latter perceived themselves professionally is thus very significant for our understanding of the history of Korean mathematics. Equally important is the perception held by *yangban sadaebu* of mathematical knowledge proper. In other words, one must take into account both the social status of *chungin* mathematicians as well as the epistemological status of mathematical knowledge in order to explain how the *tongsan* took shape in the *Choson* period. At the same time, one must take account of the institutional base of the *chungin* mathematicians which enabled them to develop their professional role—indeed, both aspects, institutional as well as professional, are equally important for understanding the social history of mathematics in Korea. Regarding the former, the contributions of King Sejong should also be remembered; his reign (1418-1450) was known as the golden age in the history of Korean science and technology.

As recorded in the *Sejong sillok* (世宗實錄) (*Annals on King Sejong*), the king promoted astronomical studies and, in particular, calendrical calculations. This activity gave practitioners the knowledge and discipline needed for the development of Korean mathematics proper ([Yeh 2002]). The Bureau of Calendrical Calculations was primarily responsible for training these mathematical practitioners. Just by glancing through the standard textbooks that were used for this training as well as for preparations to take examinations, namely the *Sanhak gemong, Yanghwi sanbop* and *Sangmyong sanbop*,[20] one easily sees that their contents went well beyond the elementary level ([Yeh 2002]; [Yeh 2004]). The authors of the first two of these Chinese texts were Zhu Shijie (朱世傑) and Yang Hui (楊輝), respectively. These two texts are among the great mathematical classics in thirteenth–century Song–Yuan China ([Li and Du 1987], 109-174).

There are not many source materials or even mathematical texts available for studying Korean mathematics in the period immediately after King Sejong's reign. By the seventeenth century both the *Sanhak gemong* and *Yanghwi sanbop* had become quite rare. This helps to explain why Kyong Song-jin was so important in the history of Korean mathematics. In the preface to the new edition of the *Sanhak gemong* (1660), Kim Si-jin (金始振) (1618-1667), the chief of Chonju (全州) prefecture, mentions that he acquired the Korean edition from Kyong Song-jin in 1657. Meanwhile, he also obtained a handwritten manuscript of the *Yanghwi sanbop* from Chong Kun-yang (鄭君瀁), the chief of Kumgu (金溝) county. This fragmentary evidence suggests that by the seventeenth century these two classic texts, the *Sanhak gemong* and *Yanghwi sanbop*, were rarely known at this time. It seems they were not easily available even to the *sadaebu*, not to mention the *chungin* mathematicians. Nevertheless, they found someone to annotate these texts, the

contrast, those candidates who had *yangban* origin could be promoted to the top class. See the website dl.vmall.com/c0sb55dig9/ for a table of *Choson* kindom officials.

[20]In the modern Chinese Pingyin sysytem, *Sangmyong sanpob* is written *Xiangming suanfa*. It was perhaps written by An Zhizhai (安止齋) in Yuan China, but some historians think the author was He Pingzi (何平子), and thus the true author is not known. The text is about practical mathematics for daily use.

county magistrate Im Chun (任濬) (1647-?), who also made graphical illustrations ([Horng 2002c]). Moreover, Kim Si-jin supplemented the *Sanhak gemong* with a chapter on sea island surveying, which he took from the *Yanghwi sanbop*.

Kim Si-jin served the chief of Chonju prefecture, just as did Pak Yul (朴繘) (1621-?), who wrote the *Chuhak ponwon* (籌學本源). This was edited and modified into the *Sanhak ponwon* (算學本原) (its new title) by Hwang Yun-sok (黃胤錫) (1729-1791) for his voluminous collected works, the *Yisu sinpyon* (理藪新編) (Practical Learning based on *Songnihak*) [Sun 2003]). According to Hwang Yun-sol, Pak Yul's *Chuhak ponwon* was written to help readers prepare for studying the *Sanhak gemong* ([Kim 1985] vol. 3, 219). Surely Pak Yul, just like Kim Si-jin, Chong Kun-yang, and Im Chum would have been well versed in the *Sanhak gemong*. Kim Yong Woon has even guessed "that he published not only an enlarged edition, but in fact a revised edition with all the chapters compiled in the light of Korean circumstances at that time" ([Kim 1973c]).

The 1660 Korean reprint of the *Sanhak gemong* was a most important event for the history of mathematics in the wider context of Sino-Korean and Sino-Japanese relations ([Horng 2002c]). Ironically, the text had been lost in China since its first publication in 1299. Seen from the Korean side, the preservation of this text offered its mathematical practitioners a viable avenue for learning to work with the system of counting rods (*chusan* (籌算)), a method of calculation that already since the early fourteenth century had been replaced by the abacus (*zhusuan*) (珠算)) in China ([Li and Du 1987], Chapter 6). Moreover, the *Sanhak gemong* contains the method of *chon wonsul* (*tianyuan shu*) (天元術), which Korean mathematicians used to advantage to solve problems involving algebraic equations in one unknown.[21] Still, the Korean version of the text represents the interplay of mathematical ideas in a kind of exchange between the *sadaebu* Kim Si-jin and the *chungin* Kyong Song-jin, through which the latter took central stage in the *Choson* mathematics of the seventeenth century. This opened the way for further interactions between the *chungin* and the *sadaebu* in the realm of mathematical studies, and Kyong Song-jin gained high respect among the *sadaebu*. As a *chungin* technocrat of rank six, he was invited to supervise a mathematics seminar for local administrators that was primarily concerned with taxation of the farmlands ([Horng and Li 2007]). The participants were local officials (also of rank six, though of *yangban* origin) as well as their associate *chungin* mathematicians.[22] Song-jin clearly played a key role in this context, despite the fact that he was a *chungin* mathematician. In fact, Kyong Song-jin was the only *chungin* mathemati-

[21]The *tianyuan shu* (technique of the celestial element) was a method dealing with algebraic equations in one unknown. It was given a systematic description in Li Ye's *Ceyuan haijing* (1248) and *Yigu yanduan* (1259). "Celestial element" refers to the unknown in the problem ([Li and Du 1987], 135-140).

[22]These local officials invariably came from the *yangban* class. After passing the central civil service examination, they were typically offered a position like the head of a county (rank six). Their most important duty then consisted in overseeing the assessment and collection of taxes. If it was found that they had made mistkaes in carrying out these duties, they were usually punished ([Horng and Li 2007]).

cian whose work was praised by his contemporary, Cho Tae-koo (1660-1723), an influential *sadaebu* scholar who promoted mathematical studies enthusiastically.

Cho Tae-koo was from the *yangban* class and held high administrative positions, including an appointment as supervisor of the Observatory. He employed the rhetoric of *kujang* (九章) (literally nine chapters) in order to raise awareness of the importance of mathematical study in the context of *sirhak*. His strategy can be seen from the last section of his *Chuso kwangyon* (籌書管見) (Humble Opinions on Mathematics), where he takes up the *Kujang mundap* (九章問答) (Q & A on the *Jiuzhang suanshu*) to engage readers in a mathematical discourse.[23] There he argued that, despite its praxis-oriented approach, the *Jiuzhang suanshu* (Nine Chapters on the Mathematical Art) nevertheless contained fundamental knowledge needed in order to pursue further learning, for example works such as Zhu Shijie's *Chon wonsul* (*tianyuan shu*) or Western plane and spherical trigonometry ([Horng 2002c]). In his postscript to the *Chuso kwangyon*, Cho Tae-koo even took up Confucius's words of encouragement to his seventy disciples that they should learn mathematics. Confucius emphasized the utility of mathematics, noting how such knowledge was indispensible for those who wished to use it to benefit society.

It is evident that the *sadaebu* Confucian Cho Tae-koo knew mathematics quite well, so he was able to help elevate the status of this discipline in the context of the *sirhak* school ([Horng 2001a]; [Wu 2007]). This represents a major turning point in Korean intellectual history, since for the Korean Confucians, just like their counterparts in Qing China, mathematics was traditionally viewed as one of the six inferior arts, and thus had nothing to do with Confucian moral philosophy. This meant that it was not seen as a respectable subject within traditional Confucianism.[24] Thus, by the seventeenth century, the *sadaebu* were beginning to view these matters from a different perspective.

Even so, it should be noted that Cho Tae-koo did not refer to his contemporary *chungin* Hong Chong-ha (1684-?). This is surprising, especially in light of the use he made of *chon wonsul*. Using this method, which had been lost in China after the early 14th century, he was to prove his superiority in mathematical expertise over the Qing Chinese astronomer He Gouzhu (何國柱) ([Kawahara 1998]; [Horng 2002a]; [Guo 2009]). In fact, Hong Chong-ha's mastery of the *chon wonsul*, alongside the innovative features of his *Kuil chip*, show that he was able to build on the tradition of the *Sanhak gemong*. This indicates that Korean mathematicians were already beginning to develop a mathematical culture of their own ([Horng 2002a]). Whether or not Hong Chong-ha was the author of the *Tongsan cho*, the *Kuil chip* and the *Tongsan cho* both represent intellectual achievements that served to establish the *tongsan*'s legitimacy ([Chen 2003]). The manner in

[23] According to Guo Shirong, the problems cited in the *Kujang muundap* mainly came from Cheng Dawei's *Suanfa tongzong* ([Guo 2009], 95-97). The Han dynasty text of the *Jiuzhang suanshu* that was brought to Korea was lost by no later than the seventeenth century, as witnessed by Cho Sok-chong (1645-1715) ([Guo 2009], 91).

[24] A brief explanation of the evolving social status of Chinese mathematicians, especially with regard to major Confucian intellectual concerns, is given in [Horng 1993b].

which the Korean mathematicians transformed Chinese mathematical knowledge by offering somewhat different interpretations of these earlier works reveals how mathematics evolved from culture to culture even though the texts were written in the same language, namely Chinese ([Horng 2002a]).

Tongsan in Context: Eighteenth-Century *Sadaebu*

It is clear that during the eighteenth century *Choson* period, both *sadaebu* Confucians as well as *chungin* mathematicians were actively engaged in mathematical studies under the influence of the *sirhak*. Among the major scholars from this period, Hwang Yun-sok (黃胤錫) (1729-1791) stands out because of his efforts to bring about a fusion of mathematics with the *songnihak* (neo-Confucianism). His *Sanhak ibmun* (算學入門) (Introduction to mathematics) provided a beginner's guide to the study of Pak Yol's *Chuhak ponwon* or *Sanhak ponwon.*[25] In fact, two manuscripts on mathematics were included in his *Yisu sinpyon* (which contains 23 volumes in total). This huge work covers such diverse subjects as *songnihak*, mathematics, astronomy, calendrical calculation, history, and geography, as well as topics relevant to the proposals of the three schools of *sirhak* (Cf. Section 2 above). Hwang Yun-sok's encyclopedic knowledge had much to do with his family background and intellectual heritage. He devoted himself to study from the time of his childhood, growing up in a family that owned a large library of books. Although he was not from the *yangban* class, he passed the military examination (*mukwa* (武科)), but could only be promoted to rank six in the Ministry of Military Affairs. However, because of his commitment to *songnihak* in his youth, he was able to study under the supervision of the master Kim Won-haeng (金元行). Among the other disciples, he became acquainted with contemporary scholars like Hong Tae-yong and So Myong-ung (徐命膺) ([Chou 2003]; [Chou 2004]). In due course, Hwang Yun-sok came to realize that the study of numbers should never be regarded as disgraceful for a Confucian scholar in the context of *songnihak*. ([Jun 2006])

Regarding the role of the *songnihak* in eighteenth-century *Choson* Korea, the *Sogye saeyo* (書計瑣錄) (Random Collection of Writing and Calculation, 1786) of Pae Sang-sol (裵相說) (1759-1789) also deserves mention.[26] This handwritten manuscript aimed to teach elementary subjects and skills such as writing and calculating.[27] Its contents were based on Chinese mathematics as adapted to include Western ideas that had been introduced into China. Still, the author's approach aimed to integrate these foreign sources into the Korean socio-cultural context. Thus, he explained the four arithmetical operations in terms of the concepts of

[25]The *songnihak* (the study of images and numbers) concerns Song-Yuan Chinese neo-Confucianism. Korean scholars essentially followed the southern Song Chinese Confucianism of Zhu Xi ([Kim 2001]).

[26]For a brief biography, see ([Guo 2009], 57).

[27]For detailed textual analysis of the *Sogye saeyo*, see [Lin 2003].

songnihak and *yin-yang*.[28] Pae's explanations followed an approach adopted earlier by the *yangban* Choe Sok-chong (崔錫鼎) (1645-1715), whose *Kusuryak* (九數略) (Summary of the Nine Branches of Numbers) has much the same flavor. In fact, Choe Sok-chong was well known for having reconstructed the calculation methods with counting rods and practical mathematics according to his philosophy of the *Book of Change* (*Yijing* (易經)). ([Kawahara 1998]) This may well explain why Pae Sang-sol showed such high respect for the *Kusuryak* whenever he referred to it ([Lin 2002]; [Lin 2004]).

Still, despite his broad intellectual pursuit, Pae Sang-sol never referred to the *Sanhak gemong* or the *Suri chongon* for reasons that remain unclear ([Guo 2009], 57-58). On the other hand, his contemporary, So Ho-su (徐浩修) (1736-1799), had a chance to read the *Suri chongon*. He then went on to write a supplementary work, the *Suri chongon pohae* (數理精蘊補解) (Supplement to the *Suri chongon*, 1787), in which he elucidated the solutions of sixteen original problems from the *Suri chongon* and gave eighteen new ones.[29] So Ho-su's father, So Myung-ung (1716-1787), was a high official under King Chonjo (正祖) and served at the *Kyujanggak* (奎章閣), the palace library.[30] As a *sadaebu*, So Ho-su was well known for his expertise in mathematics and astronomy, which was apparently due to the intellectual heritage of his family. In fact, his father had a profound knowledge not only of the *songnihak* but also of Western works ([Kim 2001]). This may well account for why So Ho-su was strongly interested in the *Suri chongon*. In addition, his *Suri chongon pohae* demonstrated the significance of the *Suri chongon* for the transmission of Sino-Korean mathematical knowledge from the late eighteenth century onward ([Oh 2004]). Perhaps it was also due to So Ho-su's comprehensive understanding of the *Suri chongon* that he made a distinction between the concept of number and that of the *sangsu* (象數) (literally phenomena underlying number). This distinction stems from metaphysical views of the *sadaebu* closely related with the *songnihak*. This example serves to remind us that some *sadaebu* Confucians were devoted to mathematical knowledge without any real concern over issues of *sanhak* versus *songnihak* ([Oh 2004]).

Hong Tae-yong (1731-1785) was another scholar who tried to liberate the study of science and mathematics study from the *songnihak* ([Hung 2003], 13-17). He is noteworthy as the author of *Chunhae suyong* (籌解需用) (Requirements for the Understanding of Mathematics, ca. 1766), in which he referred not only to *Sanhak gemong, Yanghwi sanbop, Sangmyong sanpob, Sanbop tongjong* (算法統宗), but also to *Suri chongon* as well.[31] As a *sadaebu* Confucian, Hong Tae-

[28] *Ying* and *yang*, which sometimes refer to airy substances, are two basic antitheses, but also complementary concepts in the *songnihak*.

[29] The *Shuli jinyun* was edited by the Qing Emperor Kangxi and published posthumously in 1723. It is regarded as an encyclopedia of Western mathematics transmitted to China during the early eighteenth century.

[30] So Myung-ung is one of the major *sadaebu* scholars whose work is discussed in Kim Yun Sik's study of Western science, cosmological ideas and *Yijing* studies in 17th- and 18th-century Korea. Cf. [Kim 2001].

[31] As noted above, *Sanbop tongjong* is written *Suanfa tongzong* in the Pinyin system. This is a

yong synthesized *sirhak* and mathematics in order to emphasize the applicability of mathematics and its potential benefits for the state. As noted in Section 2, Hong Tae-yong belonged to the group of the *sirhak* school concerned with issues of economic enrichment of the state. This group was also called *Pukhak* (北學) School (lit. school devoted to learning from China) because their members identified strongly with Qing China, whose Manchu rulers were long looked down upon as barbarians. Now, however, they had become quite powerful and prosperous, and Hong Tae-yong, among others, took advantage of this new situation by joining the envoy to China in order to update scientific treatises and instruments. Since the *Suri chongon* was edited by the Emperor Kangxi, ([Li and Du 1987], 216-222), whereas Qing Chinese scholars were also enthusiastic about Western studies, it was only natural that Hong Tae-yong also took this up with great interest. Needless to say, the *Suri chongon* would serve as an important reference for his mathematical studies ([Hung 2003]).

After Hong Tae-yong, another scholar in early nineteenth-century Korea, Hong Kil-chu (洪吉周) (1786-1841), undertook an extensive study of the *Suri chongon*. Hong Kil-chu's maternal grandfather, So Hyong-su (徐迴修) (1725-1779), was So Ho-su's cousin. The So family was of the *yangban* class and became well known for their pursuit of mathematics and astronomy. This may explain why Hong Kil-chu's mother, who came from the So family, was so gifted in mathematics. She mastered the *Sanhak gemong* and even invented new algorithms on her own; later she became Hong Kil-chu's first tutor in mathematics. Hong Kil-chu's family was also of noble origin. His grandfather and elder brother had been Chief State Councilor and Second High State Councilor, respectively, and his father also held a high official post. Coming from this privileged family, Hong Kil-chu had the resources and support needed so that he could commit himself entirely to the study of Confucianism, literature, and mathematics. At the age of twenty-six, he gave up the chance for a governmental career by way of the civil examination in order to pursue his scholarly interests. For this purpose, he made ample use of his large household library in order to study with the goal of becoming a Confucian scholar. This collection included books on mathematics and astronomy that had been introduced from Qing China, such as *Tongwen suanzhi* (同文算指) (edited by Matteo Ricci and Li Zhizao (李之藻)), *Jihe yuanben* (Clavius's edition of Euclid's *Elements*, translated by Matteo Ricci and Xu Guanqi), *Suli jinyun (Suri chongon), Lisuan quanshu* (曆算全書) (Collected Works of Mei Wending (梅文鼎) on Astronomy, Calendrical Science and Mathemaics), *Lixiang kaocheng* (曆象考成), and other similar works ([Jun 2006]). It thus comes as no surprise that Hong Kil-chu's two main works on mathematics, *Kiha sinsol* (幾何新說) (New Methods of Geometry) and *Hogak yollye* (弧角演例) (Explanations and Examples for Arc and Angle), reveal features of both traditional Chinese math-

Ming Chinese text devoted to practical and commercial mathematics using the abacus (珠算盤) to carry out arithmetic computations. This led to widespread use of the abacus in East Asia following transmission of the text to Korea, Japan, as well as Vietnam. For a preliminary comparative study of these different influences on Korean and Japanese mathematics, see [Horng 2002b].

ematics as well as the newly introduced Western mathematics. As historian Jun Yong Hoon has observed, in these works one finds "Western subjects and terms in geometry and spherical trigonometry, borrowed from *Jihe yuanben, Suli jinyun (Suri chongon)* and *Lixiang kaocheng*" ([Jun 2006]). He notes further that Hong Kil-chu employed new algorithms for extracting roots and other techniques not found in Chinese works, so that the *Kiha sinsol* demonstrates a new kind of ingenuity that developed into a special Korean mathematical approach to counting rod computations. This stands in sharp contrast to the contributions of Hong Tae-yong, whose *Chuhae siryong* made no substantial imprint on Korean mathematics ([Hung 2003], 115-116). Therefore, unlike Hong Tae-yong, Hong Kil-chu was a truly important transitional figure for *tongsan* at the turn of the eighteenth and nineteenth centuries.[32]

Tongsan in Context: Nam Pyong-Gil and His Team

Hong Kil-chu's other book, the *Hogak yollye*, was the first to deal with spherical trigonometry in the history of Korean mathematics ([Jun 2006]). Unfortunately, it does not seem to have had much influence on the next generation of Korean mathematicians. For example, in Cho Hui-sun's (趙羲純) *Sanhak subyu* (算學拾遺) (Gathered Remnant of Mathematics, 1860s), a text devoted to spherical trigonometry, there are no references to the *Hogak yollye*. However, Cho Hui–sun went beyond Hong Kil-chu insofar as the study of spherical trigonometry was concerned. In particular, he employed the method of logarithms to help simplify the solutions for problems in spherical trigonometry, an important innovation in Korean mathematics. As for the sources used in the *Sanhak subyu*, Cho Hui-sun referred to both the *Suri chongon* and the *Lixiang kaocheng*, as did Hong Kil-chu in his *Hogak yollye*. Yet, Cho Hui-chun also synthesized Zhu Shijie's *Siyuan yujian* (四元玉鑑) (Precious Mirror of the Four Elements) and he drew furthermore on Nam Pyong-gil's *Sanhak Chongui* (算學正義) (Rectifying the Meaning of Mathematics), which gives methods for solving systems of higher-degree equations with multiple unknowns.[33] This synthesis of various source materials, including works of Cho's contemporary mathematicians like Nam Pyong-gil, proved to be a significant feature in the Korean mathematical texts published during the second half of the nineteenth century ([Wang 2011], 103-104).

Concerning Cho Hui-sun's contributions to Korean mathematics, the question arises why he, as a military official, should have devoted himself to this type of

[32]Jun Yong Hoon argues that Hong not only had mathematical ingenuity but also that he made use of newly introduced Western mathematical resources. In addition, Jun also points out that Hong's mathematics reflected the intellectual atmosphere in Korea during the late 18th and early 19th centuries ([Jun 2006]).

[33]Since Nam Pyong-gil wrote a preface to the *Sanhak subyu* (in 1869) ([Chang 2005], 110), and since both he and Cho Hui-sun were of the *yangban* class serving in high official posts, they were undoubtedly personally acquainted.

intellectual pursuit?[34] Cho Hui-sun was a member of the *yangban* military order.[35] He presumably already held a high military post after 1866, and in 1882 he was promoted to commander in charge of protecting the royal palace ([Wang 2011], 12-13). One might imagine he had strong interests in the controversial issues that surrounded the *chon wonsul* (天元術)) and the *changunbop (jiegen fang)* (借根方) (literally, borrowing the root and power),[36] a matter that much engaged his contemporary, Nam Pyong-gil ([Horng 2000b]). Nevertheless, Cho Hui-sun adopted a flexible attitude regarding matters of methodology. If he thought the *changunbop* offered a better approach to a certain topic, such as problems of (Western) spherical trigonometry, he would choose that method. On the other hand, when it came to solving high-degree equations in several unknowns, he preferred to use the methods of the *chon wonsul*, which he deemed superior to the *changunbop* in this particular connection ([Chang 2005], 110-113).

Now to appreciate why these issues were important for Korean mathematicians at that time we must briefly turn to their Chinese counterparts to gain a sense of the historical background. The *changunbop (jiegen fang)*, introduced insto China through the *Shuli jingyun*, was a Western method for solving quadratic equations by "completing the square." Actually, the method involves nothing more than al-Khwarizmi's methods (as later refined and developed by Girolamo Cardano). In the early eighteenth century, Mei Juecheng (梅瑴成) had a chance to read Li Ye's *Ceyuan haijing* and *Yigu yanduan* at the Qing palace library. He could then compare the *tianyuan shu* with the *jiegen fang*, and in doing so he realized that they were mathematically equivalent. Mei Juecheng apparently intended to echo the views of his grandfather, Mei Wending, who argued that Western geometry was actually just a different version of Chinese *Gougu* (勾股).[37] Mei Juecheng was perhaps the first among the eighteenth-century Chinese mathematicians to make this claim, and he did so by showing how several problems can be solved by using either of the two methods ([Horng 2000b]). Since he was also one of the executive editors of the *Suli jinyun*, he could use these arguments to support Emperor Kangxi's proposal that Western learning had Chinese origins (*Xixue zhongyuan* (西學中源)) ([Horng 1993b]).

Chinese scholars and mathematicians had already proposed similar argu-

[34] It is difficult to find a counterpart figure in the history of Chinese mathematics. Zeng Goufan (曾國藩)(1811-1872), the scholar-official who led the campaign against the Taiping rebels in the middle of the 19th century, showed enthusiasm for mathematical studies. However, he regretted his own lack of mathematical competence, even though he was one of the most important patrons of mathematics in Qing China during the second half of the nineteenth century ([Su 2013], 240-250).

[35] *Yangban* (literally two orders) was a class with two distinct orders, literary and military.

[36] In handling algebraic equations of the second degree, one is instructed first by the method of *jiegen fang* to "borrow one root" (借一根) for the unknown. See *Shuli jingyun* volume 31 for more details.

[37] *Gou-gu* problems originate from the ninth chapter of the *Jiuzhang suanshu*. The ninth chapter was devoted to the Pythagorean theorem and its applications to surveying problems. Literally *gou* denotes the shortest side of a right-angled triangle, whereas *gu* denotes the other side adjacent to the right angle.

ments for the Chinese origins of the *jiegen fang (changunbop)*, which came to China through the *Shuli jinyun (Suri chongon)*. However, due to the unavailability of the *Ceyuan haijing* and *Yigu yanduan*, Qing scholars and mathematicians eventually accepted the *jiegen fang* as the proper method for solving algebraic equations (up to second degree). Thus, in 1796, Li Rui recommended the *jiegen fang* for its superiority over the *Fangcheng shu* (方程術) (for solving systems of linear equations), which was regarded as the best method that the *Jiuzhang suanshu* could provide.[38] At the same time, he regretted that Mei Juecheng was the only mathematician fully versed in the former method ([Lin 2002], 65). Only a year later, however, his views underwent a dramatic change. Working under the sponsorship of his mentor, Ruan Yuan (阮元), Li Rui had begun to annotate the *Ceyuan haijing* and *Yigu yanduan*. By studying these texts, he gained a familiarity with the *tianyuan shu*, which he now argued was superior to the *jiegen fang* ([Lin 2002], 67-72).

Concerning the larger intellectual context, the Qian-Jia (乾嘉) school, to which Li Rui was closely affiliated, played a powerful and influential role in the revival and restoration of ancient knowledge. The mathematicians of this school, like Li Rui, used the method of textual criticism to explore this knowledge, often with the intent of showing that it surpassed that of the West ([Horng 1993b]). Consequently, Li Rui's arguments for the superiority of the *tianyuan shu* over the *jiegen fang* quickly came to dominate mathematical studies during the period from 1797 to 1820 when the major works on mathematics were primarily devoted to the study of such methods ([Lin 2002], 102-103). By 1820 nearly every classical text on mathematics was annotated and explained in terms of the *tianyuan shu*. Wang Xiaotong's *Qigu suanjing* was perhaps the single best example supporting the position of Li Rui and his patron Zhang Dunren (張敦仁) regarding the power and excellence of the *tianyuan shu*. By this time, however, it became increasingly difficult to find new wood to keep this fire alive. Indeed, the pendulum began to swing back the other way, in part because the "*xifa*" (西法) (Western method) school, whose members mainly pursued mathematical studies using Western techniques, now mounted a counterattack in order to recover their former status. This school had long opposed the methodology of the Qian-Jia school, which emphasized studies based on textual criticism of classical texts (including, of course, those on mathematics) without much concern for innovation. This trend in scholarship had clearly passed its peak by 1820, and as a consequence the dominance of the *tianyuan shu* eventually came to an end ([Lin 2002]).

Let us now return to the Korean side of this story. Wang Xiaotong's *Qigu suanjing* was a major influence on Nam Pyong-gil (1820-1869). Around mid-century he published the annotated *Chipko yondan* (緝古演段) (Demonstrating Diagrams in *Chipko sanchong*, ca.1855), referring therein to the method of *changunbop* in the *Suri chongon* ([Chang 2005]; [Guo 2009], 116-121). Another similar

[38]In fact, the *fangcheng shu* is the title of the eighth chapter of the *Jiuzhang suanshu*, which is devoted to solving simultaneous systems of linear equations ([Guo, Dauben and Xu 2013], 904-1033).

endeavour was made by Nam Pyong-chol (南秉哲) (1817-1863), Nam Pyong-gil's elder brother, who in his *Haegyong sech'ohea* (海鏡細草解) used the theory of proportion in the *Suri chongon* to annotate Li Ye's *Ceyuan haijing* ([Xia 2003], 81). These works might be described as echoing Yi Shang-hyok's study of the method of *changunbop*. For in writing his *Changunbop monggu* (借根方蒙求) (Beginner's Guide of the *Changunbop*, 1854), Yi Sang-hyok relied essentially on the *Suri chongon*. His intention in writing this text was, in fact, to introduce the Western method. In this regard, one should not be surprised that Yi Sang-hyok's *Sansul kwangyon* (算術管見) (Humble Opinions on Mathematics, 1855) referred to two works by the Qing mathematician Mei Juecheng, his *Suri chongon* as well as the *Chishui yizhen* (赤水遺珍) (Pearls Recovered from the Red Rivers) ([Hsieh 2005]).

In addition to the *Suri chongon*, Yi Shang-hyok's fellow mathematicians were surely well versed in the *chon wonsul*. Like the Qing Chinese scholars, these Koreans were also interested in the ongoing dialogue regarding the *chon wonsul* versus the *changunbop*.[39] This may well explain why Nam Pyong-gil, Yi's *yangban* patron, wrote the *Muihae* (無異解) (No Difference in Methods for Solving Equations, 1855). The *Muihae* gave a critical account of Qing mathematician Li Rui's restoration of the traditional *chon wonsul* in light of the Western *changunbop* ([Horng 2000a]). Nam's conclusion was that the two methods were essentially identical, a view supported by his own *Chipko yondan* ([Chang 2005], 115).

In their early careers, Nam Pyong-gil, Yi Shang-hyok, and even Nam Pyong-chol all identified the *chon wonsul* with the *changunbop*. However, in 1867 Nam Pyong-gil published *Sanhak chongui* (算學正義) (Rectifying the Meaning of Mathematics), a work that marks a turning point in Korean discussions of this issue.[40] Nam Pyong-gil here argued that the *chon wonsul* should be regarded as superior to the *changunbop*, citing two classic sources: the method for extracting roots in the *Shushu jiuzhang* (數書九章) and the method for setting up simultaneous equations with multiple unknowns in the *Siyuan yujian* (四元玉鑑) of Zhu Shijie. A similar line of argument was set forth by Yi Shang-hyok, who in his *Iksan* (翼算) (Assistance to mathematics, 1868) distinguished clearly between these two methodological approaches in a discussion of the properties of the positive and negative signs in an algebraic equation ([Chang 2005], 115; [Guo 2009], 162). Finally,

[39]Due to Chinese influence, even before the opening of a Korean port to the outside world, Korean scholars raised such issues as *Tongdo sogi* (東道西器) (Eastern Way and Western Technology). This was a Korean counterpart to the Qing Chinese *Zhongti xiyong* (中體西用) that sought to find an accommodation with Western science and technology. However, as Pak Song-nae points out, "enthusiasm for the accommodation of Western science in the absence of any particular crisis of consciousness, and the subsequent attitude of *tondo sogi*, were unable to translate their thinking into practice on a wide scale." ([Pak 1986]) Even so, Kim Yun Sik argues, pointing to the case of So Myung-ung, that "Western cosmological knowledge could be incorporated rather easily by eighteen-century Korean scholars into the traditional cosmological system based on the *Yijing* (易經)" ([Kim 2001]).

[40]In fact, the *Sanhak chongui* began with methodology before giving an account of the underlying theory. This structure allows us to see how the author provides an encyclopaedic picture of mathematical knowledge by integrating the *tongsan*, the newly transmitted *chungsan*, as well as the *Suri chongon*. See [Chen 2007] for a detailed textual analysis of the *Sanhak chongui*.

though somewhat later, Cho Hui-sun also approved this viewpoint. ([Chang 2005], 115). Nam Pyong-gil's argument in favor of *chon wonsul* over *changunbop* was reinforced by his interpretation of the mathematical methods of the *Jiuzhang suanshu.* In his *Kujang sulhae* (九章術解) (1864), he demonstrated that the *Kujang sansul (Jiuzhang suanshu)* (九章算術) and the new mathematical canon of the *Suri chongon* did not contradict one other. As Jia-Ming Ying argues, for Nam Pyong-gil the *Kujang sansul* "was an important historical document that was to be explained in its own *contemporary* terms" ([Ying 2010]).

These four figures—Nam Pyong-gil, Nam Pyong-chol, Yi Sang-hyok, and Cho Hui-sun—brought the last chapter in the story of *tongsan* to its conclusion. All were very familiar with Li Rui's new annotated edition of the *Yigu yanduan (Okkam yondon)* and the *Ceyuan haijing (Chugwon haegyong).* Still, they already knew the *chon wonsul* through the *Sanhak gemong*, whereas they came to understand the *changunbop* by means of the *Suri chongon.* Therefore, these two works, the *Sankhak genong* and the *Suri chongon*, should be regarded as the most important mathematical texts that shaped the Korean *tongsan* tradition, which emerged through the adaptation and transformation of traditional Chinese mathematics as well as by assimilating a Chinese version of Western mathematical knowledge.[41] However, in contrast to Qing Chinese mathematicians, the Korean view of the *chon wonsul* grew out of a different kind of dialogue. Thus, Nam Pyong-gil's claim that the *chon wonsul* was superior to the *changunbop* was not driven by any ideological concerns but rather was based on methodological considerations alone. It is evident that his texts were exclusively concerned with mathematical knowledge proper. On the other hand, it should not be overlooked that the enthusiasm for mathematics of some *sadaebu* was surely due to their desire to incorporate it into the *songnihak.*

In this connection, the case of Choe Hang-gi (1803-1877) is especially illuminating because of his endeavor to integrate mathematical thought into the study of *qi (kihak* (氣學)). As historian Jun Yong Hoon indicates, Choe Hang-gi created the *kihak* "[b]y compounding resources from both Western science and Confucian natural philosophy," hoping to establish a new mechanics that would replace the Newtonian version ([Jun 2010]). He developed a comprehensive philosophy that purported to explain everything in the universe, including mathematics, as witnessed by his *Supsan chinbol* (習算津筏) (Tools of Calculation Learning, 1850). In fact, his mathematical text was mainly devoted to editing problems selected from the *Suri chongon* with only slight differences related to the numerical data posed in the problems. Moreover, Choe Hang-gi replaced the method of pen calculations used in the *Suri chongon* with that of counting rods ([Li 2010]). The reason for this, as Choe explains in his preface to the *Supsan chinbol*, stems from that fact that the method of counting rods relies on their positions, and the concept of

41 This was in striking contrast to its Chinese counterpart. In Qing China, the two most popular texts on mathematics for literati were the *Suanfa tongzong* and the *Shuli jingyun.* Regarding the wide circulation of the *Suanfa tonzong* by the early 18th century, see ([Li and Du 1987], 188).

position was of central importance in the *qi*.[42] This, in turn, reinforced his thesis regarding the primacy of the *qi*: "whereas the *qi* is the essence of number, number is a manifestation of the *qi*" ([Choe 2002] vol. 1, p. 261). In a similar vein, Choe Hang-gi even claimed that mathematics was the material from which one could forge the purest quality of the *qi* ([Choe 2002] vol. 3, p. 157]). Thus, mathematics assumed a prominent place in the *kihak*, but also in Choe's moral philosophy: it was, therefore, no longer to be regarded as one of the six inferior arts, it was indispensible for the *kila* which led to an explanation of "the best ethical conduct of human beings" ([Jun 2010], 73).

Concluding Remarks

To summarize the last three sections, the sometimes controversial debates between different parties (the advocates of *sanhak* versus *sirhak*, or the disputes over *sanhak* versus *songnihak*) proved in the long run beneficial for Korean mathematics, whose practitioners gradually enjoyed more legitimacy and respectability. Cho Tae-koo, Hong Tae-yong, and Hong Kil-chu all drew upon the issues at stake in the ongoing dialogue over *sanhak* versus *sirhak*. By the same token, the work of Kyong Song-jin and Hong Chong-ha enabled *tongsan* to develop on its own, in part by recognizing that mathematical pursuits met contemporary social needs in accordance with *sirhak* concerns. These concerns ran parallel with efforts to integrate mathematical studies into a larger intellectual framework. Thus, the legitimacy and significance of mathematical knowledge proper was in turn reinforced by the *songnihak* discourse of the *sadaebu*, including Hwang Yun-sok, Choe Sok-chong, and Choe Han-gi, all of whom shared this general interest. As for Nam Pyong-gil's team, their views on issues like *chon wonsul* versus *changunbop* were based on praxis and had little to do with ideological concerns, in striking contrast with the views of their Qing Chinese counterparts. This would seem even more significant if one takes into account that Nam Pyong-gil and Yi Shang-hyok did not at all times take the same position with regard to the methodologies of *chon wonsul* and *changunbop*.

In this article, I have discussed the role of fifteen Korean mathematicians with due attention to how their mathematical texts contributed to the larger context of *tongsan*.[43] This work spans the period from 1657, when Kyong Song-jin gave Kim Si-jin a copy of the *Sanhak gemong*, to 1868, when Yi Shang-hyok wrote his *Iksan*. It was primarily due to Kyong Song-jin and Hong Chong-ha, two *chungin* mathematicians with close family relations, that the *tongsan* was adopted as a technical term associated with the work of Korean mathematicians. This term

[42]The *qi* is basically divided into *ying qi* and *yang qi* which have "natural motion" downward and upward respectively. Using counting rods to represent polynomials or equations in terms of the celestial element (*tianyuan*) with the constant term (*tai*) (太) as an indicator, the unknown (i.e., the *tianyuan*) can be either above or below ([Li and Du 1987], 139-140). Thus, the positions of the counting rods were taken as basic mathematical concepts.

[43]The only texts not available are Hong Kil-chu's *Kiha sinsol* and *Hogak yollye.*

gained a deeper meaning through Hong Chong-ha's revision of the *tongsan* in his *Kuil jip*. Moreover, the family connections between these two *chungin* mathematicians also apparently provided an avenue for the *tongsan* to emerge as a significant development. Thus, the special status of the *chungin* mathematicians as a closed network in *Choson* society made them an indispensable factor in the history of Korean mathematics.

Unfortunately, only three of the fifteen Korean mathematicians we have studied—the *chungin* mathematicians Kyong Song-jin, Hong Chong-ha and Yi Shang-hyok—wrote texts that have survived. At any rate, it is not known whether any other major texts are still extant. This circumstance, however, does not really affect our main conclusion, namely that Kyong Song-jin and Hong Chong-ha were the two key figures who promoted the *tongsan*, a development crucial to the history of Korean mathematics. Nevertheless, one should not neglect the contributions of leading *sadaebu* Confucians, like Cho Tae-koo and Hong Tae-yong, who emphasized the importance of mathematical pursuits not only for the welfare of the state but also for cultivating personal mental discipline. On the other hand, *sadaebu* literati—like Choe Sok-chong, Hwang Yun-sok, Pae Sang-sol, and Choe Hang-gi—demonstrated mathematical expertise in their own texts, further promoting the status of this knowledge within the realm of *songnihak* concerns ([Kawahara 2010]). All of these developments over the course of two hundred years within the *Choson* period were quite unparalleled by events in contemporary China ([Horng 1993b]). Taken together, they would produce, step by step, the special characteristics of the *tongsan* tradition.

The major Korean mathematicians of the nineteenth century—figures like Hong Kil-chu, Nam Pyong-gil, Yi Shang-hyok, and Cho Hui-sun—showed little interest in ideological issues, such as whether Western learning originated from China. On the other hand, some Korean *sadaebu* did pay attention to such issues as *tongdo sogi* (東道西器) (Eastern Way and Western Technology) ([Pak 1986]). Nam Pyong-chol, the elder brother of Nam Pyong-gil, defended the Eastern Way, although he worked primarily on astronomy, calendrical science, and astronomical instruments. Unlike his younger brother, Nam Pyong-chol took this issue for granted, despite the respect he showed for the work of the Qing mathematician Jian Yong (江永), an expert on the Western method (*xifa*).[44] Nevertheless, for Nam Pyong-chol this issue was not associated with foundation myths regarding the origins of mathematical knowledge. In much the same vein, the arguments of Nam Pyong-gil and Yi Shang-hyok regarding the superiority of the *chon wonsul* over the *changunbop* suggest far more flexibility with respect to methodological comparisons than was shown by their Chinese counterparts ([Horng 1993a]). This accords with the general attitude of Korean mathematicians, who tried to make sense of mathematical knowledge on their own rather than adopting the views that prevailed in Qing China.

[44] Jian Yong and his disciple Dai Zhen were criticized for their defense of Western methods as applied to studies of astronomy and calendrical science; see [Horng 1993b].

In this regard, Nam Pyong-gil's aim in writing his *Sanhak chongui* was, according Jia-Ming Ying, to compose a new mathematical canon representative of the *tongsan*. His endeavor thus echoed an earlier intellectual concern of Kim Chong-hui, namely that *sirhak* should lead to the creation of academic works in a Korean context. From Nam Pyong-gil's perspective, the *Sanhak chongui*, as a comprehensive mathematical treatise with encyclopaedic features, should replace the *Suri chongon* to become the new mathematical canon for the Korean people ([Ying 2010], 114). In the *Kujang sulhae* (1864) and the *Sanhak chongui*, Nam Pyong-gil and his team succeded in consolidating this *tongsan* on their own terms. This was accomplished by integrating *chungsan* and *sosan* (西算) (Western mathematics) in the *Choson* context, despite the fact that rather little in either text could be regarded as original or substantial compared with the standards, Chinese or Western, of that time. It seems clear that before the Japanese colonial occupation of Korea in 1910, the *tongsan* had made its own way, starting from the cornerstone laid in the *Tongsan cho* (1718) to the completed edifice of the *Sanhak chongui* (1867).

As a final remark, we should also briefly address "the Problem of China" basically proposed by Kim Yung Sik, in terms of Korean adaptation. According to Kim Yung Sik, this problem concerns the assessment of "the significance of the Chinese scientific ideas and technical artifacts that are apparently omnipresent in traditional Korean science and technology."[45] This is important for understanding how Koreans adapted Chinese science and technology to their own needs and conditions, which leads Kim to ask how much of this "can be referred to as an 'independent' development of Korean science [leading] to new inventions and discoveries" ([Kim 1998]). On the other hand, he also relates these issues to a fundamental "why not" question: "why did Korean science, which had reached the highest level in the world in the early fifteenth century, not continue to progress but instead stagnated and declined?"[46] This question brings to mind Joseph Needham's well-known thesis regarding the case of China, which has played a major role in the historiography of Chinese science and technology.

In this connection, Kim Yung Sik formulated the following two specific challenges for future historical research on the history of Korean science and technology:

1. To examine King Sejong's personal knowledge of scientific and technical subjects and his attitude towards such specialized subjects and their practitioners.

2. To explore the relations between specialist officials, in particular *chungin* mathematicians, and generalist bureaucrats, especially

[45]Cited from [Kim 1998]. In his substantial contribution to this subject ([Guo 2009]), Guo Shirong provides detailed information about how Chinese mathematical texts influenced Korean mathematics.

[46]In the *Choson* period, Koreans made several important discoveries in science as well as innovations in technology ([Kim 1998]; [Portal 2000], 112-115).

yangban civil servants, with regard to scientific practice.

Concerning the first issue, although the *Sejong sillok* has been extensively investigated with respect to the institutionalization of mathematics,[47] and Sejong's arguments in favor of mathematical pursuits have been noted ([Yeh 2002]), nevertheless, we still do not have enough evidence to explain why the golden era in the history of Korean science passed with the end of Sejong's reign. Moreover, as argued above, in the period between the eighteenth and early nineteenth centuries, Korean *sadaebu*—including Hwang Yun-sok, Hong Tae-yong, Pae Sang-sol, So Ho-su, and Hong Kil-chu—made important contributions to mathematical studies, thereby lending prestige to the *tongsan*. These developments are significant in their own right for an appreciation of the history of Korean mathematics in this period, notwithstanding the fact that it lagged far behind contemporary European developments.

As to the second matter, the case of Kyong Song-jin shows how a *chungin* managed to earn high professional status as well as the respect of contemporary *yangban* and *sadaebu* Confucians ([Horng and Li 2007]). The close family ties between Kyong Song-jin and Hong Chong-ha also add another piece of evidence showing, as Kim Yung Sik puts it, that the *chungin* became a closed social system through intermarriage.[48] On the other hand, the first eight chapters of Hong Chong-ha's *Kuil chip* assure us of his professional expertise, especially when it came to the method of *chon wonsul*, by which he showed his mathematical superiority over the Qing Chinese astronomer He Gouzhu ([Kawahara 1998]; [Horng 2002a]; [Guo 2009]). This self-confidence would, in turn, reinforce a sense of self-esteem among the *chungin* people owing to their mathematical expertise, despite the fact that it was nearly impossible for them to elevate their social status. Nevertheless, some *sadaebu* scholars and mathematical practitioners, who, like Hwang Yun-sok and Pak Yul, were devoted to *songnihak*, argued that mathematics should not be viewed an inferior art, as had long been the case among Confucians. For these *sadaebu*, mathematics deserved serious study not merely because of its practical applications and potential utility for the administration of the state. They framed arguments within the contexts of debates over *songnihak* versus *sanhak*, and they made contributions to mathematics proper, as witnessed by their study of the *Suri chongon*. Moreover, the *sadaebu*'s dialogue with the *chungin* mathematicians sheds new light on Korean mathematics and opens doors for further explorations. It would be interesting to know whether and how the *chungin* class reacted to

[47]See [Yeh 2002] for a detailed investigation of the *Sejong Sillok*, in which passages related to Sejong's promotion of science and technology in general, and mathematics in particular, are cited. In addition, Guo Shirong summarizes Sejong's contributions to Korean science and technology during the fifteenth century in [Guo 2009].

[48]Several members of Kyong Song-jin's family were *chungin* medical doctors, whereas the father of his second wife was a *chungin* mathematician; see [Li 2003]; [Li 2004]. On the other hand, almost of all members of Hong Chong-ha's family were *chungin* mathematicians; see [Kawahara 1998]. As we have already noted, Hong Chong-ha's mother was a daughter of Kyong Yon (慶演), who was Kyong Song-jin's nephew ([Li 2003]; [Li 2004]).

contemporary discourse and rhetoric about mathematics. This could also lead to a better understanding of the role of the *chungin* in *Chonson* society.

In a similar vein, one should also take into account the roles played by such figures as Nam Pyong-gil and Yi Sang-hyok. As emphasized above, Nam Pyong-gil seems to have been unconcerned when it came to the ideological issues surrounding foundation myth and *sanhak*, issues which were highly provocative for his Qing Chinese counterparts. Nam's supervision and patronage played an important role in Yi's early career as a *chungin* mathematician.[49] Yi Shang-hyok later became a successful collaborator of Nam Pyong-gil, and they each wrote prefaces for each other's mathematical treatises. This suggests that their relationship had a special importance for the history of Korean mathematics. Eventually Yi Shang-hyok attained a high reputation of his own; his professional status not only benefited the growth of mathematical knowledge proper, but also helped *tongsan* to emerge as a respectable specialized discipline.

Acknowledgment

An early version of this article, entitled "History of Korean Mathematics in the *Choson* Period: An Overview," was presented at the 11th International Conference on the History of Science in East Asia, August 15-10, 2005, Deutsches Museum, Munich. My sincere thanks goes to Prof. Lee Sang-gu for his invitation to speak on this subject again at the 25th Anniversary Conference sponsored by the Korean Society for History of Mathematics on November 15, 2009. For help in reaching this final version, I would like to thank Jin-Fa Hsu for referring me to source materials published in Japanese. In addition, I would like to express my gratitude to Jia-Ming Yin, Jun-Hong Su, Tsang-Yi Lin, Fu-Kai Chang, and Jun-Wei Hwang for their comments on an earlier draft. Needless to say, I alone am responsible for the present version of this article.

Bibliography

[Chang 2005] Chang, Fu-Kai: *On the Dialogue between the Chon Wonsul and Changunbop: A case study of the Chipko Yanduan by Nam Pyong-gil* (1820-1869) (in Chinese). Master's thesis, National Taiwan Normal University, Taipei, Taiwan. 2005.

[Chen 2007] Chen, Chun-Ting: *Textual Analysis of Nam Pyong-gils' Sanhak Chongui* (in Chinese). Master's thesis, National Taiwan Normal University, Taipei, Taiwan. 2007.

[Chen 2003] Chen, Kwan-Liang: *Textual Analysis of Tongsan Cao* (in Chinese). Master's thesis, National Taiwan Normal University, Taipei, Taiwan. 2003.

[49] Nam Pyong-gil was serving as the adjunct chief at the Observatory where Yi Shang-hyok was one of the *chungin* astronomers, see [Wu 2003]; [Wu 2004]. However, the patronage relations require more explanation, which we are unable to clarify at the present time.

[Cho 1985] Choe, Tae-koo: *Chuso Kysuryak*. In Kim Yong Woon ed. (1985). *Hanguk Kwahak Kisul Sa Charyo Taekye: Suhak Pyon* (Collection of Source the Korean History of Science and Technology: Mathematics Section (2) (1985), 3-199. Seoul: Yigang Press.

[Choe 2002] Choe Han-gi: *Chungbo myongnamnu ch'ongso* (Expanded Collection of Choe Han-gi's Writings). 5 vols. Seoul: Song-gyun'gwan University Press. 2002.

[Choe 1985] Cho Sok-chong: *Kysuryak*. In Kim Yong Woon, ed. (1985). *Hanguk Kwahak Kisul Sa Charyo Taekye: Suhak Pyon* (Collection of Source Materials of the Korean History of Science and Technology: Mathematics Section (1) (1985), 371-704. Seoul: Yigang Press.

[Chou 2003] Chou, Chung Kuei: *Study on Hwang Yun-sok's Sanhak Rumen* (in Chinese). Master's thesis, National Taiwan Normal University, Taipei, Taiwan. 2003.

[Chou 2004] Chou, Chung-Kuei: "On Korean Confucian Mathematical Texts: A case study of Huang Yong'xi's *Suan-xue Ru-men* (Hwang Yun-esok's Sanhak Rumen)" (in Chinese), Wann-Sheng Horng *et al.*, eds., *Proceedings of the Asia-Pacific HPM 2004 Conference: History, Culture and Mathematics Education in the New Technology Era* (May 24-28, 2004, National Taichung Teachers College, Taichung, Taiwan) (2004), pp. 343-358.

[Chu 2009] Chu, Ping-yi: "A Study on the *Chongzhen lishu* Housed in the *Kyujanggak* Library and Other Related Materials" (in Korean and Chinese), *Kyujanggak* 34 (2009), 231-262.

[Guo 2009] Guo, Shirong: *Zhongguo shuxue dianji zai chaoxian bandao de liuchuan yu yingxiang* (中國數學典籍在朝鮮半島的流傳與影響) (Circulation and Influence of Rate Chinese Mathematical Texts in Korea). Jinan: Shandong Educational Press. 2009.

[Guo 2010] . Guo, Shuchun: *A History of Science and Technology in China: Mathematics Section* (in Chinese). Beijing: Kexue Chubanshe. 2010.

[Guo, Dauben and Xu 2013] Guo Shuchun, Joseph W. Dauben & Xu Yibao: *Nine Chapters on the Art of Mathematics III* Shenyang: Liaoning Education Press. 2013.

[Hong CH 1985] Hong, Chong-ha: *Kuil-chip*. In Kim Yong Woon ed. (1985). *Hanguk Kwahak Kisul Sa Charyo Taekye: Suhak Pyon*(Collection of Source Materials of the Korean History of Science and Technology: Mathematics Section (2) (1985). Seoul: Yigang Press.

[Hong TY 1985] Hong, Tae-yong: *Chuhae Hsuyong*. In Kim Yong Woon ed. *Hanguk Kwahak Kisul Sa Charyo Taekye: Suhak Pyon*(Collection of Source Materials of the Korean History of Science and Technology: Mathematics Section (3) (1985). Seoul: Yigang Press.

[Horng 1993a] Horng, Wann-Sheng: "Chinese Mathematics at the Turn of the 19th Century: Jiao Xun, Wang Lai and Li Rui", Lin, Cheng-hung and Daiwie Fu, eds., *Philosophy and Conceptual History of Science in Taiwan* (Netherlands: Kluwer Academic Publishers) (1993), pp. 167-208.

[Horng 1993b] Horng, Wann-Sheng: "*Tantian sanyou* Jiao Xun, Wang Lai and Li Rui: Interaction between Mathematics and Classical Studies in Qing Dynasty" (in Chinese), Horng, Wann-Sheng ed., *Tantian Sanyou* (談天三友). Taipei: Ming-Wen Publication Company. 1993.

[Horng 2000a] Horng, Wann-Sheng: "Sino-Korean Transmission of Mathematical Texts in the 19th Century: A case study of Nam Byung Gil's *Vu Yi He*," presented to the HPM 2000 Taipei Conference, August 9-14, 2000, Taipei: National Taiwan Normal University. 2000.

[Horng 2000b] Horng, Wann-Sheng: "Nam Pyong-gil's Comments on the 'Tian Yuan Shu' versus 'Jie Gen Fang': An HPM Perspective" (in Chinese), *Chinese Journal of Science Education* 8 (3) (2000), 215-224.

[Horng 2000c] Horng, Wann-Sheng: "Reading into Mathematical Texts from a Pedagogical Perspective: A case study of the Mei Juecheng's *Chi Shui Yi Zhen*" (in Chinese), *Bulletin of Chinese Association for the History of Science* 1 (2) (2000), 35-43.

[Horng 2001a] Horng, Wann-Sheng: "How Did *Choson Sadaebu* Scholars Read the 'Kujang' – A case study of Cho Tae-koo," *Bulletin of Academia Historica* (2) (2001), 297-324.

[Horng 2002a] Horng, Wann-Sheng: "An Eighteenth Century Sino-Korean Dialogue on Mathematics: He Gouzhu vs. Hong Jeong-Ha (Hong Chong-ha)" (in Chinese), *Chinese Studies* 12 (2) (2002), 57-80.

[Horng 2002b] Horng, Wann-Sheng: "Historical Problems of Sino-Japanese and Sino-Korean Transmission of Mathematical Cultures" (in Chinese). In Wang, Yu-Fong, ed., *Proceedings of Conference on Science, Technology, Medicine and Society* (Kaohsiung: National Science and Technology Museum) (2002), 61-72.

[Horng 2002c] Horng, Wann-Sheng: "Sino-Korean Transmission of Mathematical Texts in the 19th century: A case with Nam Pyong-gil's *Kugo Sulyo Tohae,*" *Historia Scientiarum* 12 (2) (2002), 87-99.

[Horng 2003a] Horng, Wann-Sheng: "Transmission and Transformation of Mathematical Cultures: A case study of Korean Mathematician Nam Pyong gil's *Sanhak Chongyi*" (in Chinese), *Journal of Taiwan Normal University: Humanities & Social Sciences* 48 (1) (2003), 21-38.

[Horng 2008] Horng, Wann-Sheng: "The First Visit of Kyujangguk Library's Collection of Mathematical Texts" (in Chinese), *Kyujangguk* 32 (2008), 283-293.

[Horng and Li 2007] Horng, Wann-Sheng & Chien-Tsung Li: "Kyong Song-jin and a Mathematics Workshop in 17th Century Choson Korea" (in Chinese), *Chinese Studies* 25 (1) (2007), 313-340.

[Hsieh 2003] Hsieh, Pai-Cheng: *Study on Korean Kugo Method* (in Chinese). Master's thesis, National Taiwan Normal University, Taipei, Taiwan. 2003.

[Hsieh 2005] Hsieh, San-Bao: *Preliminary Study on Yi Sang-hyok's Sansul kwanyon* (in Chinese). Master's thesis, National Taiwan Normal University, Taipei, Taiwan. 2005.

[Hung 2003] Hung, Yi-Ting: *On Hong Tae-yong's Mathematical and Sirhak Thinking: A case study of his Chuhae Hyuyong* (in Chinese). Master's thesis, National Taiwan Normal University, Taipei, Taiwan. 2003.

[Hwang 1985] Hwang Yun-sok: *Sanhak Ibmun*. In Kim, Yong Woon, ed. (1985). *Hanguk Kwahak Kisul Sa Charyo Taekye: Suhak Pyon* (Collection of Source Materials of the Korean History of Science and Technology: Mathematics Section (3) (1985), 3-216. Seoul: Yigang Press.

[Jun 2004a] Jun, Yong Hoon: *Conflicts and Consolidations of Western and Eastern Astronomy in the 17th and 18th Centuries.* Doctoral thesis, Program in History and Philosophy of Science, Seoul National University, August. 2004.

[Jun 2004b] Jun, Yong Hoon (2004). "Hong Kil-ju (洪吉周, 1768-1841) and His Studies on Mathematics," Wann-Sheng Horng *et al*, eds., *Proceedings of the Asia-Pacific HPM 2004 Conference: History, Culture and Mathematics Education in the New Technology Era* (May 24-28, 2004, National Taichung Teachers College, Taichung, Taiwan) (2004), 259-270.

[Jun 2006] Jun, Yong Hoon: "Mathematics in context: A Case in Early Nineteenth-Century Korea," *Science in Context* 19 (4) (2006), 475-512.

[Jun 2010] Jun, Yong Hoon: "A Korean Reading of Newtonian Mechanics in the Nineteenth Century," *EASTM* 32 (2010), 59-88.

[Kawahara 1998] Kawahara, Hideki: "Tongsan and Chonwonsul – Choson's Mathematics in the Mid Middle 17th through Early 18th Century," *Chosen Gakuho* (Journal of the Academic Association of Koreanology in Japan) (169) (1998), 35-71.

[Kawahara 2010] Kawahara, Hideki: *Chosen Sugakushi: Shushigakutekina tenkai to sono shuen* (The History of Choson Mathematics: Development Based on Neo-Confucianism and Its End) (in Japanese). Tokyo: University of Tokyo Press. 2010.

[Kim 1973a] Kim, Yong Woon: "Introduction to Korean Mathematical History (I)," *Korea Journal* (July) (1973), 16-23.

[Kim 1973b] Kim, Yong Woon: "Introduction to Korean Mathematical History (II)," *Korea Journal* (August) (1973), 27-32.

[Kim 1973c] Kim, Yong Woon: "Introduction to Korean Mathematical History (III)," *Korea Journal* (September) (1973), 27-32.

[Kim 1985] Kim, Yong Woon: *Hanguk Kwahak Kisul Sa Charyo Taekye:Suhak Pyon* (韓國科學技術史資料大系・數學篇) (Collection of Source Materials of the Korean History of Science and Technology: Mathematics Section). Seoul: Yigang Press. 1985.

[Kim 1986] Kim, Yong Woon: "Pan-Paradigm and Korean Mathematics in the Chosen Dynasty," *Korea Journal* (March) (1986), 25-46.

[Kim and Kim 1978] Kim, Yong Woon & Kim, Yong Guk: *Kankoku Sugakushi* (History of Korean Mathematics) (in Japanese), Tokyo: Maki Shoten. 1978.

[Kim 1998] Kim, Yung Sik: "Problems and Possibilities in the Study of the History of Korean Science," Osiris 13 (1998), 48-79.

[Kim 2001] Kim, Yung Sik: "Western Science, Cosmological Ideas, and the *Yijing* Studies in Seventeenth-and Eighteenth-Century Korea", *Seoul Journal of Korean Studies* (14) (2001), 209-334.

[Kim 2003] Kim, Yung Sik: *The Natural Philosophy of Chu Xi* (1130-1200). (Chinese translation). Shanghai: Huatong Normal University Press. 2003.

[Kuo 2007] Kuo, Shou-De: *Preliminary Study on Nam Pyong-gil's Celiang Tujie* (in Chinese). Master's thesis, National Taiwan Normal University, Taipei, Taiwan. 2007.

[Kyong 1985] Kyong, Song-jin: *Muksa jip*. In Kim, Yong Woon, ed. (1985). *Hanguk Kwahak Kisul Sa Charyo Taekye: Suhak Pyon* (Collection of Source Materials of the Korean History of Science and Technology: Mathematics Section (1) (1985) 3-368. Seoul: Yigang Press.

[Lee 1984] Lee, Ki-baik (Yi, Ki-baek): (translated by Edward W. Wagner). *A New History of Korea.* Cambridge, MA.: Harvard University Press. 1984.

[Lee 2001] Lee, Won-Sun (Yi, Won-Sun): *Study on Korean History of Western Learning* (Chinese Translation). Beijing: China Social Science Press. 2001.

[Li 2003] Li, Chien-Tsung: *Preliminary Study on Kyong Song-jin's Muksa jip* (in Chinese). Master's thesis, National Taiwan Normal University, Taipei, Taiwan. 2003.

[Li 2004] Li, Chien-Tsung: "Korean *Chungin* Mathematician Kyong Song-jin in the Seventeenth Century," Wann-Sheng Horng *et al.*, eds., *Proceedings of the Asia-Pacific HPM 2004 Conference: History, Culture and Mathematics Education in the New Technology Era* (May 24-28, 2004, National Taichung Teachers College, Taichung, Taiwan) (2004), 285-304.

[Li 2010] Li, Jian-Xun: *Study on Choe Han-gi's Philosophy of Ki and Mathematical Thought* (in Chinese). Master's thesis, National Taiwan Normal University, Taipei, Taiwan. 2010.

[Li and Du 1987] Li Yan & Du Shran: *Chinese Mathematics: A Concise History*. New York: Oxford University Press. 1987.

[Lin 2003] Lin, Ken-Hui: *Textual Analysis of Pae Sang-sol's Sogye-saeyo* (in Chinese). Master's thesis, National Taiwan Normal University, Taipei, Taiwan. 2003.

[Lin 2004] Lin, Ken-Hui: "Remarks on Pae Sang-sol's *Sogye-saeyo*" (in Chinese), Wann-Sheng Horng *et al.*, eds., *Proceedings of the Asia-Pacific HPM 2004 Conference: History, Culture and Mathematics Education in the New Technology Era* (May 24-28, 2004, National Taichung Teachers College, Taichung, Taiwan) (2004), 305-324.

[Lin 2002] Lin, Tsang-Yi: *Qing Chinese Mathematicians' Jiegen fang vs. Tianyuan shu*: 1723-1820 (in Chinese). Master's thesis, National Taiwan Normal University, Taipei, Taiwan. 2002.

[Nam PC 1985] Nam Pyong-chol: *Haekyong sechoea*. In Kim, Yong Woon, ed. (1985). *Hanguk Kwahak Kisul Sa Charyo Taekye: Suhak Pyon* (Collection of Source Materials of the Korean History of Science and Technology: Mathematics Section (5) (1985), 3-247. Seoul: Yigang Press.

[Nam 1985a] Nam, Pyon-gil: *Kujang Sulhae*. In Kim, Yong Woon, ed. (1985). *Hanguk Kwahak Kisul Sa Charyo Taekye: Suhak Pyon* (Collection of Source Materials of the Korean History of Science and Technology: Mathematics Section (6) (1985), 251-496. Seoul: Yigang Press.

[Nam 1985b] Nam, Pyon-gil: *Sanhak Chongui*. In Kim, Yong Woon, ed. (1985). *Hanguk Kwahak Kisul Sa Charyo Taekye: Suhak Pyon* (Collection of Source Materials of the Korean History of Science and Technology: Mathematics Section (7) (1985), 3-518. Seoul: Yigang Press.

[Nam 1985c] Nam, Pyon-gil: *Chipko Yondan*. In Kim, Yong Woon, ed. (1985). *Hanguk Kwahak Kisul Sa Charyo Taekye: Suhak Pyon* (Collection of Source Materials of the Korean History of Science and Technology: Mathematics Section (5) (1985), 249-364. Seoul: Yigang Press.

[Nam 1985d] Nam, Pyon-gil: *Chukryang tohae*. In Kim, Yong Woon, ed. (1985). *Hanguk Kwahak Kisul Sa Charyo Taekye: Suhak Pyon* (Collection of Source Materials of the Korean History of Science and Technology: Mathematics Section (5) (1985), 367-482. Seoul: Yigang Press.

[Nam 1985e] Nam, Pyong-gil: *Muihae*. In Kim, Yong Woon, ed. (1985). *Hanguk Kwahak Kisul Sa Charyo Taekye: Suhak Pyon* (Collection of Source Materials of the Korean History of Science and Technology: Mathematics Section (6) (1985), 191-250. Seoul: Yigang Press.

[Oh 2004] Oh, Young Sook: "*Suri Chongon Pohae* (數理精蘊補解): An 18th Century Korean Supplement to *Shuli jingyun* (數理精蘊)," Wann-Sheng Horng *et al*, eds., *Proceedings of the Asia-Pacific HPM 2004 Conference: History, Culture and Mathematics Education in the New Technology Era* (May 24-28, 2004, National Taichung Teachers College, Taichung, Taiwan) (2004), 279-284.

[Pae 1985] Pae, Sang-sol: *Sogye-saeyo.* In Kim, Yong Woon ed. (1985). *Hanguk Kwahak Kisul Sa Charyo Taekye: Suhak Pyon* (Collection of Source Materials of the Korean History of Science and Technology: Mathematics Section (4) (1985), 3-230. Seoul: Yigang Press.

[Pak 1986] Pak, Song-Nae: "Western Science and *Shihak* Scholars," *Korea Journal* (March) (1986), 4-24.

[Park 1990] Park, Seong-Rae: "History of Science in Korea: A Survey," *Korea Journal* July (1990), 4-8.

[Park 2005] Park, Seong-Rae: *Science and Technology in Korean History.* Fremont, California: Jain Publishing Company. 2005.

[Pak 1985] Pak, Yul: *Sanhak ponwon.* In Kim, Yong Woon, ed. (1985). *Hanguk Kwahak Kisul Sa Charyo Taekye: Suhak Pyon* (Collection of Source Materials of the Korean History of Science and Technology: Mathematics Section (3) (1985), 219-360. Seoul: Yugang Press.

[Portal 2000] Portal, Jane: *Korea: Art and Archaeology.* London: British Museum Press. 2000.

[So 1787] So, Ho-su: *Suri chongon pohae* (數理精蘊補解) (unknown publisher). 1787.

[Song 1986] Song, Sang-Yong: "Studies in the History of Korean Science: A Bibliographical Essay," *Korea Journal* (March) (1986), 47-64.

[Su 2013] Su, Jung-Hong: *Modern Mathematics in China (1600-1905): An approach of social history of mathematics* (in Chinese). Ph. D thesis, National Taiwan Normal University, Taipei, Taiwan. 2013.

[Sun 2003] Sun, Mei-Yin: *Preliminary Study on Pak Yul's Chuhak ponwon* (in Chinese). Master's thesis, National Taiwan Normal University, Taipei, Taiwan. 2003.

[Wang 2011] Wang. Yu-Jun: Text Analysis of Cho Hui-sun's *Sanhak subyu.* (in Chinese). Master's thesis, National Taiwan Normal University, Taipei, Taiwan. 2011.

[Wen 2009] Wen, Hong-Yuan: *Outlook of Chinese Mathematics in Yi Sang-hyok's Iksan* (in Chinese). Master's thesis, National Taiwan Normal University, Taipei, Taiwan. 2009.

[Wu 2003] Wu, Biing-Horng: *Preliminary Study on Yi Sang-hyok's Changunbop Monggu* (in Chinese). Master's thesis, National Taiwan Normal University, Taipei, Taiwan. 2003.

[Wu 2004] Wu, Bing-Horng: "On the Transmission and Transformation of Sino-Korean Mathematics: A case study of Yi Shang-hyok's text on the method of *Changunbop*)" (in Chinese), Wann-Sheng Horng *et al.*, eds., *Proceedings of the Asia-Pacific HPM 2004 Conference: History, Culture and Mathematics Education in the New Technology Era* (May 24-28, 2004, National Taichung Teachers College, Taichung, Taiwan) (2004), 325-342.

[Wu 2007] Wu, Jian-Ren: *Textual Analysis of Cho Tae-koo's Chuso kwangyon* (in Chinese). Master's thesis, National Taiwan Normal University, Taipei, Taiwan. 2007.

[Xia 2003] Xia, Wen-Jun: *How Choson Mathematician Learned from Chinese Mathematics: A case study of Nam Pyong-chol's Haekyong sechoea* (in Chinese). Master's thesis, National Taiwan Normal University, Taipei, Taiwan. 2003.

[Yeh 2002] Yeh, Chi-Hai: *Korean Mathematics in Sejong Period (1406-1450) of the Yi Dynasty* (in Chinese). Master's thesis, National Taiwan Normal University, Taipei, Taiwan. 2002.

[Yeh 2004] Yeh, Chi-Hai: "Korean Mathematics in Sejong Period (1406-1450) of the Yi Dynasty", Wann-Sheng Horng *et al.*, eds., *Proceedings of the Asia-Pacific HPM 2004 Conference: History, Culture and Mathematics Education in the New Technology Era* (May 24-28, 2004, National Taichung Teachers College, Taichung, Taiwan) (2004), 359-370.

[Yi 1985a] Yi Sang-hyok: *Changunbop Monggu.* In Kim, Yong Woon ed. (1985). *Hanguk Kwahak Kisul Sa Charyo Taekye: Suhak Pyon* (Collection of Source Materials of the Korean History of Science and Technology: Mathematics Section (4) (1985), 349-557. Seoul: Yigang Press.

[Yi 1985b] Yi Sang-hyok: *Iksan.* In Kim, Yong Woon, ed. (1985). *Hanguk Kwahak Kisul Sa Charyo Taekye: Suhak Pyon* (Collection of Source Materials of the Korean History of Science and Technology: Mathematics Section (4) (1985), 561-686. Seoul: Yigang Press.

[Yi 1996] Yi Song-nu: *Study on Yangban Class in Early Choson Period* (Chinese Translation). Taipei: Society of Korean Study Press. 1996.

[Ying 2010] Ying, Jia-Ming: *Nam Pyong-Gil's* (1820-1869) *Reinterpretation of Classical Mathematics* (in Chinese). Ph. D thesis, National Taiwan Normal University, Taipei, Taiwan. 2010.

[Ying 2011] Ying, Jia-Ming: "The *Kujang shlhae* 九章術解: Nam Pyong-Gil's reinterpretation of the mathematical methods of the *Jiuzhang suanshu,"* *Historia Mathematica* 38(1) (2011), 1-27.

[Ying 2013] Ying, Jiia-Ming: "A Survey of Geometrical Diagrams in Korean Mathematical Texts from the 17th to the 19th Century," *Historia Scientiarum* 23 (1) (2013), 38-58.

Transmission of Probability Theory into China at the end of the Nineteenth Century

SIU Man Keung and LIH Ko Wei

In celebration of the seventieth birthday of Joseph Dauben, an esteemed scholar and a good friend

Introduction

In spite of the occurrence of many uncertain events in human experience in different civilizations since antiquity, be it in the East or the West, a quantitative approach to probability was not developed until the sixteenth and seventeenth centuries in Western Europe. This peculiar "miss" is particularly notable in the history of Chinese mathematics, even though knowledge and skill in numerical calculation had long been well developed in ancient and medieval China. Probability first came to China in the form of a book, *Jueyi Shuxue* (literally meaning: "mathematics to resolve uncertainty"). It was translated around 1880 through the collaboration of the British missionary John Fryer (1839-1928) and the Chinese mathematician HUA Heng-fang (1833-1902) from an article on probability written by Thomas Galloway (1796-1851) for the *Encyclopaedia Britannica*; this article was published separately as the book *A Treatise on Probability.*

This paper concerns this episode and related issues: it tells the story of how the theory of probability was first introduced into China in the late-nineteenth century. In a way this is not a story of success, but not one of complete failure either. This first book on probability, *Jueyi Shuxue*, introduced a new branch of mathematics to Chinese mathematicians and scientists towards the last years of the old Imperial China, though it seems to have exerted little influence on the subsequent development of probability and statistics which took place in the Chinese Republic. The serious study of probability and statistics, in the modern sense of the subject, only got underway in China in the 1930s, from which point onwards its development was similar to that in the Western world (For an account of the history of statistics in China see for instance [Li, Mo 1993]). Why are we interested in this "unsuccessful" attempt? A main reason is because we look at the episode from the viewpoint of incompatibility of the cultural, social and linguistic contexts. We thus see this "failure" as an instance of a far more general phenomenon, for when a foreign scientific discipline is implanted into an indigenous environment in which the concepts and technical means are foreign and largely incompatible with the scientific language of the time, one should hardly be surprised when this new discipline does not bear fruit immediately.

Many researchers have brought forth scholarly studies of the development of probability theory in the West. Some have analyzed why the subject arose as late as the sixteenth and seventeenth centuries in Europe, even though the notions of

D.E. Rowe and W.-S. Horng (eds.), *A Delicate Balance: Global Perspectives on Innovation and Tradition in the History of Mathematics*,
Trends in the History of Science, DOI 10.1007/978-3-319-12030-0_17

"probable" or "chance" were by no means foreign to ancient civilizations both in the East and the West, in which games were a familiar part of life since early antiquity. In the case of China, the absence of interest in the theory of probability until the subject was transmitted through *Jueyi Shuxue* in the late-nineteenth century raises a particularly interesting question, especially in view of the many accomplishments of Chinese mathematicians from antiquity to medieval times. We thus turn to address this issue with the benefit of perceptive views offered by other scholars who have been interested in this tantalizing question.

Jueyi Shuxue

The book entitled *Jueyi Shuxue* (決疑數學) [Jueyi Shuxue 1896] has been studied by several historians of mathematics. They have concluded this is a translation based on an article of Thomas Galloway (1796-1851) in the eighth edition of *Encyclopaedia Britannica* of 1859 (see [Bréard 2008], [Guo 1989], [Wang 2006], [Wang, Bréard 2010], [Yan 1990]). Thomas Galloway was a Scottish mathematician who taught at the Royal Military College at Sandhurst beginning in 1823. He later became an actuary with the Amicable Life Assurance Company of London from 1833 to the end of his life. Galloway's text first appeared as an article in the seventh edition of the *Encyclopaedia Britannica* in 1839 and was also published separately as a book [Galloway 1839]. The translation was accomplished through the collaboration of the British missionary John Fryer (傅蘭雅 1839-1928) and the Chinese mathematician HUA Heng-Fang (華蘅芳 1833-1902). This was accomplished in the traditional manner of those days, namely, the Western translator would explain the text in detail to the Chinese translator, who would then study the content as deeply as possible and then write out a Chinese text in consultation with the Western translator. The translation was published in 1896 at the initiation of ZHOU Xue-Xi (周學熙 1866-1947), a well-known Chinese official, politician and industrialist. Although it was already announced in the Chinese magazine *Gezhi Huibian* (格致彙編) that the translation had been finished by the autumn of 1880, it took more than fifteen years before it was finally published. According to what ZHOU's nephew, the mathematician ZHOU Da (周達 1879-1949), said in the preface of a reprinted edition of the book in 1909, only a few copies were printed and their circulation was not very wide (see Section 3) [Guo 1989].

Jueyi Shuxue reproduces the original English text [Galloway 1839] quite faithfully. The content essentially reflects that of a text on probability after the style of Laplace to which the work of Poisson has been added. The introduction begins with the sentence: "The doctrine of probability is an extensive and very important branch of mathematical science, the object of which is to reduce to calculation the reasons which we have for believing or expecting any contingent event, or for assenting to any conclusion which is not necessarily true." After some general discussion of the subject a historical account of the theory from the Pascal-Fermat correspondence to the fundamental treatises of Huygens, Bernoulli and others are given including the landmark classics *Ars Conjectandi* by Jakob Bernoulli (called by the name of James Bernoulli in Galloway's book), followed by the works of Montmort, De Moivre, Bayes, Laplace and Poisson, with the work of De Morgan of 1837 mentioned at the end. By the end of the nineteenth century probability theory had developed well beyond Laplace's work [Schneider 1987], but the translation reflects the enthusiasm the translators still held for Laplace's work.

A conscientious Chinese reader would, through this faithful translation, gain knowledge of the development of probability theory in Europe since the seventeenth century from a comprehensive historical account spanning some ten pages of the book. Still, it is questionable to what extent a Chinese reader in the late-nineteenth century would have been able to comprehend the logical and philosophical context of a subject that was so new and foreign to him. We cite here a representative passage from this historical account:

> The foundations of the mathematical theory of probabilities were laid by Pascal and Fermat about the middle of the 17th century. [...] and a correspondence on the subject took place between these two illustrious geometers, which is preserved in their respective works, and throws some light on the history of mathematics in that age.
>
> About the same period Huygens composed his tract *De Ratiociniis in Ludis Aleæ*, which was first published in the *Exercitationes Geometricæ* of Schooten in 1658. This was the first systematic treatise which appeared on the doctrine of chances. It contained an analysis of the various questions which had been solved by Pascal and Fermat, and at the end five new questions were proposed, the solutions of which, simple as they may now appear, were then attended with considerable difficulty. The analysis of two of them was in fact given for the first time by Montmort, half a century after their publication. [...]
>
> James Bernoulli appears to have been the first who perceived that the theory of probability may be applied to much more important purposes than to regulate the stakes and expectations of gamesters, and that the phenomena, both of the moral and physical world, anomalous and irregular as they appear when viewed in detail, exhibit, when considered in large numbers, a constancy of succession which renders their occurrence capable of being submitted to numerical estimation. The *Ars Conjectandi*, published in 1713, seven years after the death of the author, contains a number of interesting questions relative to combinations and infinite series; but the most remarkable result which it contains is a theorem respecting the indefinite repetition of events, which may be said to form the basis of all the higher applications of the theory. It consists in this, that if a series of trials be instituted respecting an event which must either happen or fail in each trial, the probability becomes greater and greater, as the number of trials is increased, that the ratio of the number of times it happens, to the whole

number of trials, will be equal to its *a priori* probability in a single trial; and that the number of trials may be made so great as to give a probability, approaching as nearly to certainty as we please, that the difference between the ratio of its occurrences to the number of trials, and the fraction which measures its *a priori* probability, will be less than any assigned quantity. Bernoulli informs us, that the solution of this important theorem had engaged his attention during a period of twenty years.

In the interval between the death of Bernoulli and the appearance of the *Ars Conjectandi*, Montmort published his *Essai d'Analyse sur les Jeux de Hazard*. The first edition was in 1708; the second, which is considerably extended, and enriched by several letters of John and Nicolas Bernoulli, appeared in 1713. [...]

About the same time, Demoivre began to turn his attention to the subject of probability, and his labours, which were continued during a long life, contributed greatly to the advancement of the general theory, as well as the extension of some of its most interesting applications. [...]

The objects and important applications of the theory of probabilities having been made known by the works now mentioned, the subject has ever since been regarded as one of the most curious and interesting branches of mathematical speculation, and accordingly has received more or less attention from almost every mathematician of eminence. A great variety of questions connected with it and especially relating to lotteries, are interspersed in the volumes of the *Paris* and *Berlin Memoirs*, (particularly the latter,) by John and Nicolas Bernoulli, Euler, Lambert, Beguelin, and others. D'Alembert has likewise treated of the theory in several of the volumes of his *Opuscula*; and it is not a little remarkable, that in some instances its first principles should have been misunderstood by so ingenious and profound a writer. [...]

[...] The *Transactions of the Royal Society* for the years 1763 and 1764, contain two papers by the Rev. Mr. Bayes, with additions to the latter by Dr. Price, which deserve to be noticed, inasmuch as the principles on which the probability of an event is determined, when the event depends on causes of which the existence and influence are only presumed from experience, are there for the first time correctly laid down. The question proposed and solved by Bayes was this: a series of experiments having been made relative to an event, to determine the presumption there is, that the fraction which measures its probability falls within given limits.

One of the earliest applications of the theory of probability was to determine, from observations of mortality, the average duration of human life, and the value of pecuniary interests depending on its continuance or failure. [...]

The application of the theory of probability to the subject of jurisprudence, and the verdicts of juries and decisions of tribunals, has been discussed by the Marquis Condorcet in various articles in the *Encylopédie Méthodique*; [...]

> The analysis of probability has been applied with signal advantage in many researches of Natural Philosophy, but especially in appreciating the mean errors of observations. [...] The mean errors of observations were treated as questions of probability by Lagrange in the Turin *Memoirs* for 1773; but it is to Laplace that the theory owes its principal extension and most important results. [...]
>
> Laplace's great work, the *Théorie Analytique des Probabilités* , first published in 1812, is one of the most remarkable productions that has ever appeared in abstract science. The principles of the calculus, as well as the peculiar methods of analysis which it requires, and the most interesting and difficult questions which it presents, are here discussed in a far more general manner than had been attempted by any former writer on the subject; and it may be said, accordingly, to have placed the theory under an entirely new aspect. [...]
>
> Next to the *Théorie Analytique* of Laplace, the most important work which hitherto appeared on the subject of probability is the recent one of Poisson, entitled, *Recherches sur la Probabilité des Judgements*, (Paris 1837.) [...]
>
> It is in these two works of Laplace and Poisson that the higher and more abstruse parts of the theory of probabilities must be studied. A very clear exposition of the principles, accompanied with many interesting remarks on the uses and applications of the theory, is given by Lacroix in his valuable little work, *Traité Elémentaire du Calcul des Probabilités*, Paris 1822.
>
> Since the time of Demoivre, the English treatises on the general theory of probability have neither been numerous, nor, with one or two exceptions, very important. [...] but by far the most valuable work in the language is the Treatise in the *Encyclopedia Metropolitana*, by Professor De Morgan, 1837. In this very able production, Mr. De Morgan has treated the subject in its utmost generality, and embodied, with a moderate compass, the substance of the great work of Laplace.

It is interesting to note a passage in the introduction that has this to say about probability and gambling:

> In fact, most of the questions of this class to which the calculus can be applied, are connected with lotteries and games of hazard. The results obtained from the analysis of such questions cannot be considered as being of any great value in themselves, but they frequently throw light on subjects of far higher importance which present analogous combinations. It is true that the mathematical theory comes in aid of moral considerations, and demonstrates the ruinous tendency of gambling even

> when the conditions of the play are equal, mathematically speaking; but, unfortunately, those who indulge a passion for this vice are seldom capable of appreciating the force of such arguments.

This expression of a negative attitude towards gambling perhaps had an appeal for the Chinese translator, who was brought up in a Confucian tradition in which gambling was censured by public opinion as a vice that led to corruption of moral standards.

Historical Background of the Translation

The translation of Western books on science and mathematics went on as a fervent activity in Imperial China in the latter half of the nineteenth century. One has to look at it in the historical context of what the country went through during that period, which the historian Immanuel HSÜ (徐中約 1923-2005) describes as a "search for a way to survival in the new world that had been forcibly thrust upon China by the West after the mid-19th century" [Hsü 1970/1995]. The Chinese, "burdened by tradition and heritage, and as yet ignorant of the nature of the Western world, groped in the dark." HSÜ maintains that "the dynamics of change suggests that modern Chinese history is not characterized by a passive response to the West, but by an active struggle of the Chinese to meet the foreign and domestic challenges in an effort to regenerate and transform their country from an outdated Confucian universal empire to a modern national state, with a rightful place in the family of nations" [Hsü 1970/1995].

To see this in a wider historical context one has to look at the three waves of transmission of Western science into China. The translation of Euclid's *Elements* by XU Guang-qi and Matteo Ricci in 1607 led the way during the first wave [Siu 2011]. As pointed out by Siu [Siu 1995/1996]: "The gain of this first wave seemed momentary and passed with the downfall of the Ming Dynasty. Looking back we can see its long-term influence, but at the time this small window which opened onto an amazing outside world was soon closed again, only to be forced open as a wider door two hundred years later by Western gunboats that inflicted upon the ancient nation a century of exploitation and humiliation, thus generating an urgency to know more about the Western world."

The second wave came in the wake of the first wave and lasted from the mid-seventeenth century to the mid-eighteenth century. Instead of Chinese scholar-officials the chief promoter was Emperor Kangxi of the Qing Dynasty (reigned 1662-1722). Instead of Italian and Portuguese Jesuits the Western partners were mainly French Jesuits, the so-called "King's Mathematicians" sent by Louis XIV, the "Sun King" of France (reigned 1643-1715), in 1685 [Du, Han 1992]. Emperor Kangxi learnt assiduously Western mathematics and astronomy from these French Jesuits. Later in 1713 a group of selected Chinese officials and scholars was summoned to form the Office of Mathematics that was established in *Mengyangzhai* (Studio for the Cultivation of the Youth 蒙養齋), a place situated in the garden

Changchunyuan (Garden of the Exuberant Spring 暢春園) inside the Imperial Palace. Besides serving as a school for taking lessons in mathematics, astronomy and science, the main task of this Office was to compile the monumental treatise *Lüli Yuanyuan* (Origin of Mathematical Harmonics and Astronomy 律曆淵源), comprising three parts: *Lixiang Kaocheng* (Compendium of Observational Computational Astronomy 曆象考成), *Shuli Jingyun* (Collected Basic Principles of Mathematics 數理精蘊), and *Lülü Zhengyi* (Exact Meaning of Pitchpipes 律呂正義). The treatise was completed in 1722/1723. The second book *Shuli Jingyun* includes both traditional Chinese mathematics, the part that was still extant and was understood at the time, as well as Western mathematics, most likely from the "lecture notes" prepared by the missionaries for Emperor Kangxi during his ardent study of mathematics in the 1680's. Interested readers will find a more in-depth discussion of this second wave in a paper of Catherine Jami [Jami 2002], as well as in [Jami 2012]

The third wave came in the last forty years of the nineteenth century in the form of the so-called "Self-strengthening Movement" after the country suffered from foreign exploitation during the First Opium War (1839-1842) and the Second Opium War (1856-1860). This time the initiators were some officials of high rank led by Prince Gong (Yixin 恭親王奕訢 1833-1898) with contributions from Chinese scholars and Protestant missionaries coming from England or America [Chan, Siu 2012, Siu 2009, Xiong 1994].

The theme and mood of the three waves of transmission of European science into China were reflected in the respective slogans prevalent in each period. In the first part of the seventeenth century the idea was: "In order to surpass we must try to understand and to synthesize (欲求超勝必須會通)." In the first part of the eighteenth century it became: "Western learning has its root in Chinese Learning (西學中源)." In the latter part of the nineteenth century the slogan took on a very different tone: "Learn the strong techniques of the '[Western] barbarians' in order to control them (師夷長技以制夷)." In a paper on European science in China, Catherine Jami says: "[...] the cross-cultural transmission of scientific learning cannot be read in a single way, as the transmission of immutable objects between two monolithic cultural entities. Quite the contrary: the stakes in this transmission, and the continuous reshaping of what was transmitted, can be brought to light only by situating the actors within the society in which they lived, by retrieving their motivations, strategies, and rationales within this context." [Jami 1999].

Headed by Prince Gong and supported strongly by some officials of high rank, schools were established to teach Western science and mathematics, and offices were set up to translate Western texts in science and mathematics. In particular, an establishment known by the name of *Tong Wen Guan* (College of Foreign Languages 同文館) was set up in 1862 by decree, with the section on mathematics and astronomy established in 1866. The establishment of *Tong Wen Guan* was at first intended as a language school to train interpreters but later developed into a college of Western learning, along with other colleges of similar nature that sprouted in other cities like Shanghai, Guangzhou, Fuzhou, Tianjin, along with the establishment of arsenals, shipyards and naval schools.

At the beginning of the nineteenth century a new wave of Christian missionaries came to China, starting with Robert Morrison (1782-1834) of the London Missionary Society of London in 1807. As pointed out by Benjamin Elman [Elman 2005]: "Although their enthusiasm for China was often fired by evangelicalism back home, British and American Protestants in China swiftly recognized that Chinese literati were interested in the sciences that the missionaries accepted as part of their Christian heritage. Like their Jesuits predecessors, British and American missionaries viewed science as emblematic of their superior knowledge systems."

Once the missionaries had this view they expanded their non-evangelical activities to include education, medical work and publication. In 1868 the English missionary John Fryer was recruited to work in the translation department of the Jiangnan Arsenal (江南製造局) in Shanghai. Fryer was the most prolific translator who took part in completing more than a third of the total output in this department. During the period 1868-1896, he collaborated with the Chinese mathematician HUA Heng-Fang, who was his colleague at the Jiangnan Arsenal. (See [Bennett 1967] and [Horng 1993] for a detailed account of Fryer and HUA respectively.) The translation of *Jueyi Shuxue* was accomplished during this period in which the Chinese government, as well as many Chinese scholars, harboured a fervent and urgent desire to learn from the Westerners in order to resist foreign exploitation of their country.

After *Jueyi Shuxue* was published in 1896, Fryer issued it again in 1897 through his *Gezhi Shushi* (Chinese Scientific Book Depot, literally: bookshop for investigating things and extending knowledge 格致書室), which he set up in 1884 as a retail outlet of the Jiangnan Arsenal [Elman 2005]. The book was reprinted in 1909 by the Chinese mathematician ZHOU Da, who had great interest in the book, a copy of which he inherited from his uncle who first printed it (see Section 2). Still, ZHOU Da lamented that it was not given adequate attention so that circulation was not widespread. Despite this admirable effort of ZHOU Da, who stressed in a preface to the reprinting how important this theory was to mankind, the translated book did not seem to capture much attention at the time. The fact that the book remained unpublished for fifteen years after it was translated may reflect doubts of the translators themselves about its suitability for the indigenous scientific community. Or, it may possibly reflect their lack of confidence in having a strong enough grasp of this new branch of mathematics. This may explain why the book does not present an introduction written by Fryer nor by HUA, whereas the two other well-known and popular translated books by them contained such explanatory introductions. This famous pair of collaborators produced *Daishu Shu* (Method of Algebra 代數術) in 1873 and *Weiji Suyuan* (The Origins of the Differential and Integral Calculus 微積溯源) in 1874. These were, respectively, translations of the articles on "Algebra" and "Fluxions" in the eighth edition of the *Encyclopaedia Britannica*, which were written by the Scottish mathematician William Wallace (1768-1843). Both texts were influential in transmitting these new subjects, algebra and calculus, into China at the time (see [Hu 1998] for more on the translation of those two books). In contrast, the new subject of probability theory was not met with similar enthusiasm from the scientific community, partly it seems because the translators themselves had doubts about how to synthesize the subject with the Chinese mathematical tradition!

Reception of the Translation at the Time

In 1896 a leading Chinese intellectual from this period, LIANG Qi-Chao (梁啟超 1873-1929), compiled a list of Western books for recommended reading. He published this in the periodical *Shiwu Bao* (The China Progress 時務報), which he had helped to establish in that same year. *Jueyi Shuxe* was included on this list, even though the translation was not yet published, as LIANG Qi-Chao pointed out: "The translation of *Jueyi Shushu* (決疑數術 which should be *Jueyi Shuxue*) was completed but the book is not yet published [by the Translation Bureau of the Jiangnan Arsenal]. The Westerners apply this method to insurance and other matters. There was once an article published in *Gezhi Huibian* (格致彙編) to introduce the subject. The article is brief and cannot give the details." This magazine is the 1891 Spring issue of *Gezhi Huibian* [literally: compendium for investigating things and extending knowledge] with the English title *The Chinese Scientific and Industrial Magazine* [Gezhi Huibian], a Chinese magazine published from 1876 to 1892 by the English missionary John Fryer (1839-1928) and the Chinese scholar XU Shou (徐壽 1818-1884) in Shanghai, and was modelled after an earlier publication of a similar nature, the *Peking Magazine.*

In the latter part of the nineteenth century some Western missionaries published periodicals to propagate Western learning in old imperial China as a means of spreading Christian faith. Thus, the periodical *Zhongxi Wenjian Lu* (Record of News in China and the West 中西聞見錄) with the English title *Peking Magazine* was founded in August of 1872 by the American missionary William Alexander Parsons Martin (1827-1916) and the English missionary Joseph Edkins (1823-1905), supported by the Society for the Diffusion of Useful Knowledge in Beijing [Peking]. At the back of the title-page of the first issue the objective in publishing this periodical was clearly stated in a short note, further elaborated in the ensuing preface. The short note says, "*Zhongxi Wenjian Lu* [*Peking Magazine*] adopts the practice and format of newspapers in the Western world in publishing international news and recent happenings in different countries, as well as essays on astronomy, geography and *gewu* [science, literally meaning " investigating things"]. The magazine will be published once every month. Any gentleman, Chinese or Westerner who gathers new information or has his own views to express, is invited to submit it to the editors at the *Shi Yiyuan* (Charity Hospital) of *Mishi* [*Street*] (Rice Market Street). The editors will select those items that are considered to be fit for print. In this way, new information will be attained through collective effort to benefit more people so as to enable them to become more and more knowledgeable" [Zhongxi Wenjian Lu] .Upon the closing down of the Society for the Diffusion of Useful Knowledge in August of 1875, the periodical was also terminated after thirty-six issues, to be revived in 1876 in the form of another periodical, *Gezhi Huibian*.

The aforementioned article in the 1891 Spring issue of *Gezhi Huibian* titled "Using mathematics to resolve uncertainty" was probably the only other mentioning of the subject besides *Jueyi Shuxue* at the time. No author's name is attached to the article. It was likely to have come from the pen of Fryer himself. Twenty-six years later in 1917 another article on probability appeared in the journal *Kexue* (Science 科學), which was the official publication of a newly established organization *Zhongguo Kexue She* (Science Society of China 中國科學社) founded in 1914 by a group of Chinese students studying in the United States who were devoted to bringing about progress and development of their motherland and who shared the view that science would be the key to move the country forward. The journal *Kexue* was established with the intention of promoting science among the Chinese people, to "promote science, encourage industry, authorize terminologies, and spread knowledge" [Buck 1980]. The author of this article on probability is HU Minfu [HU Ming-fu] (胡明復), who was the first Chinese to earn a doctorate in mathematics abroad (from Harvard in 1917) as well as the first Chinese mathematician to publish a paper in an international journal of fame [Linear integro-differential equation with a boundary condition, *Trans. Amer. Math, Soc.*, 19(4) (1918), 1-43.] HU Ming-fu returned to China after 1917 and made great contributions to the scientific development of the country before he met a tragic untimely death while swimming in Wuxi in 1927 [Zhang 1991]. Going further than the 1891 article on probability in *Gezhi Huibian*, HU's article on probability not only explained what the subject was about but also touched upon philosophical aspects and the role it plays in scientific thinking [Hu 1917]. Actually, the article was published as the third part under the general heading of methodology in science.

In a different direction, LIANG Qi-Chao also applied his knowledge of probability to another area of human activity. In one of his political essays, entitled "Nü xue (Women's schooling 女學)" and published in the March issue of *Shiwu Bao* in 1897, he again referred to the book *Jueyi Shuxe*:

> *Mengzi* (孟子) says, "To dwell in idleness without instruction is a life akin to that of the birds and beasts." [...] In general, the people in a state are bound to let each one have an occupation. If everyone can support himself, then the state can be governed extensively. If there are those who cannot do this, then its strength and weakness varies with the number of people who have no occupation. What is the reason for this? Those persons who have no occupation, must be supported by those who have an occupation. If they are not supported, then those who have no occupation are in a perilous situation, if they are supported, then those who have an occupation are in danger. This is what it means. What the Westerners in their translations called 'production and distribution' is what is expressed in the *Great Learning* (大學) with

> "Those who produce are numerous, those who eat are few." *Guanzi* (管子) said: "If a single peasant is not engaged in farming, someone will suffer hunger; if a single woman does not engage in weaving, someone will suffer cold." These are not empty words. The above is like taking the entire population of a state and its material production, and by the *Mathematical Art of Probability* (*jueyi shushu* 決疑數術), increasing and decreasing, dispersing and adding it. The obtained ratio is indeed like this: in China, those who sponge on others are half of those who create profit. [...] There are two hundred million women who belong entirely to the consumers and of whom not a single one is productive. Because they cannot support themselves and have to be supported by others, men elevate them like dogs, horses, or slaves, which makes women's life even more miserable." (This English translation is taken from [Bréard 2008], with slight modifications; the full essay is in [Liang 1936].)

Even though LIANG Qi-Chao referred to the notion of probability in his political essay on reform in advocating literacy for women and hence in defending women's rights, it seems that he did not entirely grasp the concept of probability and regarded probability just as simply a proportion! This kind of superficial understanding of probability by a well-educated scholar of the time is not surprising. The actual influence of the book *Jueyi Shuxue* seems to have been rather minimal, and subsequent studies and works on probability and statistics suggest little direct impact from the book. It turns out that most of the terms the translators coined in probability theory did not survive so that Chinese terminologies adopted in later textbooks were different. An obvious example is "*jueyi shuxue*" (決疑數學), which was later replaced by "*gailü lun*" (概率論) after the temporary adoption of various other translated terms.

Lack of Influence of the Translation

A major obstacle for Chinese translations of modern mathematics texts in the nineteenth century was the lack of existing terminology. This meant not only that new Chinese terms had to be coined but also that Western expressions had to be adopted in a mathematical symbolism that was foreign to Chinese readers. Thus, one reason for the lack of influence of *Jueyi Shuxue* might have to do with the mathematical language adopted in the translation. This, after all, was not a complete symbolic format already commonly adopted in Western mathematical literature but rather a hybrid between the Chinese format adopted in traditional Chinese mathematical texts and some newly coined symbols with a Chinese flavour. To our modern eyes this would indeed look awkward and impede facility in computation or even ease in comprehension. Let us look at some excerpts from the *Jueyi Shuxue*:

(1) Section II, Article 13: Application of the binomial theorem.

This deals with the expansion of the expression $(p+q)$ to the power h. The original text says:

In order to place the proposition now demonstrated in a clearer light, let us consider separately the different terms of the development of $(p+q)^h$, namely,

$$\begin{aligned} p^h &+ hp^{h-1}q + \frac{h(h-1)}{1 \cdot 2}p^{h-2}q^2 + \cdots \\ &+ \frac{h(h-1)(h-2)\cdots(h-n+1)}{1 \cdot 2 \cdot 3 \cdots n}p^{h-n}q^n \\ &+ \cdots + q^h. \end{aligned}$$

The Chinese translation rendered the formula in the following form:

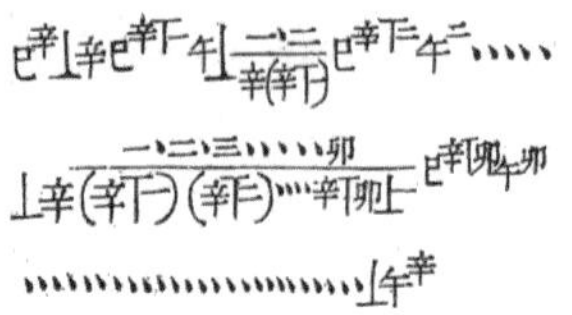

The original text goes on to say:

The first term p^h expresses the probability that the event E will occur in every one of the h trials. The second term $hp^{h-1}q$ expresses the probability that E will occur $h-1$ times, and F once, without distinction of order; that is to say F may happen at the first or last or any intermediate trial.

The Chinese translation rendered the second term in the following form:

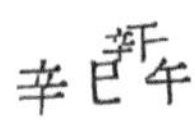

(2) Section VIII, Article 96: Method of computing the value of the integral which expresses the probability.

This deals with the computation of the famous integral of e to the power minus x squared. The original text says:

The integral $\int e^{-t^2}dt$ is computed as follows. Developing the exponential e^{-t^2} in a series of the ascending powers of t^2, and integrating the successive terms between the limits $t=0$ and $t=\tau$, we find

$$\int_0^\tau e^{-t^2}dt = \tau - \frac{\tau^3}{1 \cdot 3} + \frac{\tau^5}{1 \cdot 2 \cdot 5} + \frac{\tau^7}{1 \cdot 2 \cdot 3 \cdot 7} + \text{\&c.};$$

a series which converges rapidly when τ is less than unity.

The Chinese translation rendered the integral in the following form:

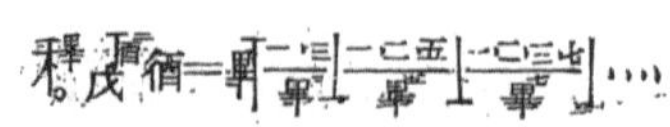

This kind of Chinese-Western mixture of symbolic language was adopted in all the translated mathematical texts at the time. For a student of mathematics today this hybrid certainly looks awkward, however, with sufficient practice familiarity can still be attained. Still, Chinese mathematicians in the nineteenth century might not have regarded this hybrid style as a crucial difficulty in learning the mathematics. Thus, it seems that hindrance to the development of probability theory in ancient and medieval China arose more for cultural and social reasons, to which we now turn.

As HUA Heng-Fang might have anticipated, the concept of probability was not readily received by the indigenous scientific community of his time. With algebra there was the tradition of solving equations in ancient and medieval China, in fact at a level that was quite advanced when compared with that of the Western world in the corresponding period. Similarly, with calculus there was the notion of infinitesimals already used in the computation of areas and volumes as explained by LIU Hui (劉徽) in the third century. But the quantitative notion of uncertainty was totally absent in all Chinese mathematical classics up to that point.

This peculiar "miss" in the history of Chinese mathematics presents a baffling question in two respects. The first concerns an "internal" issue, since the needed mathematical knowledge and skill in numerical calculation had long been developed in ancient and medieval China (see Section 6). The second aspect reflects an "external" issue, since probability theory has so many applications. It is commonly agreed that traditional Chinese mathematics paid much attention to demands of solving real world problems. Indeed, this exemplified a basic tenet of traditional Chinese philosophy of life shared by the class of *shi* (intellectuals 士), namely, self-improvement and social interaction (經世致用). Had the Chinese mathematicians become aware of the fundamental significance and applicability of the notion of probability they would doubtless have devoted their efforts to its investigation. One might, of course, argue that the actual "real-life" problems in China of the nineteenth century were not yet developed to the stage that gave rise to probability theory in the European world of the seventeenth and eighteenth centuries.

The Notion of Probability in Ancient and Medieval China

Perhaps this question of the peculiar "miss" has to be viewed not just within the Chinese context alone but also against the full background of developments in the West as well.[1] As noted at the outset of this essay, despite the role of uncertain events throughout human history, a quantitative approach to probability was not developed until the sixteenth and seventeenth centuries in Western Europe. For instance, gambling as a human activity has a very long history, an early example of risk taking. According to Florence Nightingale David "the real problem which

[1]For relevant studies of this, see [Bernstein 1996], [Daston 1988], [David 1962], [Hacking 1975/2006], [Hald 1990], [Hald 1998], [Maistrov 1974], [Meusnier 1996], [Plato 1994], [Sheynin 1974], [Stigler 1986].

confronts the historian of the calculus of probabilities is its extremely tardy conceptual growth– in fact one might almost say, its late birth as an offspring of the mathematical sciences" [David 1962].

Mark Elvin points out [Elvin 2002/2010, Elvin 2004]–following Alistair Crombie's six styles of scientific thinking in the European world (namely, "postulational," 'experimental', "hypothetical modelling," "taxonomy," "probabilistic," and "historical derivation") [Crombie 1994]–that probabilistic and statistical thinking was the last that came to maturity. One might for this reason regard this style as perhaps the least natural kind of scientific thinking, thus an approach that needs time to incubate, be it in the West or the East. In his account of the prehistory of the theory of probability, Oscar Sheynin mentions a saying of Aristotle: "Evidently, none of the traditional sciences busies itself about the accidental, [...] this (the accidental) none of the recognized sciences considers, but only sophistic [...]." Sheynin points out that Aristotle was the first to attempt an explanation of chance [Sheynin 1974]. In a paper that discusses the possible and the probable [Sambursky 1956], Shmuel Sambursky points out an interesting feature of Greek philosophy in general and its mathematics in particular that caused their "miss" in understanding the intriguing and important concept of probability, namely, the cosmic outlook of the ancient Greeks which "saw precision restricted to the heavens and failed to see laws in the fluctuations and irregularities on earth." [Sambursky 1956] A breakdown of this barrier between "heavenly" and "terrestrial" phenomena in the sixteenth/seventeenth centuries freed the human mind so that the investigation of chance events was on the agenda. The ancient Greeks perceived mathematics as an idealistic world of "mathematical reality" for pure intellectual pursuit and presented mathematics as a beautiful and perfect form that embodied truth with certainty. In Plato's dialogue *Theaetetus* [Plato 1921] Socrates said, "If Theodorus, or any other geometrician, should base his geometry on probability, he would be of no account at all." Drawing definite conclusions via deductive logic in Greek mathematics is to be contrasted with what Plato defined in another dialogue, *Phaedrus* [Plato 1914], as "accepted by the people because of its likeness to truth", which reminds one of the German term for probability —"Wahrscheinlichkeit".

In the Chinese scene there were likewise instances in philosophy, in mathematics and in social practices that could have hinted at probabilistic thinking, but this had not germinated and bloomed into a mathematical theory as it did in the West. No recognizable mathematical account on the notion of probability appeared in the mathematical literature in China before its transmission from the West through *Jueyi Shuxue*. Mathematically speaking it is known that Chinese mathematicians developed much earlier than Pascal what is commonly known as Pascal's triangle, solved problems related to permutations and combinations, and for administrative reasons treated problems of a statistical nature since early times.

Chinese mathematicians in the Song Dynasty and the Yuan Dynasty developed the arithmetical triangle, known in the Western world as Pascal's triangle (but actually the triangle was not first discovered by Pascal either). They used this initially as a means for extracting square and cube roots, and they later applied it to solving polynomial equations, working with finite differences and summing finite arithmetical series. The triangle can be dated back to the beginning of the eleventh century in the work of JIA Xian (賈憲 early 11th century), who employed it in the extraction of square and cube roots. It is fortunate that the work of JIA Xian, which was lost, was reported and appeared first in print in the mathematical treatise *Xiangjie Jiuzhang Suanfa* (Detailed Explanations of the Nine Chapters on the Mathematical Method 詳解九章算法) of YANG Hui (楊輝 1238-1298) in 1261, and the triangle of binormial coefficients appeared again in another mathematical treatise *Siyuan Yujian* (Jade Mirror of Four Elements 四元玉鑑) of ZHU Shi-Jie (朱世傑 1249-1314) in 1303, who applied it to polynomial equations, finite differences and finite series [Li 1999, Shen 2000].

Folklore wisdom depicted in some anecdotes embraced an awareness of chance events. In the old book *Thirty-six Stratagems* (三十六計) the Stratagem Number 27 known as "Faked madness but not insane (假痴不癲)" [Ma, Tong 2007] tells the anecdote of how General DI Qing (狄青 1008-1057) of the Northern Song Dynasty in the 11th century boosted the morale of his army before the battle against NONG Zhi-Gao (儂智高 Nungz Ciqgau 1025-1055), the chieftain of the ethnic group of *Zhuang* (壯族) in the Southwestern part of China, by tossing 100 coins all with heads up! The whole army hailed this omen of success and marched into battle with full confidence. General DI ordered that the 100 coins be nailed to the ground and covered up by a silk cloth to await a triumphant return. The army did return in triumph, only to find that all the one hundred specially prepared coins had heads on both sides! This anecdote indicates that people in the old days knew that the chances for certain events were slim.

In Book 18 of the *Mengxi Bitan* (Dream Pool Essays 夢溪筆談), compiled by SHEN Kuo (沈括 1031-1095) in 1088, there is recorded the following problem in counting the number of possible situations in chess:[2]

> The story-tellers say that I-Hsing (一行) once calculated the total number of possible situations (局) in chess, and found he could exhaust them all. I thought about this a good deal, and came to the conclusion that it is quite easy. But the numbers involved cannot be expressed in the commonly used terms for numbers. I will only briefly mention the large numbers which have to be used. With two rows and four pieces the number of probable situations will be 81 different kinds. With three rows and nine pieces the number will be 19,683. Using four rows and sixteen pieces the number will be 43,046,721. For five rows and twenty-five pieces, the number will be 847,288,609,443 $\cdots$. Above seven rows we do not have any names for the large numbers involved. When the whole 361 places are used, the number will come to some figures (of the

[2]This English translation is taken from [Needham 1959/1970].

order of) ten thousand written fifty-two times ($10,000^{52}$ 連書萬字五十二).

One problem in *Shushu Jiuzhang* (Mathematical Treatise in Nine Sections 數書九章) of QIN Jiu-Shao (秦九韶 c.1202-1261) in 1247 made use of the idea of sampling in collecting grains for taxation [Xu, Zhang 1995]:

> When a peasant paid tax to the government granary in the form of 1534 *dan* (石) of rice, it was found out on examination that a certain amount of rice with husks was present. A sample of 254 grains was taken for further examination. Of these 28 grains were with husks. How many genuine grains of rice were there, given that one *shao* (勺) contains 300 grains?" (Note: In the mensuration system of the Song Dynasty, 1 *dan* = 10 *dou* (斗) = 100 *sheng* (升) = 1000 *ge* (合) = 10,000 *shao*. According to tradition recorded in *Jiuzhang Suanshu* (Nine Chapters on the Mathematical Art 九章算術), a grain of rice with husk was counted as half a grain of rice.) [The answer is given as: 4,348,346,456 grains, out of the original 1534 x 10,000 x 300 = 4,602,000,000 grains.]

In the late-seventeenth century CHEN Hou-Yao (陳厚耀 1660-1722) treated in his *Cuozong Fayi* (The Meaning of Methods for Alternation and Combination 錯綜法義) [Chen late 17th century] problems of permutations and combinations in connection with divination (trigrams, hexagrams), the ten heavenly stems (天干) and the twelve earthly branches (地支) to form the astronomical sexagesimal cycles, as well as with dice throwing and card games. (See [Bréard 2012] for a more detailed discussion.) In the foreword to his treatise, CHEN Hou-Yao pointed out that even though all necessary mathematical techniques had already been provided by the famous ancient mathematical treatise *Jiuzhang Suanshu* (believed to be compiled between the second century B.C.E. and the second century), there was no discussion in the treatise of any method to deal with "alternation and combination." The author was among the group of scholars summoned by Emperor Kangxi to work on the compilation of *Lüli Yuanyuan*; in fact, it was CHEN himself who proposed to Emperor Kangxi that such a treatise ought to be compiled. This book is perhaps the first systematic account of combinations and permutations in China, and its contents are very original. However, it is not known whether the author conceived of it by himself, or whether the material was already known at the time in China, or whether the author might have learnt the subject from Emperor Kangxi, who in turn might have learnt it from the French Jesuits during the second wave of transmission of Western science into China. Judged by the fact that the content was not included in the mathematical treatise *Shuli Jingyun*, the latter circumstance seems rather unlikely. That makes the question even more intriguing as to whether CHEN Hou-Yao came upon this subject by himself.

The Status of the Notion of Probability in China

Games of chance were never unfamiliar to the Chinese, even though gambling practice was officially banned by different rulers throughout the long history of China [Guo, Xiao 1995]. This practice of an official ban might explain the lack of interest of the scholarly class which did not deign to investigate and discuss problems relating to chance as a scholarly pursuit. In the Song Dynasty a practice known by the name of *guan pu* or *pu mai* (gambling-based sale 關撲, 撲賣) [Luo, Xu 1994] was not uncommon to boost the sale of goods by shop-owners. If a customer won a coin-tossing game, then he got back his stake together with the goods for free, while if he lost the game then he lost his money but got no goods. It was apparent that those shop-owners had a good notion of the odds (even if they might not know how to calculate the odds theoretically) and the strategy in playing the games in order to stay in business, but that they kept the knowledge to themselves for their own advantage.

Divination (占卜) and fortunetelling may lead to probabilistic thinking but again such knowledge might be covered up in secrecy by those who practiced it as a profession. (On the other hand, divination might also lead to the opposite by thinking that human affairs are destined by the deities so that it is meaningless to investigate the notions of chance and risk!) Physiognomy (面相學), which had long been practiced in China, can be regarded as related to some kind of statistical thinking. The idea of insurance, and thence the related idea of quantification of probabilistic occurences, might have arisen in businesses, for example, in *yin hao* (銀號) or *qian zhuang* (錢莊) or *piao hao* (票號) (Chinese versions of private banks). Another possible business sector in medieval China might have been *biao ju* (鏢局) or *biao hang* (鏢行) (Chinese versions of courier services with military escort). Unfortunately, detailed documentary evidence of such transactions is not easy to come by.

In his very thorough analysis of the situation of probabilistic thinking in premodern China, Mark Elvin [Elvin 2002/2010] points out that there may have been more than what is now known from writings. As a kind of "internal" obstacle he points out that "the cosmos was seen by not a few as an immense system of numbers, and there was a general preoccupation with chance and hazard; but even in the thought of Wang Chong (王充 27-97), who seems to have been the most lucid of the philosophically obsessed in this domain, the notion of an inherent character of luckiness or unluckiness in individuals blocked off easy access to the idea of objective equiprobability as between different players" [Elvin 2002/2010]. In another detailed and comprehensive study on probability and statistics in China, Andrea Bréard suggests that the *Jueyi Shuxue* was "a linguistic and epistemological failure in attempting to translate unfamiliar phenomena into an existing taxonomic system" [Bréard 2008]. Because the classical doctrines of probability theory were transmitted into China "as a body of knowledge without the same time-sequence they had in the West," Chinese readers could not comprehend the theory in its logical and philosophical context. Andrea Bréard also refers to the discussion by

Mark Elvin and concluded that further scrutiny is needed [Bréard 2008].

Having mentioned all these factors we have to admit that in the history of Chinese mathematics the kind of probing into mathematical concepts, purely for the curiosity of intellectual pursuit, was not as marked as in the West. The axiomatic approach to the study of geometry is one example; the investigation of the solvability by radicals of polynomial equations instead of just obtaining highly accurate methods of solving them is another; and (probably!) the study of probability starting from games of chance is a third. It is rather ironic to find that an essential feature of Greek mathematics, which caused the "miss" of a development of probabilistic thinking in the ancient world (as described in Section 6), ultimately evolved in Europe and blossomed into an intellectual pursuit for curiosity's sake as well as for the "mathematization" of the physical world until in the sixteenth/ seventeenth centuries probability theory developed into an important branch of mathematics. The absence of such a feature in Chinese mathematics during ancient and medieval times turned out to be, oddly enough, one of the reasons for this "miss." In any event, the development of probabilistic thinking failed to take root within the Chinese setting, so that the subject had to wait a long time before it was first transmitted from the West toward the end of the nineteenth century!

Bibliography

[Bennett 1967] Bennett, A.A. *John Fryer: The Introduction of Western Science and Technology into Nineteenth-Century China*, Cambridge: Harvard University Press.

[Bernstein 1996] Bernstein, P. L. *Against the Gods: The Remarkable Story of Risk*, New York: John Wiley & Sons.

[Bréard 2008] Bréard, Andrea, *Reform, Bureaucratic Expansions and Production of Numbers: Statistics in Early 20th Century China*, Habilitationsthesis, Technical University of Berlin, Berlin.

[Bréard 2012] Bréard, Andrea, Divination with hexagrams as combinatorial practice. *Zhouyi Studies* (English Version) 8(1): 157-174 .

[Buck 1980] Buck, P. *American Science and Modern China: 1876-1936*,Cambridge: Cambridge University Press.

[Chan, Siu 2012] Chan, Y.C., Siu, M.K. (陳葉祥, 蕭文強), Facing the change and meeting the challenge: Mathematics curriculum of *Tongwen Guan* in China in the second half of the nineteenth century. *Zentralblatt für Didaktik der Mathematik-The International Journal of Mathematics Education*, 44(4): 461-472.

[Chen late 17th century] Chen Hou-Yao (陳厚耀) [in Chinese] *The Meaning of Methods for Alternations and Combinations* (《錯綜法義》)with a synopsis by Han, Q., in Guo, S.C. (郭書春), ed., *Collection of Chinese Classics in*

Science and Technology (*Mathematics*) (《中國科學技術典籍通彙・數學卷》), vol. 4, Zhengzhou: Henan Educational Press (河南教育出版社), 1993, 685-688.

[Crombie 1994] Crombie, Alistair, *Styles of Scientific Thinking in the European Tradition: The History of Argument and Explanation Especially in the Mathematical and Biomedical Sciences and Arts*, London: Duckworth.

[Daston 1988] Daston, L.orraine, *Classical Probability in the Enlightenment*, Princeton: Princeton University Press.

[David 1962] David, F.N., *Games, Gods and Gambling: The Origins of History of Probability and Statistical Ideas from the Earliest Times to the Newtonian Era.* London: Charles Griffin & Co.

[Du, Han 1992] Du, S.R., Han, Q. (杜石然, 韓琦), The contribution of French Jesuits to Chinese science in the seventeenth and eighteenth centuries, *Impact of Science on Society*, 167: 265-275.

[Elman 2005] Elman, B. A., *On Their Own Terms: Science in China 1550-1900*, Cambridge: Harvard University Press.

[Elvin 2002/2010] Elvin, M., Personal luck: why premodern China — probably — did not develop probabilistic thinking, in *Concepts of Nature: A Chinese-European Cross-Cultural Perspective.* ed. Vogel, U., Günter, D., Leiden: Brill, pp. 400-468; Chinese translation in *Chinese Science and Scientific Revolutions: Selected Research Essays on Joseph Needham's Problem and Related Questions* (《中國科學與科學革命：李約瑟難題及相關問題研究論著選》), ed. Liu, D., Wang, Y.C. (劉鈍, 王揚宗), Shenyang: Liaoning Educational Press (遼寧教育出版社), 2002, 426-496.

[Elvin 2004] Elvin, M., Some reflections on the use of 'styles of scientific thinking' to disaggregate and sharpen comparisons between China and Europe from Song to mid-Qing times (960-1850 CE), *History of Technology*, 25: 53-103.

[Galloway 1839] Galloway, T., *A Treatise on Probability: Forming the Article under That Head in the* 7th *Edition of the Encyclopaedia Britannica*, Edinburgh: Adam and Charles Black.

[Gezhi Huibian] , *Gezhi Huibian*, [*Chinese Scientific and Industrial Magazine*] (《格致彙編》) [in Chinese] sixty issues (1876-1892) reprinted and collected in six volumes, Nanjing Antique Bookshop (南京古籍書店),Nanjing.

[Guo, Xiao 1995] Guo, S.L., Xiao, M.H. (郭雙林, 蕭梅花) [in Chinese] *A History of Gambling in China* (《中華賭博史》), Beijing: Chinese Social Sciences Press (中國社會科學出版社).

[Guo 1989] Guo, S.R. (郭世榮), [in Chinese] *Jueyi Shuxue*, the first book on probability introduced into China from the West, *China Historical Materials of Science and Technology* (《中國科技史料》), 10(2): 90-95.

[Hacking 1975/2006] Hacking, Ian, *The Emergence of Probability: A Philosophical Study of Early Ideas about Probability, Induction and Statistical Inference*, 2nd ed., (1st ed. 1975), Cambridge: Cambridge University Press.

[Hald 1990] Hald, Anders, *A History of Probability and Statistics and Their Applications before 1750*, New York: Wiley.

[Hald 1998] Hald, Anders, *A History of Mathematical Statistics from 1750 to 1930*, New York: Wiley.

[Horng 1993] Horng, W.S. (洪萬生), Hua Hengfang (1833-1902) and his notebook on learning mathematics, *Xue Suan Bi Tan. Taiwanese Journal on Philosophy and History of Science*, 2(2): 27-76.

[Hsü 1970/1995] Hsü, C.Y. (徐中約) *The Rise of Modern China*. 5th ed. (1st ed. 1970), Oxford: Oxford University Press.

[Hu 1917] Hu, M.F. (胡明復), [in Chinese] Probability theory, *Kexue* (《科學》), 3(3): 261-269.

[Hu 1998] Hu, M.J. (胡明杰), *Merging Chinese and Western Mathematics: The Introduction of Algebra and the Calculus in China, 1859-1903*, Ph.D. dissertation, Princeton University.

[Jami 1999] Jami, Catherine, "European science in China" or "Western Learning"? Representations of cross-cultural transmission, 1600-1800, *Science in Context*, 12(3): 413-434.

[Jami 2002] Jami, Catherine, Imperial control and western learning: The Kangxi Emperor's performance, *Late Imperial China*, 23(1): 28-49.

[Jami 2012] Jami, Catherine, *The Emperor' s New Mathematics. Western Learning and Imperial Authority During the Kangxi Reign (1662–1722)*, Oxford: Oxford University Press.

[Jueyi Shuxue 1896] *Jueyi Shuxue*, [Mathematics to Resolve Uncertainty] (《決疑數學》) [in Chinese] translated by Fryer, J., Hua, H.F. (傅蘭雅, 華蘅芳).

[Li 1999] Li, D. (李迪), ed., [in Chinese] *A Complete History of Chinese Mathematics, Volume 6* (*Xi Xia, Jin, Yuan, Ming Dynasties*) (《中國數學史大系》第六卷 (西夏金元明)》), Beijing. Beijing Normal University Press (北京師範大學出版社).

[Li, Mo 1993] Li, H.C., Mo, Y.T. (李惠村, 莫曰達), [in Chinese] *A History of Statistics in China* (《中國統計史》), Beijing: Statistics Press (統計出版社).

[Liang 1936] Liang, Q.C. (梁啟超), [in Chinese] *Collected Works From the Ice-Drinker's Studio* (《飲冰室合集》), 149 vols., Beijing: Zhonghua Books (中華書局).

[Luo, Xu 1994] Luo, X.B., Xu, R.S. (羅新本, 許蓉生), [in Chinese] *Gambling Customs in Ancient China* (《中國古代賭博習俗》), Xian: Shanxi Peoples' Press (陝西人民出版社).

[Ma, Tong 2007] Ma, J.Y., Tong, X.G. (馬繼業, 仝晰綱), [in Chinese] *Thirty-six Stratagems* (《三十六計》), Wuhan: Chongwen Books (崇文書局).

[Maistrov 1974] Maistrov, L .E., *Probability Theory: A Historical Sketch*, translated from Russian by S. Kotz, S., New York: Academic Press.

[Meusnier 1996] Meusnier, N., L'émergence d'un mathématique du probable au XVIIe siècle, *Revue d'histoire des mathématiques*, 2: 119-147.

[Needham 1959/1970] Needham, Joseph, *Science and Civilization in China, Volume 3: Mathematics and the Sciences of the Heaven and the Earth*, 2nd ed, (1st ed. 1959), Cambridge: Cambridge University Press.

[Plato 1914] *Plato: Euthyphro, Apology, Crito, Phaedo, Phaedrus*, trans. H.N. Fowler, Cambridge :Harvard University Press.

[Plato 1921] Plato *Plato: Theaetetus, Sophist*, translated in English by H.N. Fowler, Cambridge: Harvard University Press.

[Plato 1994] Plato, Jan von, *Creating Modern Probability: Its Mathematics, Physics and Philosophy in Historical Perspective*, Cambridge: Cambridge University Press.

[Sambursky 1956] Sambursky, Samuel, On the possible and the probable in ancient Greece, *Osiris*, 12: 35-48.

[Schneider 1987] Schneider, Ivo, Laplace and thereafter: The status of probability calculus in the nineteenth century, in *The Probabilistic Revolution, Volume I: Ideas in History*, Krüger, L., Daston, L.J., Heidelberger, M., eds., Cambridge: MIT Press, pp. 191-21.

[Shen 2000] Shen, K.S. (沈康身), ed., [in Chinese] *A Complete History of Chinese Mathematics, Volume 5* (*Northern and Southern Song Dynasties*) (《中國數學史大系》第五卷 (兩宋)》), Beijing: Beijing Normal University Press (北京師範大學出版社).

[Sheynin 1974] Sheynin, O.B., On the prehistory of the theory of probability, *Archive for History of Exact Sciences*, 12: 97-141.

[Siu 1995/1996] Siu, M.K. (蕭文強), Success and failure of XU Guang-qi: Response to the first dissemination of European science in Ming China, *Studies in History of Medicine and Science, New Series*,14(1-2): 137-178.

[Siu 2009] Siu, M.K. (蕭文強), Mathematics education in East Asia from antiquity to modern times, in *Dig Where You Stand: Proceedings of a Conference on On-going Research in the History of Mathematics Education, Gardabaer, June 20-24, 2009*, Bjarnadottir, K., Furinghetti, F., Schubring, G., eds, University of Iceland, Reykjavík, pp. 197-208.

[Siu 2011] Siu, M.K. (蕭文強), 1607, a year of (some) significance: Translation of the first European text in mathematics - the *Elements* - into Chinese, in *History and Epistemology in Mathematics Education: Proceedings of the 6th European Summer University*, Barbin, E., Kronfellner, M., Tzanakis, C., eds., Vienna: Verlag Holzhausen, pp. 573-589.

[Stigler 1986] Stigler, S.M., *The History of Statistics: The Measurement of Uncertainty before 1900*, Cambridge: Belknap Press.

[Wang 2006] Wang, Y.J. (王幼軍), [in Chinese] *Jueyi Shuxue*: The earliest probability book in Chinese with Laplacian probability style, *Studies in the History of Natural Sciences* (《自然科學史研究》), 25(2): 159-169.

[Wang, Bréard 2010] Wang, Y.J., Bréard, A.(王幼軍, 白安雅), [in Chinese] The evolution of statistics in modern China—From social statistics to mathematical statistics, *Journal of Shanghai Jiaotong University* (*Philosophy and Social Sciences*) (《上海交通大學學報（哲學社會科學版）》), 18(3): 61-67, 74.

[Xiong 1994] Xiong, Y.Z. (熊月之) [in Chinese], *The Dissemination of Western Learning and the Late Qing Society* (《西學東漸與晚清社會》), Shanghai: Shanghai Peoples' Press (上海人民出版社).

[Xu, Zhang 1995] Xu. K., Zhang, J.H. (許康, 張杰恒), [in Chinese] An earliest known example in statistical sampling—Analysis of the problem on "rice-grain composition" by Qin Jiushao, in *Collection of Research Papers on History of Mathematics, Volume 6* (《數學史研究文集第六輯》), ed. Li, D. (李迪), Huhehaote: University of Inner Mongolia, pp. 53-55.

[Yan 1990] Yan, D.J. (嚴敦傑), [in Chinese] A postscript to the tens books of *Jueyi Shuxue*, in *Collection of Papers on the History of Mathematics in the Ming-Qing Period* (《明清數學史論文集》), ed. Mei, R.Z. (梅榮照), Nanjing: Jiangsu Educational Press (江蘇教育出版社), pp. 421-444.

[Zhang 1991] Zhang, Z.G. (張祖貴), [in Chinese] Hu Mingfu, the first doctor of mathematics in modern China, *China Historical Materials of Science and Technology* (《中國科技史料》), 12(3): 46-53.

[Zhongxi Wenjian Lu] *Zhongxi Wenjian Lu* a [*Peking Magazine*] (《中西聞見錄》) [in Chinese] thirty-six issues (1872-1875), reprinted and collected in four volumes. Nanjing: Nanjing Antique Bookshop (南京古籍書店).

Selected Bibliography of Joseph W. Dauben

Bibliography

Books and Monographs

Georg Cantor: His Mathematics and Philosophy of the Infinite (Cambridge, Mass.: Harvard University Press, 1979). Abridged Chinese translation by Zheng Yu-xin, 1989. Reprinted with a paperback edition by Princeton University Press, 1990; a new Chinese translation by Zheng Yu-xin, 2008.

Editor, *Mathematical Perspectives: Essays on Mathematics and its Historical Development* (New York: Academic Press, 1981).

The Intersection of History and Mathematics, J. Dauben, Chikara Sasaki, and Mistuo Sugiura, eds. (Basel: Birkhäuser, 1994).

Abraham Robinson. The Creation of Nonstandard Analysis, A Personal and Mathematical Odyssey (Princeton: Princeton University Press, 1995; paperback ed., 1998; Chinese translation by Wang Qian, Beijing: Science Press, 2005).

States of the Art. Flores quadrivii: Studies in Honor of Christoph J. Scriba, J. Dauben, Menso Folkerts, Eberhard Knobloch, and Hans Wußing, eds. (San Diego: Academic Press, 1996).

Editor, *The History of Mathematics from Antiquity to the Present* (NY: Garland Press, 1985; rev. ed., Albert Lewis, ed., CD-ROM version, Providence, RI: The American Mathematical Society, 2000).

Writing the History of History of Mathematics (an Historiography Project of the International Commission on History of Mathematics), J. Dauben and C.J. Scriba, eds. (Basel: Birkhäuser, 2002).

From China to Paris: 2000 Years of Mathematical Transmission. Proceedings of a Conference Held at the Rockefeller Foundation Research and Conference Center, Bellagio, Italy, May, 2000, J. Dauben, Y. Dold-Samplonius, M. Folkerts, and Benno van Dalen, eds., in *Boethius* (Stuttgart: Steiner Verlag, 2002).

(with GUO Shuchun and XU Yibao) 九章筭術, *Nine Chapters on the Art of Mathematics.* A Critical Edition and English Translation based upon a New Collation of the Ancient Text and Modern Chinese Translation. Library of Chinese Classics, 3 vols., (Shenyang: Liaoning Education Press, 2013).

康托尔: 家史和他的青年时代 (Cantor: His Family History and Youth), Zheng Yu-xin, trans. (Nanjing: Nanjing University Press, 2013).

D.E. Rowe and W.-S. Horng (eds.), *A Delicate Balance: Global Perspectives on Innovation and Tradition in the History of Mathematics*,
Trends in the History of Science, DOI 10.1007/978-3-319-12030-0_18

Chapters in Books

"The Development of Cantorian Set Theory." Chapter V of *From the Calculus to Set Theory, 1630-1910*, ed. Ivor Grattan-Guinness (London: Duckworths, 1977): 181-219.

"Mathematics in Germany and France in the Early 19th Century: Transmission and Transformation." In *Epistemological and Social Problems of the Development of the Sciences in the Early 19th Century*, ed. M. Otte (Dordrecht: D. Reidel, 1981): 371-399.

"Peirce's Critique of Cantorian Set Theory." *Proceedings of the C.S. Peirce Bicentennial International Congress, Amsterdam, The Netherlands* (Lubbock, Texas: Texas Tech University Press, 1981): 93-98.

"Conceptual Revolutions and the History of Mathematics: Two Studies in the Growth of Knowledge." In E. Mendelsohn, ed. *Transformation and Tradition in the Sciences* (Cambridge, England: Cambridge University Press, 1984): 81-103.

"Abraham Robinson and Nonstandard Analysis: History, Philosophy and Foundations of Mathematics." In P. Kitcher and W. Aspray, eds., *New Perspectives on the History and Philosophy of Mathematics* (Minneapolis: University of Minnesota Press, 1987): 177-200.

"Abraham Robinson: Les Infinitesimaux, l'Analyse Non-Standard, et les Fondements des Mathématiques." In H. Barreau, ed., *La Mathématique Non-Standard* (Fondements des Sciences; Paris: Editiions du CNRS, 1989): 157-184.

"Are There Revolutions in Mathematics?" In J. Echeverría, A. Ibarra and T. Mormann, eds., *The Space of Mathematics* (Berlin: De Gruyter, 1992): 203-226.

"The 'Pythagorean Theorem' and Chinese Mathematics. Liu Hui's Commentary on the 句股 *Gou Gu* Theorem in Chapter Nine of the 九章算術 *Jiu Jang Suan Shu*." *Amphora. Festschrift in Honor of Hans Wussing* (Leipzig: B.G. Teubner, 1992): 133-155.

"Conceptual Revolutions and the History of Mathematics: Two Studies in the Growth of Knowledge" and "Revolutions Revisited," Chapters 4 and 5 of D. Gillies, ed., *Revolutions in Mathematics* (Oxford: Clarendon Press, 1992; issued in paperback, 1995): 49-82.

"Mathematics: An Historian's Perspective." In Sasaki Chikara, ed., *The Intersection of History and Mathematics. Proceedings of the International Symposium on History of Mathematics* (Tokyo, 1990; Basel: Birkhäuser, 1994): 1-13.

"Mathematical Exchanges between the United States and China" (with Zhang Dian-Zhou, East China Normal University, Shanghai, China). In E. Knobloch and D. Rowe, eds., *History of Modern Mathematics* (Orlando: Academic Press, 1994): 263-297.

"Peirce and History of Science." In *Peirce and Contemporary Thought: Philosophical Inquiries. Proceedings of the Plenary Lectures delivered at the Peirce Sesquicentennial Congress* (Harvard University, 1989; New York: Fordham University Press, 1995): 146-195.

"Paradigms and Proofs: How Revolutions Transform Mathematics." *Paradigms and Mathematics*, Elena Ausejo and Mariano Hormigón, eds. (Madrid: Siglo XXI de Espanña, 1996): 117-148.

"Arguments, Logic and Proof: Mathematics, Logic and the Infinite." *History of Mathematics and Education: Ideas and Experiences. Proceedings of the Essen Symposium on Foundations and 19th-century Mathematics*, H.N. Jahnke, N. Knoche and M. Otte, eds. (Göttingen, Vandenhoeck & Ruprecht, *Studien zur Wissenschafts-, Sozial- und Bildungsgeschichte der Mathematik*, vol. 11, 1996): 113-148.

"Mathematics at the University of Toronto: Abraham Robinson in Canada (1951-1957)." In *States of the Art. Flores quadrivii: Studies in Honor of Christoph J. Scriba* (San Diego: Academic Press, 1996): 93-136.

"*Historia Mathematicae*: Journals of the History of Science, in honor of the 10th anniversary of the founding of the Italian journal *Nuncius* (Firenze, June 5-6, 1997). In *Journals and History of Science*, Marco Beretta, Claudio Pogliano, and Pietro Redondi, eds. (Florence: Olschki Editore, 1998): 1-30.

"The United States" and (with Liu Dun) "China." Chapters in *Writing the History of History of Mathematics* (an Historiography Project of the International Commission on History of Mathematics), J. Dauben und C.J. Scriba, eds. (Basel: Birkhäuser, 2002): 263-285 and 297-306.

"Internationalizing Mathematics East and West: Individuals and Institutions in the Emergence of a Modern Mathematical Community in China." In *Mathematics Unbound: The Evolution of an International Mathematical Community*, K.H. Parshall and A. Rice, eds. (Providence, R.I., and London, England: The American Mathematical Society and the London Mathematical Society, 2002): 253-286.

"History of Mathematics in the 19th Century: An Historiographic Overview." Chapter 5 in *From Natural Philosophy to the Sciences: Historiography of Nineteenth-*

Century Science, David Cahan, ed. (Chicago: Chicago University Press, 2003): 129-162.

"算數書 *Suan Shu Shu* (A Book on Numbers and Computation): Preliminary Investigation." In *Form, Zahl, Ordnung - Studien zur Wissenschafts- und Technikgeschichte. Ivo Schneider zum 65. Geburtstag*, Rudolph Seising, Menso Folkerts, and Ulf Hashagen, eds. Boetheus: Texte und Abhandlungen zur Geschichte der Mathematik und der Naturwissenschaften (Stuttgart: Franz Steiner Verlag, 2004): 151-168.

"Georg Cantor, Paper on the 'Foundations of a General Set Theory' (1883)." Chapter 46 of *Landmark Writings in Western Mathematics 1640-1940*, Ivor Grattan-Guinness, ed. (London: Routledge, 2005): 600-612.

"Chinese Mathematics." In *The Mathematics of Egypt, Mesopotamia, China, India, and Islam. A Sourcebook*, Victor Katz, ed. (Princeton, NJ: Princeton University Press, 2007): 187-384.

"算數書 *Suan Shu Shu* (A Book on Numbers and Computation): What is the Proper Figure of a 鄆都 *yun du*? *Problemgeschichte der Chinesischen Mathematik.* Two Problems on the Computation of Volumes from the Most Ancient Mathematical Text Yet Known from Ancient China." *Mathematics Celestial and Terrestrial. Festschrift für Menso Folkerts zum 65. Geburtstag, Acta Historica Leopoldina* 54 (2008): 343-360.

"Modern Mathematics in China: The Role of Scientific Exchanges East and West; From the Jesuits and Matteo Ricci to the Cultural Revolution," 東西科技的對話: 國際研討會 *Dongxi keji de duihua: guoji de yantaohui* (Dialogues between East and West Science and Technology: International Workshop) (Hsinchu: National Tsinghua University, 2012): 1-52.

(with Karen Hunger Parshall), "The History of Mathematics Education in North America to 1800" and (with Xu Yibao), "Mathematics Education in Modern China" in *Handbook of Mathematics Education*, G. Schubring and A. Karp, eds. (Heidelberg: Springer-Verlag, 2014).

Articles in Journals and Periodicals

"The Trigonometric Background to Georg Cantor's Theory of Sets," *Archive for History of Exact Sciences* 7 (1971): 181-216.

"Georg Cantor's Philosophy of Mathematics: the Irrational and the Transfinite Numbers," *Proceedings of the XIIIth International Congress of the History of*

Science 5 (Moscow, 1974): 86-93.

"Denumerability and Dimension: The Origins of Georg Cantor's Theory of Sets," *Rete* 2 (1974): 105-134.

"The Invariance of Dimension: Problems in the Early Development of Set Theory and Topology," *Historia Mathematica* 2 (1975): 273-288.

"Georg Cantor and Pope Leo XIII: Philosophical and Theological Dimensions of Transfinite Set Theory," *Journal of the History of Ideas* 38 (1977): 85-108.

"Georg Cantor: The Personal Matrix of his Mathematics," *Isis* 69 (1978): 534-550.

"Georg Cantor: The Roles of Personality and Psychology in the Creation of Transfinite Set Theory," *Annals of the New York Academy of Sciences: Papers in Mathematics* 321 (New York: The New York Academy of Sciences, 1979): 27-45.

"Mathematics and World War I: The International Diplomacy of G.H. Hardy and Gösta Mittag-Leffler," *Festschrift in Honor of Erwin H. Hiebert, Historia Mathematica*, 7 (1980): 261-288.

"Peirce's Place in Mathematics," *Festschrift in Honor of Carolyn Eisele, Historia Mathematica* 9 (1982): 311-325.

"Georg Cantor and the Origins of Transfinite Set Theory," *Scientific American* 248 (June, 1983): 122-131.

"La filosofía de C.S. Peirce sobre los conjuntos infinitos," *Mathesis* 2 (2) (1986): 191-212.

"Charles S. Peirce, Evolutionary Pragmatism and the History of Science," *Centaurus. International Magazine of the History of Mathematics, Science and Technology*, 38 (1996): 22-82.

"Ancient Chinese Mathematics: The 九章算術 *Jiu Zhang Suan Shu* (Nine Chapters of the Art of Mathematics) *versus* Euclid's *Elements*. Aspects of Proof and the Linguistic Limits of Knowledge." Chandrasekhar Memorial Issue, *International Journal of Engineering Science* 36 (1998): 1339-1359.

"Marx, Mao and Mathematics: The Politics of Infinitesimals." *Documenta Mathematica. Journal der Deutschen Mathematiker-Vereinigung.* Proceedings of the International Congress of Mathematicians, Berlin, August 18-27, 1998, vol. 3,

Plenary Lectures and Ceremonies (1998): 799-810.

"*Historia Mathematica*: 25 Years/Context and Content." In honor of the 25th anniversary of the journal's founding, *Historia Mathematica* 26 (1999): 1-23.

"Abraham Robinson. 1918-1974." *Biographical Memoirs of the National Academy of Sciences* 82 (2002): 1-44.

"Mathematics, Ideology, and the Politics of Infinitesimals: Mathematical Logic and Nonstandard Analysis in Modern China." *Festschrift in Honor of Ivor Grattan-Guinness*, John Dawson, ed. *History and Philosophy of Logic* 24 (4) (2003):327-363.

"Mathematics and Ideology: The Politics of Infinitesimals. In Memory of Mariano Hormigón," *Llull. Revista de la Sociedad Española de Historia de las Ciencias y de las Técnicas* 27 (59)(2004): 291-355.

"Marx, Mao and Mathematics: Nonstandard Analysis and the Cultural Revolution. In Memory of Mariano Hormigón," *Acta Historica Leopoldina* 48 (2007) 159-171.

"算數書 *Suan shu shu* (A Book on Numbers and Computations). English Translation with Commentary," *Archive for History of Exact Sciences*, 62 (2) (Heidelberg: Springer Verlag, 2008): 91-178.

Essay Review Articles

Michael Hallett, *Cantorian Set Theory and Limitation of Size* (Oxford: Clarendon Press, 1984); reviewed in *British Journal of Philosophy of Science* 39 (1988): 179-188.

Jean-Claude Martzloff, *Histoire des mathématiques chinoises*, Paris: Masson, 1988; reviewed in *Historia Mathematica* 20 (1993): 437-446.

"Searching for the Glassy Essence: Recent Studies on Charles Sanders Peirce," *Isis* 86 (1995): 290-299.

Elena Ausejo and Mariano Hormigón, eds., *Messengers of Mathematics: European Mathematical Journals (1800-1946)*; reviewed in *Llull* 18 (1995): 303-327.

Gert König, ed., *Konzepte des mathematisch Unendlichen im 19. Jahrhundert* (Göttingen: Vandenhoeck & Ruprecht, 1990); reviewed for the *Archives internationales d'histoire des sciences* 47 (1997): 173-177.

Jean-Claude Martzloff, *A History of Chinese Mathematics*, Berlin: Springer-Verlag, 1997; reviewed in *Nieuw Archief voor Wiskunde* 17 (1999): 491-498.

Georges Ifrah, *The Universal History of Numbers. From Prehistory to the Invention of the Computer*; reviewed in the *Notices of the American Mathematical Society*, (Volume I), 49 (1) (January, 2002): 32-38; (Volume II), 49 (2) (February, 2002): 211-216.

W. Tait, *The Provenance of Pure Reason. Essays in the Philosophy of Mathematics and Its History*, New York: Oxford University Press, 2005; reviewed in *History and Philosophy of Logic* 29 (2) (2008): 183-193.

Catherine Jami: *The Emperor's New Mathematics. Western Learning and Imperial Authority During the Kangxi Reign (1662-1722).* Oxford: Oxford University Press, 2012, reviewed for *Sino-Western Cultural Relations Journal*, XXXIV (2012): 70-81.

Jean-Claude Martzloff: *Le calendrier chinois: structure et calculs (104 av. J.-C.-1644). Indétermination céleste et réforme permanente. La construction chinoise officielle du temps quotidien discret à partir d'un temps mathématique caché, linéaire et continu.* Paris: Honoré Champion, 2009, reviewed in EASTM (East Asian Science, Technology, and medicine), 33 (2011): 90-114.

Contributors

Tom Archibald
Department of Mathematics
Simon Fraser University
8888 University Drive
Burnaby BC V5A 1S6
Canada
e-mail: tarchi@sfu.ca

Elena Ausejo
Universidad de Zaragoza
Facultad de Ciencias (Matemáticas)
Ciudad Universitaria
Pedro Cerbuna 12
E 50009 Zaragoza
Spain
email: ichs@unizar.es

Craig Fraser
Institute for the History and
Philosophy of Science and Technology
Victoria College
73 Queen's Park Crescent
University of Toronto
Toronto, Ontario M5S 1K7
Canada
e-mail: craig.fraser@utoronto.ca

Ivor Grattan-Guinness
Middlesex University
Queensway, EN3 4SA, Enfield
Greater London
UK
e-mail: eb7io6gg@waitrose.com

Jeremy Gray
Department of Mathematics and Statistics
Open University
Walton Hall,
Milton Keynes
Buckinghamshire MK7 6AA
UK
e-mail: j.j.gray@open.ac.uk

Niccolò Guicciardini
Università degli Studi di Bergamo
via Pignolo, 123
24121 Bergamo
Italy
e-mail: niccolo.guicciardini@unibg.it

GUO Shirong
Institute for the History
of Science and Technology
Inner Mongolia Normal University
Hohhot 010022
China
e-mail: gsr@imnu.edu.cn

Wann-Sheng HORNG
Department of Mathematics
National Taiwan Normal University
88 Sec. 4, Ting Chou Road
Taipei 11677
Taiwan
e-mail: horng@math.ntnu.edu.tw

JI Zhigang
School of History and Culture of Science
Shanghai Jiao Tong University
800 Dongchuan Road
Shanghai 200240
China
e-mail: jizhig@gmail.com

LI Wenlin
Academy of Mathematics
and Systems Science
Chinese Academy of Sciences
P.O. Box 2734
Beijing 100080
China
e-mail: wli@math.ac.cn

D.E. Rowe and W.-S. Horng (eds.), *A Delicate Balance: Global Perspectives on Innovation and Tradition in the History of Mathematics*,
Trends in the History of Science, DOI 10.1007/978-3-319-12030-0

LIH Ko Wei
Institute of Mathematics
Academia Sinica
Roosevelt Road
Taipei 10617
Taiwan
e-mail: makwlih@sinica.edu.tw

LIU Dun
Institute for the History of Natural Sciences
Chinese Academy of Sciences
Tsinghua University
Beijing 100084
China
e-mail: liudun@ustc.edu.cn

Jean-Claude Martzloff
Centre de recherche sur les civilisations de l'Asie orientale
Collège de France
52, rue du Cardinal Lemoine
75005 Paris
France
e-mail: jeanclaude.martzloff@sfr.fr

Karen Hunger Parshall
Departments of History and Mathematics
P. O. Box 400137
University of Virginia
Charlottesville, Virginia 22904-4137
USA
e-mail: khp3k@virginia.edu

Walter Purkert
Hausdorff Edition
Hausdorff Center for Mathematics
Endenicher Allee 62
53115 Bonn
Germany
e-mail: edition@math.uni-bonn.de

David E. Rowe
Institut für Mathematik
Staudingerweg 9
Johannes Gutenberg Universität
55099 Mainz
Germany
e-mail: rowe@mathematik.uni-mainz.de

SIU Man Keung
Department of Mathematics
The University of Hong Kong
Pokfulam
Hong Kong
e-mail: mathsiu@hku.hk

MIX
Papier aus verantwortungsvollen Quellen
Paper from responsible sources
FSC® C105338

If you have any concerns about our products,
you can contact us on
ProductSafety@springernature.com

In case Publisher is established outside the EU,
the EU authorized representative is:
Springer Nature Customer Service Center GmbH
Europaplatz 3, 69115 Heidelberg, Germany

Printed by Libri Plureos GmbH
in Hamburg, Germany